Sustainable Solutions for E-Waste and Development

Rajesh Kumar K. V.
Woxsen University, India

Hemachandran Kannan
Woxsen University, India

Dmitry Spodarets
Data Phoenix, USA

Parvez Alam Khan
Universiti Teknologi PETRONAS, Malaysia

Bikash Kumar Pradhan
Woxsen University, India

IGI Global
PUBLISHER of TIMELY KNOWLEDGE

A volume in the Practice, Progress, and Proficiency in Sustainability (PPPS) Book Series

Published in the United States of America by
IGI Global
Engineering Science Reference (an imprint of IGI Global)
701 E. Chocolate Avenue
Hershey PA, USA 17033
Tel: 717-533-8845
Fax: 717-533-8661
E-mail: cust@igi-global.com
Web site: http://www.igi-global.com

Library of Congress Cataloging-in-Publication Data

CIP Data in progress

Title: Sustainable Solutions for E-Waste and Development

ISBN: 9798369310182

This book is published in the IGI Global book series Practice, Progress, and Proficiency in Sustainability (PPPS) (ISSN: 2330-3271; eISSN: 2330-328X)

British Cataloguing in Publication Data
A Cataloguing in Publication record for this book is available from the British Library.

For electronic access to this publication, please contact: eresources@igi-global.com.

Practice, Progress, and Proficiency in Sustainability (PPPS) Book Series

Ayman Batisha
International Sustainability Institute, Egypt

ISSN:2330-3271
EISSN:2330-328X

Mission

In a world where traditional business practices are reconsidered and economic activity is performed in a global context, new areas of economic developments are recognized as the key enablers of wealth and income production. This knowledge of information technologies provides infrastructures, systems, and services towards sustainable development.

The **Practices, Progress, and Proficiency in Sustainability (PPPS) Book Series** focuses on the local and global challenges, business opportunities, and societal needs surrounding international collaboration and sustainable development of technology. This series brings together academics, researchers, entrepreneurs, policy makers and government officers aiming to contribute to the progress and proficiency in sustainability.

Coverage

- Global Content and Knowledge Repositories
- Global Business
- Strategic Management of IT
- Technological learning
- Knowledge clusters
- Socio-Economic
- Outsourcing
- Green Technology
- Eco-Innovation
- E-Development

IGI Global is currently accepting manuscripts for publication within this series. To submit a proposal for a volume in this series, please contact our Acquisition Editors at Acquisitions@igi-global.com or visit: http://www.igi-global.com/publish/.

The Practice, Progress, and Proficiency in Sustainability (PPPS) Book Series (ISSN 2330-3271) is published by IGI Global, 701 E. Chocolate Avenue, Hershey, PA 17033-1240, USA, www.igi-global.com. This series is composed of titles available for purchase individually; each title is edited to be contextually exclusive from any other title within the series. For pricing and ordering information please visit http://www.igi-global.com/book-series/practice-progress-proficiency-sustainability/73810. Postmaster: Send all address changes to above address.

Titles in this Series

For a list of additional titles in this series, please visit:
www.igi-global.com/book-series/practice-progress-proficiency-sustainability/73810

Natural Resource Management Issues in Human-Influenced Landscapes
Dilini Hemachandra (Faculty of Agriculture, University of Peradeniya, Sri Lanka) and Jeevika Weerahewa (Faculty of Agriculture, University of Peradeniya, Sri Lanka)
Engineering Science Reference • copyright 2024 • 300pp • H/C (ISBN: 9781668470510) • US $225.00 (our price)

Water-Soil-Plant-Animal Nexus in the Era of Climate Change
Ahmed Karmaoui (Moulay Ismail University, Meknès, Morocco & Moroccan Center for Culture and Science, Morocco)
Engineering Science Reference • copyright 2024 • 478pp • H/C (ISBN: 9781668498385) • US $240.00 (our price)

Operational Research for Renewable Energy and Sustainable Environments
Joshua Thomas (UOW Malaysia KDU, Malaysia) Gerhard Wilhelm Weber (Poznan University of Technology, Poland) Roman Rodríguez Aguilar (Universidad Panamericana, Mexico) Elias Munapo (North West University, South Africa) and Pandian Vasant (Modeling Evolutionary Algorithms Simulation and Artificial Intelligence (MERLIN), TDTU, Vietnam)
Engineering Science Reference • copyright 2024 • 300pp • H/C (ISBN: 9781668491300) • US $240.00 (our price)

The Role of Women in Cultivating Sustainable Societies Through Millets
Raghvendra Kumar (GIET University, India) and Ishaani Priyadarshini (University of California, Berkeley, USA)
Engineering Science Reference • copyright 2024 • 293pp • H/C (ISBN: 9781668498194) • US $265.00 (our price)

Opportunities and Challenges for Women Leaders in Environmental Management
Perfecto Gatbonton Aquino, Jr. (The University of Cambodia, Cambodia) Revenio Cabanilla Jalagat, Jr. (Al-Zahra College for Women, Oman) and Mercia Selva Malar Justin (Xavier Institute of Management and Entrepreneurship, India)
Engineering Science Reference • copyright 2024 • 259pp • H/C (ISBN: 9781668459867) • US $250.00 (our price)

Quality of Life and Climate Change Impacts, Sustainable Adaptation, and Social-Ecological Resilience
Kasturi Shukla (Symbiosis International University) Yogesh B. Patil (Symbiosis International University, India) Ronald C. Estoque (Forestry and Forest Products Research Institute (FFPRI), Japan) and Pedro Antonio López de Haro (Universidad Autónoma Indígena de México, Mexico)
Engineering Science Reference • copyright 2024 • 300pp • H/C (ISBN: 9781668498637) • US $250.00 (our price)

IGI Global
PUBLISHER of TIMELY KNOWLEDGE

701 East Chocolate Avenue, Hershey, PA 17033, USA
Tel: 717-533-8845 x100 • Fax: 717-533-8661
E-Mail: cust@igi-global.com • www.igi-global.com

Table of Contents

Detailed Table of Contents

This chapter studies Indian chemical sector efficiency and productivity from a sustainable development perspective. This study explores Indian chemical companies' innovative potential, operational efficiency, financial and human resource management techniques, and successes and challenges. The analysis focuses energy efficiency in production to reduce environmental harm. It emphasises the need for good waste management to reduce electronic waste and promote recycling. The report also emphasises the necessity of competent labour and solid financial management for corporate growth. Finally, it stresses how technology and new industrial methods boost efficiency. Using academic and empirical data, the chapter provides policymakers, industry executives, and researchers with valuable ideas to improve the sustainability and efficiency of the Indian chemical sector. It emphasises the need for collaboration to develop and implement sustainable economic and environmental solutions.

Electronic materials/components that have reached the end of their functional life cycle are considered electronic waste (e-waste). Electronics contain potentially hazardous or toxic elements, which many people are unaware of; thus, they must be handled with care when not required anymore. Every product created leaves harmful effects on environment. Copper, gold, silver, platinum etc. are rich resources found in e-waste that can be recovered and utilized again. Loss of these resources, however, results from incorrect disposal, which accelerates the depletion of natural resources. Landfill and burning of e-waste

Green computing can be described as efficient resource management in distributed computing environments such as cloud, fog, and edge computing environments. In green computing, green refers to the eco-friendly environment with environmental responsibility to efficiently manage the computing resources. Mainly, two significant reasons are associated with global warming issues from the distributed computing perspective. They are high power consumption of cloud datacenters and high CO2 emission rate. According to many survey reports, every year, cloud datacenters alone produce nearly 90 million metric tons of CO2 into the environment, and now, this high power consumption of datacenters has become one of the primary reasons for global warming issues. Also, it stated that this CO2 emission rate would increase by 8% every year if it did not identify proper control measurements. Therefore, it is crucial to enhance e-waste management by increasing the efficiency of computing resource usage and minimizing high power consumption, high CO2 emission rate, inefficient recycling policies, etc.

Continuous e-waste generation causes harms to the environment and the economy. This study aims to explore e-waste management systems in the private commercial banks in Bangladesh and ways to reduce e-waste generation and sustainable solutions of e-waste management. The rapid economic growth and establishment of commercial private banks has created a market for computers, consumer electronics, and household appliances in Bangladesh. This study looks into the generation of e-waste growth rate and remedial measures. The e-waste is harmful to human health and for the environment. Barium and lithium are examples that affect brain, heart, liver, kidneys, and skeletal system. It's critical to raise awareness about the dangers of inappropriate e-waste disposal. E-waste must be recycled properly to have a low environmental impact. It was found that banking generated 2176.30 tons of e-waste. Results showed statistically significant growth rate. Buy back guarantee and bring your own device may be options for e-waste reduction.

Information and telecommunications technology (ITT) has expanded into many aspects of modern life and has positively affected human life even in the most remote areas of developing countries. The rapid growth in ITT has led to an improvement in the capacity of computers but simultaneously to a decrease in the product's lifetime as a result of which increasingly large quantities of e-waste are generated annually. It contains hazardous components that if it is not properly managed can have adverse environmental and health effects. The management of e-waste poses significant challenges in developing countries. Due to the lack of adequate infrastructure, e-waste is burned or dumped in open areas. Informal and inefficient recycling practices are generally employed which further exacerbate pollution and health risks. So, the implementation of e-waste management, adequate recycling, and waste disposal facilities have become

The exponential surge in global e-waste has emerged as a critical social, ecosystem, and public health matter. Within this chapter, the authors delve into the multifaceted economic, social, educational, legal, and technological challenges entwined with electronic equipment waste management. It accentuates the imperativeness of devising effective strategies to overcome these hurdles. Insufficient management practices culminate in the emission of perilous substances, thus exacerbating pollution levels and health risks. Numerous nations have adopted regulations and policies to govern the people and proper disposal of electronic equipment waste. In this chapter, the authors explore the utmost significance of managing electronic equipment waste in connection with the Sustainable Development Goals (SDGs). The discourse encompasses exemplary practices and strategies. Conclusively, this chapter emphasizes the paramount importance of robust policies and continuous innovation in achieving sustainable electronic equipment waste management and making valuable contributions towards the realization of the SDGs.

Nanotechnology is one of the most relevant technical factors that have applications in different types of industrial sectors, and it seems to have a consistent scope of technical innovation that focuses on technical derivation with sustainable outcomes. In this study, the perception of nanotechnology and its prominent usage with potential outcomes in different industrial activities has been derived. In addition to this, the study has also reflected on the advantages along with some of the negative factors that come with nanotechnology and its integration. The study helps to evaluate the perception of understanding nanotechnology along with the purpose of evaluating the correlation between the social specifications and commercial attributes.

With the current global overexploitation and depletion of coal resources, it has become necessary to shift toward renewable energy sources to fulfil the world's energy requirements. Among these alternatives, solar energy stands out as the most promising solution. The sun, known for its life-enhancing properties, showers the Earth with an abundant amount of energy. In the case of India, for instance, the country receives an impressive 5000 trillion kWh of energy annually, with a substantial portion of the nation enjoying daily solar radiation ranging from 4 to 7 kWh per square meter. India has demonstrated worthy expertise in harnessing solar energy. Following coal extraction, solar energy developed as the most dependable and widespread energy source capable of meeting the world's energy demands. The study

data regarding material concentrations in specific electronic equipment and the absence of universally applicable standard methods for evaluating elemental concentrations in e-waste. The chapter underscores the value of integrating technical support and financial incentives to bolster e-waste management efforts. Furthermore, it highlights the pivotal roles played by the private sector, non-governmental organizations, and government bodies in the effective implementation of these strategies. Collaboration between businesses, governments, and consumers is underscored as essential for establishing a robust circular economy.

Preface

In the ever-evolving landscape of technology and digital innovation, our reliance on electronic devices has reached unprecedented levels, ushering in an era of connectivity and convenience. However, this rapid surge in technological progress has given rise to a formidable environmental challenge: electronic waste, or e-waste. "Sustainable Solutions for E-Waste and Development" stands as a collective endeavor to comprehend and address the intricate interplay between sustainable development and the escalating issue of e-waste.

As editors, we embarked on this journey with a shared commitment to unravel the complexities surrounding e-waste and to present a comprehensive exploration of sustainable solutions. The authors, each a distinguished expert in their field—Rajesh Kumar K V, Hemachandran Kannan, Dmitry Spodarets, Parvez Alam Khan, and Bikash Kumar Pradhan—have brought forth their unique perspectives to create a tapestry of knowledge that transcends disciplinary boundaries.

This book navigates the multifaceted impacts of e-waste, transcending traditional boundaries to illuminate its environmental, social, and economic repercussions. It serves as a beacon, shedding light on the alarming rates of e-waste generation, the hazardous components embedded in electronic devices, and the far-reaching effects on both ecosystems and human health. Through this exploration, we emphasize the urgent imperative for innovative and practical strategies, unveiling insights into effective e-waste management, recycling initiatives, circular economy models, policy frameworks, and technological advancements.

Our intended audience spans a diverse spectrum, including environmentalists, policy makers, industry professionals, researchers, academics, and conscientious individuals interested in the intersection of sustainability and the digital world. We also envision this book as a valuable resource for educators striving to instill knowledge about e-waste and sustainability in their students.

The recommended chapters, thoughtfully curated by our esteemed authors, serve as pillars supporting the overarching theme of sustainable e-waste management. From the foundational "Introduction to E-waste and Sustainable Development" to the forward-looking "Future Trends and Emerging Solutions in Sustainable E-waste Management," each chapter is a crucial building block in understanding and addressing the e-waste challenge.

Chapter 1 initiates our exploration by scrutinizing the efficiency and productivity of the Indian chemical sector through the lens of sustainable development. The study delves into the innovative potential, operational efficiency, and management techniques of Indian chemical companies, emphasizing the need for energy-efficient production to mitigate environmental harm. It underscores the significance of waste management to reduce e-waste and promote recycling. Furthermore, the chapter highlights the essential role of competent labor and sound financial management for corporate growth. It concludes by emphasizing how technology and new industrial methods can enhance efficiency, offering policymakers, industry executives, and researchers valuable insights to enhance the sustainability of the Indian chemical sector through collaborative efforts.

Chapter 2 sheds light on the crucial issue of electronic waste (e-waste) and its environmental impact. Focusing on the hazardous components present in electronic devices, it stresses the importance of proper handling and disposal to prevent environmental harm. The chapter explores the potential recovery of valuable resources from e-waste, such as copper, gold, silver, and platinum. It addresses the consequences of incorrect disposal methods, including the depletion of natural resources and various types of pollution that impact human health. The integration of new technologies for cost-effective and sustainable e-waste management is discussed, along with the challenges faced in waste management. This chapter aims to classify various types of e-waste and their environmental impact, providing a comprehensive understanding of e-waste management and its challenges.

As electronic equipment usage surges, Chapter 3 focuses on the growing issue of e-waste, emphasizing the challenges posed by non-biodegradable electronic products. It discusses the best practices and innovative approaches in e-waste collection and recycling, aiming to address environmental concerns, conserve resources, and promote sustainable waste management. The chapter highlights the release of toxic chemicals from e-waste, leading to environmental and health issues, especially in low-income areas where recycling facilities are often located. The involvement of women and children in e-waste recycling and the associated health risks are explored. This chapter provides insights into effective e-waste collection and recycling strategies to mitigate environmental and health concerns.

In the digital age, the proliferation of electronic devices has given rise to a pressing global issue – e-waste. Chapter 4 explores the social dimensions of e-waste, emphasizing the health risks associated with improper management and the equity concerns in its production, disposal, and recycling. Drawing from various studies, the chapter advocates for a multi-pronged approach to address the e-waste challenge, emphasizing global collaboration, regulatory frameworks, public awareness, and technological interventions. It serves as a call to action, urging readers to recognize the societal repercussions of e-waste and work collectively towards responsible e-waste management for a sustainable and equitable future.

Green computing, a crucial aspect of sustainable development, is explored in Chapter 5. It defines green computing as efficient resource management in distributed computing environments, such as Cloud, Fog, and Edge computing. The chapter highlights the environmental impact of high power consumption and CO2 emissions from datacenters, underscoring the need for efficient e-waste management to minimize these effects. It discusses the role of green computing in reducing the carbon footprint of electronic devices, emphasizing the importance of efficient resource usage, recycling policies, and minimizing environmental impact. This chapter offers valuable insights for industry professionals, researchers, and policymakers aiming to enhance the sustainability of computing environments.

Chapter 6 investigates the challenges posed by continuous e-waste generation in the private commercial banks of Bangladesh. It explores the rapid economic growth and the resultant increase in the generation of e-waste from computers, consumer electronics, and household appliances. The study addresses the harmful effects of e-waste on human health and the environment, focusing on elements like barium and lithium that can affect various bodily systems. The chapter emphasizes the critical need for awareness about the dangers of inappropriate e-waste disposal and advocates for proper recycling to minimize environmental impact. It concludes by highlighting the importance of sound e-waste policies for sustainable development, offering insights into potential solutions for reducing e-waste generation in the banking sector.

Information and telecommunications technology (ITT) has positively impacted modern life but has also led to a surge in e-waste. Chapter 7 delves into the challenges of managing e-waste in developing countries, where inadequate infrastructure often results in improper disposal practices. It highlights the adverse environmental and health effects of e-waste, especially due to informal and inefficient recycling methods. The chapter stresses the crucial need for e-waste management, adequate recycling facilities, and waste disposal infrastructure. By reviewing e-waste generation, challenges, disposal methods, and their impact on human health and the environment, this chapter provides valuable insights for policymakers, practitioners, and researchers involved in effective e-waste management.

Electronic trash, or e-waste, presents both opportunities and challenges in modern times. Chapter 8 explores the potential for recycling and recovering valuable resources from e-waste, leading to cost savings, reduced demand for new materials, and increased employment opportunities in the recycling and waste management sectors. However, the chapter also highlights the challenges in managing e-waste, given the toxic compounds present in electronic gadgets, such as lead and mercury. Adequate regulatory frameworks and efficient e-waste policy enforcement are deemed essential to address the environmental degradation associated with e-waste. The study analyzes the economic aspects of managing e-waste, offering insights into promoting sustainable practices.

Chapter 9 addresses the escalating problem of e-waste generation in India and worldwide. It highlights the damage to public health and the environment caused by toxic substances and improper disposal of e-waste. Despite policy measures and initiatives, e-waste management remains a significant challenge. The chapter emphasizes the need for a systematic waste collection system, accurate estimations, and early forecasting to achieve sustainable e-waste management. By presenting existing e-waste scenarios in India, the chapter aims to raise awareness about the potential harms of e-waste and promote global sustainability. It serves as a foundation for understanding the challenges of e-waste in the context of sustainable development goals.

The exponential increase in e-waste poses significant environmental concerns. Chapter 10 explores the crucial role of Extended Producer Responsibility (EPR) in addressing the negative social and environmental effects of e-waste. Entrepreneurs are provided with a detailed roadmap for navigating the challenges of EPR adoption in e-waste management. The study presents EPR as both a difficulty and an opportunity for entrepreneurs, showcasing how embracing its principles can minimize environmental impact, promote resource conservation, and contribute to a more sustainable future. By analyzing the potential of EPR in the context of e-waste management, this chapter offers valuable insights for entrepreneurs and policymakers alike.

IoT-driven solutions take center stage in Chapter 11, addressing the global e-waste challenge and emphasizing health risks and equity issues. The chapter justifies the relevance of IoT in addressing e-waste's social impacts, discussing its role in real-time data collection, precise tracking, and assessment of health risks associated with e-waste. The discussion navigates through the disparities in developing nations, highlighting the role of IoT in tailored awareness campaigns, policy development, and enforcement. The chapter concludes by emphasizing the imperative of awareness, regulation, responsible recycling, and global collaboration in tackling e-waste's societal repercussions, offering crucial insights for policymakers, practitioners, and researchers.

The unpredictable chemical components of e-waste are scrutinized in this chapter, revealing severe impacts on ecosystems and human health. Chapter 12 analyzes the potential threat posed by e-waste to sustainable societies and development goals, conducting an extensive literature review and content analysis. It emphasizes the importance of e-waste management methods, such as recycling, to reduce the exorbitant amount of e-waste. By focusing on the contribution of developed countries to e-waste generation, the chapter provides insights into the challenges of sustainable development goals in the context of e-waste.

Chapter 13 explores the multifaceted economic, social, educational, legal, and technological challenges associated with electronic equipment waste management. It emphasizes the need for effective strategies to overcome these challenges, addressing the emission of perilous substances and their impact on pollution levels and health risks. The chapter discusses regulations and policies implemented by nations for the proper disposal of electronic equipment waste, providing insights into exemplary practices and strategies. Conclusively, the chapter underscores the paramount importance of robust policies and continuous innovation in achieving sustainable electronic equipment waste management and contributing to the realization of Sustainable Development Goals (SDGs).

Nanotechnology takes center stage in Chapter 14, evaluating its applications in different industrial sectors and its potential outcomes. The study reflects on the advantages and negative factors associated with nanotechnology and its integration. By evaluating the correlation between social specifications and commercial attributes, the chapter offers a comprehensive understanding of nanotechnology's role in sustainable outcomes. It explores the perception of nanotechnology and its usage, contributing valuable insights to the ongoing discourse on the sustainable application of advanced technologies.

With a focus on renewable energy sources, Chapter 15 explores the role of solar energy in meeting the world's energy requirements. It highlights the significant potential of solar energy, especially in countries like India, and addresses the essential facts of solar waste management. The chapter discusses the challenges related to managerial policy in solar waste management, emphasizing the importance of waste management as solar panel installations expand. By exploring the opportunities and challenges associated with solar waste, the chapter contributes to the broader conversation on transitioning towards sustainable energy sources.

Chapter 16 examines strategies and approaches designed to reduce the production of electronic waste. It outlines methodologies, such as developing electronic goods with enhanced durability, facilitating consumer education, implementing circular economy initiatives, establishing take-back and recycling programs, and enacting legislative measures. These methods collectively contribute to the reduction of electronic waste by prolonging product lifespan, fostering responsible consumption, advocating for reuse and recycling, and establishing a regulatory framework. The chapter emphasizes the importance of these strategies in mitigating the environmental consequences associated with electronic waste, contributing to a more sustainable and resource-efficient electronic industry.

E-waste generation has increased exponentially with the widespread use of electronic devices, posing significant threats to the environment and human health. Chapter 17 explores the current situation, emerging trends, and the future of smart e-waste management. It discusses the challenges faced by countries in implementing advanced methods, such as the Internet of Things (IoT) and artificial intelligence (AI), in e-waste management. The study aims to portray the current state of smart e-waste management, offering insights into the potential benefits and challenges of incorporating IoT and AI technologies to address the growing issue of e-waste.

Chapter 18 emphasizes the importance of engineers in revolutionizing e-waste management through cutting-edge technology. It explores innovative strategies and technologies, including AI-driven diagnostics, repair guidance, component identification, and value assessment, as the future of efficient, effective, and sustainable e-waste recycling. The research aims to explore ways in which technology can play a significant positive role in society without requiring on-site involvement. By examining advanced technology's role in a multi-step resource recovery system, the chapter contributes to the ongoing discourse on sustainable e-waste management.

Chapter 19 delves into the most recent technological advances in electronic waste (e-waste) management approaches and models. It explores various recycling processes, including mechanical, chemical, and biological methods, and discusses the increased operational efficacy of e-waste recycling through the use of artificial intelligence and machine learning algorithms. The study highlights the popularity of urban mining as a method of obtaining precious metals from technological trash. Additionally, the chapter explores the implementation of Extended Producer Responsibility (EPR) rules to promote correct e-waste management and disposal. These technological advances collectively contribute to more sustainable and effective e-waste management, decreasing environmental impact and aiding the recovery of valuable materials.

Chapter 20 provides a comprehensive exploration of the complexities and recommendations associated with the management of electronic waste, advocating for a transition towards a circular economy. It underscores the significance of adopting sustainable practices, including eco-friendly product designs, extended producer responsibility, recycling, and remanufacturing. The chapter identifies challenges such as the scarcity of data on material concentrations and the absence of universally applicable standard methods for evaluating elemental concentrations in e-waste. It emphasizes the value of integrating technical support and financial incentives to bolster e-waste management efforts and underscores the pivotal roles played by the private sector, non-governmental organizations, and government bodies. The collaborative efforts of businesses, governments, and consumers are highlighted as essential for establishing a robust circular economy and addressing the challenges associated with electronic waste management.

We extend our gratitude to the contributing authors for their dedication and expertise, without which this work would not have been possible. As editors, we are honored to present "Sustainable Solutions for E-Waste and Development" as a testament to the collective effort needed to pave the way for a more sustainable and responsible digital future.

May this book inspire thoughtful consideration, informed action, and a shared commitment to building a world where technological progress harmonizes with ecological responsibility.

Rajesh Kumar K. V.
Woxsen University, India

Hemachandran Kannan
Woxsen University, India

Dmitry Spodarets
Data Phoenix, USA

Parvez Alam Khan
Universiti Teknologi PETRONAS, Malaysia

Bikash Kumar Pradhan
Woxsen University, India

Chapter 1
Assessing the Operational Effectiveness of Chemical Companies in Advancing E-Waste Management and Development

Sahini Reddy Pulla
Woxsen University, India

Prakhyath Kandula
Woxsen University, India

ABSTRACT

This chapter studies Indian chemical sector efficiency and productivity from a sustainable development perspective. This study explores Indian chemical companies' innovative potential, operational efficiency, financial and human resource management techniques, and successes and challenges. The analysis focuses energy efficiency in production to reduce environmental harm. It emphasises the need for good waste management to reduce electronic waste and promote recycling. The report also emphasises the necessity of competent labour and solid financial management for corporate growth. Finally, it stresses how technology and new industrial methods boost efficiency. Using academic and empirical data, the chapter provides policymakers, industry executives, and researchers with valuable ideas to improve the sustainability and efficiency of the Indian chemical sector. It emphasises the need for collaboration to develop and implement sustainable economic and environmental solutions.

INTRODUCTION

India is currently facing the growing problem of electronic trash (e-waste) because of fast industrialization and widespread adoption of technology. With its position as the second-largest mobile phone

DOI: 10.4018/979-8-3693-1018-2.ch001

producer and the fifth-largest user of electronic devices worldwide, the country is currently seeing an unprecedented increase in outdated gadgets, which is a major factor in the growing e-waste problem. Within this framework, the chemical industry assumes a crucial role, not only as a purchaser of electronic components, but also as a significant influencer in establishing sustainable methods for the handling of electronic waste. As per the research of Ambitionbox.com (2022), the efficiency prospect of a chemical company can be measured by the products and their relative quantities of feedstocks that consider how well it can utilize energy.

The electronic waste situation in India is characterised by its extensive magnitude and complex composition. The chemical sector, heavily involved in the industrial processes that contribute to the production of electronic waste, is at the centre of this changing problem. This chapter seeks to examine the present condition of e-waste management in the Indian chemical sector, providing insight into the existing legal frameworks, the distinct obstacles encountered, and the inventive solutions and exemplary methods being utilised.

India faces the challenge of dealing with more than 3 million metric tonnes of e-waste produced every year, highlighting the urgent requirement for a comprehensive and strategic strategy to its management. In this context, the chemical industry plays a crucial role in managing the intricate materials and processes involved in the creation of electrical devices. Our objective is to analyse the present condition of electronic waste in the Indian chemical industry, in order to understand the complexities of this issue and provide guidance on how to adopt sustainable and responsible methods.

This investigation involves an examination of the regulatory framework that governs the management of electronic waste, the difficulties faced by the chemical sector, and the creative ideas and exemplary methods that have the potential to transform the current situation. The Indian chemical industry's approach to addressing the e-waste challenge is comprehensive, encompassing adherence to current legislation, incorporation of circular economy ideas, and use of advanced technologies.

Through conducting this extensive analysis, our aim is to make a valuable contribution to the ongoing discussion on e-waste management in India. The knowledge acquired from this investigation not only provides a brief overview of the current difficulties but also lays the groundwork for well-informed suggestions and tactics targeted at promoting a sustainable, circular economy within the chemical industry. India's technological progress is accompanied by the need to address electronic waste, and the chemical industry will have a crucial role in handling this trash to create a more ecologically aware and responsible industrial environment.

As per the concentration by Edge (2020), the harmfulness of the utilization of the synthetic substances can forestall word-related diseases and advance future possibilities in labourers' well-being.

LITERATURE REVIEW

E-Waste Management in the Indian Chemical Industry

The increasing problem of electronic trash (e-waste) has led to a significant amount of written works, which demonstrate the worldwide need to tackle the environmental and health risks linked to incorrect disposal and recycling. Thiounn and Smith (2020) asserted the importance of efficiency in chemical companies lies in the process where the company can able to recycle their chemical processing in a systematic manner. In the Indian environment, where the market for electronic devices is growing rapidly, it is emphasised in the literature that a systematic and comprehensive approach to managing electronic waste is crucial, especially in the chemical sector.

E-Waste Reality in India

The literature on the issue commences by vividly depicting the e-waste scenario in India. Reports continuously emphasise the country's rapid rise as a major contributor to the worldwide e-waste production. Research indicates a consistent and ongoing rise, with forecasts of a significant surge in the production of electronic waste, positioning India as the focal point of the worldwide e-waste problem.

The Role and Contribution of the Chemical Industry

The literature highlights the essential role of the chemical industry in the creation of e-waste, as it is extensively involved in the manufacturing of electronic devices. Chemical processes, which encompass the creation of semiconductor materials as well as the development of adhesives and coatings, play a substantial role in the intricate makeup of electronic waste. Samsami *et al.* (2020) asserted that chemical discharges can be negative for the environment due to poor management of resources and the lack of raw materials. Companies that deal with chemicals require a constant supply of resources and raw materials, like oil, natural gas, and minerals, to process the production of their products. Recognising this contribution is essential when formulating efficient e-waste management strategies.

Legal and Administrative Structures Governing Regulations

A significant portion of the literature review is focused on examining the regulatory frameworks that govern the management of electronic waste in India. The E-Waste (Management and Handling) Rules, 2016, play a crucial role in defining the legal duties and obligations of producers, consumers, and recyclers in the context of electronic waste. The literature examines the effectiveness of these regulations, emphasising both accomplishments and difficulties in their implementation within the chemical industry.

Obstacles Encountered by the Chemical Industry

The literature frequently explores the various difficulties encountered by the chemical industry in efficiently handling electronic waste. The wide range of materials utilised in electronic gadgets presents a substantial challenge, requiring sophisticated technology to effectively separate and recycle them. The financial obligations, particularly for smaller businesses, present a hindrance, highlighting the necessity for customised solutions that can accommodate the diverse capabilities of different participants in the chemical industry.

Novel Solutions and Optimal Practices

The literature presents a range of inventive solutions and exemplary practices implemented by chemical businesses in India. Circular economy ideas, involving the design of easily recyclable electronic components, are examined as a possible route. Collaborations between chemical businesses, electronics manufacturers, and waste management bodies are recognised as promoting holistic approaches to e-waste disposal. The integration of modern technologies, such as artificial intelligence and machine learning, emerges as a revolutionary force in boosting and lowering the environmental imprint.

Community Engagement and Corporate Social Responsibility

Literature on e-waste management within the chemical industry extends beyond technology solutions to incorporate community engagement and corporate social responsibility (CSR). Initiatives aiming at raising public awareness about e-waste hazards and creating collection sites through CSR activities are regarded as essential components in building a culture of accountability and compliance.

Future Outlook and Recommendations

The literature finishes with a forward-looking viewpoint, presenting insights into the future of e-waste management in the Indian chemical sector. Recommendations range from improved regulatory enforcement to higher spending in research and development. Collaborative efforts, both within the business and between sectors, are emphasized as important to establishing comprehensive and sustainable e-waste management solutions.

In integrating the many strands of literature, it becomes obvious that e-waste management in the Indian chemical industry is a complicated challenge requiring a nuanced and coordinated response. The literature review sets the stage for our in-depth exploration of the current situation, challenges, and potential solutions in the subsequent sections of this chapter, contributing to the ongoing discourse on sustainable industrial practices in the context of e-waste.

METHODS

India is currently seeing a substantial surge in electronic garbage (e-waste) as a result of its expanding population and rapid technological advancements. The chemical industry, which is crucial for numerous industrial procedures, plays a pivotal role in addressing the challenges posed by electronic waste (e-waste). This chapter analyses the current state of e-waste management in the Indian chemical business, including its existing condition, the challenges it faces, and potential options for adopting sustainable practices.

The Current State of Electronic Trash (E-Waste) in India

India has become a significant contributor to the global e-waste issue, mostly due to rapid urbanisation and extensive technology adoption. The electronic devices market is seeing substantial expansion, leading to a proportional increase in the quantity of obsolete equipment. According to recent figures, India generated over 3 million metric tonnes of electronic waste (e-waste) in 2022, and this figure is expected to rise in the next years. The chemical industry in India is a major buyer of electronic components for its manufacturing processes. The accumulation of electronic waste poses substantial environmental and health hazards, necessitating a comprehensive approach to its management and elimination. The challenge lies not only in the substantial volume of electronic waste, but also in its complex composition, comprising multiple components that require specialised processing.

The Legal and Administrative Framework That Governs Rules

India has established regulatory frameworks to regulate the management of electronic trash, recognising the urgency of addressing this matter swiftly. The E-Waste (Management and Handling) Rules, 2016, offer guidelines for the proper disposal and recycling of electronic waste. All stakeholders, encompassing manufacturers, consumers, and recyclers, are obligated to adhere to these criteria, which emphasise the necessity of collective accountability. Complying with these laws is crucial in the chemical industry. It is incumbent upon companies to adopt environmentally sustainable practices for disposing of electronic equipment, ensuring that hazardous substances are managed with meticulous care. However, there are persistent challenges in ensuring uniform application of these standards throughout the diverse landscape of the Indian chemical industry.

Challenges in the Chemical Sector's Treatment of Electronic Waste

The chemical sector faces specific challenges in managing electrical waste in India. A significant impediment arises from the diverse array of materials employed in electrical devices, which includes both precious metals and hazardous substances. Chemical enterprises must commit substantial financial resources to invest in advanced technology that facilitates efficient segregation and recycling. A further challenge stems from the lack of standardised protocols for managing electronic waste in the chemical industry. While larger businesses have the requisite resources to implement comprehensive e-waste management plans, smaller firms may encounter challenges in meeting regulatory obligations. Addressing this gap is essential for a cohesive and effective approach to e-waste management. Furthermore, there is a dearth of general awareness among the public on the perils linked to improper disposal of electronic waste. It is crucial to educate individuals in the chemical industry and the general public in order to foster a strong sense of responsibility and compliance with e-waste laws.

Innovative Solutions and Best Practices

Multiple chemical companies in India have implemented innovative strategies and optimal approaches to address the challenges posed by electronic waste. To achieve optimal results, it is crucial to integrate circular economy concepts into industrial processes. Companies are continually researching ways to create electronic components that have improved recyclability, hence reducing the environmental impact of their products. The collaboration of chemical enterprises, electronics manufacturers, and waste management organisations is fostering an all-encompassing approach to electronic waste treatment. These partnerships aim to improve recycling technologies and minimise the release of toxic substances into the environment by sharing information and combining resources. In addition, the chemical industry is progressively adopting advanced technology, such as artificial intelligence (AI) and machine learning. These solutions enhance the efficiency of sorting and recycling electronic waste, thereby reducing the environmental impact and improving overall operational effectiveness.

Active Involvement With the Community and Dedication to Business Social Responsibility

Effective e-waste management extends beyond mere compliance with rules; it necessitates active community involvement and the practice of corporate social responsibility (CSR). Indian chemical businesses are increasingly recognising the importance of actively engaging local communities to improve awareness about the hazards of e-waste and promote proper disposal practices. CSR initiatives are being employed to establish e-waste collection centres in collaboration with local authorities. These centres serve as both collection points and central locations for public on the importance of recycling and the potential economic benefits of proper disposal of electronic waste.

RESEARCH AND DEVELOPMENT MODELS

Research and development (R&D) are essential for fostering innovation, particularly in the Indian chemical industry's efforts to manage e-waste. Strong R&D models play a critical role in developing sustainable solutions. This section examines the many research and development approaches utilised by chemical firms in India as they address the difficulties presented by electronic waste.

Technologies for the Recovery of Materials

The primary objective of R&D models is to develop cutting-edge technology for the recovery of sophisticated materials. Chemical companies conduct methodical research to develop procedures that effectively recover valuable substances from electronic waste. This encompasses the retrieval of valuable resources such as precious metals, rare earth elements, and other valuable materials that are present in abandoned electronic gadgets. Companies strive to enhance the extraction of these minerals by employing inventive chemical procedures, with the goal of reducing waste and maximising the utilisation of resources.

Principles of Eco-Design

Another notable research and development model revolves around the ideas of eco-design. Chemical corporations are allocating resources towards doing research aimed at redesigning electrical components in order to enhance their recyclability and reusability. This entails the creation of materials that are ecologically sustainable, readily separable, and harmonious with current recycling procedures. Through the incorporation of eco-design principles into the process of product development, corporations actively contribute to the production of electronic gadgets that possess a diminished ecological footprint over their entire lifespan.

Joint Endeavours

The chemical companies that face barriers and problems that contribute to their inefficiency can be caused by the presence of insufficient resources. In this context, Mallapragada *et al.* (2023) asserted that the chemical industry faces obstacles due to the increased rate of greenhouse gas emissions and extreme use of steel and cement. Collaborative research and development projects are becoming a potent approach for tackling the intricate issues of e-waste management. Chemical firms are progressively establishing strategic alliances with electronics manufacturers, waste management agencies, and research institutions. These alliances enable the consolidation of resources, skills, and ideas, resulting in more comprehensive and efficient solutions for managing electronic waste. Collaborative research centres and joint projects allow for a combined and coordinated effort to address the complex problems related to electronic waste.

Integration of Circular Economy

R&D approaches in the chemical sector are shifting towards incorporating circular economy ideas. This entails conducting research and devising procedures that adhere to the principle of a closed-loop system, in which materials are consistently recycled and reused. Companies are actively seeking new methods to effectively complete their production cycles, guaranteeing that the disposal of electronic items is not an endpoint but rather a starting point for further resource utilisation.

Technological Advancement

Technological innovation is a fundamental element of research and development models when it comes to managing electronic trash. Chemical businesses are utilising advanced technologies like artificial intelligence (AI) and machine learning to improve the effectiveness of e-waste sorting and recycling procedures. AI-driven solutions are applied for the automatic identification and separation of distinct items in electronic trash, speeding recycling operations and eliminating errors.

Life Cycle Assessment (LCA) is a method used to evaluate the environmental impact of a product or process throughout its entire life cycle, from raw material extraction to disposal. As proposed by Charnes, Cooper and Rhodes (1978), in order to examine the dynamic proficiency can be created through assistance of practical viability inside the process. R&D models frequently integrate Life Cycle Assessment (LCA) approaches to thoroughly assess the ecological footprint of electronic devices. This entails evaluating the complete life cycle, encompassing the extraction of raw materials, the manufacturing process, usage, and the disposal at the end of its life. Through the implementation of Life Cycle Assessments (LCAs), chemical businesses acquire valuable knowledge regarding the specific areas where their products have the greatest negative environmental effects. This information allows them to focus their research and development endeavours on reducing these impacts, therefore promoting a comprehensive approach to sustainability.

Centres for Innovation and Business Start-Ups

Creating specialised innovation hubs and incubators focused on e-waste management functions as a proactive research and development (R&D) framework. These hubs offer a cooperative setting for researchers, scientists, and industry specialists to investigate innovative solutions. Chemical firms have a crucial role in promoting innovation and entrepreneurship, which in turn helps create a dynamic environment dedicated to the advancement and application of sustainable techniques in e-waste treatment.

DISCUSSIONS

The future of e-waste management in the Indian chemical industry hinges on the efficient enforcement of rules, the advancement of novel technologies, and the collaboration among diverse stakeholders. To ensure the establishment of a long-lasting framework, it is crucial for policymakers, industry stakeholders, and the general public to collaborate in a harmonic manner.

Regulatory authorities must first prioritise improving the effectiveness of enforcement operations and ensuring consistent compliance throughout the chemical sector. This involves meticulous monitoring of e-waste disposal techniques and penalties for non-compliance. Simultaneously, the government might offer incentives to businesses that adopt environmentally conscious practices, so encouraging a shift towards sustainable management of electronic waste.

Also, it is imperative to spend financial resources towards research and development in the chemical industry. Indian chemical firms have the potential to become global leaders in sustainable practices, particularly in the field of e-waste recycling, by developing innovative chemical processes or adopting environmentally friendly materials.

Moreover, it is crucial to foster collaboration not only within the organisation but also across diverse sectors. As indicated by the concentrate by Abanades et al. (2021), the basic audit of chemical businesses incorporates a few factors, for example, hydrogenation techniques, drug producing arising energy applications. Partnerships between the government, academic, and business sectors have the potential to expedite the development of comprehensive solutions for the management of electronic waste. Knowledge-sharing platforms and industry forums can improve the spread of exemplary methods and technological advancements.

Ultimately, it is crucial to implement a heightened level of education and awareness campaigns. Concentrating on teaching professionals in the industry and the general population about the risks associated with improper disposal of electronic waste, the benefits of recycling, and the participation of the chemical sector in sustainable practices could potentially foster a shift in culture towards responsible management of electronic waste.

CONCLUSION

In essence, efficiently handling electronic waste in the chemical industry in India is a multifaceted endeavour that necessitates a cohesive and collaborative approach. Regulations establish a foundation, but their effective implementation and enforcement are vital. The chemical industry, leveraging its technological expertise, should take the lead in developing innovative recycling methods and materials to achieve a sustainable and circular economy. In the opinion of Chen et al. (2022), various methodologies to effectively accomplish a carbon-impartial society keeping up with worldwide industrial enterprise including various negative outflows in the compound business will be given more assistance from top to bottom investigation.

In order to have an efficient e-waste management plan, it is crucial to include community participation, company social responsibility, and public awareness campaigns. India's current surge in technology is being paralleled by the chemical industry's aggressive initiatives to address the e-waste challenge. This not only fosters ecological sustainability but also positions the country as a global pioneer in conscientious and innovative industrial methods.

REFERENCES

Abanades, S., Abbaspour, H., Ahmadi, A., Das, B., Ehyaei, M. A., Esmaeilion, F., El Haj Assad, M., Hajilounezhad, T., Jamali, D. H., Hmida, A., & Ozgoli, H. A. (2021). A critical review of biogas production and usage with legislations framework across the globe. *International Journal of Environmental Science and Technology*, 1–24. PMID:34025745

Ambitionbox.com. (2022). *462 Chemicals Companies in India | AmbitionBox*. AmbitionBox. Available at: https://www.ambitionbox.com/chemicals-companies-in-india

Baumann, M., Moody, T. S., Smyth, M., & Wharry, S. (2020). A perspective on continuous flow chemistry in the pharmaceutical industry. *Organic Process Research & Development*, *24*(10), 1802–1813. doi:10.1021/acs.oprd.9b00524

Braverman, H. (2020). Scientific management. In *Industrial Work and Life* (pp. 111–127). Routledge. doi:10.4324/9781003085706-9

Charnes, A., Cooper, W. W., & Rhodes, E. (1978). Measuring the of decision making units. *European Journal of Operational Research*, *2*(6), 429–444. doi:10.1016/0377-2217(78)90138-8

Chen, L., Msigwa, G., Yang, M., Osman, A. I., Fawzy, S., Rooney, D. W., & Yap, P. S. (2022). Strategies to achieve a carbon neutral society: A review. *Environmental Chemistry Letters*, *20*(4), 2277–2310. doi:10.100710311-022-01435-8 PMID:35431715

Farrell, M. J. (1957). The measurement of productive efficiency. *Journal of the Royal Statistical Society. Series A (General)*, *120*(3), 253–281. doi:10.2307/2343100

Forbes. (2023). *What is Organizational Behavior?* https://www.forbes.com/advisor/business/what-is-organizational-behavior

HR, G., & Aithal, P.S. (2022). Why is it Called Doctor of Philosophy and Why Choosing Appropriate Research Philosophical Paradigm is Indispensable During Ph. D. Program in India? *International Journal of Philosophy and Languages*, *1*(1), 42–58.

Leng, J., Ruan, G., Jiang, P., Xu, K., Liu, Q., Zhou, X., & Liu, C. (2020). Blockchain-empowered sustainable manufacturing and product lifecycle management in industry 4.0: A survey. *Renewable & Sustainable Energy Reviews*, *132*, 110112. doi:10.1016/j.rser.2020.110112

Livemint. (2023). *Indian chemical industry: A growing force and its prospects mint*. Available at: https://www.livemint.com/market/stock-market-news/indian-chemical-industry-a-growing-force-and-its-prospects-11695551490833.html#:~:text=Chemical%20industry%20revenue%20has%20been,in%20the%20last%205%20years.&text=The%20Indian%20chemicals%20sector%20stands,of%20%24178%20billion%20in%202021

Luu, D. H. (2021). The impact of capital structure on firm value: A case study in Vietnam. *The Journal of Asian Finance. Economics and Business*, *8*(5), 287–292.

Mallapragada, D. S., Dvorkin, Y., Modestino, M. A., Esposito, D. V., Smith, W. A., Hodge, B. M., Harold, M. P., Donnelly, V. M., Nuz, A., Bloomquist, C., Baker, K., Grabow, L. C., Yan, Y., Rajput, N. N., Hartman, R. L., Biddinger, E. J., Aydil, E. S., & Taylor, A. D. (2023). Decarbonization of the chemical industry through electrification: Barriers and opportunities. *Joule*, *7*(1), 23–41. doi:10.1016/j.joule.2022.12.008

Oglu Karimli, V. M., & Oglu Baghirov, H. S. (2020). *Administrative management theory at high educational institutions and its following stages.* Administrative Management Theory at High Educational Institutions and Its Following Stages.

Panjaitan, T. W., Dargusch, P., Wadley, D., & Aziz, A. A. (2023). A study of management decisions to adopt emission reduction measures in heavy industry in an emerging economy. *Scientific Reports*, *13*(1), 1–15. doi:10.103841598-023-28417-2 PMID:36697460

Ratti, R. (2020). Industrial applications of green chemistry: Status, Challenges and Prospects. *SN Applied Sciences*, *2*(2), 263. doi:10.100742452-020-2019-6

Rim, K. T. (2020). In silico prediction of toxicity and its applications for chemicals at work. *Toxicology and Environmental Health Sciences*, *12*(3), 191–202. doi:10.100713530-020-00056-4 PMID:32421081

Ruggiano, N., & Perry, T. E. (2019). Conducting secondary analysis of qualitative data: Should we, can we, and how? *Qualitative Social Work: Research and Practice*, *18*(1), 81–97. doi:10.1177/1473325017700701 PMID:30906228

Saeed, B. B., Afsar, B., Hafeez, S., Khan, I., Tahir, M., & Afridi, M. A. (2019). Promoting employee's proenvironmental behavior through green human resource management practices. *Corporate Social Responsibility and Environmental Management*, *26*(2), 424–438. doi:10.1002/csr.1694

Samsami, S., Mohamadizaniani, M., Sarrafzadeh, M. H., Rene, E. R., & Firoozbahr, M. (2020). Recent advances in the treatment of dye-containing wastewater from textile industries: Overview and perspectives. *Process Safety and Environmental Protection*, *143*, 138–163. doi:10.1016/j.psep.2020.05.034

Sileyew, K. J. (2019). *Research design and methodology*. Cyberspace.

Sun, X., Atiyeh, H. K., Huhnke, R. L., & Tanner, R. S. (2019). Syngas fermentation process development for production of biofuels and chemicals: A review. *Bioresource Technology Reports*, *7*, 100279. doi:10.1016/j.biteb.2019.100279

Suri, H. (2020). Ethical considerations of conducting systematic reviews in educational research. Systematic reviews in educational research: Methodology, perspectives and application, 41-54. doi:10.1007/978-3-658-27602-7_3

Syaifullah, D. H., Tjahjono, B., McIlhatton, D., & Zagloel, T. Y. M. (2022). The impacts of safety on sustainable production performance in the chemical industry: A systematic review of literature and conceptual framework. *Journal of Cleaner Production, 366*, 132876. doi:10.1016/j.jclepro.2022.132876

Thiounn, T., & Smith, R. C. (2020). Advances and approaches for chemical recycling of plastic waste. *Journal of Polymer Science*, *58*(10), 1347–1364. doi:10.1002/pol.20190261

Vega, F., Baena-Moreno, F. M., Fernández, L. M. G., Portillo, E., Navarrete, B., & Zhang, Z. (2020). Current status of CO2 chemical absorption research applied to CCS: Towards full deployment at industrial scale. *Applied Energy*, *260*, 114313. doi:10.1016/j.apenergy.2019.114313

Chapter 2
Environmental Impacts of E-Waste: Pollution and Resource Depletion

Rajeev Kumar
https://orcid.org/0000-0002-0820-5970
Manav Rachna International Institute of Research and Studies, Faridabad, India

Arti Saxena
Manav Rachna International Institute of Research and Studies, Faridabad, India

Jyoti Chawla
Manav Rachna International Institute of Research and Studies, Faridabad, India

Vijay Kumar
Manav Rachna International Institute of Research and Studies, Faridabad, India

ABSTRACT

Electronic materials/components that have reached the end of their functional life cycle are considered electronic waste (e-waste). Electronics contain potentially hazardous or toxic elements, which many people are unaware of; thus, they must be handled with care when not required anymore. Every product created leaves harmful effects on environment. Copper, gold, silver, platinum etc. are rich resources found in e-waste that can be recovered and utilized again. Loss of these resources, however, results from incorrect disposal, which accelerates the depletion of natural resources. Landfill and burning of e-waste are responsible for various types of pollution and a lot of impact on human health. New technologies have been applied to manage these wastes at low cost and sustainable ways. However, there are numerous difficulties and challenges in managing various types of waste. The aim of this chapter is to classify various types of waste and their impact on the environment. It also explores the management of waste by new technologies and challenges faced at the time of management of waste.

DOI: 10.4018/979-8-3693-1018-2.ch002

1. INTRODUCTION

Electric and electronic items now play a more important role in our daily activities than any time before. Due to fast growing technological and increasing demands of new gadgets (laptop, phone, tablet, inverter, printers etc.), lot of used electric and electronic materials have been throwing without any concerned. The volume and numbers of these items are increasing day by day which is very alarming conditions for the globe. Electronic products have integral role in our routine work more than ever before. These habits increase financial burden as well as pose hazardous effects on the environment if they are not disposed/recycled in a proper manner and contribute to E-waste (Jain et al., 2023). Various hazardous elements/compounds including lead, cadmium, mercury, lithium, zinc, nickel, barium, poly-brominated compounds etc. may enter the environment from discarded electronic devices. When electronic debris is disposed of in landfills, harmful compounds leach into groundwater, harming both aquatic and terrestrial species. This may have an impact on the wellness of people in developing nations where the majority of electronic garbage is disposed. However, it is also possible to recover the elements like silver, gold, copper, aluminum etc. if the discarded products are recycled after proper disposal (Manikandan et al., 2023). If not disposed/handled properly these may have adverse effects on human, plants and animals through various exposure routes. Figure 1 depicts the different ways to handle the electronic waster and the consequences associated with each approach. Recycling, reusing or refurbishing the devices is the safest approach to reduce their impact on environment and health. Other options of electronic waste treatment pose many occupational hazards as well as release many toxins into the environment which may further lead to many health implications. To ensure the safe handling of these devices after their useful life, awareness is also required at consumer end and their disposal should be prohibited along with municipal solid waste. Moreover, E-waste disposal units for formal management of E-waste should be available and proper guidance manual should be available within the waste handling organizations for processing of E-waste. Forti et al., 2020 estimated that by 2030 E-waste generation will exceed 74 million tons globally but the recycling is being done at much lesser pace. Mor et al., 2021 in a study related to awareness of E-waste management practices. It is discussed that the users are aware that hazardous materials are present in E-waste, but not aware about their adverse on health and environment. There is a need to adopt sustainable approach for production, utilization and disposal of E-waste (Sinha, 2011; Bekabil, 2020; Yabe et al., 2010; Zeng & Li, 2016).

Since October 2019, only 71% of global population was under national E-waste policy or regulation, which was only 44% in 2014. Countries like India and China have national legislative laws in place, but only 78 of the 193 countries have an E-waste national policy. The European Union has produced guidelines for hazardous substances and waste electrical and electronic equipment (WEEE). (Murphy et al., 1998). Good policies and legal regulations can only be made when you have good E-waste data. The data includes quantity and flow of E-waste, monitoring, controlling, illegal dumping and improper treatment. Good data minimize the generation of E-waste and promote recycling and refurnishing of E-waste creates green jobs.

Ananno et al 2021 have been studied the analysis and management of E-waste in Bangladesh. A government agency ought to oversee the entire waste management system. Given its attention to environmental welfare, the Ministry of Forests and Environment appears to be the most pertinent ministry in this situation. There should be an E-waste management and regulatory board (EMRB). This board will supervise and address the nation's progress in E-waste management. Additionally, the board will finance waste management initiatives across the country, encourage widespread adoption of waste management practises, apply the legislative framework established by the parliament, and have the authority to justify and certify waste management systems. The public will hear from the communication and information group about the advantages and significance of trash management. In order to raise public awareness, it will also host meetings, seminars, and campaigns. Additionally, this agency will actively promote trash management initiatives and applicable local regulations. The current waste management practices will be analyzed, improved, and modified by the technical and certification group. In order to maximize output, this department will investigate current procedures and devise fresh approaches to inject more realism and significance throughout the entire process. This team can also look at potential methods for using E-waste to produce sustainable energy.

Khatiwada et al 2023 have been studied management of E-waste in Nepal by using machine learning.

An IoT sensor-based alert system was added to a Waste Management System (WMS), which is purely manual and can notify the appropriate individuals when waste is overflowing so they may take prompt action. Several deep learning techniques are used for the picture processing, which improves classification and prediction accuracy. A CNN algorithm is used in the deep learning model to extract features, and an RBM model is used to process the extracted features for increased accuracy. The suggested system is validated against an open-source dataset, and the outcomes are contrasted with those of other methods currently in use. The comparison demonstrates that the suggested method offers higher E-waste forecast accuracy with 96% than the other.

Bagwan, 2024 has been investigated the management of E-waste using Auto Regressive Integrated Moving Average (ARIMA) models in Maharashtra, India. According to the report, the average capacity for processing E-waste between 2023 and 2030 would be 163563.15 metric tonnes. The projected numbers are expected to increase gradually over time, with 248 recyclers by 2030. Analyzing the pace of change in the ability to process E-waste from 2023 to 2030 will reveal opportunities for 6.86 percent yearly growth. This also shows the extent of entrepreneurship in the E-waste recycling sectors by indicating an average annual growth rate of 7.23 percent for recyclers.

Governments all over the world are creating national E-waste legislation but this legislation can be achieved if the disposal policy is enacted at the municipal or local level. According to certain research, the bulk of E-waste is shipped from developed countries to developing or under developed countries but the quantity flow of E-waste is hard to measure. It is brought illegally under pretence of reuse. This raises environmental burden in the destination places and poses significant health issues.

Increasing use of electrical and electronics items all around the world is only one reason of increasing E-waste. Motivation of this chapter is that how to explore various sources of E-waste, their environmental impacts, precious metals being used in electronic gadgets and their recovery and management of E-wastes. Also this chapter explores how to reduce these E-waste, how to utilize unused or dumped electrical or electronics item and how to aware public regarding E-waste and their management. Thus the aim of this chapter is to summarize the data of E-waste, their classification, and their environmental impacts, recovery of precious metals and management of these E-wastes.

2. CLASSIFICATION OF E-WASTE

There are wide range of electrical and electronics products containing circuits, electrical components with a battery. It could be a basic kitchen appliance, toy, tools to music, mobile phone, laptops, LED TV etc. Also, sensors are being deployed to the advanced devices are the part of the product. These devices

Figure 1. Ways to handle the electronic waster and the consequences associated with each approach

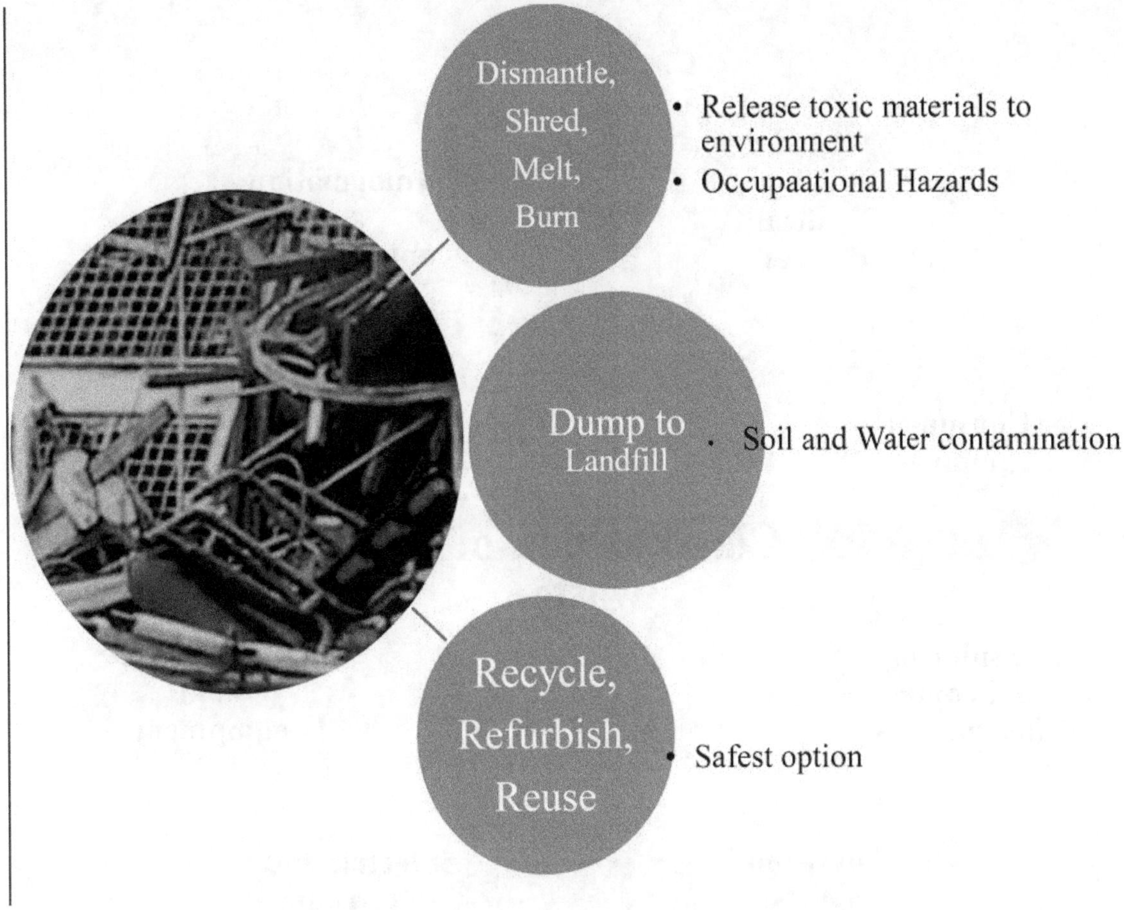

once discarded, become E-waste. Since each gadget has different material and therefore has different recycling method but their impact is harmful to the environment and human health.

There is no universal system for classifying e-waste since it differs from nation to nation. E-products can be classified according to similar functions, comparable material composition i.e. valuable/hazardous materials, and life time of the products. Furthermore, goods in the same category have to have a uniform distribution of average weight and lifespan, as this might make quantitative evaluation of related goods easier. Lastly, distinct assignments should be made for large or environmentally significant E-waste products, for which a wealth of data may be accessible. As of right now, only one classification scheme meets those requirements: the UNU's classification (Wang et al., 2012). As per UNU's mainly, electrical and electronics products are classified into 54 different types of product categories (Balde et al., 2022). Again, these 54 product categories are grouped into various general categories and are given as (Figure 2) (Forti et al., 2020):

Figure 2. Classification of e-waste

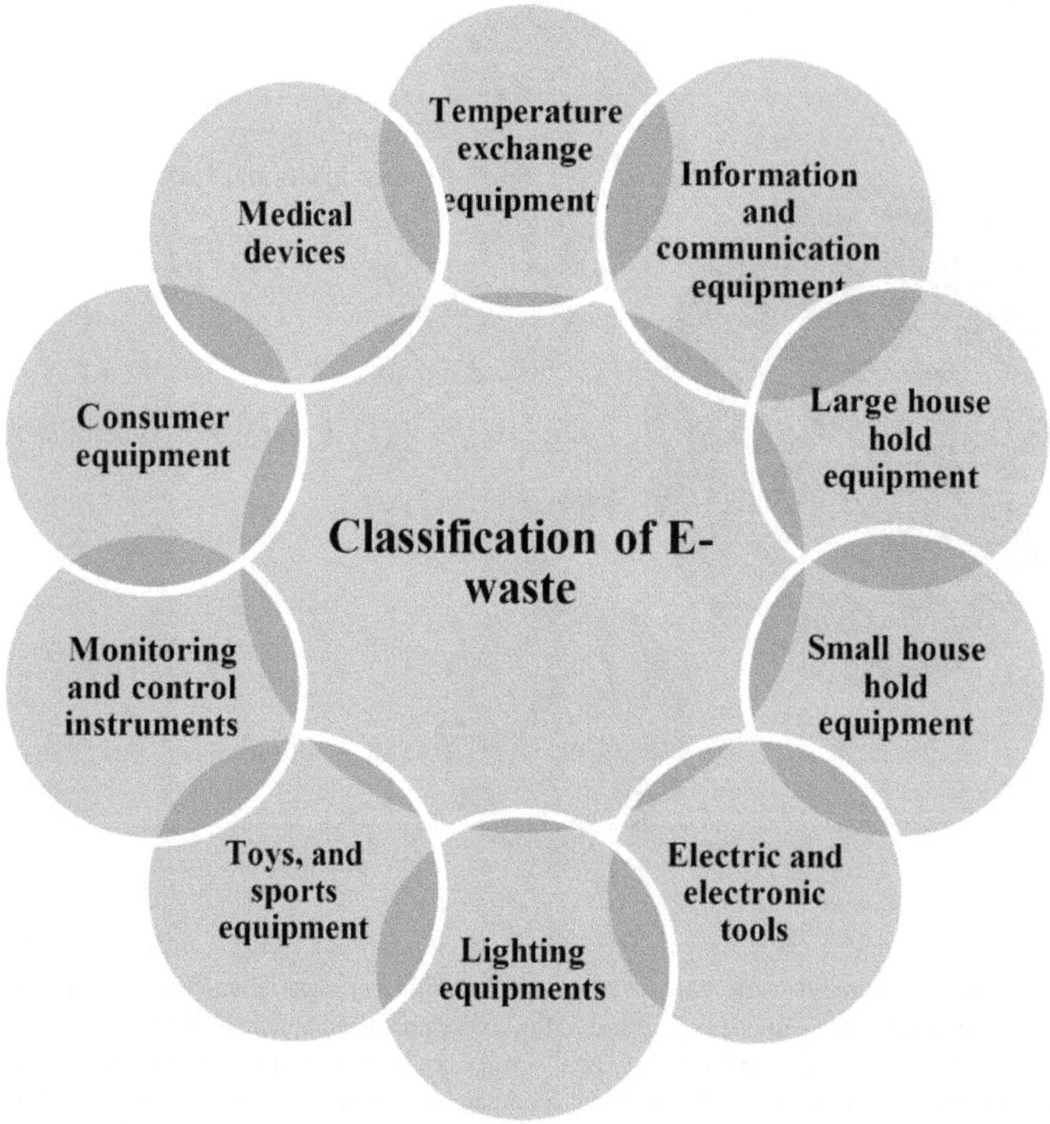

2.1. Temperature Exchange Equipment

Depending on the application, many types of temperature exchanger equipment are employed in the various industries, but their common function is to transfer heat from a hot area to a cold area. Heat energy is transferred via conduction, convection, or radiation in temperature exchanger equipment. Cooling and freezing equipment such as refrigerators, freezers, air conditioners, and heat pumps etc are the very common examples of temperature exchange equipment.

2.2. Information and Communication Equipment

The information and communications technology (ICT) sector is one of the fastest expanding industries globally. It is also the leading cause of the global e-waste catastrophe. ICT equipment refers to everyday electrical and electronic products for personal use, such as laptops, CPUs, monitors, mice, printers, keyboards, audio amplifiers, CDs, pen-drives, TV sets, cameras, DVDs etc

2.3. Large Household Equipment

Around 98% E-waste (household) generated by big appliances is disposed of in landfills each year. Our negligent dumping, like every other item on our list, costs not only in terms of environmental effect but also in terms of precious metals loss. This category includes major household appliances such as washing machines, freezers, dryers, refrigerators, electric stoves, printing machines, microwave and various panels etc.

2.4. Small House Hold Equipment

Most of the countries provide convenient recycling solutions for minor household equipment but it is very large in numbers and generated E-waste. Small household gadgets, commonly known as consumer electronics such as calculators, watches, toasters, electric shavers, hair dryers, radio sets, video cameras, toys, tools, vacuum cleaners, fryers, small medical devices, and control instruments etc.

2.5. Electric and Electronic Tools

This category includes tools used to prepare materials such as wood, metal, and others. They might be huge or small in size and, like other types of e-waste, are harmful to the environment. Drill and welding machines, lawn mowers, detectors, sewing machines, flash lights, and soldering irons are the best examples of this category.

2.6. Lighting Equipment

Lighting technology has become so common place in modern life that we no longer notice it. You turn on the light without thinking every time you enter in a room. Turning on the light when you wake up in the night is also typical subconscious behavior. What are the most common forms of lighting equipment that have the greatest potential for producing e-waste? Bulb, discharge lamps, tube lights, LED lamps, switches, wires, fluorescent lighting, and sodium vapor lamps are commonly used lighting equipment at everywhere.

2.7. Toys, and Sports Equipment

This category generates most rapidly increasing E-waste. The reason for this is that children and youngsters outgrow their toys rapidly or they become faulty after extensive use. Electric toys, video games, racing items are widely purchased toys by the children.

2.8. Monitoring and Control Instruments

Heating regulator, thermostat, and smoke detector are used in labs, industries and homes are the monitoring and control instruments. From the last one decade these devices are widely used in different areas and responsible for E-waste generation.

2.9. Consumer Equipment

Consumer equipment is one of the highly useable electronic gadgets in home. Washing machines, televisions, LED, smart TV, telephones, refrigerator, calculators, smart phones, personal computers, laptop, digital cameras and MP3 players are very few examples of consumer equipment. People are using these equipments on regular basis and lots of equipments are damage or discarded daily due to any reasons which are responsible of lot of E-waste daily.

2.10. Medical Devices

Pacemakers, hip implants, contraceptive intrauterine devices, drug releasing pumps, defibrillators, and diagnostic equipment are the few examples of electronic medical devices. Moreover, high frequency surgical equipment, therapeutic and diagnostic ultrasound equipment, medical lasers, infant incubators and warmers, and medical beds are few examples of electrical medical devices. These devices are more widely used in hospitals. The likelihood of producing a large volume of e-waste every day rises as a result of everyday use and the ongoing development of new technology.

Global E-waste generation in 2019 on the basis of their categorization is given in the following table 1. The decline in the category of screen and monitors is due to the fact that screen/monitor is slighter flat screen displays have replaced CRT monitors, although no. of pieces continues to grow.

3. ENVIRONMENTAL IMPACTS AND POLLUTION BY E-WASTE

Technological advancement is the root cause of E-waste as it leads to the increased usage of available advanced electronics and tossing of old ones. When this E-waste is out in the open, toxic chemicals are released into the air that harms the environment. This is one of the biggest environmental impacts of E-waste and that is why E-waste is most dangerous specifically during summers. Those toxic substances can also seep into the soil and groundwater, affecting human health and both, land and marine animals. In this manner, E-waste is significantly contributing towards air pollution, water pollution and

Table 1. Categories of e-waste generated in 2019

Category	Small Equipment	Large Equipment	Temp. Exchange Equipment	Screens and Monitors	Small IT and Telecommunication Equipment	Lamps	Ref.
E-waste generated (in Metric ton)	17.4	13.1	10.8	6.7	4.7	0.9	Forti et al. 2020
E-waste increased since 2014 (in %)	+4	+5	+7	-1	+2	+4	Forti et al. 2020

soil pollution. According to the World Health Organization, lots of harmful health effects are caused by direct or indirect exposure to E-waste i.e. breathing problem, premature birth, and metal distress. The Global E-waste monitor 2020 reported a record more than 53 million metric tons of E-waste worldwide in 2019, more than 20% in five years (Forti et al., 2020). By 2030, it will reach 74 million metric tons, virtually doubling in 16 years. The United Nations (UN) warned that because of the ongoing consumer demand for digital devices, there will be time that will lead to a "tsunami of E-waste" if E-waste is not handled efficiently. The major pollutants and hazardous materials from E-waste are shown in figure 3.

3.1. Air Pollution

Developing countries find it difficult to practice formal E-waste recycling procedures because of the costly operational expenses involved. Hence, they focus on practicing informal e-waste practices. Though informal e-waste recycling practices are successful in extracting valuable parts from e-waste, the way in which they are practiced or conducted causes harm to the workforce involved. The dismantling of electronic products release various types of toxins such as dioxins into the air and particulate matters of various sizes causing respiratory health issues to the surrounding mankind and involved workforce.

Figure 3. Major pollutants and hazardous materials from e-waste

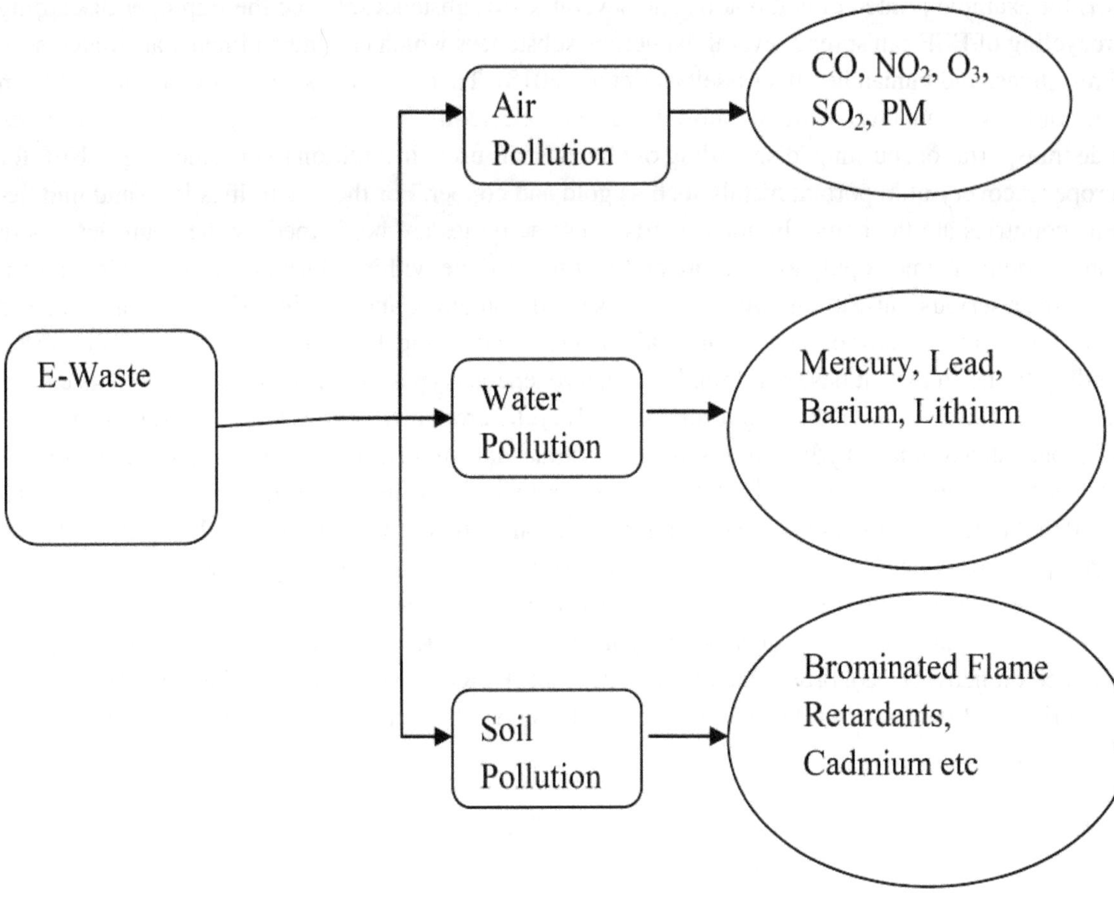

Also, for removing highly valuable raw materials like gold and silver, acids, de-soldering and other chemicals are used which release harmful fumes in areas resulting in severe respiratory and breathing issues (Gangwar et al., 2019, Awasthi, 2018). For example, getting copper from electronic devices, toxic e-waste materials are burned which releases fine particles that pollute the air and result in chronic diseases and cancers affecting human health badly. The chronically polluted air results in damaging the water quality, soil, animal, and plant species.

Polybrominated diphenyl ethers and polybrominated biphenyls are the principal toxic chemical substances that get released when any of the E-waste is burned (Zeng et al., 2010). PBDE and PBBs are used as flame retardants in televisions, computers, and other electronic devices. At the time of disposal of PBDE and PBBs, when it is burnt it releases CO_2 into the air. The release of CO_2 contributes to greenhouse gas emissions and changes in climatic conditions, which ultimately give rise to human-generated global warming. Even the manufacturing of electronic devices releases a large amount of CO_2 into the atmosphere.

In 2019, the E-waste reported (53.6 Mt) was found to have more than 50 tons of Hg and 70 thousand tons of hazardous brominated flame retardants (Lebbie et al., 2021). Having an elevated amount of rare-earth-element (REE), could also be among the important reason of REE pollution (Brewer et al., 2022). At the dumping sights of E-Waste, it is obvious to get heavy metal pollution as metals are being used in electrical and electronic equipments (EEE) as a must. For instance, copper is used in electric wiring and cadmium is utilized in batteries (Kumar & Fulekar, 2019). There are various important components used in EEE for example printed circuit boards and several toxic substances, hence the improper discarding and recycling of EEE can spread several dangerous substances which can inturn impact air, water, soil, food resources, and human health (Abdelbasir et al., 2018). There is always a threat to air quality from such E-waste as it contains hazardous substances and toxic additives. The recycling methods of e-waste include transportation, burning, dismantling of all materials used, in addition to the smelting of EEE for the proper recovery of important metals such as gold and copper. For these activities low- and middle-income countries are the home. In other words, these activities are performed by the countries where low and middle income population is more and that too is done within informal settings which in turn can become a serious cause of air pollution because of the smelting and burning of e-waste, as it release airborne pollutants. Volatile organic compounds are produced during the dismantling process of E-waste. According to the studies, it has been found that heterogeneous types of hazardous air pollutants can be spread during e-waste combustion as it releases polycyclic aromatic hydrocarbons, hydrogen chloride, poly-halogenated aromatic hydrocarbons, dioxins, furans and significant amounts of particulate matter (PM). For the separation of copper from wires in e-waste, open burning is used in Thailand, which can pose a threat to the employees via the inhalation of PM and other pollutants (Bungadaeng et al., 2019). According to a research, The significant source of PM was found to be the illegal e-waste recycling and this can exceed acceptable limits in near future. Unsuitable e-waste recycling can also release various inorganic and organic pollutants such as heavy metals, polycyclic aromatic hydrocarbon (PAH), polychlorinated biphenyl (PCB), brominated flame retardant (BFR), polychlorinated dibenzofuran (PCDF), polychlorinated dibenzodioxin (PCDD), perfluoroalkyl and polyfluoroalkyl substance (PFAS) (Ahirwar et al., 2021).

3.2. Water Pollution

70% of planet earth is water. Deposition of hazardous E-waste into the water supplies like ponds, streams and lakes results in severe acidification and toxification effects alarming biodiversity. Acidification kills marine and freshwater organisms in turn damaging the ecosystem where revitalization is not viable. Nuclear power plants and mining also releases a lot of radioactive elements in water and hence contributing into water pollution.

Unsuitable disposal of EEE items can give rise to water pollution due to the toxic metals present in it. A significant amount of water i.e. approximately 600k litres can be polluted by a single mobile phone battery. The most important compound which is found in all three forms and can pollute water is Mercury. It is a voluminous compound found in e-waste and can last for centuries. According to a recent research, uncontrolled and improper methods of E-waste recycling had a major impact on rice, fish, sea food and vegetables with heavy metals and livestock with constant air pollutants (Lin et al., 2022). The main source of E-waste, either discarding or shipping, is found to be developed countries to other developing countries in Asia such as in Guiyu, where almost seventy percent of E-waste ends up. Studies revealed that approximately 12.5 percent of this waste is recycled via prehistoric procedures, which concludes to the discharge of toxic materials into the eco-system. Apart from constant organic pollutants, there exists various other heavy metals in the groundwater and rivers of Guiyu, which causes its water incompatible for household activities like drinking and cooking (Li et al., 2020). Also, besides polluting the groundwater, E-waste also plays a significant role in producing ample amounts of waste water during the procedure of collecting copper particles, for example, during the used hydraulic shaking bed separation procedure.

3.3. Soil Pollution

When electronic waste is improperly disposed of, it either goes into the landfills abruptly or is dumped illegally. Both of these are clearly not healthy environmental practices and cause potential harm to the underground water and soil.

Electronic waste contains heavy metals and flame retardants which end up contaminating the underground water and the crops. When the soil and water get contaminated due to the presence of toxic heavy metals such as lead, mercury, lithium, barium, and cadmium, they become vulnerable to absorbing these toxins causing dangerous health hazards to mankind. Not just this, but the release of large and heavy particles from the burning, shredding, or dismantling of electronic waste gets quickly re-deposited into the groundwater and soil. And resultantly it starts adversely impacting the soil quality, leading to reduced plant growth, production of contaminated soil and potentially harming the ecosystem.

Various air and water pollutants of E-waste also contribute significantly towards soil pollution. For example, Acid Rain etc. During weathering of certain rocks, soil gets toxic chemicals. Radioactive elements from mining and nuclear power plants also come into the soil and hence reduce soil fertility due to increase in alkalinity, salinity or pH.

E-waste also has a foremost affect on soil and its biological components. The collection of E-waste in particular sites like mountains or the outskirts of cities is affecting the microbial community present at those sites, especially in countries such as China, Pakistan and India. This impact on microbial population in turn affects the soil-ecological function. For example, the heavy metals such as mercury, lead, cadmium, nickel, arsenic, chromium, and consistent organic pollutants present in e-waste can cause a reduction in the normal soil microbial biota (Salam et al 2019). For instance, a research in southeastern China revealed that this kind of soil was badly contaminated with cadmium and copper as a consequence of improper e-waste's dismantling methods (Fang et al., 2022). The foremost cause for spreading vicious substances and heavy metals into the soil environments is open burning. The unethical and improper practices involved in discarding, recycling and dismantling of E-wastes became the foremost cause for the lethal impacts on soil ecosystems in several countries. The metal pollution is also caused by the recycling activities such as getting copper from wires, taking out gold by the use of acidic substances, plastic melting and burning of circuit. All these activities can also cause surface soil pollution because of the presence of heavy metals in e-waste (Preeti & Sayali 2021).

4. RESOURCE DEPLETION DUE TO GENERATION OF E-WASTE

In both developed and developing countries, electrical and electronic equipment has a shorter lifespan because of advancement in technology, resulting in tones of electronic waste. Natural resources are being used more and more in electric and electronic technologies, which are causing their depletion to occur at an alarming rate. Urban mining will immediately stimulate secondary production since there aren't enough metals from the core of the earth to go around. The electronic sector has been using a variety of metals, raw materials, chemicals and polymers in the manufacturing of various components and other parts of the electronics materials (Ramesh et al., 2023, Li et al., 2022). Metals including cobalt, copper, gold, nickel, aluminium, lithium, silver, and tin are among the most popular ones. Some of these materials shield the inner workings of electrical equipment, facilitate cleaning, or enhance their usability. Some are employed in laying out wiring for the gadgets because they are very conductive and perform admirably at moving electricity. There is a significant amount of copper wiring needed to make circuit boards. In the same ways many more metals have been used significantly for making important parts of the electronic gadgets. Up to 69 elements from the periodic table can be found in E-waste are shown in figure 4.

More precious metals are found in IT and communication system chips than in home products. For instance, a cell phone contains more than 40 elements, including basic metals such as copper and tin as

Figure 4. Elements present in the e-waste

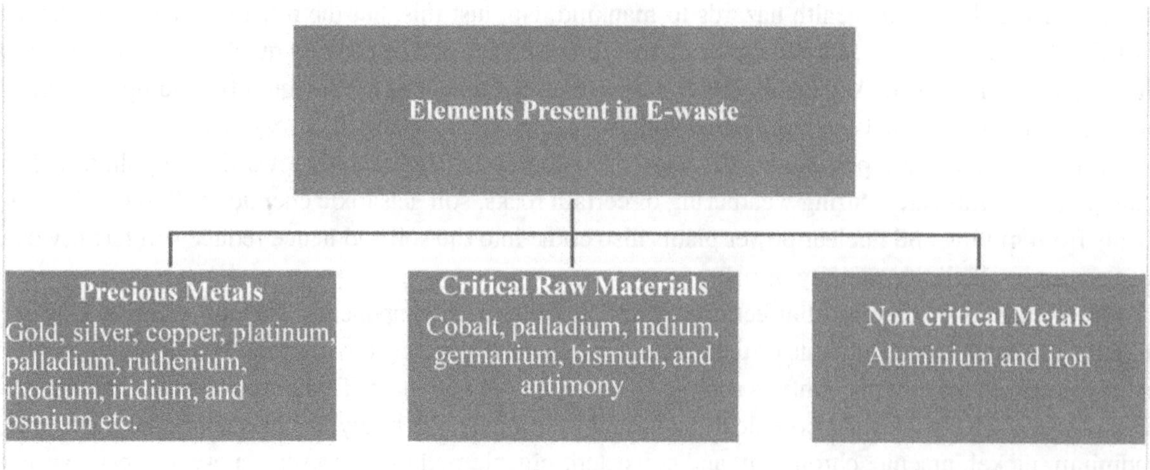

well as some precious metals such as gold, palladium, and silver (Gautam et al., 2022). Many valuable metals, including gold and copper, as well as dangerous substances, including lead, are present in the waste printed circuit boards (PCBs) (Priya & Hait, 2018). To create indium-tin-oxide (ITO), which is utilized in LCD screens, indium is used (Savvilotidou, et al., 2015). In lithium ion batteries, which are frequently used in portable electronic devices including mobile phones, laptops, computers, cameras, and players, Co and Li are most commonly used. LEDs contain poly-metallic materials such as gallium, indium, and gold (Zhan et al., 2015) (Figure 5).

The primary source of metals used in electrical and electronic devices is mining. Information and communication technologies (ICTs) require many natural resources extracted from vast territories, which also depletes other valuable resources like water and energy during manufacture, producing tones of CO_2 emissions. There are typically 10,000 tonnes of CO_2 emissions produced to create one ton of gold, platinum or palladium (Cayumil et al., 2016). The use of precious metals in computers ensures that digital information is transmitted quickly and accurately. More than any other metal, gold best satisfies these requirements. Therefore, there is an average yearly need of 300 tones for gold in electrical and electronic equipments. More than 5 million tones of CO_2 are produced by this extraction alone. The annual CO_2 emission level from other metals such as copper, indium, silver, platinum, ruthenium cobalt, tin

Figure 5. Precious and base metals present in e-waste

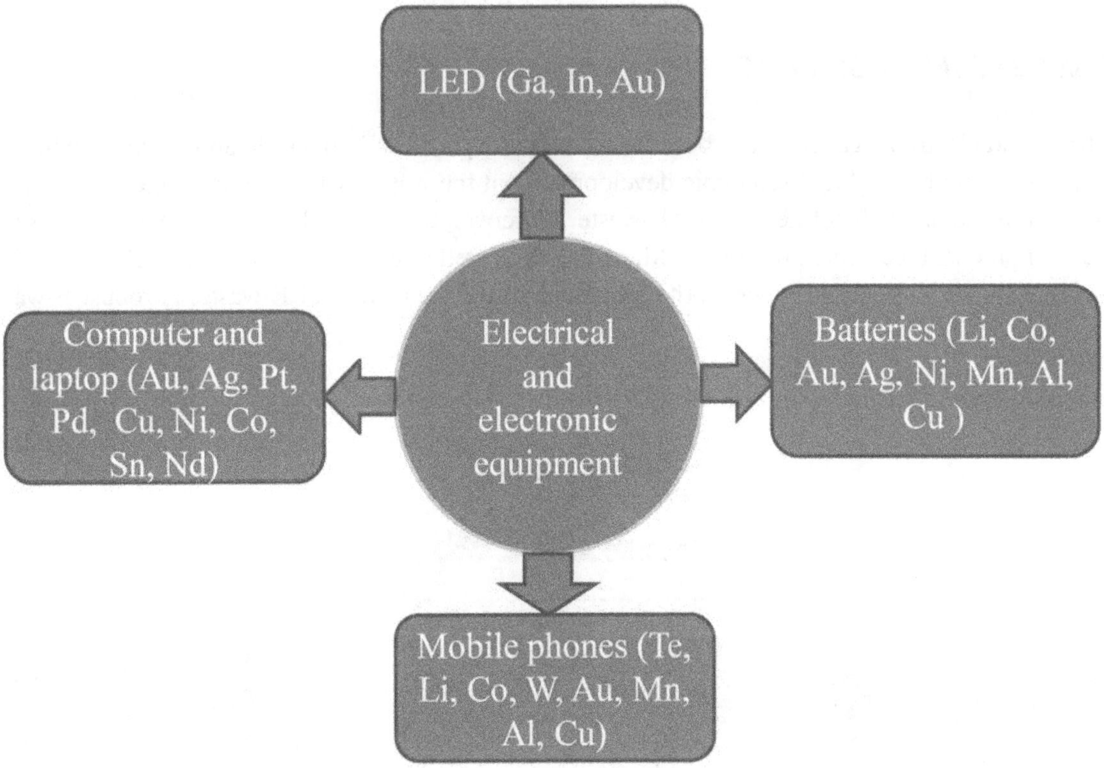

and palladium, used in electrical and electronic devices is more than 20 million tones, which is around 0.001 of the total CO_2 emissions worldwide (Devika, 2010).

It is urgently necessary to keep an eye on the amount and flow of E-waste and to make sure that these monitoring efforts are successful and ongoing in order to safeguard the environment and promote a sustainable society. Recycling is more cost effective than mining for precious metals from E- wastes (He et al., 2018). E-waste can be recycled formally or informally to recover important metals. Open burning to recover precious metals like copper, gold, mercury, platinum, and palladium is known as informal recycling. Recovering of precious metals from E-waste by using of various chemicals such as nitric acid, cyanide and sulfuric acid is known as formal recovery.

Three key phases are involved in recycling of E-waste for formal recovery. The first phase is dismantling, which focuses on locating valuable or hazardous components in the E-waste, second phase is upgrading, which focuses on metallurgical/mechanical processing protocols to get the desirable materials ready for refining, and the last phase is refining, which focuses purifying/retreating the recovered materials to make them suitable for their further use (Atia et al., 2019) (Figure 6).

After being heated, discarded PCBs produced two distinct types of solid products, carbonaceous non-metallic fractions (NMFs) and metallic products made of foils and/or droplets rich in copper, tin-lead droplets rich in copper, and some wires. Two different types of metallic phases were found to include significant amounts of Pt, Au, Pd, and Ag with only very small amounts being retained by the NMFs. Several valuable metals, including Pt, Au, Pd, and Ag were successfully concentrated in a small volume fraction using this procedure, which also resulted in volume reduction of up to 75% for additional processing or refinement (Cayumil et al., 2016).

5. E WASTE MANAGEMENT

Electrical and electronic gadgets have become the part and parcel of human life and its consumption is directly linked with the global economic development but the way in which these gadgets are disposing off is questionable, therefore creating E-waste. Moreover, the disposal is unsustainable because of the lack of practical recycling procedures. Many health as well as environmental related challenges has been faced by various countries across the globe due to the deposition of E-waste. A robust E-waste

Figure 6. Recycling methodology of e-waste

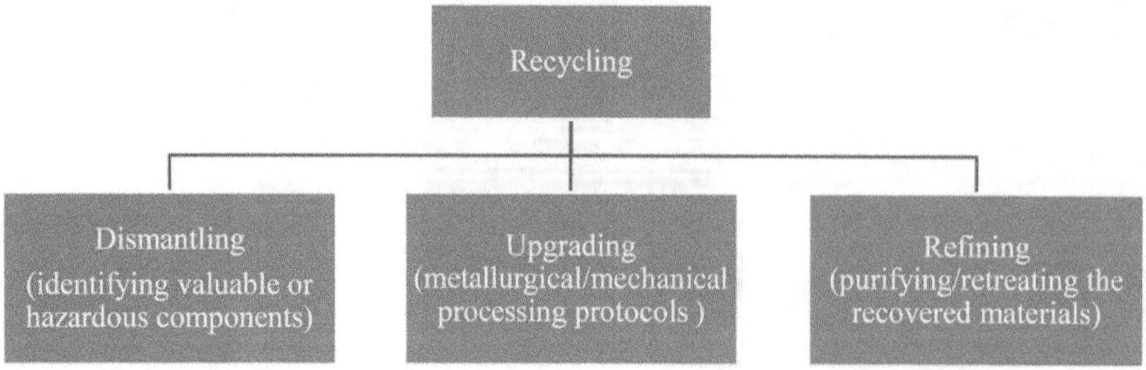

management system is required to monitor the quantities and the flows of E-waste with an aim to provide sustainable disposal solution. Therefore, there is a need for the development of system which comprises of good recycling infrastructure, sound policies and implementable legal procedures for the disposal or management of E-waste.

Growing urbanization, mobility, industrialization etc. leads to the growth of E-gadgets. Globally 53.6 Mt of e-waste has been generated with an average of 7.3 kg per capita, which is projected to be 74.7 Mt by 2030 with an increase of annual average of 2 Mt (Forti et al., 2020) (Figure 7). The generated E-waste is expected to be 110 Mt by 2050, which is the fastest growing waste globally (Parajuly et al., 2019). The estimated value of raw material in e-waste generated in 2019 is 57 bn. USD contains aluminum, copper and iron as main raw material. Out of this, only 17.4% of the global E-waste is documented with a worth of 10 bn. USD and 4 Mt of raw material is available for recycling. Out of these, 8% of the e-waste stays in waste bins leads to landfills or incinerated. These discarded products can be refurnished and reused. The biggest irony is that this E-waste is illegally deported to some low or middle income countries under the guise of being reused but are dumped in big volumes ranging from 7% to 20% generation of E-waste in that country. The main causes of the growing amount of E-waste are mainly by higher consumption rate of E-gadgets, shorter life cycles and few repair options of gadgets.

In 2019, the continent in which collection and recycling of E-waste took place is given in table 2. E-waste systems do not cover batteries or electrical components of vehicles.

Figure 7. Facts about the e-waste generated in 2019

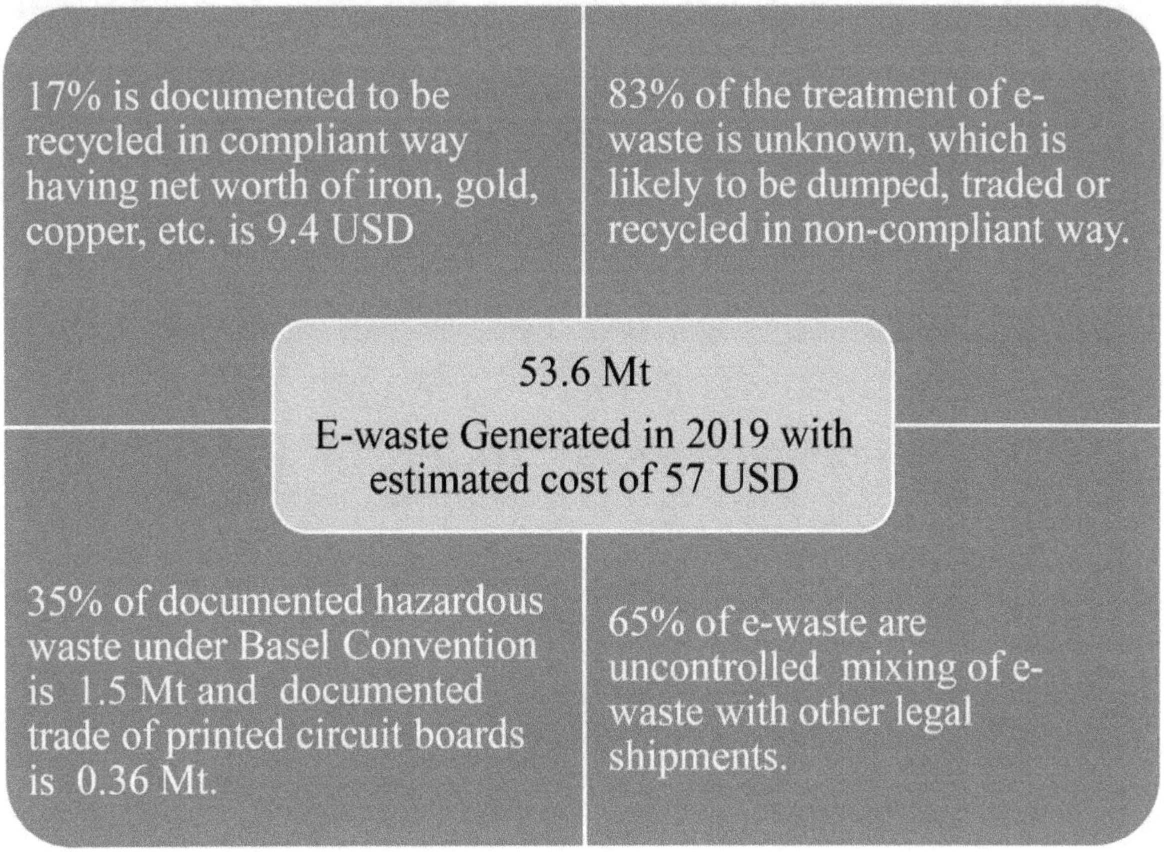

Table 2. Status of e-waste generation in continents

Continent	Europe	Asia	America	Oceania	Africa	Ref.
Generation of E-waste (in Mt)	12	24.9	13.1	0.9	2.9	Andeobu et al. 2021
Per capita generation of E-waste (in kg.)	16.2	5.6	13.3	16.1	2.5	Andeobu et al. 2021
Collection and recycling of E-waste (in %)	42.5	11.7	9.4	8.8	0.9	Andeobu et al. 2021

E-waste management involves a combination of policies and methods from start to finish. These strategic actions should be viewed through the lens of technical availability, affordability, economic, and environment friendly. A sustainable path with strategic intervention based on current management practices is crucial for management of E-waste in effective manners. Figure 8 shows the strategic plan for collection, dumping and management of E-waste in well mannered ways.

It has been studied within the European Union that high-income countries such as US, Australia, Japan, etc. have developed complaint E-waste management infrastructure for collecting E-waste locally through various municipalities and recovering good amount of recyclable components of the collected

Figure 8. Management of e-waste

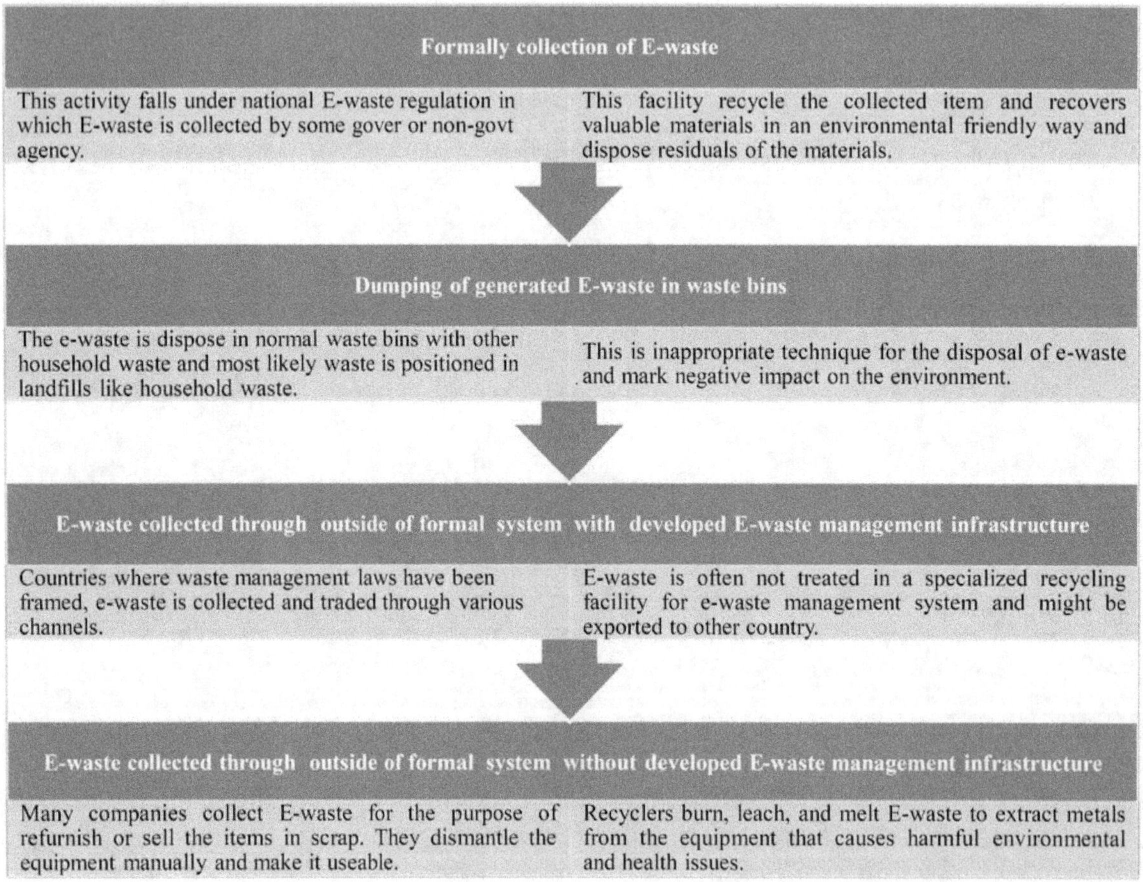

E-waste and disposal of residuals in a complaint way without putting additional burden on the environment. Target studies conducted by various researchers on some countries revealed that significant amount of generated E-waste still unmanaged and disposed without following proper de-pollution steps (Magalini et al., 2019: Balde et al., 2020: Forti et al., 2020). Pello, recycling robots, pneumatic waste pipes, solar-powered trash compactors, E-waste kiosks, recycling apps are the new technologies have been applied to manage these wastes at low cost and sustainable ways. The internet of things (IOT) is implemented by many of these new, emerging debris management technologies to encourage sustainability and accelerate waste collection.

There is some highlighting which help to reduce the E-waste:

- Recycling of the waste is the best ways to reduce waste
- Educate everyone about the impact of E-waste on environment as well as on human beings
- Teach the young generation and child about the reduction of E-waste
- Store your data in the cloud to avoid storage devises
- Purchase very few electronic items and repair if possible
- Sell unused electronic items rather than discarding
- Donate your unused electronic items rather than discarding

6. CHALLENGES AND ISSUES FOR MANAGING E-WASTE

There are many challenges for effective implementation of E-waste management strategies. The amount of E-waste generation is too large to be handled with available infrastructure. Collection and process mechanism for E-waste management is not defined. Moreover, maximum people are not aware about the environmental and health implications related to improper disposal of electronic/electrical products (Liu et al., 2009). The common approach for disposal that has been followed by consumers includes the disposal of small/low-cost items like cables, chargers, batteries etc. along with household waste and selling of products like LCD, washing machine, mobile phones, laptops etc. in informal sector (Rajesh et al., 2022). There is a need for appropriate estimation of e waste generated/imported within a country in order to finalize the requirement of infrastructure for proper handling of waste. In addition to this technological guidance regarding e waste recycling should be provided to the concerned manpower working in the recycling units. Improper processes may also lead to many occupational hazards in workers. Figure 9 depicts the challenges associated with sustainable e waste management.

There are numerous difficulties and challenge in managing various types of waste. Lack of awareness, lack of technological guidance, specified regulations, environmental implications, unorganized waste collection and management, estimation of waste import, lake of infrastructure for sustainable management, occupational hazards, insufficient funds for sustainable recycling methods are the major challenges for sustainable management of E-waste. The infrastructure that is currently in place cannot handle the large volume of E-waste generated. E-waste management collection and processing mechanisms are not yet established properly. Additionally, the majority of people are unaware of the negative effects of improper disposal of electronic and electrical items.

In sustainable E-waste management, there are countless challenges and associated health risks are exponentially increasing. A holistic framework is required to fill the gaps between stakeholders and regulatory bodies in order to solve the issues related to waste management of discarded products. For

Figure 9. Challenges associated with sustainable e-waste management

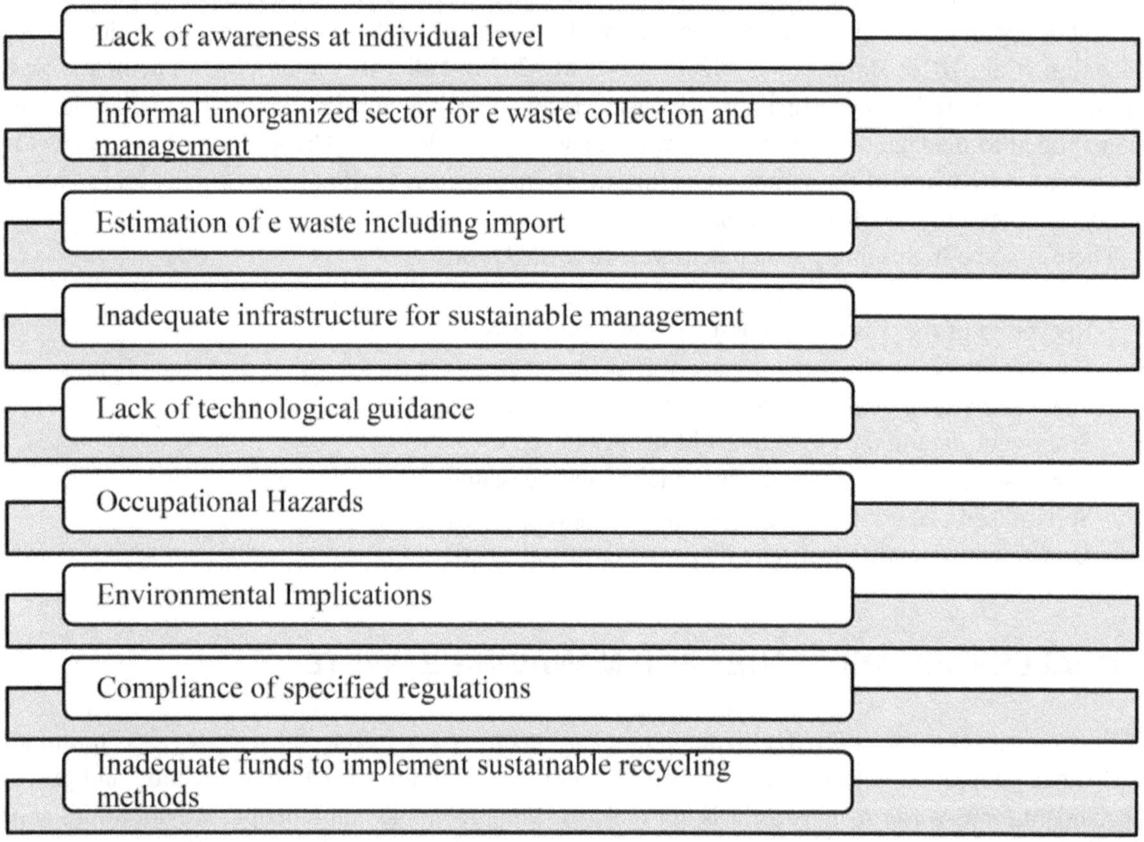

appropriate handling, skilling of workers need to be ensured in order to avoid occupational health hazards and for environment protection. Awareness of consumer has also very integral role and should be taken care of with the help of media or special campaigns. Another important step that can be taken is to minimize the level of toxic materials used in electronic products.

7. CONCLUSION

This study aimed to assess the environmental impact of E-waste and valuable resource depletion. The development of various methods for recycling valuable materials for use in renewable energy systems is being pursued by scientific committee and organizations. Many precious metals that were recovered from E-waste are more valuable than those found in ore. E-waste management rules have started to be implemented by policymakers, allowing technical transfer to recover high-quality metals and other valuable materials from E-waste. Sustainable development goals would be achieved by using recent E-waste management policies. Ag, Pd, Pt, and Au among other precious metals found in waste PCBs, demonstrated a definite propensity to combine with the primary metallic components of waste PCBs, notably Cu, Pb, and Sn, to produce solutions for their recovery in heat treatment process.

REFERENCES

Abdelbasir, S. M., Hassan, S. S., Kamel, A. H., & El-Nasr, R. S. (2018). Status of electronic waste recycling techniques: A review. *Environmental Science and Pollution Research International*, *25*(17), 16533–16547. doi:10.100711356-018-2136-6 PMID:29737485

Ahirwar, R., & Tripathi, A. K. (2021). E-waste management: A review of recycling process, environmental and occupational health hazards, and potential solutions. *Environmental Nanotechnology, Monitoring & Management*, *15*, 100409. doi:10.1016/j.enmm.2020.100409

Ananno, A. A., Masud, M. H., Dabnichki, P., Mahjabeen, M., & Chowdhury, S. A. (2021). Survey and analysis of consumers' behaviour for electronic waste management in Bangladesh. *Journal of Environmental Management*, *282*, 111943. doi:10.1016/j.jenvman.2021.111943 PMID:33465720

Andeobu, L., Wibowo, S., & Grandhi, S. (2021). An assessment of e-waste generation and environmental management of selected countries in Africa, Europe and North America: A systematic review. *The Science of the Total Environment*, *792*, 148078. doi:10.1016/j.scitotenv.2021.148078 PMID:34147806

Atia, T. A., Elia, G., Hahn, R., Altimari, P., & Pagnanelli, F. (2019). Closed-loop hydrometallurgical treatment of end-of-life lithium ion batteries: Towards zero-waste process and metal recycling in advanced batteries. *Journal of Energy Chemistry*, *35*, 220–227. doi:10.1016/j.jechem.2019.03.022

Awasthi, A. K., Wang, M., Awasthi, M. K., Wang, Z., & Li, J. (2018). Environmental pollution and human body burden from improper recycling of e-waste in China: A short-review. *Environmental Pollution*, *243*, 1310–1316. doi:10.1016/j.envpol.2018.08.037 PMID:30268981

Bagwan, W. A. (2024). Electronic waste (E-waste) generation and management scenario of India, and ARIMA forecasting of E-waste processing capacity of Maharashtra state till 2030. *Waste Management Bulletin*, *1*(4), 41–51. doi:10.1016/j.wmb.2023.08.002

Baldé, K., D'Angelo, E., Luda, V., Deubzer, O., & Kühr, R. (2022). *Global transboundary e-waste flows monitor 2022*. United Nations Institute for Training and Research (UNITAR).

Bekabil, U. T. (2020). Industrialization and environmental pollution in Africa: An empirical review. *J. Resour. Dev. Manag*, *69*, 18–21.

Brewer, A., Dror, I., & Berkowitz, B. (2022). Electronic waste as a source of rare earth element pollution: Leaching, transport in porous media, and the effects of nanoparticles. *Chemosphere*, *287*, 132217. doi:10.1016/j.chemosphere.2021.132217 PMID:34826916

Bungadaeng, S., Prueksasit, T., & Siriwong, W. (2019). Inhalation exposure to respirable particulate matter among workers in relation to their e-waste open burning activities in Buriram Province, Thailand. *Sustainable Environment Research*, *29*(1), 1–12. doi:10.118642834-019-0030-7

Cayumil, R., Khanna, R., Rajarao, R., Mukherjee, P. S., & Sahajwalla, V. (2016). Concentration of precious metals during their recovery from electronic waste. *Waste Management (New York, N.Y.)*, *57*, 121–130. doi:10.1016/j.wasman.2015.12.004 PMID:26712661

Devika, S. (2010, November). Environmental impact of improper disposal of electronic waste. In *Recent Advances in Space Technology Services and Climate Change 2010 (RSTS & CC-2010)* (pp. 29–31). IEEE. doi:10.1109/RSTSCC.2010.5712793

Fang, J., Zhang, L., Rao, S., Zhang, M., Zhao, K., & Fu, W. (2022). Spatial variation of heavy metals and their ecological risk and health risks to local residents in a typical e-waste dismantling area of southeastern China. *Environmental Monitoring and Assessment*, *194*(9), 604. doi:10.100710661-022-10296-1 PMID:35867165

Forti, V., Balde, C. P., Kuehr, R., & Bel, G. (2020). *The Global E-waste Monitor 2020: Quantities, flows and the circular economy potential.* Academic Press.

Gangwar, C., Choudhari, R., Chauhan, A., Kumar, A., Singh, A., & Tripathi, A. (2019). Assessment of air pollution caused by illegal e-waste burning to evaluate the human health risk. *Environment International*, *125*, 191–199. doi:10.1016/j.envint.2018.11.051 PMID:30721825

Gautam, P., Behera, C. K., Sinha, I., Gicheva, G., & Singh, K. K. (2022). High added-value materials recovery using electronic scrap-transforming waste to valuable products. *Journal of Cleaner Production*, *330*, 129836. doi:10.1016/j.jclepro.2021.129836

He, P., Wang, C., & Zuo, L. (2018). The present and future availability of high-tech minerals in waste mobile phones: Evidence from China. *Journal of Cleaner Production*, *192*, 940–949. doi:10.1016/j.jclepro.2018.04.222

Jain, M., Kumar, D., Chaudhary, J., Kumar, S., Sharma, S., & Verma, A. S. (2023). *Review on E-waste management and its impact on the environment and society*. Waste Management Bulletin. doi:10.1016/j.wmb.2023.06.004

Khatiwada, B., Jariyaboon, R., & Techato, K. (2023). E-waste management in Nepal: A case study overcoming challenges and opportunities. *e-Prime-Advances in Electrical Engineering, Electronics and Energy*, *4*, 100155.

Kumar, P., & Fulekar, M. H. (2019). Multivariate and statistical approaches for the evaluation of heavy metals pollution at e-waste dumping sites. *SN Applied Sciences*, *1*(11), 1–13. doi:10.100742452-019-1559-0

Lebbie, T. S., Moyebi, O. D., Asante, K. A., Fobil, J., Brune-Drisse, M. N., Suk, W. A., Sly, P. D., Gorman, J., & Carpenter, D. O. (2021). E-waste in Africa: A serious threat to the health of children. *International Journal of Environmental Research and Public Health*, *18*(16), 8488. doi:10.3390/ijerph18168488 PMID:34444234

Li, W., & Achal, V. (2020). Environmental and health impacts due to e-waste disposal in China–A review. *The Science of the Total Environment*, *737*, 139745. doi:10.1016/j.scitotenv.2020.139745 PMID:32516663

Li, Z., Qiu, F., Tian, Q., Yue, X., & Zhang, T. (2022). Production and recovery of tellurium from metallurgical intermediates and electronic waste-A comprehensive review. *Journal of Cleaner Production*, *366*, 132796. doi:10.1016/j.jclepro.2022.132796

Lin, S., Ali, M. U., Zheng, C., Cai, Z., & Wong, M. H. (2022). Toxic chemicals from uncontrolled e-waste recycling: Exposure, body burden, health impact. *Journal of Hazardous Materials*, *426*, 127792. doi:10.1016/j.jhazmat.2021.127792 PMID:34802823

Liu, Q., Li, K. Q., Zhao, H., Li, G., & Fan, F. Y. (2009). The global challenge of electronic waste management. *Environmental Science and Pollution Research International*, *16*(3), 248–249. doi:10.100711356-009-0131-7 PMID:19308473

Magalini, F., Thiébaud, E., & Kaddouh, S. (2019). Quantifying WEEE in Romania 2019 vs 2015. Sofies UK.

Manikandan, S., Inbakandan, D., & Nachiyar, C. V. (2023). Towards sustainable metal recovery from e-waste: A Mini Review. *Sustainable Chemistry for the Environment*, 100001.

Mor, R. S., Sangwan, K. S., Singh, S., Singh, A., & Kharub, M. (2021). E-waste management for environmental sustainability: An exploratory study. *Procedia CIRP*, *98*, 193–198. doi:10.1016/j.procir.2021.01.029

Murphy, D., Meyers, C., Oelson, S., McKean, K., & Custer, S. (1998). *Exceptions*. Sopris West CO.

Ni, H. G., Zeng, H., Tao, S., & Zeng, E. Y. (2010). Environmental and human exposure to persistent halogenated compounds derived from e-waste in China. *Environmental Toxicology and Chemistry*, *29*(6), 1237–1247. doi:10.1002/etc.160 PMID:20821565

Parajuly, K., Kuehr, R., Awasthi, A. K., Fitzpatrick, C., Lepawsky, J., Smith, E., Widmer R., & Zeng, X. (2019). *Future e-waste scenarios*. Academic Press.

Preeti, M., & Sayali, A. (2021). Scientometric analysis of research on end-of-life electronic waste and electric vehicle battery waste. *J. Scientometr. Res*, *10*(1), 37–46. doi:10.5530/jscires.10.1.5

Priya, A., & Hait, S. (2018). Comprehensive characterization of printed circuit boards of various end-of-life electrical and electronic equipment for beneficiation investigation. *Waste Management (New York, N.Y.)*, *75*, 103–123. doi:10.1016/j.wasman.2018.02.014 PMID:29454818

Rajesh, R., Kanakadhurga, D., & Prabaharan, N. (2022). Electronic waste: A critical assessment on the unimaginable growing pollutant, legislations and environmental impacts. *Environmental Challenges*, *7*, 100507. doi:10.1016/j.envc.2022.100507

Ramesh, M. V., Paramasivan, M., Akshay, P., & Jarin, T. (2023). A review on electric and electronic waste material management in 21st century. *Materials Today: Proceedings*. Advance online publication. doi:10.1016/j.matpr.2023.01.057

Savvilotidou, V., Hahladakis, J. N., & Gidarakos, E. (2015). Leaching capacity of metals–metalloids and recovery of valuable materials from waste LCDs. *Waste Management (New York, N.Y.)*, *45*, 314–324. doi:10.1016/j.wasman.2015.05.025 PMID:26087646

Sinha, S. (2011). *Waste Electrical and Electronic Equipment: The EU and India, Sharing Best Practices*. Toxics Link.

Wang, F., Huisman, J., Baldé, K., & Stevels, A. (2012, September). *A systematic and compatible classification of WEEE. In 2012 Electronics Goes Green 2012*. IEEE.

Yabe, J., Ishizuka, M., & Umemura, T. (2010). Current levels of heavy metal pollution in Africa. *The Journal of Veterinary Medical Science*, *72*(10), 1257–1263. doi:10.1292/jvms.10-0058 PMID:20519853

Zeng, X., & Li, J. (2016). Measuring the recyclability of e-waste: An innovative method and its implications. *Journal of Cleaner Production*, *131*, 156–162. doi:10.1016/j.jclepro.2016.05.055

Zhan, L., Xia, F., Ye, Q., Xiang, X., & Xie, B. (2015). Novel recycle technology for recovering rare metals (Ga, In) from waste light-emitting diodes. *Journal of Hazardous Materials*, *299*, 388–394. doi:10.1016/j.jhazmat.2015.06.029 PMID:26150281

Chapter 3
E-Waste Collection and Recycling Best Practices and Innovations

Pathapati Preethi
Woxsen University, India

Thotakura Deepak
Woxsen University, India

Siddadapu Venkata Anusha
Woxsen University, India

D. Bhagya
https://orcid.org/0000-0002-1685-5930
Sapthagiri College of Engineering, India

ABSTRACT

Over the last two decades, the usage of electronic equipment in human life has been increasing dramatically and the life span of electronic products is short. E-waste is growing in the world, because they do not biodegrade easily, and most of them are not reused or recycled. In this chapter, the authors have explained the best practices and innovative approaches in e-waste collection and recycling, aiming to address environmental concerns, conserve resources, and promote sustainable waste management. The main problem is the release of toxic chemicals like lead, chromium, manganese, and polybrominated diphenyl ethers (PBDEs) from e-waste leads to many environmental and health issues as most recycling facilities are located in low-income areas. In those mainly women and children often participate in e-waste recycling as a form of income and are frequently exposed to hazardous pollutants. Some of the health effects include impaired learning and memory functions; altered thyroid, estrogen, and hormone systems; and neurotoxicity.

DOI: 10.4018/979-8-3693-1018-2.ch003

1. INTRODUCTION

The concept of E-waste collection and recycling has come to light due to the increasing use of electronic items, the amount of pollution they exposed to air, land, and soil, and the hazardous compounds they contain such as lead, chromium, mercury, polyvinyl chloride, gallium arsenide, cadmium, beryllium, and others. When these compounds eventually reach our bodies. This finally harms us. There should be suitable procedures in place to manage E-waste.

The impact of technological advancement and incorporation in day-to-day living gadgets has been seen globally in recent years, empowering individuals in the form of electronic appliances or devices. Appliances, such as refrigerators, washing machines, stoves, televisions, battery-powered toys, and mobile phones, are prevalent in houses regardless of necessity or economic level. Personal cameras, laptops, cell phones, air conditioners, computers, and security systems have become commonplace in upper-income families. All of these technologies have found an irreplaceable place in our lives and have both positive and negative impacts. The darker side of technology developments in the electronics industry referred to as the poison of the digital age, has been e-waste created from EOL electronic gadgets utilized in every household (Bhaskar & Turaga, 2018).

E-waste is detrimental to our environment. Assume you go to a mobile shop to exchange your phone. Instead of recycling the merchandise, some retailers often burn them. This allows them to extract copper, silver, and gold. When these technologies are burned, hazardous gases and chemicals are released. Hydrocarbons, brominated dioxins, and other heavy particles are among the harmful chemicals. Because these particles persist in the air, humans ultimately inhale them. Animals inhale these as well.

The Quantity of electronic trash created and reported by various organizations frequently differs. The main cause is that the estimating techniques differ and the units they use to describe themselves are not standard. The lack of adequate guidelines by ISO or any other international authority organization for the assessment of e-waste creation is another factor (Bhaskar & Turaga, 2018).

The optimum number of vehicles and the proper routing are necessary for efficient collection to minimize unnecessary collection costs and potential profit losses. A flexible and reliable communication system might also increase the amount of rubbish that is legitimately collected between the community and the garbage collection company. The field of e-waste recycling and emerging technologies have seen new trends emerge, and internet-based garbage collection services have already been investigated in China and India. Investigated the primary factors and challenges that affect Chinese individual decisions to recycle e-waste through e-commerce. The desire to recycle via e-commerce generation was shown to be favorably correlated with attitudes, personal standards, and the simplicity of e-commerce, despite the fact that residents did not all agree to use it (Nowakowski & Pamuła, 2020).

Electronic waste recycling is a gold mine from an economic standpoint since it may produce a wide range of priceless materials, including palladium, copper, iron, gold, aluminium, platinum, and silver. Technically the majority of these metals are recoverable and recyclable. Recycling of e-waste also adds to the issue of unemployment in several nations, including China, which employs around 100,000 recyclers at one of its facilities. Hazardous substances including lead, mercury, and cadmium, as well as gases and chemicals like flame retardants and chlorofluorocarbons, may all be avoided through recycling, Global e-waste is estimated by the United Nations University to be worth more than $62 billion in raw materials (Aboelmaged, 2021).

1.1 Purpose of Chapter

This chapter will also explain innovative ways of collecting E-waste which include installing reverse vending machines that collect pieces of e-waste and deliver incentives or rewards in return is a great way to encourage people to recycle their electronic devices. Mobile E-waste collection units, such as specifically constructed trucks or trailers, move to different neighborhoods or events to collect E-waste directly from the community. For example, these mobile collecting units may collect E-waste during an event in a different part of the city. Initiatives in schools and places of employment encourage students and workers to recycle their outdated electronic equipment by establishing E-waste collecting programs at schools and places of employment. E-waste drop-off kiosks in public locations such as shopping malls, libraries, and transit hubs would make it easier for individuals to dispose of their electronic trash. Platforms for the online trading of e-waste should be developed. These platforms should allow users to trade or sell their outdated electronic equipment to other people who may find such gadgets valuable, hence prolonging the usable life of these devices. Competitions or schools to see who can collect the most E-waste will develop a feeling of community engagement and responsibility.

Certainly, there are methods for recycling electronic waste that have the potential to improve sustainability, resource recovery, and environmental protection. Scientists and researchers are encouraging the use of non-harmful microorganisms like bacteria, fungi, and algae to carry out a "clean" recycling of electronic trash. The idea of "E-waste to energy" is a game-changer. It is a way to deal with the growing problem of electronic garbage and to use the promise of renewable energy production. Make machines that can take apart and sort electronic devices quickly and efficiently, sorting valuable parts from things that can't be recycled. Automation can make recycling work better and protect people from dangerous chemicals (Andeobu et al., 2021). Using the latest chemical processes to dissolve and collect metals from E-waste, with solvents that are good for the environment and less need for ways that use a lot of energy. Urban mining is the idea of getting valuable metals out of old technology that has been thrown away in cities. By taking valuable metals out of old electronics.

2. LITERATURE REVIEW

The global demand for smarter electrical and electronic equipment (EEE) is increasing every year, resulting in massive amounts of obsolete EEE ending up in landfills. Its unsustainable methods of collecting, treating, and disposing of waste E-waste, pose enormous threats to the environment. With this rising e-waste, the formal sectors lack the infrastructure, technology, manufacturing, and expertise of e-waste in an environmentally responsible manner. This article expresses the global best practices in the field of E-waste management to shed light on the importance of policy implementation, technology requirements, and social awareness to achieve it (Murthy & Ramakrishna, 2022). The research proposes strategies for establishing long-term e-waste recycling systems, including potential processing routes and scenarios that integrate the factors involved in e-waste collection, pre-treatment, and metal extraction, and mentions the alternative method of electronics to green electronics which can reduce the e-waste (Kurniawan et al., 2022). The various disposal methods for potential pollutants of lead, mercury, lead, cadmium, etc. are Landfilling, Incineration, Recycling, and Re-use of e-waste, where the article expresses the methods of informal recycling i.e., chemical recycling, mechanical and Thermal recycling methods. Any recycling process involves the dismantling, or removal of various parts of e-waste containing hazardous substances such as Polychlorinated biphenyls (PCBs) and Hg, as well as the separation of plastic, removal of cathode ray tubes (CRTs), the separation of ferrous and non-ferrous metals, printed circuit boards. Strong acids are used to remove precious metals and electronics like computers, printers, and cell phones can be recycled by using filling or slight modifications, which is one of the alternative methods to reduce the e-waste generation (Annamalai, 2015). In this, the e-waste chemicals are disposed of by using the methods of selecting samples and strong chemical disposal methods to reduce the e-waste (Wong et al., 2007). Informal recyclers obtain e-waste from local waste and recover the targeted metals such as copper, aluminum, iron steel, etc. using rudimentary and primitive methods, which result from heavy environmental pollutants in the atmosphere. These metals recover through open burning and acid leaching, which are harmful to the environment, instead of these traditional methods, new innovative methods are discovered which levels of treatment are Decontamination, Dismantling, Segregation, Hammering, Shredding, Special treatment process, Pulverization, and electronic separation methods are suggested to reduce the e-waste materials (Vats & Singh, 2014). Lifecycle assessment (LCA) is now a widely used tool in e-waste management because it can estimate the impact of material consumption as well as evaluate the environmental and economic aspects. Extended Producer Responsibility (EPR), on the other way, deals with e-waste on a national scale and holds manufacturers accountable for their end-of-life products. Some developed countries like Japan and Switzerland, have adopted EPR principles, which are well supported by electronic companies. To comprehensively understand the combination of LCA and EPR principles may be the best model for managing e-waste in the future. To achieve sustainable development, electronic companies are encouraged to invest capital in establishing reverse supply chains to handle their end-of-life products (Doan et al., 2019). This article suggests when designing and implementing incentive schemes, firms consider operating transparency in e-waste management and allowing variation in collection activity and treatment process. It also suggests to manage successfully e-waste, it is vital to implement the development of eco-design equipment, proper e-waste collection, recycling of valuable materials using appropriate techniques, disposing of e-waste appropriately, preventing illegal trade of used electronic devices to developing countries and raise the awareness about the negative impacts of e-waste. By using EPR channel techniques, e-waste can be dismantled or recycled in a proper handling manner and disposed of such waste.

3. COLLECTION METHODS

3.1 E-Waste in the Residential Sector

Residential E-waste management involves the proper processing, collecting, recycling, and disposal of domestic E-waste. E-waste is a broad category of electronic and electrical products that have outlived their usefulness or are no longer needed. This category may include ancient computers, laptops, mobile phones, televisions, refrigerators, washing machines, and other electrical equipment and gadgets.

Curbside collection: In certain areas, the government schedules curbside collection days during which homeowners can deposit E-waste products, such as outdated laptops and small gadgets, alongside their usual garbage for pickup.

Drop-off locations: Many municipalities have designated drop-off locations or collecting events where citizens may bring their E-waste for appropriate disposal or recycling.

Take-back programs from electronic stores: some electronic stores provide take-back programs that allow customers to return old gadgets while purchasing new ones.

One of the most creative techniques for collecting E-waste from the residential sector is the deployment of a "Smart E-waste collection system". This approach combines technology and convenience to encourage and expedite domestic electronic waste collection. Create a mobile app that locals may use on their cell phones. The software acts as a central platform for scheduling and tracking E-waste collections. Residents may use the app to arrange a convenient time for E-waste removal from their houses. They can select a certain collection day and time. Encouraging sustainability by ensuring that collected E-waste is recycled or repaired effectively, with an emphasis on reducing environmental impact and optimizing resource recovery (Wang et al., 2016).

The smart E-waste collection system makes E-waste disposal simple and enjoyable for residents by combining technology, convenience, and data-driven insights. This strategy may considerably increase participation rates and assure responsible E-waste management by offering a user-friendly mobile app and combining contemporary and inventory management tools.

3.2 E-Waste in the Commercial and Industrial Sectors

The appropriate treatment, recycling, and disposal of electronic waste created by businesses, organizations, and industrial facilities is referred to as E-waste management in the commercial and industrial sectors. Because of the usage of electronic equipment and gadgets in different commercial and industrial processes, their sector generates a substantial amount of E-waste. Effective E-waste management in this industry is critical for environment preservation, resource conservation, regulatory compliance, and long-term business operations.

Programs for internal collection: Internal E-waste collection initiatives are frequently established within larger corporations' premises. They may have designated collection bins or spaces where staff may discard outdated gadgets.

E-waste recycling companies: To handle their E-waste, many organizations collaborate with specialist E-waste recycling companies or suppliers. These businesses offer pickup services and make certain that everything is in order.

One of the most creative approaches for collecting E-waste from the commercial and industrial sectors is the establishment of an "E-waste exchange platform". This platform uses technology and market dynamics to help companies, recyclers, and manufacturers exchange E-waste effectively and sustainably. Create an online marketplace dedicated to E-waste trades. This portal connects E-waste generators with qualified recyclers and firms in need of raw materials. Businesses publish their E-waste goods on the site, offering extensive information about the kind, amount, and condition of the equipment. They can upload images and designate the status of data sanitization or deletion. Make transparent bidding and price methods available so that purchasers may submit bids for the listed E-waste goods. Businesses can select the most appealing deals. Integrate logistics and transportation services into the platform certified recyclers and purchasers may schedule pickups directly through the website, making the collection process more efficient (Widmer et al., 2005).

The E-waste exchange platforms establish a dynamic and transparent E-waste marketplace, promoting a circular economy by encouraging enterprises to regard E-waste as a resource rather than a waste. It improves sustainability, decreases landfill waste, and encourages commercial and industrial electronic equipment recycling and reuse.

3.3 E-Waste in Educational Institutions

The appropriate treatment, recycling, and disposal of E-waste created by educational institutions such as schools, colleges, universities, and research institutes are referred to as E-waste management in the educational sector. Because they rely on electronic technology for teaching, research, and administration, educational institutions frequently generate considerable volumes of E-waste. Effective E-waste management in this industry is critical for environmental preservation, regulatory compliance, and the promotion of sustainable practices.

E-waste collection drives: E-waste collection drives are regularly organized by schools, colleges, and universities to encourage students and employees to bring in their old electronic gadgets for recycling or appropriate disposal.

Initiatives in schools and workplaces that encourage students and employees to recycle outmoded electronic equipment by establishing E-waste collecting programs are crucial for encouraging sustainability and proper e-waste disposal. The effort starts with awareness initiatives in schools and workplaces. These ads inform students and staff about the environmental effects of E-waste and advantages of proper disposal. Schools and workplaces have designated collection stations. These collection locations are often located in public places or near building entrances.

These are clear criteria in place for what sorts of electronic trash are allowed. This includes old laptops, desktop computers, displays, keyboards, mouse, mobile phones, chargers, cables, and other peripherals. Guidelines also describe the state of objects (for example, whether or not they must be in functioning order) and may exclude dangerous products. To minimize overflow and guarantee smooth operations, regular schedules for collecting and emptying the bins are created.

Collaborations with recycling companies: educational institutions may form collaborations with E-waste recycling companies to handle E-waste in an economically friendly manner.

Transportation of collected E-waste to approved recycling centers or facilities has been planned. Participant's feedback is sought to discover areas for improvement to make them more successful and engaging (Borthakur & Govind, 2017).

3.4 E-Waste in Manufacturing and Technology Companies

E-waste management in manufacturing and technology refers to the complete procedures and techniques used by these businesses to treat, recycle, and dispose of E-waste created during their operations. Due to the employment of electronic components, equipment, and gadgets in the manufacturing and product development processes, manufacturing and technology organizations frequently generate a significant amount of E-waste. E-waste is critical in various areas for environmental protection, regulatory compliance, resource conservation, and sustainability.

Inventory and Evaluation: Take a full inventory of electronic equipment gadgets, and components involved in manufacturing operations. Examine the condition, functioning, and lifetime of each item to determine which should be discarded or replaced.

Reuse and refurbishment: Consider reusing or renovating electronic equipment that is still functioning but no longer required wherever possible. This method can assist in increasing gadget lifespan and minimize E-waste creation.

Hazardous waste management: Hazardous materials such as lead, mercury, and chemicals may be present in some electrical equipment used in production. To avoid environmental pollution, strictly follow legislation controlling the disposal of hazardous E-waste.

One of the most inventive approaches for collecting E-waste from industrial and technology organizations is the establishment of a "Reverse supply chain E-waste collection network". This network combines cutting-edge technology with sustainability principles to produce a streamlined, ecologically responsible, and effective approach to collecting electronic waste from various businesses. Implement blockchain technology to develop a transparent and secure E-waste monitoring system. On the blockchain, each electric gadget and component is issued a unique digital identification. This slows for real-time tracking of E-waste throughout its full course, from disposal to recycling or reuse. Install Internet of Things (IoT) enabled smart E-waste collecting containers. These bins can check their fill levels automatically and give alarms when they need to be emptied. IoT sensors can also detect and classify E-waste by kind, making the sorting process more efficient (Schwarzer et al., 2005).

3.5 E-Waste in Environmental Organizations

In the environmental business, E-waste management refers to the methods and tactics used by environmental organizations and agencies to handle the issues connected with E-waste. These groups are critical in encouraging ecologically appropriate E-waste disposal, recycling, and long-term practices.

Campaigns for public awareness: Environmental groups frequently organize public awareness campaigns to educate the public on the necessity of proper E-waste disposal and to offer information on various collection options.

Collaborations: They may work with other industries, such as retail or government, to help with E-waste collection and recycling.

One of the most creative techniques for collecting E-waste from the environmental business is the creation of a "smart E-waste collection network". This method makes use of cutting-edge technology and data-driven tactics to simplify the collection of E-waste from various environmental organizations and agencies. Install Internet of Things-enabled E-waste collecting bins at key places on the grounds of environmental groups, recycling facilities, and government institutions. These bins are outfitted with sensors that continuously check their fill levels. Use smart routing algorithms to improve E-waste collection routes using data from IoT devices. These algorithms may prioritize overflowing bins, guaranteeing efficient and timely collection. Using historical data, use predictive analytics to forecast E-waste creation patterns. This information aids in the proactive scheduling of collections and efficient allocation of resources. Publish sustainability indicators such as carbon footprint reduction, resource conservation, and landfill diversion to demonstrate the environmental advantages of participating in the smart E-waste collecting network (Zeng et al., 2017).

4. RECYCLING METHODS

4.1. By Using Harmful Micro-Organisms

The use of non-harmful microorganisms, like bacteria, fungi, and algae, in electronic waste recycling involves a sustainable and environmentally friendly approach to break down and recover useful materials from discarded E-waste. This process is often referred to as "bioremediation" or "biological recycling".

To begin, non-harmful microorganisms known for their capacity to metabolize or interact with E-waste components are chosen. The microorganisms used are selected based on the components in the electronic trash and the anticipated recycling outcomes.

First, the E-waste is prepared for bioremediation. Disassembling electronic equipment to reveal its components, removing hazardous items such as batteries, and ensuring that the trash is clean and free of pollutants are all examples of what it entails.

The processed E-waste is subsequently introduced to the microorganisms of choice. Bacteria, fungi, and algae are employed because they naturally degrade or interact with certain elements used in electrical equipment. Some bacteria, for example, may remove precious metals like old, silver, or copper from electronic circuit boards via a technique known as bioleaching. Plastics and organic components included in E-waste can be broken down by fungi. Heavy metals and other toxins can be absorbed by algae (Wu et al., 2019).

Microorganisms metabolize, convert, or accumulate particular components of E-waste over time. They may, for example, release enzymes or acids that break down things into simpler forms.

4.2. E-Waste to Energy

The process of transforming waste material into energy in the form of electricity or heat is known as waste to energy. This is a low-cost energy generation method. This technique is also known as a sustainable waste management chain. This method also yields numerous biofuels, including CBG and Biodiesel. CBG is a novel type of biomass fuel. NextGen Energia is one of the companies that is effectively revitalizing the energy industry by giving franchise opportunities to invest in CBG/CNG pumps as well as CBG/CNG-producing facilities or projects.

The shredded E-waste is incinerated or burned in specially built facilities known as waste-to-energy plants. High temperatures are generated during the combustion process, which breaks down the organic components of E-waste, creating heat and gases.

E-waste to Energy advantages: E-waste to energy helps remove electronic trash from landfills, lowering the environmental strain and saving valuable landfill space.

Resource recovery: After combustion, valuable metals such as copper and aluminum can be recovered, helping to resource conservation.

Renewable energy: Because it uses waste products as a fuel source, electricity generated from E-waste to energy processes is often considered a kind of renewable energy.

Environmental advantages: E-waste to energy plants that are properly managed can lessen the environmental effect of electronic trash and the emission of dangerous chemicals associated with typical waste disposal techniques (Ahirwar & Tripathi, 2021).

4.3. Chemical Process

The most recent chemical procedures for dissolving and collecting metals from e-waste promise to create ecological effects, efficient, and energy-efficient alternatives. These techniques make use of solvents and technologies that are intended to have a low environmental effect while optimizing metal recovery. Environmentally friendly solvents are used in these new techniques to selectively dissolve the metals found in E-waste. These solvents are carefully chosen to do as little harmful to the environment and human health as possible while still recovering target metals.

E-waste is often pre-processed before using chemical techniques to remove non-metallic and hazardous components. This guarantees that the emphasis is on obtaining precious metals while minimizing environmental hazards.

In a controlled atmosphere, the E-waste is subsequently exposed to the solvent of choice. The solvent interacts with the metals, selectively dissolving them based on their chemical characteristics. Different metals may require different solvents or extraction conditions.

The solvent-metal interaction causes chemical reactions in a solution that result in the production of metal complexes. These chemicals are then separated from the rest of the E-waste.

The dissolved metals are recovered from the metal-rich solution. Precipitation, electrolysis, or other separation processes may be used. The ultimate objective is to obtain pure complexes.

Efforts are undertaken to collect and recycle the solvent used in the process to reduce waste environmental effects. This minimizes solvent usage and the overall environmental footprint (Kaya, 2016).

4.4. Urban Mining

The process of extracting valuable resources, resources, such as metals and minerals, from E-waste, discarded electronics, and other urban waste streams is referred to as urban mining. It is an environmentally beneficial method of recycling that strives to eliminate the need for traditional mining and the environmental effects connected with it.

The process of extraction of valuable minerals from E-waste, notably metals like gold, silver, copper, and rare earth elements, is the focus of urban mining.

Electronic items like computers, cell phones, televisions, and home appliances are examples of E-waste.

Households, corporations, and electronic recycling facilities all contribute to the collection of E-waste.

The gathered E-waste is sorted and classified to identify the materials and components.

Before urban mining can begin, E-waste is generally disassembled, and non-metallic or dangerous components such as plastic, batteries, and circuit boards are removed.

Preprocessing helps to speed up the urban mining process while also lowering environmental hazards.

To recover metals from E-waste, urban mining involves a variety of metallurgic processes. Leaching, smelting, and solvent extraction are examples of this method.

The process used is determined by the individual metal and its shape in E-waste.

Chemical techniques are sometimes used to selectively dissolve and recover metals from E-waste. These procedures are intended to be both effective and ecologically beneficial.

To reduce environmental damage, environmentally friendly solvents are frequently used.

As a result, metals are extracted and subjected to purification and refining procedures. To get high-purity metals, this may entail melting and casting.

Metals recovered from urban mining may be utilized to make new electrical devices, lowering the demand for virgin resources (Zeng et al., 2018).

5. CONCLUSION

The study concludes that E-waste is determinantal to human beings and also the environment, which requires adequate treatment to gain environmental sustainability. E-waste management is made more difficult by weak management systems that lack official collection, classification, and a recycling mechanism. This will eventually influence environmental sustainability and harm health and the natural ecology. When not properly managed, E-waste poses a tremendous danger to our environment, public health, and natural resources. We can avoid these hazards and utilize the unique resources inherent in E-waste by implementing systematic collecting and recycling activities.

Recycling E-waste not only saves valuable metals, minerals, and energy, but it also minimizes the discharge of dangerous pollutants and greenhouse gases into the environment.

Furthermore, it promotes job development, economic growth, and technical innovation in the recycling and sustainable electronic sector.

To have a real impact, governments, corporations, communities, and people must work together to raise awareness, create effective recycling systems, and promote responsible electronic usage. By doing so, we can get closer to a more sustainable and environmentally friendly future (Chandrasekaran et al., 2018).

REFERENCES

Aboelmaged, M. (2021). E-waste recycling behavior: An integration of recycling habits into the theory of planned behavior. *Journal of Cleaner Production*, *278*, 124182. doi:10.1016/j.jclepro.2020.124182

Ahirwar, R., & Tripathi, A. K. (2021). E-waste management: A review of recycling process, environmental and occupational health hazards, and potential solutions. *Environmental Nanotechnology, Monitoring & Management*, *15*, 100409. doi:10.1016/j.enmm.2020.100409

Andeobu, L., Wibowo, S., & Grandhi, S. (2021, August 27). A Systematic Review of E-Waste Generation and Environmental Management of Asia Pacific Countries. *International Journal of Environmental Research and Public Health*, *18*(17), 9051. doi:10.3390/ijerph18179051 PMID:34501640

Annamalai, J. (2015). Occupational health hazards related to informal recycling of E-waste in India: An overview. *Indian Journal of Occupational and Environmental Medicine*, *19*(1), 61. doi:10.4103/0019-5278.157013 PMID:26023273

Bhaskar, K., & Turaga, R. M. R. (2018). India's E-Waste Rules and Their Impact on E-Waste Management Practices: A Case Study. *Journal of Industrial Ecology*, *22*(4), 930–942. doi:10.1111/jiec.12619

Borthakur, A., & Govind, M. (2017). How well are we managing E-waste in India: Evidences from the city of Bangalore. *Energy, Ecology & Environment*, *2*(4), 225–235. doi:10.100740974-017-0060-0

Chandrasekaran, S. R., Avasarala, S., Murali, D., Rajagopalan, N., & Sharma, B. K. (2018). Materials and energy recovery from e-waste plastics. *ACS Sustainable Chemistry & Engineering*, *6*(4), 4594–4602. doi:10.1021/acssuschemeng.7b03282

Doan, L. T. T., Amer, Y., Lee, S. H., & Phuc, P. N. K. (2019). Strategies for E-waste management: A literature review. *International Journal of Energy and Environmental Engineering*, *13*(3), 157–162.

Kaya, M. (2016). Recovery of metals and nonmetals from electronic waste by physical and chemical recycling processes. *Waste Management (New York, N.Y.)*, *57*, 64–90. doi:10.1016/j.wasman.2016.08.004 PMID:27543174

Kurniawan, K., Soefihara, M. D. A., Nababan, D. C., & Kim, S. (2022). Current status of the recycling of e-waste in Indonesia. *Geosystem Engineering*, *25*(3-4), 83–94. doi:10.1080/12269328.2022.2142856

Murthy, V., & Ramakrishna, S. (2022). A review on global E-waste management: Urban mining towards a sustainable future and circular economy. *Sustainability (Basel)*, *14*(2), 647. doi:10.3390u14020647

Nowakowski, P., & Pamuła, T. (2020). Application of deep learning object classifier to improve e-waste collection planning. *Waste Management (New York, N.Y.)*, *109*, 1–9. doi:10.1016/j.wasman.2020.04.041 PMID:32361385

Schwarzer, S., De Bono, A., Giuliani, G., Kluser, S., & Peduzzi, P. (2005). E-waste, the hidden side of IT equipment's manufacturing and use. *Environment Alert Bulletin, 5*.

Vats, M. C., & Singh, S. K. (2014). Status of e-waste in India-A review. *transportation, 3*(10), 16917-16931.

Wang, Z., Guo, D., & Wang, X. (2016). Determinants of residents' e-waste recycling behavior intentions: Evidence from China. *Journal of Cleaner Production*, *137*, 850–860. doi:10.1016/j.jclepro.2016.07.155

Widmer, R., Oswald-Krapf, H., Sinha-Khetriwal, D., Schnellmann, M., & Böni, H. (2005). Global perspectives on e-waste. *Environmental Impact Assessment Review*, *25*(5), 436–458. doi:10.1016/j.eiar.2005.04.001

Wong, M. H., Wu, S. C., Deng, W. J., Yu, X. Z., Luo, Q., Leung, A. O. W., Wong, C. S. C., Luksemburg, W. J., & Wong, A. S. (2007). Export of toxic chemicals–a review of the case of uncontrolled electronic-waste recycling. *Environmental Pollution*, *149*(2), 131–140. doi:10.1016/j.envpol.2007.01.044 PMID:17412468

Wu, Q., Du, Y., Huang, Z., Gu, J., Leung, J. Y., Mai, B., Xiao, T., Liu, W., & Fu, J. (2019). Vertical profile of soil/sediment pollution and microbial community change by e-waste recycling operation. *The Science of the Total Environment*, *669*, 1001–1010. doi:10.1016/j.scitotenv.2019.03.178 PMID:30970449

Zeng, X., Duan, H., Wang, F., & Li, J. (2017). Examining environmental management of e-waste: China's experience and lessons. *Renewable & Sustainable Energy Reviews*, *72*, 1076–1082. doi:10.1016/j.rser.2016.10.015

Zeng, X., Mathews, J. A., & Li, J. (2018). Urban mining of e-waste is becoming more cost-effective than virgin mining. *Environmental Science & Technology*, *52*(8), 4835–4841. doi:10.1021/acs.est.7b04909 PMID:29616548

Chapter 4
E-Waste Dilemma:
Unravelling the Heath and Equity Issues

Dibya Nandan Mishra
https://orcid.org/0000-0002-9918-115X
Department of Management Studies, National Institute of Technology, Silchar, India

Bhairab Chandra Patra
Indian Institute of Management Sambalpur (IIM Sambalpur), India

ABSTRACT

The proliferation of electronic devices in today's digital age, while ushering in technological advancements, has given rise to a pressing global issue: the exponential accumulation of electronic waste (e-waste). Not only is this a grave environmental concern, but its ramifications also spill over into health and equity domains. Delving into the social dimensions of e-waste, this chapter seeks to spotlight the pressing health risks associated with improper e-waste management and the equity concerns inherent in its production, disposal, and recycling. Drawing from various studies, the chapter posits that addressing the e-waste conundrum requires a multi-pronged approach, emphasizing the importance of global collaboration, regulatory frameworks, public awareness, and technological interventions.

1. INTRODUCTION

The rise of the digital age, underpinned by rapid technological advancements and innovations, has catalyzed the proliferation of electronic devices across the globe. With this boom, however, has come an unintended consequence: electronic waste, more commonly termed as e-waste. This mounting crisis, often shadowed by other pressing environmental issues, carries with it profound implications that ripple across ecological, socio-economic, and health spheres.

1.1 Historical Overview

The genesis of the e-waste problem can be traced back to the mid-20th century, when the first wave of electronic consumer goods began permeating households. Radios, televisions, and eventually personal

DOI: 10.4018/979-8-3693-1018-2.ch004

computers became symbols of modernity and progress (Wilson, J., 2009). As manufacturers produced more advanced, compact, and affordable versions of these devices, the lifespan of electronic goods started to diminish. The trend towards built-in obsolescence, spurred by rapid innovations and the consumer's desire for the latest technology, amplified the volume of discarded devices (Kang, H., & Schoenung, J. M., 2005).

1.2 Defining E-Waste

At its core, e-waste encompasses a vast array of discarded electronic devices. This includes, but is not limited to, computers, smartphones, televisions, refrigerators, and even toys embedded with electronic components. Notably, these discarded electronics are a potent mix of materials – some valuable like gold and copper, and others hazardous like mercury and lead (Kiddee, P., Naidu, R., & Wong, M. H., 2013).

1.3 The Global Scale of the Issue

The sheer volume of e-waste generated worldwide is staggering. As of 2020, it was estimated that 53.6 million metric tons of e-waste was generated, with only a fraction being appropriately recycled (Baldé, C. P., Forti, V., & Kuehr, R., 2020). This global discrepancy in e-waste generation and its management creates hotspots, particularly in countries with lax regulations and oversight.

1.4 Environmental Ramifications

The environmental impacts of improperly managed e-waste are multifaceted. Landfills brimming with discarded electronics leach toxic substances into the soil and groundwater, resulting in long-term contamination. Moreover, informal e-waste recycling hubs, particularly prevalent in parts of Asia and Africa, employ rudimentary methods such as open burning. This releases toxic fumes laden with dioxins, furans, and heavy metal particulates into the atmosphere, which not only degrades air quality but also exacerbates the greenhouse effect (Chan, J. K., & Wong, M. H., 2013).

1.5 Socio-economic and Health Dimensions

Beyond environmental concerns, the e-waste conundrum has grave socio-economic implications. Informal recycling sectors, often devoid of health and safety standards, employ marginalized communities, including children. These workers, in their bid to extract valuable metals, are exposed to a cocktail of hazardous chemicals, leading to an array of health issues ranging from skin and respiratory ailments to more severe conditions like neurotoxicity and carcinogenic effects (Grant, K., Goldizen, F. C., Sly, P. D., et al., 2013).

Moreover, the e-waste trade has economic ramifications. While it provides a source of income for many in the informal sector, the externalized health and environmental costs far outweigh the immediate economic benefits. Furthermore, the shipment of e-waste from developed to developing countries under the guise of "second-hand goods" further complicates the issue, creating a disparity in e-waste responsibilities (Widmer, R., Oswald-Krapf, H., Sinha-Khetriwal, D., Schnellmann, M., & Böni, H., 2005).

As the curtain rises on the e-waste narrative, it's evident that the issue is not just about discarded electronics. It's a tale of rapid technological progression, societal consumption patterns, environmental

Figure 1. Outlook on global e-waste generation 2019-2030
Source: Statista 2023

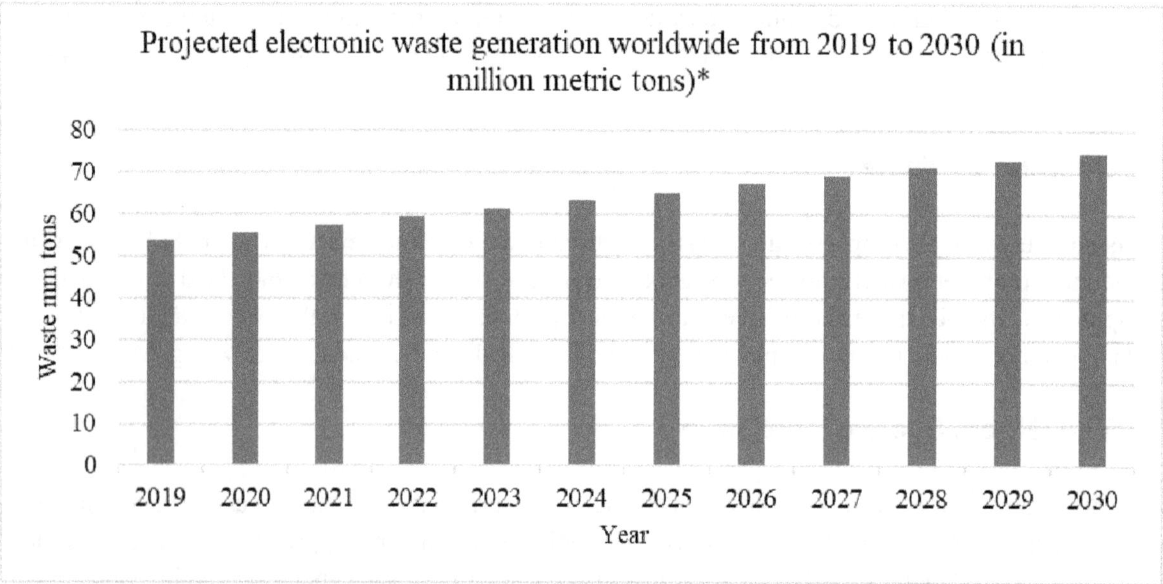

degradation, and the often-overlooked human cost. Tackling this behemoth requires a multi-pronged approach, encompassing stringent regulations, technological innovations in recycling, and a shift in societal perceptions towards consumption and disposal. Figure 1 shows the global e-waste generation in million metric tons) and the forecast till 2030.

This chapter aims to solve the following research questions:

- How have historical and technological factors contributed to the global rise of e-waste and shaped current consumption and disposal patterns?
- In what ways do global disparities in e-waste generation impact the environment, socio-economic conditions, and health of diverse populations?
- What specific health risks are associated with toxic substances in e-waste, and how do these risks affect vulnerable populations across regions?
- How do equity issues manifest in e-waste management, considering global responsibilities, the exploitation of vulnerable communities, and environmental justice?

To solve the above framed research questions the objective formed for the chapter are:

1. Conduct a literature review to understand the historical and technological factors influencing the global rise of e-waste and their impact on consumption and disposal patterns.
2. Identify patterns and disparities in generation, management practices, and their effects on the environment, socioeconomics, and health.
3. Investigate the health risks associated with toxic substances in e-waste, considering demographic vulnerabilities and regional variations.

4. Develop recommendations for policymakers, industries, and communities based on findings to address the social impacts of e-waste, emphasizing strategies for more equitable and sustainable electronic waste management.

2. HEALTH RISKS OF E-WASTE

2.1 Toxic Substances in E-Waste

E-waste is composed of a myriad of materials, many of which are hazardous to human health and the environment. The disposal of e-waste, when done improperly, releases these toxic materials into our surroundings (Oteng-Ababio, 2012).

Lead is one of the most prevalent heavy metals found in e-waste, especially in soldering of printed circuit boards and in cathode ray tube (CRT) monitors. Prolonged exposure to lead can result in damage to the nervous and reproductive systems, and it can impair child development (Needhidasan, Samuel, & Chidambaram, 2014).

Mercury is another dangerous element commonly found in flat-screen monitors, thermostats, and printed circuit boards. Mercury ingestion or inhalation can lead to brain, lung, and kidney damage (Schluep et al., 2013).

Cadmium, frequently used in rechargeable batteries and semiconductors, is carcinogenic and can lead to kidney failure (Kiddee, Naidu, & Wong, 2013).

Brominated flame retardants are used to reduce the flammability of electronic products. However, they can disrupt endocrine activity, leading to neurodevelopmental disorders (Herat & Agamuthu, 2012).

2.2 Health Impacts on Communities

The communities surrounding informal e-waste recycling sites face severe health consequences. Most of these sites employ rudimentary methods such as burning cables to extract copper, releasing dioxins and furans, which are highly toxic (Sepúlveda et al., 2010).

Children in these communities often have elevated levels of toxins in their blood. A study in Guiyu, China, a major e-waste recycling hub, found that children had significantly higher levels of lead in their blood than those from a control town (Huo et al., 2007).

Pregnant women exposed to e-waste toxins risk having children with birth defects and lowered IQs (Zheng et al., 2013).

2.3 Occupational Hazards

Workers in informal e-waste sectors face a high risk of exposure to toxic chemicals. Many are unaware of the hazards associated with e-waste and lack the necessary protective equipment (Asante et al., 2012). Inhalation of toxic fumes and direct contact with hazardous materials can lead to skin rashes, nausea, headaches, and respiratory problems (Wong, Wu, & Deng, 2007).

Chronic exposure can lead to severe health complications, including lung disease, reproductive disorders, and even certain types of cancers (Ogunseitan, Schoenung, Saphores, & Shapiro, 2009).

Figure 2. E-waste components affecting human health
Source: Impacts of E-Waste on Health—Blog. (n.d.)

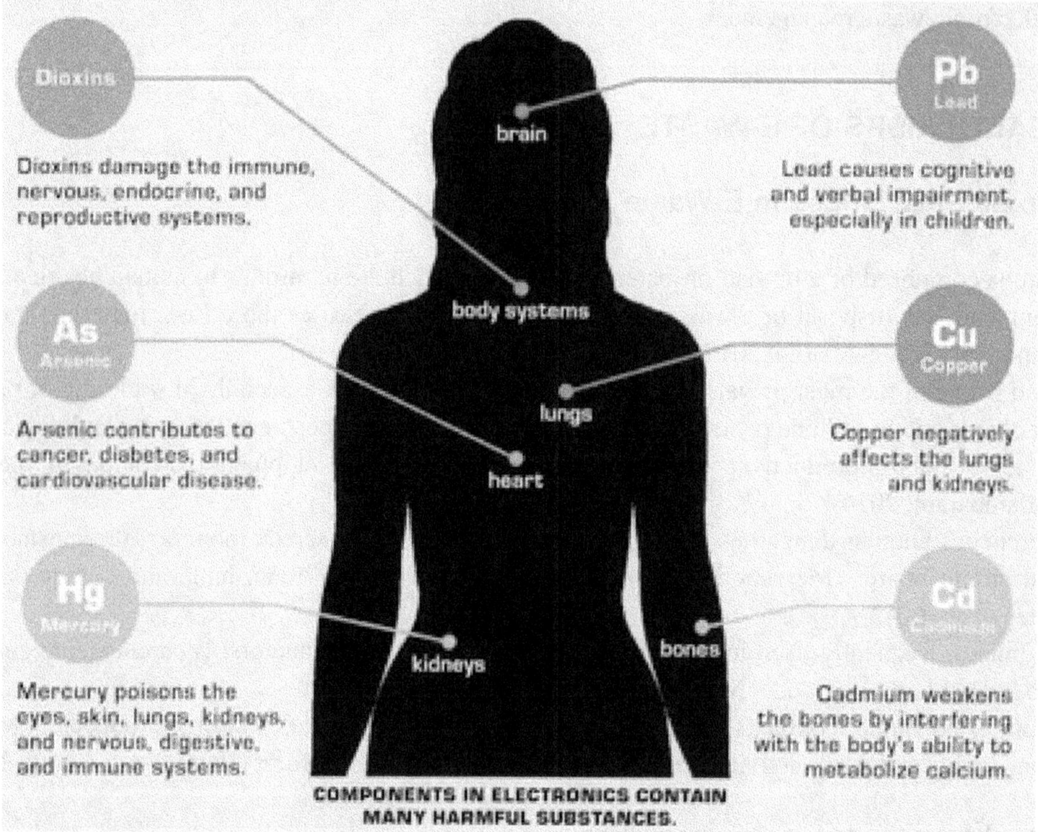

3. ENVIRONMENTAL IMPACTS OF E-WASTE

3.1 Soil Contamination

Improper disposal of e-waste significantly affects the soil. Heavy metals such as lead, cadmium, and mercury can leach into the soil, affecting its quality and leading to contamination of the food chain. This contamination can affect both flora and fauna, leading to a decline in crop yields and affecting the organisms that rely on the soil for survival (Liu, Tang, & Chen, 2013).

Soil contamination can also affect groundwater, which poses risks to both human and ecological health. Communities relying on such contaminated groundwater sources can suffer from various ailments and diseases (Tsydenova & Bengtsson, 2011).

3.2 Air Pollution

The informal recycling practices, especially the burning of e-waste to recover metals, release toxic fumes and chemicals into the atmosphere. Pollutants such as dioxins, furans, and polychlorinated biphenyls

(PCBs) are highly toxic and can cause respiratory diseases and other health complications. The release of these chemicals also contributes to global warming and climate change (Chan & Wong, 2013).

3.3 Water Pollution

E-waste dumped near water bodies or in landfills can contaminate surface and groundwater. The heavy metals and chemicals leaching from e-waste can make their way into rivers and seas, affecting marine life. Fish exposed to such pollutants can suffer from reproductive issues, and humans consuming such fish can ingest these harmful substances (Widmer et al., 2005).

3.4 Impact on Biodiversity

Biodiversity can suffer significant losses due to e-waste contamination. Species exposed to the harmful chemicals present in e-waste can face reduced reproductive rates, shortened lifespans, and even extinction in extreme cases. Moreover, the loss of one species can have a cascading effect on the ecosystem, impacting the food chain and affecting various other organisms (Manomaivibool, 2009).

Figure 3 shows different categories of pollution because of e-waste.

4. EQUITY ISSUES IN E-WASTE MANAGEMENT

4.1 Global Disparities in E-Waste Generation and Management

Developed countries, due to their higher rates of technology usage and rapid technological turnover, generate a substantial portion of the world's e-waste. However, a concerning practice is the export of

Figure 3. Environmental and health effects
Source: Adanu et al. (2020)

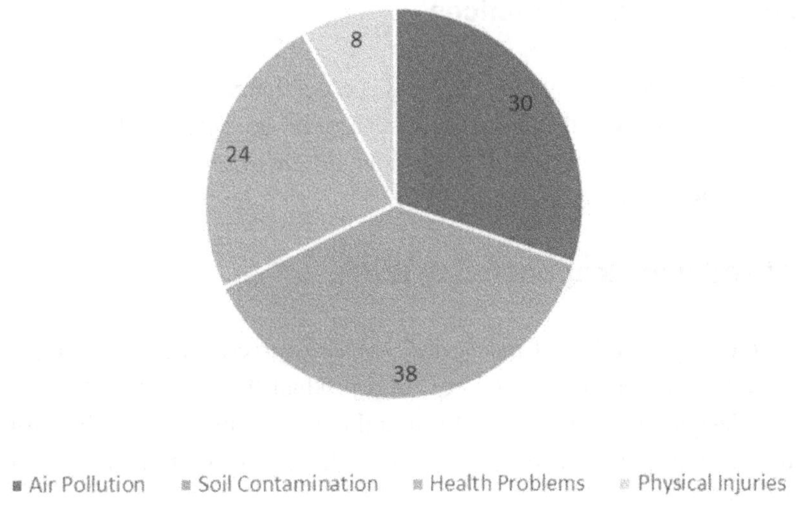

e-waste from these developed nations to developing countries under the guise of 'used electronics' or charitable donations (Schluep et al., 2013). This can create an undue burden on the receiving countries, often leading to unsafe recycling practices and environmental degradation, highlighting the disparities in e-waste management responsibilities (Ongondo, Williams, & Cherrett, 2011).

4.2 The Exploitation of Vulnerable Populations

In many developing nations, e-waste recycling has become a source of livelihood for marginalized communities. Yet, the lack of proper facilities and awareness means that these individuals often engage in risky recycling processes, endangering their health. Moreover, they may not receive fair compensation for their work, underscoring the exploitation inherent in the e-waste trade (Pickren, 2014).

4.3 Environmental Injustice and Marginalized Communities

Marginalized communities often bear the brunt of e-waste dumping and the consequent environmental degradation. Without adequate resources or political power, these communities find it challenging to resist e-waste imports or demand better recycling practices, perpetuating a cycle of environmental injustice (Lepawsky, 2015).

4.4 The Digital Divide and Its Implications

The digital divide, a term signifying the gap in access to information and communication technologies, exacerbates the e-waste problem. As affluent nations continue to update their electronic equipment, items deemed 'obsolete' in one country may still be valuable in another, less-developed nation. Addressing this divide can reduce e-waste generation and encourage a more equitable distribution of technology (Perkins, Kjellén, & Nehler, 2017).

5. ADDRESSING THE SOCIAL IMPACTS OF E-WASTE

5.1 Safe E-Waste Recycling Practices

The establishment of safe e-waste recycling facilities is crucial in mitigating the hazards associated with e-waste disposal. Governments and international organizations must establish and enforce stringent guidelines to ensure recycling processes are both safe and environmentally friendly (Grant, Goldizen, Sly, Brune, Neira, van den Berg, & Norman, 2013).

5.2 Extended Producer Responsibility (EPR)

EPR policies shift the responsibility of managing e-waste from consumers or governments to the producers. By incorporating the environmental cost of a product throughout its lifecycle, EPR promotes sustainable product design and environmentally sound recycling practices. It incentivizes manufacturers to develop products with a longer lifespan and fewer hazardous materials (Manomaivibool, 2009).

5.3 Promoting Circular Economy

The circular economy model prioritizes product reuse, refurbishment, and recycling. By designing products to be repurposed at the end of their lifecycle, manufacturers can significantly reduce the generation of e-waste. Furthermore, the adoption of a circular economy would necessitate collaboration between governments, businesses, and consumers to instill environmentally friendly consumption habits (Ghisellini, Cialani, & Ulgiati, 2016).

5.4 Public Awareness and Advocacy

Educating the public about the societal and environmental impacts of e-waste is essential for fostering responsible electronic consumption. Awareness campaigns can be spearheaded by governments, NGOs, and community organizations, emphasizing the importance of recycling electronics and advocating for safer e-waste management policies (Oswald & Reller, 2010).

5.5 Policy Harmonization across Borders

Given the transboundary nature of e-waste, harmonizing policies across countries can significantly improve management. Regional agreements can foster collaboration, setting universally accepted guidelines and promoting shared technological solutions. The Basel Convention serves as a framework, guiding the transboundary movements of hazardous waste and setting criteria for its environmentally sound management (Balde, Wang, Kuehr, & Huisman, 2015).

5.6 Technological Innovations in E-Waste Management

Investing in technological advancements can lead to more efficient recycling processes. Developing technologies can separate and recover valuable materials from e-waste, reducing the environmental impact and improving economic returns. Collaboration between academia, industry, and governments can spur innovations in e-waste recycling technologies (Kellner & Goe, 2016).

5.7 Financial Mechanisms and Incentives

Government incentives can stimulate better e-waste management. Subsidies or tax breaks can be provided to companies that adopt green manufacturing processes or that establish advanced recycling facilities. Similarly, imposing penalties or taxes on products that are not environmentally friendly can act as a deterrent, pushing manufacturers towards sustainable practices (Saphores, Nixon, Ogunseitan, & Shapiro, 2006).

5.8 Stakeholder Engagement

Engaging multiple stakeholders, including manufacturers, retailers, consumers, and NGOs, can promote holistic solutions to the e-waste dilemma. Workshops, conferences, and public forums can be organized to brainstorm solutions, share knowledge, and develop collaborative strategies (Song, Yang, & Liu, 2015).

6. CONCLUSION

As the global e-waste dilemma intensifies, understanding its multifaceted nature is paramount. The management of electronic trash has evolved beyond mere environmental concerns, firmly entrenching itself in the realms of public health, social justice, and equity. The detrimental health impacts, especially in disadvantaged communities, and the deep-seated issues of social disparity underscore the urgency of robust intervention.

To begin, the crux of the matter lies in comprehending the vast health ramifications. As illustrated, toxins emanating from improperly handled e-waste pose acute and chronic health threats, with the vulnerable sections of society—like children, pregnant women, and the elderly—bearing the brunt (Puckett, Westervelt, Gutierrez, & Davis, 2002). Further exacerbating the scenario is the occupational exposure faced by workers in informal recycling sectors, where rudimentary methods and lax safety measures heighten health risks (Chi, Streicher-Porte, Wang, & Reuter, 2011).

Equity issues in e-waste management paint a similarly grim picture. Developed countries, while being prolific producers of e-waste, frequently resort to exporting their electronic trash to less developed regions. This unsavory practice not only shirks responsibility but burdens already strained systems in recipient nations, leading to environmental degradation and heightened health risks (Widmer et al., 2005). Such exportation of waste creates power imbalances and intensifies environmental injustices.

Yet, amidst this bleak setting, there's a beacon of hope. Solutions like proper recycling practices, the promotion of a circular economy, and public advocacy are already proving transformative in some regions. The key is collaboration—be it between nations via policy harmonization, industries through technological innovations, or communities rallying for change. Instruments like Extended Producer Responsibility (EPR) push the envelope, promoting eco-friendliness right from product design to its eventual disposal (Ongondo, Williams, & Cherrett, 2011).

In conclusion, e-waste, with its inherent complexities, demands a cohesive and multifaceted approach. Addressing the intertwined issues of health risks and equity is a formidable challenge, but with concerted effort and global unity, it's a challenge humanity can surmount. The path to a more equitable and sustainable future, while littered with obstacles, is navigable, and every step taken in the right direction has the potential to forge a lasting impact.

REFERENCES

Adanu, S., Francis, G., & Attah, M. (2020). Challenges of adopting sustainable technologies in e-waste management at Agbogbloshie, Ghana. *Heliyon*, *6*(8), e04548. doi:10.1016/j.heliyon.2020.e04548 PMID:32802976

Asante, K. A., Agusa, T., Biney, C. A., Agyekum, W. A., Bello, M., Otsuka, M., Itai, T., Takahashi, S., & Tanabe, S. (2012). Multi-trace element levels and arsenic speciation in urine of e-waste recycling workers from Agbogbloshie, Accra in Ghana. *The Science of the Total Environment*, *424*, 63–73. doi:10.1016/j.scitotenv.2012.02.072 PMID:22446112

Baldé, C. P., Forti, V., & Kuehr, R. (2020). *The Global E-waste Monitor 2020: Quantities, flows, and the circular economy potential. United Nations University (UNU)/International Telecommunication Union (ITU)/International Solid Waste Association.* ISWA.

Balde, C. P., Wang, F., Kuehr, R., & Huisman, J. (2015). The global e-waste monitor – 2014. United Nations University, IAS – SCYCLE. Bonn, Germany.

Chan, J. K., & Wong, M. H. (2013). A review of environmental fate, body burdens, and human health risk assessment of PCDD/Fs at two typical electronic waste recycling sites in China. *The Science of the Total Environment*, *463*, 1112–1122. doi:10.1016/j.scitotenv.2012.07.098 PMID:22925483

Chi, X., Streicher-Porte, M., Wang, M. Y., & Reuter, M. A. (2011). Informal electronic waste recycling: A sector review with special focus on China. *Waste Management (New York, N.Y.)*, *31*(4), 731–742. doi:10.1016/j.wasman.2010.11.006 PMID:21147524

Ghisellini, P., Cialani, C., & Ulgiati, S. (2016). A review on circular economy: The expected transition to a balanced interplay of environmental and economic systems. *Journal of Cleaner Production*, *114*, 11–32. doi:10.1016/j.jclepro.2015.09.007

Global e-waste generation outlook 2030. (n.d.). *Statista*. Retrieved December 9, 2023, from https://www.statista.com/statistics/1067081/generation-electronic-waste-globally-forecast/

Grant, K., Goldizen, F. C., Sly, P. D., Brune, M. N., Neira, M., van den Berg, M., & Norman, R. E. (2013). Health consequences of exposure to e-waste: A systematic review. *The Lancet. Global Health*, *1*(6), e350–e361. doi:10.1016/S2214-109X(13)70101-3 PMID:25104600

Herat, S., & Agamuthu, P. (2012). E-waste: A problem or an opportunity? Review of issues, challenges, and solutions in Asian countries. *Waste Management & Research*, *30*(11), 1113–1129. doi:10.1177/0734242X12453378 PMID:22851536

Huo, X., Peng, L., Xu, X., Zheng, L., Qiu, B., Qi, Z., Zhang, B., Han, D., & Piao, Z. (2007). Elevated blood lead levels of children in Guiyu, an electronic waste recycling town in China. *Environmental Health Perspectives*, *115*(7), 1113–1117. doi:10.1289/ehp.9697 PMID:17637931

Impacts of E-Waste on Health—Blog. (n.d.). Retrieved December 9, 2023, from https://www.greenit.co.in/impacts-of-e-waste-on-health.html

Kang, H., & Schoenung, J. M. (2005). Electronic waste recycling: A review of U.S. infrastructure and technology options. *Resources, Conservation and Recycling*, *45*(4), 368–400. doi:10.1016/j.resconrec.2005.06.001

Kellner, R., & Goe, M. (2016). Technological trends in e-waste management. *Waste Management & Research*, *34*(7), 570–577.

Kiddee, P., Naidu, R., & Wong, M. H. (2013). Electronic waste management approaches: An overview. *Waste Management (New York, N.Y.)*, *33*(5), 1237–1250. doi:10.1016/j.wasman.2013.01.006 PMID:23402807

Kumari, N., & Lenka, U. (2023). Employment and Retention of Differently-abled People in the Workplace Through Assistive Technologies. *International Journal of Digital Technologies*, *2*(1).

Kumari, N., & Lenka, U. (2023). Scientometric review of disability inclusion in the workplace.

Lepawsky, J. (2015). The changing geography of global trade in electronic discards: Time to rethink the e-waste problem. *The Geographical Journal*, *181*(2), 147–159. doi:10.1111/geoj.12077

Liu, Q., Tang, Y., & Chen, S. (2013). E-waste recycling induced polybrominated diphenyl ethers, polychlorinated biphenyls, and polycyclic aromatic hydrocarbons contamination in the ambient air. *Journal of Environmental Sciences (China)*, *25*(1), 83–92.

Manomaivibool, P. (2009). Extended producer responsibility in a non-OECD context: The management of waste electrical and electronic equipment in India. *Resources, Conservation and Recycling*, *53*(3), 136–144. doi:10.1016/j.resconrec.2008.10.003

Mishra, D. N., & Panda, R. K. (2021). How delightful is Indian wellness tourism? A netnographic study. [AHTR]. *Advances in Hospitality and Tourism Research*, *9*(1), 132–156. doi:10.30519/ahtr.784232

Mishra, D. N., & Panda, R. K. (2023). Evaluating visitor–therapist relationship in Indian spa and wellness resorts. *Journal of Hospitality and Tourism Insights*, *6*(5), 2433–2461. doi:10.1108/JHTI-06-2022-0246

Needhidasan, S., Samuel, M., & Chidambaram, R. (2014). Electronic waste–an emerging threat to the environment of urban India. *Journal of Environmental Health Science & Engineering*, *12*(1), 36. doi:10.1186/2052-336X-12-36 PMID:24444377

Ogunseitan, O. A., Schoenung, J. M., Saphores, J. D., & Shapiro, A. A. (2009). The electronics revolution: From e-wonderland to e-wasteland. *Science*, *326*(5953), 670–671. doi:10.1126cience.1176929 PMID:19900918

Ongondo, F. O., Williams, I. D., & Cherrett, T. J. (2011). How are WEEE doing? A global review of the management of electrical and electronic wastes. *Waste Management (New York, N.Y.)*, *31*(4), 714–730. doi:10.1016/j.wasman.2010.10.023 PMID:21146974

Oswald, I., & Reller, A. (2010). Electronic waste–a challenge for sustainable development. *Environmental Sciences Europe*, *22*(1), 1–9.

Oteng-Ababio, M. (2012). Electronic waste management in Ghana-issues and practices. *Sustainable Development*, *20*(4), 293–307.

Patra, B. C., & Lenka, U. (2021). An AHP analysis of scientometrically derived factors of entrepreneurial intentions of women and constructing a conceptual research framework. *Journal of Research in Marketing and Entrepreneurship*, *23*(1), 139–158. doi:10.1108/JRME-02-2020-0017

Patra, B. C., & Lenka, U. (2022). Barriers to entrepreneurial intentions of women: Nominal group technique, analytic hierarchy process, and scientometric approach instigating the necessity of policy intervention. *Journal of Public Affairs*, *22*(2), e2429.

Patra, B. C., & Lenka, U. (2023). Scientometric, fuzzy NGT and DEMATEL analysis for determining sustainable business practices for entrepreneurial firms. *Benchmarking*.

Patra, B. C., & Mishra, D. N. (2023). Boosting entrepreneurial intentions: A novel EEG-guided protocol. *MethodsX*, *11*, 102358. doi:10.1016/j.mex.2023.102358 PMID:37701733

Perkins, D. N., Kjellén, M., & Nehler, T. (2017). The digital divide: Information technology, computerization, obsolescence, and electronic waste. *Cognitive Technologies*, 173-210.

Pickren, G. (2014). Political ecologies of electronic waste: Uncertainty and legitimacy in the governance of e-waste geographies. *Environment & Planning A*, *46*(1), 26–45. doi:10.1068/a45728

Puckett, J., Westervelt, S., Gutierrez, R., & Davis, S. (2002). *Exporting harm: The high-tech trashing of Asia*. The Basel Action Network.

Saphores, J. D., Nixon, H., Ogunseitan, O. A., & Shapiro, A. A. (2006). Household willingness to recycle electronic waste: An application to California. *Environment and Behavior*, *38*(2), 183–208. doi:10.1177/0013916505279045

Schluep, M., Hagelüken, C., Kuehr, R., Magalini, F., Maurer, C., Meskers, C., & Wang, F. (2013). *Recycling–from e-waste to resources*. United Nations Environment Programme & United Nations University.

Sepúlveda, A., Schluep, M., Renaud, F. G., Streicher, M., Kuehr, R., Hagelüken, C., & Gerecke, A. C. (2010). A review of the environmental fate and effects of hazardous substances released from electrical and electronic equipments during recycling: Examples from China and India. *Environmental Impact Assessment Review*, *30*(1), 28–41. doi:10.1016/j.eiar.2009.04.001

Song, X., Yang, J., & Liu, G. (2015). Stakeholder coordination in municipal solid waste management: A critical review. *Waste Management (New York, N.Y.)*, *38*, 157–168.

Tsydenova, O., & Bengtsson, M. (2011). Chemical hazards associated with the treatment of waste electrical and electronic equipment. *Waste Management (New York, N.Y.)*, *31*(1), 45–58. doi:10.1016/j.wasman.2010.08.014 PMID:20869229

Widmer, R., Oswald-Krapf, H., Sinha-Khetriwal, D., Schnellmann, M., & Böni, H. (2005). Global perspectives on e-waste. *Environmental Impact Assessment Review*, *25*(5), 436–458. doi:10.1016/j.eiar.2005.04.001

Wilson, J. (2009). *Electronics: A history of everyday things*. Cambridge University Press.

Wong, C. S., Wu, S. C., Deng, W. J., Yu, X. Z., Luo, Q., Leung, A. O. W., Wong, C. S. C., Luksemburg, W. J., & Wong, A. S. (2007). Export of toxic chemicals: A review of the case of uncontrolled electronic-waste recycling. *Environmental Pollution*, *149*(2), 131–140. doi:10.1016/j.envpol.2007.01.044 PMID:17412468

Zheng, G., Xu, X., Li, B., Wu, K., Yekeen, T. A., & Huo, X. (2013). Association between lung function in school children and exposure to three transition metals from an e-waste recycling area. *Journal of Exposure Science & Environmental Epidemiology*, *23*(1), 67–72. doi:10.1038/jes.2012.84 PMID:22854517

Chapter 5
E-Waste Management: A Significant Solution for Green Computing

K. Dinesh Kumar
https://orcid.org/0000-0003-0843-1561
Amrita Vishwa Vidyapeetham, India

Dipalee Divakar Rane
D.Y. Patil College of Engineering, India

Appalaraju Muralidhar
https://orcid.org/0000-0002-0730-3309
Vellore Institute of Technology, Chennai, India

Sam Goundar
https://orcid.org/0000-0001-6465-1097
RMIT University, Australia

P. Viswanatha Reddy
Viswam Engineering College, India

ABSTRACT

Green computing can be described as efficient resource management in distributed computing environments such as cloud, fog, and edge computing environments. In green computing, green refers to the eco-friendly environment with environmental responsibility to efficiently manage the computing resources. Mainly, two significant reasons are associated with global warming issues from the distributed computing perspective. They are high power consumption of cloud datacenters and high CO2 emission rate. According to many survey reports, every year, cloud datacenters alone produce nearly 90 million metric tons of CO2 into the environment, and now, this high power consumption of datacenters has become one of the primary reasons for global warming issues. Also, it stated that this CO2 emission rate would increase by 8% every year if it did not identify proper control measurements. Therefore, it is crucial to enhance e-waste management by increasing the efficiency of computing resource usage and minimizing high power consumption, high CO2 emission rate, inefficient recycling policies, etc.

DOI: 10.4018/979-8-3693-1018-2.ch005

INTRODUCTION

E-waste, often known as electronic garbage, is a rising worldwide issue in our increasingly digital and technologically oriented society. E-waste is made up of abandoned electronic equipment and gadgets, which can include everything from obsolete cellphones and laptops to smartwatches and other home appliances. In the environment, decomposition is one of the most important factors unless the substances are moved into different forms. The electronic equipment is mainly made of metal and other chemical materials. Without a proper management strategy, it's quite impossible to prevent upcoming environmental threats and disasters. The world's technological dependency is increasing daily; parallelly, e-waste is also growing 400 metric tons of CO_2 were released into the atmosphere by the (ICT) industry worldwide in 2007. Personal computers and other gadgets, such as computer displays, have had a significant negative influence on the environment Airehrour et al. (2019). The production of 41.8 million metric tons of electronic garbage was seen all over the world a few years ago Debnath, Roychoudhuri, and Ghosh (2016a). As we move forward in time, the most recent research from the United Nations University (UNU) paints a troubling picture: the volume of electronic trash is expected to rise to a staggering 52.2 million metric tons by the year 2021. This trend emphasizes the frightening rate at which our digital consumption practices are causing the ever-escalating e-waste problem Prasad et al. (2021). Electronic waste poses a significant threat to both human health and the environment due to its composition of heavy metal ions like Cadmium, Lead, Bromine, Mercury, Chromium, and various flame retardants. When humans and the environment come into direct contact with such waste, it can lead to highly hazardous situations Rautela et al. (2021). These materials have the potential to wreak havoc on human health by directly harming vital systems, including the cardiovascular, respiratory, digestive, and neurological systems Kumar, Holuszko, and Espinosa (2017) Joshi et al. (2023). So overall, E-waste creates an imbalance in the ecological system. Through its lifecycle and emitting greenhouse gases, E-waste also causes global warming. While the production of electronic devices and other gadgets requires a high amount of energy. During the manufacturing, gases like CO_2, Sulphur, and CO_2 are uncovered in the air layer.

The 20th century was revolutionary because of several technological inventions. The present and future also depend on technological advancement. Alongside the more industrial extension, there is a necessity for that. The ecosystem of this planet is becoming more unsuitable for using unguided and excessive energy. On a worldwide scale, governments and academic experts agree that carbon dioxide and other pollutants are the primary drivers of global warming and climate change. These pollutants include a variety of indicators, such as sulfur dioxide, suspended particle matter, and nitrogen oxides. This collective acknowledgment emphasizes these pollutants' crucial role in influencing the Earth's climate and the necessity for comprehensive solutions to limit their effectsAmri, Zaied, and Lahouel (2019). As it's a concerning topic in the present world, research took place in this field to find the solution. Green computing takes place by fulfilling the necessity and reducing the excessive usage of resources. Green computing involves reducing energy consumption, managing resources, and adapting to renewable energy. As it is a pressing issue in the modern world, researchers dug deep into this area to discover the answer. "Green computing" is implemented by decreasing excessive resource use while satisfying needs. Green computing involves conserving energy, managing resources, and incorporating renewable energy. Green computing promotes the responsible handling and disposal of electronic equipment, serving as a deterrent to the sneaky introduction of harmful e-waste into our delicate ecosystems. Electronic equipment's proper stewardship protects the land, water, and air from the insidious incursion of dangerous contaminants through rigorous end-of-life treatment, preventing contamination and ecological damage Panda (2013a).

To achieve our objective, we have structured the article into distinct sections. Section 2 will Discuss the challenges and current practices in managing electronic waste. Section 3 will explain the concept of Green Computing and how it can mitigate EWaste-related issues. Section 4 will explore the connection between E-Waste and global warming, highlighting the environmental issues associated with electronic waste. Section 5 presents solutions to address E-Waste from a global warming perspective. Section 6 will discuss how E-Waste management and its environmental impact are specific to India.

E-WASTE MANAGEMENT

Waste management involves organizing and systematically handling different waste materials, including solid, liquid, and hazardous. It includes processes such as collection, transportation, processing, recycling, and disposal, all of which should be carried out in an environmentally responsible manner. Appropriate waste management is essential for maintaining public health, preventing environmental pollution, preserving resources, and encouraging sustainable living.

E-waste is a growing problem in our modern world. Electronic waste, which includes discarded devices like computers, smartphones, televisions, medical and industrial equipment, refrigerators, and other appliances, can pose significant environmental and health risks if not disposed of properly. That's why e-waste management must ensure that these materials are safely and responsibly recycled and their hazardous components don't harm the environment or human health. In the context of prosperous nations, e-waste accounts for approximately 1% of the entire solid waste generated. E-waste generated from computers, flat-screen monitors, capacitors, transformers, and batteries contains harmful components like lead, cadmium, mercury, and polychlorinated biphenyls. In India alone, about 14 million PCs, 16 million mobile phones, and 80 million TVs are currently in use Panda (2013b). Countless wastes currently being used uneconomically or left unutilized are posing significant risks to the human environment Mukta and Ahmed (2020).

The following reasons highlight why e-waste is such a pressing issue:

- A remarkable amount of e-waste is constantly growing and shows no signs of slowing down.
- E-waste has significant environmental consequences that significantly impact the planet.
- A large portion of e-waste is shipped overseas, which can further harm the environment and human health.
- E-waste presents potential health hazards to those who come into contact with it.
- electronic waste disposal must be handled carefully to prevent data security breaches.

Present Situation of E-Waste Management

The present situation of E-waste management is quite complex and challenging. With the rapid growth of technology, the amount of electronic waste generated is increasing alarmingly. Unfortunately, most of it is in landfills or illegally exported to developing countries, causing severe environmental and health hazards. Many countries have implemented laws and regulations to control and manage E-waste, but the lack of proper infrastructure and resources makes it difficult to handle this problem effectively. Therefore, there is a need for a more comprehensive and sustainable approach to managing E-waste, which includes proper disposal, recycling, and recovery of valuable materials from electronic devices.

Figure 1 illustrates the Projected e-waste generation worldwide from 2019 to 2030. It's interesting to note that electronic equipment started becoming a part of our daily lives as early as the 1990s Balde et al. (2015). However, around 2005, e-waste became an emerging problem Widmer et al. (2005), and by 2014, it had already become a mini catastrophe. The current state of e-waste management is concerning, and projected trends from 2019 to 2030 based on available data paint a grim picture. The statistics speak for themselves - e-waste generation has steadily increased. In 2019, approximately 53.6 million metric tons of e-waste were generated worldwide. By 2020, this figure had already climbed to 55.5 million metric tons. The upward trajectory continued in 2021, with 57.4 million metric tons; in 2022, it reached 59.4 million. The projection for 2030 is particularly alarming a staggering 74.7 million metric tons. The reasons behind this exponential growth are multifaceted. Rapid technological advancements, shorter product lifecycles, and a culture of consumerism have all contributed to the proliferation of electronic devices. The world increasingly relies on electronics, from smartphones and laptops to household appliances and industrial machinery. However, with this dependence comes a significant environmental cost.

Lifecycle of Electronic Equipment

It's crucial to think about an electrical device's lifetime to comprehend it better. The lifetime of EEE in India is divided into three tiers, per Wath et al. (2010). We've taken a different approach and classified it into four phases Debnath, Roychoudhuri, and Ghosh (2016b). You can see a schematic lifecycle of e-waste in Figure 2.

- EEE generation
- WEEE generation
- WEEE re-processing

Figure 1. Projected e-waste generation worldwide from 2019 to 2030

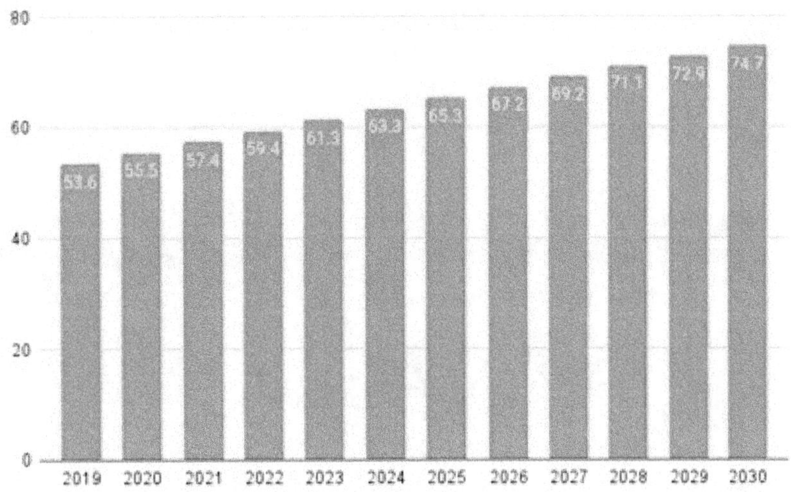

Phase I – Production of EEE

Manufacturing EEE initiates its lifecycle within the EEE value chain. These products can exist as stand-alone units or as components of larger assembled entities that represent entirely new products. They may be produced and assembled within India or imported from various countries, including China, Europe, and Korea. Despite not being the end consumers, these products reach customers through distributors and retailers. It's worth noting that they are categorized as 'Associated Elements' rather than contributing to Waste Electrical and Electronic Equipment (WEEE) generation. However, there has been a recent shift in this practice. This change reflects evolving approaches to product lifecycle management and sustainability.

Phase II – Generation of UEEE

The subsequent phase involves the generation of Used Electronic and Electrical Equipment (UEEE). Notably, Electrical and Electronic Equipment (EEE) consumers extend beyond just offices and institutions, encompassing domestic users as well. Launching the latest electronic devices, such as smartphones or tablets, has increasingly become a highly anticipated consumer event. However, amidst the glitzy advertisements and long store queues, there's an underlying narrative: Many older devices are cast aside with each new product release. Technological advancements primarily drive UEEE. Additionally, reasons for UEEE generation include the desire to upgrade equipment and replace older items with newer ones.

Phase III – Decision for Fate

At this stage, an important decision is made regarding the end-of-life of any electronic and electrical equipment (EEE). The decision is based on whether the equipment is repairable and can be refurbished

Figure 2. Life cycle of e-waste

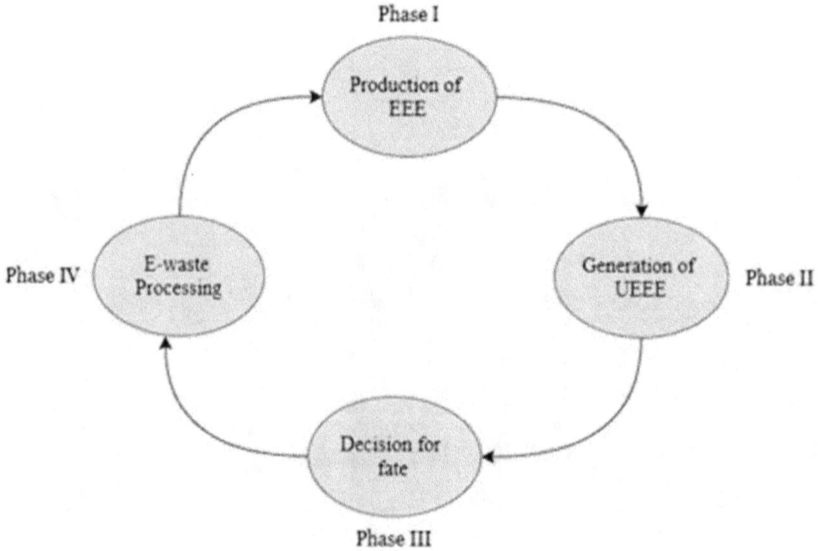

for resale in the secondhand market or should be WEEE. One of the notable characteristics of the Indian E-waste Supply Chain Management system is the presence of specialized industries dedicated to refurbishing and upgrading used personal computers. These industries have a significant role in the overall system.

Phase IV – E-waste Processing

The E-waste processing phase is categorized into three parts: Pre and post processing, and proper treatment.

Pre-processing: During the pre-processing stage of e-waste, the collected waste is sorted and dismantled. The plastics are sold to recyclers while the rest goes for e-waste treatment.

Proper Treatment: At e-waste treatment plants, different techniques and processes are utilized to extract valuable materials, like precious metals, and to recycle usable components from the e-waste.

Post Processing: As part of e-waste processing, the post-processing phase involves the responsible handling of waste materials, commonly known as "Zero Value Waste." However, in India, managing Zero Value Waste is still a challenge, and proper disposal of it remains a significant concern.

E-Waste Impact on Human and Environment

E-waste negatively affects both people and the environment. This kind of garbage consists of outdated electronics that still contain dangerous elements, including cell phones, laptops, televisions, and other electronic equipment. These materials can leach into the soil and groundwater if not disposed of properly. It is vital to emphasize the serious and widespread consequences of improper electrical waste disposal. When ewaste is improperly disposed of in landfills, it poses a significant time bomb for human and environmental health. The potential for toxic metals to seep into groundwater is a weighty problem as they can slowly infiltrate our crucial resources, posing a marked threat to all areas of life, including humans, plants, animals, and more. Sadly, ewaste is one of the ultimate contributors to the alarming expansion in air and water pollution, releasing toxic elements like copper, lead, mercury, cadmium, and others that cause comprehensive damage to our health. These substances can lead to various health problems, from heart issues to birth defects, and can affect nearly every system in the human body. The need for responsible e-waste management is extremely urgent as it is closely tied to our well-being and the maintenance of our environment. Amateurishness, restricted infrastructure, insufficient funding, and passive community involvement are challenges in handling E-waste Rautela et al. (2021).

Recycling of E-Waste

The process of e-waste recycling is a crucial and highly productive endeavor that involves several important steps. Figure 3 displays the e-waste recycling.

- The first step, picking, relies on manual labor to meticulously sort and identify items or components that should not be crushed or removed from the e-waste stream.
- After sorting, the next step is disassembly, where all components are carefully isolated and categorized into core materials. These core materials are destined for further recycling and reuse.

The third step is size reduction, where the e-waste is broken down into small fragments and refined into smaller pieces to facilitate the separation process and minimize dust and environmental impact.

- The fourth step is using an over-band magnet to release all magnetic materials, including steel and iron fragments, from the refined material. This step helps separate magnetic metals from non-magnetic ones.
- The fifth step involves separating metallic components such as aluminum, copper, and brass from non-metallic components like glass or plastic. This separation can be accomplished using magnets or optical identification techniques.
- Finally, in the last step of the recycling process, water separation is used to segregate non-metallic components, particularly the separation of plastic from glass. Water is often utilized in this process to help segregate these materials effectively. Once all components are separated, they are resold as raw materials for reuse in various manufacturing processes.

This process of extracting valuable materials from e-waste helps recover valuable resources and contributes to solving the growing problem of e-waste accumulation. By recycling e-waste systematically and environmentally friendly, we reduce the strain on natural resources and mitigate the harmful effects of improper disposal on human health and the environment.

GREEN COMPUTING

Green computing is an approach to computing that prioritizes efficient resource use and reduces power consumption to reduce the environmental impact of computing and IT-related activities. It's an important initiative promoting environmentally responsible behavior in the tech industry. The ever-growing technological advancements have increased the need for infrastructure and facilities to support it. We live

Figure 3. Recycling of electronic waste

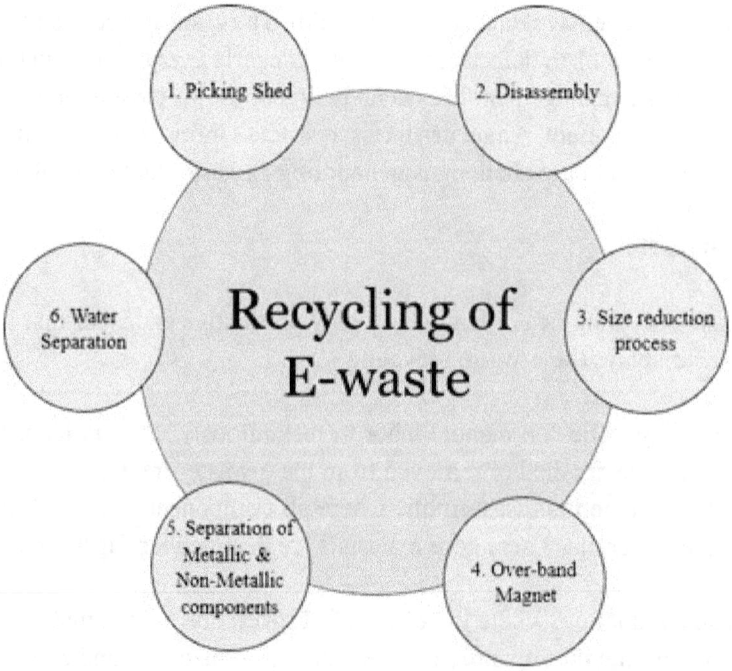

in an abundant information era, and it requires essential equipment and infrastructure to create, process, store, and utilize this data Ranbhise (2014). This is where green computing comes in, which is all about utilizing computing resources efficiently and promoting an environmentally responsible approach to computing. Its primary focus is on reducing power consumption and minimizing the associated environmental impact. Green computing is an essential step towards combating global warming and ensuring the sustainability of our planet. By adopting green computing practices, individuals and organizations can reduce their carbon footprint and contribute to environmental sustainability. This not only protects the environment but also results in tangible benefits such as reduced energy consumption and paper costs. Overall, embracing green computing is vital for a more sustainable and eco-friendly future.

In essence, green computing consists of the four core areas shown in Figure 5 and is discussed as follows Sadiku, Ampah, and Musa (2018), Srivastava and Sharma (2015). Green computing comprises four core areas: Green Use, Green Design, Green Disposal, and Green Manufacturing.

- The Green Use area involves using environmentally proper resources while reducing energy consumption.
- Green Design refers to developing energy-efficient and eco-friendly objects and services.
- Green Disposal involves recycling e-waste with the tiniest or no effect on the environment.
- Green Manufacturing involves manufacturing electronic devices with the tiniest or no effect on the environment.

The main factors causing green computing include Nazir, Majeed, and Birjais (2018):

- The consumption of a lot of electricity directly impacts natural resources and the environment. In addition, when we update our computers and dispose of outdated hardware devices and peripherals, we produce hazardous toxic waste that damages the environment.
- When we update our computers and dispose of outdated hardware devices and peripherals, we produce a hazardous toxic waste that damages the environment.

Figure 4. Key approaches to green computing for sustainable technology

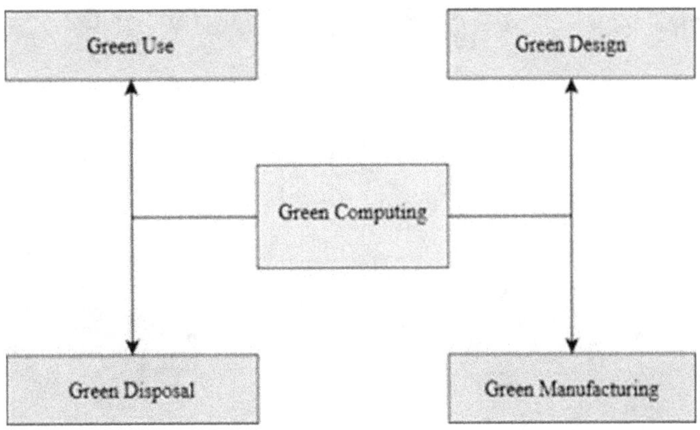

GLOBAL WARMING

The world is presently facing a significant challenge in the form of environmental and energy-related enterprises. The most burning among these issues is global warming, causing a gradational increase in the earth's average temperature over a long period. This miracle has far-reaching consequences for climate and overall environmental changes, with e-waste being the most concerning factor that plays a major part in global warming. The rising temperature is causing trouble to life on Earth, from land to ocean. The emigration of feasts like CH_4, CO_2, CO, SO_2, and other substances triggers the Greenhouse effect, leading to further temperature increases. In recent decades, worldwide energy demand has increased mainly. According to the International Energy Agency (IEA), if current trends continue, global energy demand will increase by 1.2 by 2035Khan et al. (2021). Although there has been a minor decline in conventional reactionary energies similar to coal, oil painting, and natural gas, they remain the world's primary energy sources. Oil painting, natural gas, and coal accounted for nearly 90% of electricity generation sources in 2016Norouzi and Fani (2020). It means the situation worsens, and resource management and energy reduction are necessary to neutralize the heat. In the following Figure 3 demonstrates the global warming effect.

Global Warming Issues

The effect of global warming is a multiplex challenge, and its influences are farreaching. With the incremental increase in Earth's average surface temperature, we're facing many problems. Let's take a closer look at some of the critical issues:

Raising Temperatures: The most obvious and crucial global warming situation is the rise in global temperatures. The more superheated climate is causing more frequent and severe warmth waves, making it unsupportable for humans and other animals.

Figure 5. Global warming effect from e-waste

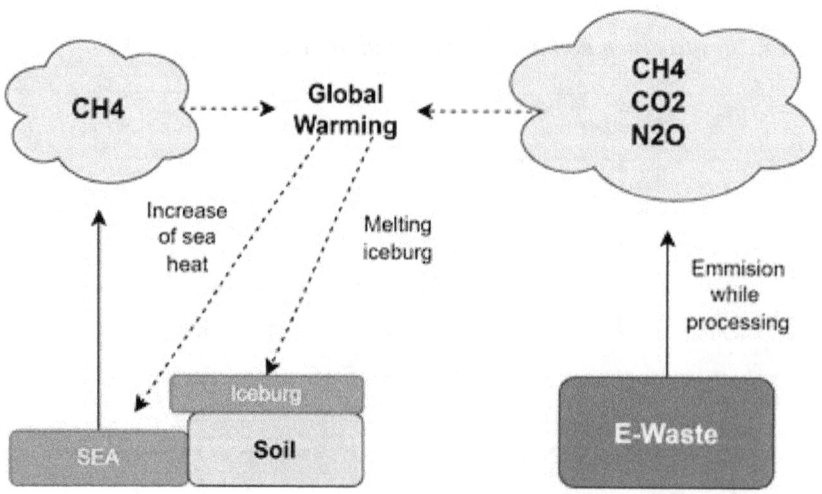

- **Shrinking Ice and Glaciers:** As the temperature is increasing, the ice caps and glaciers are melting at a rough rate, contributing to the rise in ocean conditions.
- **Trouble for Coastal Areas:** The melting of ice and warming of seawater are guiding to a rise in ocean situations, which could be disastrous for coastal residents, structures, and ecosystems.
- **Extreme Weather Patterns:** Global warming generates more frequent and damaging intense rainfall events, such as hurricanes, famines, cataracts, and backfires. These circumstances put millions of lives at risk and forced widespread impairment of property and structures.
- **Decline in Biodiversity:** Climate change and altered territories are making it difficult for colorful species to survive, leading to a steady decline in biodiversity. It's high time we take a step forward and address these issues before it's too late.

SOLUTION OF E-WASTE MANAGEMENT IN GLOBAL WARMING PERSPECTIVE

The management of e-waste is a crucial aspect of the fight against global warming. Effective solutions must be implemented to prevent electronic waste's hazardous environmental effects.

Proposed Solutions

Table 1 displays the analysis that indicates the usage of various methods or technologies in the existing work. In recent times, the industry of electrical waste has grown significantly, reflecting the widespread use of electronic devices in homes and businesses Parmar, Pandya, and Kumbharana (2011). The emergence of electronic waste has become increasingly complex, which demands efficient technologies and processes for proper management? Regrettably, inadequate disposal systems have made e-waste a significant environmental burden Jayarajan et al. (2018). However, the challenges posed by e-waste provide an exciting opportunity, as it is a combination of toxic and valuable materials, which has the potential to become a substantial business prospect Tiwari and Dhawan (2014). With various studies highlighting the potential advantages of harnessing electrical waste for green computing, managing it responsibly and sustainably is essential. With the proliferation of electronic devices, the issue of e-waste management has become increasingly pressing. Regulations such as WEEE and RoHS are positive steps towards creating a more sustainable society. It is essential to develop safer electronic devices that use eco-friendly materials. Manufacturers need to prioritize innovation towards sustainable materials and processes. Currently, electronic equipment design prioritizes energy efficiency and power consumption. Still, we must shift our focus towards developing safer electronic devices that use biodegradable, less toxic, and eco-friendly raw materials. Manufacturers play a crucial role in this transition, and innovation must be directed towards sustainable materials and manufacturing processes. To create a greener future, computer users and businesses must take responsibility and make small but effective changes to their work habits to reduce their environmental impact. Here are some practical steps that we can all take to make computing more sustainable:

- Prioritizing important printing.
- Utilizing recycled paper.
- Printing both-sided.
- Investing in high-efficiency, energy-saving equipment and top-rated energy stars is wise.

- Turning off systems instead of leaving them in standby mode.
- It's essential to acquire new equipment only when necessary, not just because a newer model is available. This helps to avoid unnecessary expenses and ensures that resources are utilized efficiently.
- Opting for smaller systems with fewer attachments.
- Unplugging unused peripherals (printer, audio system, scanner, modem, etc.).
- It is better to optimally charge UPS batteries instead of leaving them connected all day.

Various methods and tools have been developed to address this challenge. However, all these approaches share some standard requirements, such as investing in new technology and increasing awareness and knowledge.

This article will discuss an approach that prioritizes, these are the three fundamental principles of e-waste management.

- Eco-friendliness E-waste Management Approach
- Cost-effectiveness E-waste Management Approach
- Minimal impact on the environment and natural resources

Eco-Friendly E-Waste Management Approach

Ideal strategies and rules of green computing involve:

- Using power efficiently.
- Reducing paper consumption.
- Promoting eco-friendly equipment.
- Safely recycling old devices.

In Europe, several environmental regulations address waste management, recycling, and pollution control. Electronics companies are also developing a range of green computers, such as eco-friendly desktops and laptops, to reduce e-waste. They are trying to ensure that all electronic hardware adheres to the restricted use of hazardous substances, such as brominated flame-retardants, PVCs, and heavy metals like lead, cadmium, and mercury. However, the challenge lies in creating reliable green computer materials. Even the front-runner, a tin/copper/silver alloy, requires higher melting temperatures, affecting chip life. Let's break down some key points of eco-friendly e-waste management approach and elaborate on them:

- **Smart Power Usage:** Green computing is all about intelligent power consumption management, meaning devices can enter low-power states when unused. It is essential to optimize energy consumption to promote green computing.
- **Reduction of Paper Consumption:** The shift towards a paperless society is a great way to promote efficiency and environmental responsibility. By embracing digital documentation and communication, we can conserve trees and reduce the waste generated by paper production and disposal. It's a simple and effective way to impact our planet positively.

Table 1. Analysis on e-waste management existing work

Reference	Methods or Technologies That Have Been Utilized
Tsitomeneas, Kokkosis, and Charitopoulos (2014)	E-waste legislating, designing, and managing involves the application of various principles, theories, and practices.
Srivastava and Sharma (2015)	The Interpretive Structural Modeling (ISM) methodology is employed to evolve mutual relationships among the factors.
Jayarajan et al. (2018)	An electric device is proposed for implementing a monitoring system in e-waste dump yards.
Jian and Shanshan (2010)	Implement an ERP system to standardize the market for recycling electrical products.
Mangaonkar, Sabhani, and Sirsat (2014)	Using GIS, identifying the primary source of e-waste
Anwar et al. (2018)	Reduce electronic waste by utilizing virtualization, its feature and its role as a crucial component of green computing.
Tiwari and Dhawan (2014)	Create a time-series, multiplelifetime end-of-life model.
Mukta and Ahmed (2020)	Reuse, improve, or recycle the electronic waste. Create an electronic device that uses non-traditional energy sources for minimal power.
Debnath et al. (2015)	Quality Function Deployment (QFD) is a powerful analytical tool that can be used to identify various parameters from primary research data. These parameters can then be used to understand and improve e-waste recycling practices as a green computing approach.
Masuduzzaman, Amit, and Alauddin (2018)	Recycling e-waste as aggregate in concrete can increase the potential use of this waste material and have a positive environmental impact.

- **Recommendation of Environment-Friendly Equipment:** Energy Star-rated devices and other green certifications encourage manufacturers to design sustainable products, driving the industry towards more environmentally friendly solutions.
- **Safe Recycling of Old Machines:** Improper handling of e-waste can pose a significant environmental threat. That's why it's crucial to implement safe and efficient e-waste recycling programs. Taking responsibility for our e-waste can help protect our environment and create a more sustainable future.
- **Eco-Friendly Computers:** Electronics giants are making significant strides in producing eco-friendly computers. These devices are engineered to be free of hazardous materials, such as lead, cadmium, and mercury, commonly used in traditional computer manufacturing.
- **Challenges of Green Materials:** Replacing hazardous materials with eco-friendly alternatives is essential for environmental protection, but it's not without its challenges. One of the biggest challenges is ensuring the reliability and performance of electronic components with greener materials, such as solder. Despite the challenges, it's essential to continue pushing forward in this area to create a more sustainable and responsible future for all.
- **Energy-Efficient Manufacturing:** A critical aspect of green computing is minimizing the energy-intensive manufacturing of computer components. Manufacturers can achieve this by optimizing their processes, adopting cleaner energy sources, and reducing waste throughout production. Manufacturers must continue to prioritize eco-friendliness in their production processes to minimize the impact on our planet.

- **Bio-Plastics:** Bio-plastics, derived from plant-based polymers, present an excellent opportunity to reduce our reliance on petroleum-based plastics. However, the challenge lies in maintaining the thermal stability of these materials to prevent electronics from overheating. By continuing to explore and develop these materials, we can create a more sustainable future and reduce our environmental impact.

Cost-Effective E-Waste Management Approach

While it's important to prioritize eco-friendliness, we also need to consider cost effectiveness when it comes to e-waste management solutions. To ensure sustainability, these must be economically feasible. Heightened costs can prevent individuals, businesses, and governments from adopting responsible practices. To execute cost-effectiveness, we need to discover a proportion between the cost of recycling and the possible revenue developed from salvaged materials. This may involve government subsidies, business incentives, or educating consumers about the long-term savings associated with proper e-waste disposal. We can also enable the development of innovative recycling technologies to decrease the overall cost of e-waste management.

Preserving the Environment and Natural Resources

E-waste management is essential for maintaining the environment and natural resources. It is crucial to embrace practices that decrease the depletion of raw materials, like material recovery and reuse. Recycling and repurposing electronic components can restrict the need for new resources, eventually easing the hurdle on ecosystems. It is equally important to ensure that e-waste management does not disrupt local ecosystems or harm natural resources during the collection and processing stages. Hence, it is essential to choose responsible disposal methods that prevent soil and water contamination. Proper waste disposal facilities should have the necessary infrastructure to ensure that e-waste does not become an environmental hazard.

E-WASTE PERSPECTIVE: THE INDIAN SCENARIO

India is one of the biggest producers of e-waste all over the world. It has been reported that India produces approximately 18.5 lakh metric tonnes (MT) of electronic waste annually, with Mumbai and Delhi-NCR being the biggest contributors. By 2018, this figure is predicted to rise to 30 lakh MT. Other cities that produce significant amounts of e-waste include Bangalore, Chennai, Kolkata, Ahmadabad, Hyderabad, Pune, Surat, and Nagpur. Among India's eight most populous states, Maharashtra is the biggest e-waste producer, followed by Tamil Nadu, Andhra Pradesh, Uttar Pradesh, Delhi, Gujarat, Karnataka, and West Bengal. It is worth noting that industrialized countries have established dedicated facilities for e-waste recycling. However, poorer nations like India often resort to manual recycling, which puts employees at risk of harmful exposure to e-waste materials. More than half of the e-waste generated in affluent countries is exported to underdeveloped nations, with China, India, and Pakistan being the primary destinations. In these countries, the e-waste is recycled, and metals such as copper, iron, silicon, nickel, and gold are extracted. Figure 6 illustrates India's state-wise e-waste generation.

Figure 6. State-wise e-waste generation in India

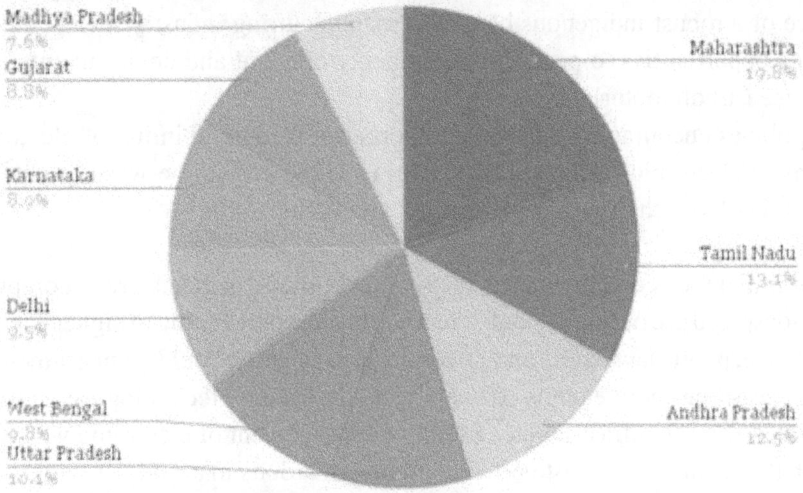

In the 2018 Environmental Performance Index, which included 180 countries, India came in at 177. This poor ranking can be attributed to inadequate environmental health policies and the high number of deaths caused by air pollution. India is also the fifth-largest producer of electronic waste worldwide, generating over two million tonnes of it annually. However, fewer than 2% of this e-waste is properly recycled, which raises questions about public health and groundwater pollution. E-waste management in India is predominantly controlled by the informal sector, which operates without regulation and often fails to recover the total value of materials. This informal sector also poses risks to the environment and workers' safety due to improper handling of toxins. A dismantling facility for electronic waste is located in Seelampur, Delhi, where people, including children, extract valuable components using harmful methods like open incineration and acid-leaching. Raising awareness, improving recycling infrastructure, and reforming existing policies are necessary to address these challenges. A significant portion of discarded electronics is collected by informal channels such as repair shops, used product dealers, and ecommerce vendors for reuse and part cannibalization. Enhancing these practices can contribute to better e-waste management in India.

Green Computing's Critical Role in Combating Global Warming

The issue of global warming has brought attention to the practices of the IT sector, which have been found to have implications for climate change. One way to positively contribute to global efforts in reducing greenhouse gas emissions and addressing climate change is through adopting green computing principles and developing a sustainable hardware industry in India. This serves as a reminder of the interconnectedness of economic development, technological advancement, and environmental responsibility as we strive to tackle the challenges posed by global warming. From a global warming perspective, several insights and implications exist:

- Dependence on foreign equipment involves energy-intensive manufacturing processes and long transportation distances, contributing to carbon emissions and exacerbating global warming.
- The absence of a robust indigenous hardware manufacturing industry means India needs to take advantage of opportunities to produce more energy-efficient and environmentally friendly hardware to reduce carbon footprints.
- Tax relief policies encouraging hardware imports can lead to an influx of electronic devices, resulting in more electronic waste and contributing to the growing e-waste problem, which poses environmental risks, including greenhouse gas emissions.

Moreover, the initial focus on low-cost infrastructure building neglects green computing principles, resulting in the widespread use of energy-inefficient hardware, contributing to higher energy consumption and emissions. Hence, public awareness and government legislation could be instrumental in promoting green computing, adopting more energy-efficient and eco-friendly technologies, and reducing carbon emissions associated with IT infrastructure. Effective management of electronic waste is crucial in the fight against global warming, as e-waste often contains hazardous materials. If not handled properly, it can release greenhouse gases and other pollutants, making legislation to manage e-waste responsibly vital.

Finally, emphasizing the importance of long-term sustainability in IT infrastructure development aligns with the global goal of mitigating climate change. Sustainable practices, including energy-efficient hardware and responsible e-waste management, can help reduce the ICT (Information and Communication Technology) sector's carbon footprint.

CONCLUSION

E-waste, or electronic waste, is a major concern for the well-being of both humans and animals. As the world continues to grapple with the consequences of global warming, we must prioritize effective management strategies for handling e-waste. Implementing these strategies can help mitigate the potential environmental harm caused by e-waste while promoting energy efficiency and responsible waste processing methods. Green computing is key to creating a sustainable and eco-friendly modern society. The constant development of technology has led to progress in the fields of green computing and e-waste management. Nowadays, it's more important than ever for businesses and organizations to incorporate e-waste management as an essential component of their green computing practices. Neglecting this responsibility could lead to numerous difficulties and complexities for our planet. The great news is that virtualization and reduced carbon footprint are the exclusive priorities of government and academic research. We anticipate that our study will help progress the field and open up exciting new potential for green computing.

REFERENCES

Airehrour, D., Cherrington, M., Madanian, S., & Singh, J. (2019). Reducing ICT carbon footprints through adoption of green computing. Academy of Economic Studies in Bucharest. Department of Economic Informatics. doi:10.12948/ie2019.04.17

Amri, F., Ben Zaied, Y., & Ben Lahouel, B. (2019). ICT, total factor productivity, and carbon dioxide emissions in Tunisia. *Technological Forecasting and Social Change*, *146*, 212–217. doi:10.1016/j.techfore.2019.05.028

Anwar, S., Ghaffar, M., Razzaq, F., & Bibi, B. (2018). E-waste reduction via virtualization in green computing. *American Scientific Research Journal for Engineering, Technology, and Sciences*, *41*(1), 1–11.

Balde, C. P., & Kuehr, R. (2015). E-waste statistics-Guidelines on classification, reporting and indicators. United Nations University.

Debnath, B., & Baidya, R. (2015). E-waste recycling as criteria for green computing approach: analysis by QFD tool. *Computational Advancement in Communication Circuits and Systems: Proceedings of ICCACCS 2014*, 139–144.

Debnath, B., Roychoudhuri, R., & Ghosh, S. K. (2016). E-waste management–a potential route to green computing. *Procedia Environmental Sciences*, *35*, 669–675. doi:10.1016/j.proenv.2016.07.063

Dinesh Kumar, K., & Umamaheswari, E. (2023). An efficient proactive VM consolidation technique with improved LSTM network in a cloud environment. *Computing*, 1–28. doi:10.100700607-023-01214-5

Jayarajan, P., Thenmozhi, S., Maheswar, R., Malathy, S., & Udaiyakumar, R. (2018). Smart cloud enabled E-waste management system. *2018 International Conference on Computer Communication and Informatics (ICCCI)*, 1–5.

Jian, L., & Zhang, S. (2010). Study e-waste management based on EPR system. *2010 International Conference on E-Business and E-Government*, 820–823. 10.1109/ICEE.2010.213

Joshi, S., Sharma, M., Ekren, B. Y., Kazancoglu, Y., Luthra, S., & Prasad, M. (2023). Assessing Supply Chain Innovations for Building Resilient Food Supply Chains: An Emerging Economy Perspective. *Sustainability (Basel)*, *15*(6), 4924. doi:10.3390u15064924

Khan, K., Su, C.-W., Umar, M., & Yue, X.-G. (2021). Do crude oil price bubbles occur? *Resources Policy*, *71*, 101936. doi:10.1016/j.resourpol.2020.101936

Kumar, A., Holuszko, M., & Denise, C. R. E. (2017). E-waste: An overview on generation, collection, legislation and recycling practices. *Resources, Conservation and Recycling*, *122*, 32–42. doi:10.1016/j.resconrec.2017.01.018

Kumar, K. D., & Umamaheswari, E. (2018). Prediction methods for effective resource provisioning in cloud computing: A survey. *Multiagent and Grid Systems*, *14*(3), 283–305. doi:10.3233/MGS-180292

Kumar, K. D., & Umamaheswari, E. (2018). Efficient Cloud Resource Scaling Based on Prediction Approaches. *International Journal of Engineering & Technology, 7*(4.10).

Kumar, K. D., & Umamaheswari, E. (2019). Ewptnn: An efficient workload prediction model in cloud computing using two-stage neural networks. *Procedia Computer Science*, *165*, 151–157. doi:10.1016/j.procs.2020.01.097

Kumar, K. D., & Umamaheswari, E. (2020). Hpcwmf: A hybrid predictive cloud workload management framework using improved LSTM neural network. *Cybernetics and Information Technologies, 20*(4), 55–73. doi:10.2478/cait-2020-0047

Kumar, K. V. R., Kumar, K. D., Poluru, R. K., Basha, S. M., & Reddy, M. P. K. (2020). Internet of things and fog computing applications in intelligent transportation systems. In *Architecture and security issues in fog computing applications* (pp. 131–150). IGI Global. doi:10.4018/978-1-7998-0194-8.ch008

Mangaonkar, N., Sabhani, P., & Sirsat, S. (2014). Green Technology for avoiding E-waste with the help of GIS. *2014 International Conference on Advances in Communication and Computing Technologies (ICACACT 2014)*, 1–6. 10.1109/EIC.2015.7230750

Masuduzzaman, M., Shishir, K. S. A., & Alauddin, M. (2018). Utilization of E-waste in Concrete and its Environmental Impact-A Review. In *2018 international conference on smart city and emerging technology (ICSCET)* (pp. 1–4). IEEE. doi:10.1109/ICSCET.2018.8537301

Mukta, T. A., & Ahmed, I. (2020). Review on E-waste management strategies for implementing green computing. *Int. J. Comput. Appl, 177*, 45–52.

Nazir, Majeed, & Birjais. (2018). Contribution of green computing towards it for providing sustainable environment. *International Journal of Advanced Research in Computer Science, 9*(2).

Norouzi, N., & Fani, M. (2020). Black gold falls, black plague arise-An Opec crude oil price forecast using a gray prediction model. *Upstream Oil and Gas Technology, 5*, 100015. doi:10.1016/j.upstre.2020.100015

Panda, R. (2013). E-waste management: A step towards green computing. *International Journal of Environmental Engineering and Management, 4*(5), 2231–1319.

Parmar, Pandya, & Kumbharana. (2011). Optimization of Energy Usage for Computer Systems by Effective Implementation of Green Computing. *International Journal of Advanced Networking Applications,* 87–90.

Prasad, K. D., Krishna Murthy, P., Hima Gireesh, Ch., Prasad, M., & Sravani, K. (2021). Prioritization of e-waste management strategies towards green computing using AHP-QFD approach. *Proc. Eng, 3*, 33–40.

Ranbhise, P. (2014). Green computing a way towards environmentally sustainable future. *2014 International Conference on Contemporary Computing and Informatics (IC3I)*, 1094–1100. 10.1109/IC3I.2014.7019596

Rautela, R., Arya, S., Vishwakarma, S., Lee, J., Kim, K.-H., & Kumar, S. (2021). E-waste management and its effects on the environment and human health. *The Science of the Total Environment, 773*, 145623. doi:10.1016/j.scitotenv.2021.145623 PMID:33592459

Sadiku, Ampah, & Musa. (2018). Green computing: a primer. *Journal of Scientific and Engineering Research, 5*(4), 247–251.

Srivastava, R., & Sharma, D. (2015). Factors affecting e-waste management: an interpretive structural modeling approach. *2015 Fifth international conference on communication systems and network technologies*, 1307–1312. 10.1109/CSNT.2015.158

Tiwari, D., & Dhawan, N. G. (2014). E-waste management: An emerging challenge to manage and recover valuable resources. *International Journal of Environmental Research and Development*, *4*(3), 253–260.

Tsitomeneas, Kokkosis, & Charitopoulos. (2014). *Legislation, design and management of the electrical and electronic waste (e-waste) procedures.* Academic Press.

Wath, S. B., Vaidya, A. N., Dutt, P. S., & Chakrabarti, T. (2010). A roadmap for development of sustainable E-waste management system in India. *The Science of the Total Environment*, *409*(1), 19–32. doi:10.1016/j.scitotenv.2010.09.030 PMID:20951410

Widmer, R., Oswald-Krapf, H., Sinha-Khetriwal, D., Schnellmann, M., & B¨oni, H. (2005). Global perspectives on e-waste. *Environmental Impact Assessment Review*, *25*(5), 436–458. doi:10.1016/j.eiar.2005.04.001

Chapter 6
E-Waste Management for Sustainable Development:
A Case of Private Commercial Banks in Bangladesh

Md. Shafiqul Islam
International University of Business Agriculture and Technology, Bangladesh

A. K. Md. Jawadul Haque
Fast Security Islami Bank Limited, Bangladesh

ABSTRACT

Continuous e-waste generation causes harms to the environment and the economy. This study aims to explore e-waste management systems in the private commercial banks in Bangladesh and ways to reduce e-waste generation and sustainable solutions of e-waste management. The rapid economic growth and establishment of commercial private banks has created a market for computers, consumer electronics, and household appliances in Bangladesh. This study looks into the generation of e-waste growth rate and remedial measures. The e-waste is harmful to human health and for the environment. Barium and lithium are examples that affect brain, heart, liver, kidneys, and skeletal system. It's critical to raise awareness about the dangers of inappropriate e-waste disposal. E-waste must be recycled properly to have a low environmental impact. It was found that banking generated 2176.30 tons of e-waste. Results showed statistically significant growth rate. Buy back guarantee and bring your own device may be options for e-waste reduction.

1. INTRODUCTION

E-waste is a growing concern as well as a significant commercial potential, considering the volume of e-waste created and the harmful and valuable components included inside. Iron, copper, aluminum, gold, and other metals account for more than 60 percent of e-waste, while contaminants account for 2.70 percent. These pollutants' produces high toxicity, particularly when burnt or recycled in uncontrolled

DOI: 10.4018/979-8-3693-1018-2.ch006

circumstances. The Basel Convention designated e-waste as hazardous and established a framework for regulating the Trans boundary movement of such trash (Widmer *et al.,* 2005). Electronic garbage or e-waste is one of the world's fastest increasing concerns. E-waste is composed of a variety of components, some of which include dangerous compounds that, if not properly disposed of, can have a negative influence on human health and the environment. Electronic garbage, or e-waste, is a growing environmental concern in a number of developing nations, including Bangladesh. It is one of the most recent categories of garbage to be classified as a waste material. Bangladesh, in line with worldwide trends and in response to its continued rapid expansion, is expanding its reliance on electronics-based appliances, equipment, and tools in people's homes, offices, industrial units, cars, and communication systems. This has resulted in an increase in the rate of e-waste creation (BUET, 2018). Many developing countries and countries in transition have an informal or less-organized small-scale sector that acquires discarded or obsolete computing equipment for repair and resale. Numerous informal and unskilled groups collect and recover elements such as copper, gold, and plastics from electronic and electrical garbage (e-waste), unaware of the harmful health and environmental consequences (Quinn *et al.,* 2021). E-waste defined as all types of electrical and electronic devices and their parts that have been thrown away as garbage by the owner with no intention of reuse. According to a European Union study, e-waste is growing at a rate of 5.0 percent per year, roughly three times faster than other waste streams in the solid waste sector (Schwarzer *et al.,* 2015). The use of electronic components and microchips in domestic appliances, office equipment, transportation systems, and industrial instruments and gadgets has increased exponentially during the last few decades. As these products reach the end of their useful lives, they are accumulating in ever-increasing quantities at various waste sites in practically every country on Earth. Commercial banks are no exception, and the digitalisation of the industry, as well as the emergence of e-commerce, internet banking, and cloud computing, have accelerated the trend. As a result, the amount of electronic garbage generated by commercial banks in Bangladesh has increased many folds. Electronic waste (e-waste) or waste from Electrical and Electronic Equipment (WEEE) contains a number of hazardous compounds, including lead, chromium, and plastic additives. Often containing valuable metals and other reusable materials such as plastics, e-waste has been viewed as a source of additional revenue by an increasing number of Bangladesh's low-income and unemployed residents. Recycling scrap and used electrical equipment is a lucrative business in Bangladesh, where there is little or no environmental regulation or occupational health legislation to protect workers who are directly exposed to health risks associated with waste recycling activities. In our country, the informal sector recycles a relatively tiny percentage of used electrical equipment; the remainder is dumped in landfills, rivers, ponds, sewers, lakes, channels, and open places. Improper recycling and recovery practices can have a significant negative influence on the environment. Crude methods of deconstruction frequently result in toxic emissions, which contaminate the environment and expose workers and the general public to hazardous elements. Assessment of e-waste generation is critical for addressing the difficulties of e-waste, setting targets, and identifying optimum management strategies. In this research, an attempt has been made to quantify the electronic waste generated by commercial banks in Bangladesh. Electronic trash (e-waste) management is undergoing tremendous change. Apart from a fast-growing global population, globalisation is pushing product demand, resulting in increased costs for a variety of resources. The planet is becoming more crisscrossed with used electrical items and recyclable resources. It means both opportunities and challenges of e-waste management (William *et al.,* 2008). E-waste policies and laws are emerging to address the worldwide challenge of rising e-waste. Even assessing the creation of e-waste is proving difficult in a number of nations. The primary impediments to implementing current approaches for e-waste evaluation are data availability, dependability, and completeness (Kumar & Rawat 2018). All electronic e-waste include pollutants including lead, cadmium, beryllium, or brominated flame retardants. Even in developed nations, recycling and disposing of e-waste can pose major risks to workers and communities, and extreme caution must be used to minimize harmful exposure during recycling operations and leaching of material such as heavy metals from landfills and incinerator ashes (Wath *et al.*, 2011). The majority of industrialized nations have enacted legislation requiring electronic producers and importers to collect old electronic items at the end of their useful lives (EoL), based on the extended producer responsibility principle (Nnorom & Osibanjo, 2008). However, consumers and producers are generally unaware of the long-term negative consequences of improperly disposed of e-waste. As a result, users must either keep EOL equipment in their possession or hand it over to corner side *Kabaris*, as there is no curbside collection mechanism for these EEEs (Electrical and Electronic Equipment's), with informal recyclers disposing of e-waste in an unorganized manner and depositing the remainder in municipal waste bins. In India, a large amount of e-waste is accessible, and sadly, 90 percent of e-waste is processed by informal agencies such as kabaris, recyclers, and so on. They are greedy, extracting the desired materials such as copper, aluminums, and steel and leaving the remainder of the e-waste behind or allowing it to burn in the open. India had 4.64 million desktop computers, 431 thousand notebook computers, and 89 thousand servers in 2005-06 (CPCB, 2011). Mostly informal recyclers are using primary e-waste disposal methods including open burning to gather metals (copper, aluminum, and steel) and to deposit the rest of the materials (plastics, ceramics, and asbestos) all which are hazardous by nature and explode in the atmosphere during incineration/open dump burning (Vats & Singh 2014). It is feasible to recover precious metals with a high degree of efficiency using the correct procedure (Zhang *et al.*, 2011). The adverse consequences of informal e-waste recycling on the environment are most severe for workers who handle this material, although contamination may travel thousands of kilometers beyond recycling centers. E-waste has a greater impact on air pollution and animal species and also on the biodiversity. As mentioned previously, electronic trash contains hazardous components that are harmful to human health and for the environment. Additionally, it can have a significant effect on the neurological and reproductive systems of humans, resulting in sickness and birth abnormalities. To minimize these hazardous impacts of e-waste, it is critical to properly recycle, refurbish, resell, or reuse goods. The enormous stream of e-waste will only deteriorate if the public is not educated on proper disposal methods (Gaidajis *et al.*, 2010).

1.1 Problem Statement

Bangladesh demonstrates significant economic growth and a developing market for electric, electronic, and home appliance customers. These needs have resulted in an increase in the production of locally manufactured electronics and an increase in the demand for reusing these items. In the semi-formal and informal industries, equipment is typically repaired and repurposed. Nowadays, reusing or recycling equipment, as well as disposing, creates risk. These e-wastes may pose environmental and health risks. Currently, there is a lack of awareness and proper information on the danger of e-waste in Bangladesh.

1.2 Rationale of the Study

Bangladesh has over 61 commercial banks, each having a sizable network of branches, sub branches, and agent outlets. Almost all branches and outlets operate with a significant number of electronic equipment. Over the product's life span, which is typically five years, these electronic devices are considered to have a zero/one taka value after depreciation and sold to recyclers/informal collectors via brokers, thereby entering the supply chain for recycling, dismantling, and extraction of precious materials, among other things.

1.4 Objective of the Study

The main objective of this study was to estimate e-waste generation of PCBs, growth rate, present scenarios and the remedial measures of e-waste management. The study aims the following objectives:

- Determine the volume of e-waste generated by the PCBs
- Estimate the growth rate of e-waste during 2016 -2020 of PCBs
- Explore the remedial measures in light of new enactment of Rules in June2021
- Know the present scenarios of e-waste management by the PCBs

2. LITERATURE REVIEW

Electrical and electronic devices are now a necessary commodity for a variety of daily activities, and they are growing five times faster than humans, such as mobile phones, which have grown from zero to 7.2 billion in just three decades. Improper dispose can result in environmental dangers that are harmful to human health, marine life, groundwater contamination, and soil. Bangladesh has elementary solid waste management facilities and difficult to manage this massive incursion of electrical and electronic waste. Inadequate public knowledge, policies, and funding in the field of waste management are a few of the major issues that have contributed to this dire situation. A set of sustainable E-waste management systems has been proposed, as well as the obstacles that may arise during their implementation. The successful deployment of these proposed methods would improve the quality of E-waste management in Bangladesh, boosting the present 35 percent overall E-waste recycling rate and providing tremendous energy from waste (Karim *et al.,* 2014; Khalil *et al.,* 2014). A study mentioned that Bangladesh accounts for 7 percent of global e-waste disposal on a percentage basis (Islam, 2016). E-waste is a term that refers to large and small household appliances, information and telecommunications equipment, lighting equipment, electrical and electronic gadgets, toys, sports equipment, and medical devices (Pariatamby, 2013).

2.1 Bangladesh Perspective

Electrical and electronic devices are now a necessary commodity for a variety of daily activities, and they are growing five times faster than humans. The image of electrical and electronic waste production in Bangladesh, as well as modern E-waste management technologies, has been extensively given in this study. The e-waste management strategies and systems are conventional and disconnected from contemporary technology capability. Bangladesh proposed a set of sustainable e-waste management systems, as well as the challenges that may arise during implementation. The successful deployment of these proposed methods would improve the quality of e-waste management in Bangladesh, boosting the present 35 percent overall E-waste recycling rate and providing tremendous energy from waste (Karim *et al.,* 2014; Khalil *et al.,* 2014). Customers utilize non-brand televisions based on their income capability, and these sets have a lifespan of no more than 3-5 years. In most cases, 60 percent of repairers work on non–brand TVs, while the remaining 20 percent work on roughly 170 brand TVs per year. As a result, non-brand TV sets are more likely to generate e-waste. About half of the waste generated by assembler businesses is sold, 30 percent is dumped, and 20 percent is held for a long time. 30 percent of repairers kept the old and damaged TV sets for repair and future use, 15 percent threw them away, and 5 percent refused to notify. Customers sold 40 percent of rejected sets to repairers, 10 percent threw them away, and 20 percent reused them after they were repaired. The results of the survey allow us to readily depict the dumping management landscape of the entire city of Dhaka. Electronic waste, or e-waste, is a term that refers to materials that are created when electronic equipment reaches the end of their useful life and are discarded due to their inability to perform their intended purpose. E-waste is a term that refers to large and small household appliances, information and telecommunications equipment, lighting equipment, electrical and electronic gadgets, toys, sports equipment, and medical devices (Pramila *et al.,* 2012).

2.2 E-Waste Recycling

The term "e-waste recycling" refers to the process of recovering and reusing these electronic wastes. It's straightforward. It is a method for recovering materials from electronic waste. This allows you to incorporate them into new technological devices. The electronic wastes derived from air conditioners, televisions, electric cookers, heaters, DVD players, fans, microwaves, and radios. They may also be found in information technology equipment such as computers, laptops, smartphones, batteries, hard drives, circuit boards, and displays. Due to the potential for reducing environmental dangers and pollution, e-waste recycling is one of the most discussed concerns in the world today. Additionally, it has the potential to defend our lives as humans and other living forms that exist in our globe. Recycling e-waste is a developing trend that was sparked by the extensive environmental contamination caused by e-waste. Additionally, millions of technological devices are used every day. When they approach the end of their useful lives, they largely decompose in landfills. What's more, just 12.5 percent of e-waste gets recycled (Zhang *et al.,* 2011). Several metals may also be recovered and recycled to create new steel and metal goods. These metals are more valuable like gold (Au) derived from printed circuit boards, Central processing Units (CPU), connectors / fingers, connector plating, pins, contacts, EPROM, ROM, RAM, integrated circuits, transistors, diodes, and switches. You can remove the glass from computer displays and televisions' CRTs (Cathode Ray Tubes). CRTS includes a number of potentially harmful chemicals, including lead and this is hazardous to human health as well as the immediate environment. Metals should be removed using over-band magnets. It helps you in removing ferrous and non-ferrous components from the glass.

2.3 Toxic Elements in E-Waste

E-waste comprises about 1000 different chemicals that are harmful, resulting in significant pollution upon disposal. Electronic garbage components, such as central processing units (CPUs) and cathode ray tubes found in televisions and computer displays, may include hazardous materials such as lead, mercury, cadmium, beryllium, and brominated flame retardants. A mobile phone/smartphone is made up of over 50 distinct components, including base metals (such as copper and tin), non-ferrous special metals (such as cobalt, indium, and antimony), and precious metals (such as silver, gold, palladium). Lead, mercury, arsenic, cadmium, selenium, hexavalent chromium, and flame retardants present in excess of threshold levels render e-waste dangerous in nature (Sthiannopkao & Wang, 2013).

2.4 Environmental Impact

The consequences of e-waste are numerous. Pollutants are released into the atmosphere; they have a detrimental effect on living organism. Volumes of e-waste material and process residues are never recycled and are instead thrown in open fields, riverbanks, ponds, and wetlands, rivers, and irrigation ditches. Destruction of drinking water sources has been compounded by indiscriminate dumping and even landfills. Additionally, when waste is thrown in open areas, it pollutes the soil, crops, plants and microorganisms. After burning of e-waste several gases, fly ash, and fine particles and harmful poisons are released into the air, which have a detrimental effect on human and animal life when inhaled. Recyclable e-waste is classified as a commodity, but non-recyclable e-waste is classified as waste. Both categories of e-waste have classified in to hazardous and non-biodegradable. Land filled e-waste produces toxic leachates, which eventually affect groundwater. When acids and sludge from melted computer chips are dumped on the ground, they contribute to soil acidification. Electronic trash incineration can produce harmful fumes and gases, damaging the nearby air (Gaidajis *et al.,* 2010). It is also impacting on the air. The air pollution caused by e-waste is more detrimental to certain animal species than others, posing a hazard to both these species and the biodiversity. Over time, air pollution damages soil and water quality, as well as plant species, wreaking havoc on ecosystems (Sivaramanan, 2013). After burning, shredding, or deconstructing e-waste, massive particles are released, and the soil becomes contaminated. Soil pollution depends on a number of variables, including temperature, pH level, and soil makeup (Agbeshie *et al.,* 2020). Acidification caused by metals has the ability to harm both marine and freshwater organisms, hence disrupting biodiversity and destroying ecosystems over time. Acidification also has a detrimental effect on ecosystems, making recovery more difficult (Vasanthi *et al.,* 2008). Heavy metal and organic contaminants have been reached, the biota of neighbouring waterways; it assumes that they are causing harmful impacts. Huang *et al.,* (2014) suggested future research to assess both contamination levels in biota and deleterious impacts on biota in surrounding water bodies.

2.6 Health Impact

Many metals including mercury, cadmium, lead, barium, and lithium are all toxic components derived from electronic waste. It has detrimental health effects and affects human brain cells, the heart, the liver, the kidneys, and even the skeletal system. It causes a substantial effect on neurological and reproductive systems, and birth defects (Orisakwe *et al.,* 2019).

3. METHODOLOGY

The study's major focus is on the generation of e-waste by private commercial banks in Bangladesh and how are they managing e-waste. There are total 5085 branches of 41 private commercial banks in Bangladesh. As a consequence, each district or Upazila city in Bangladesh has been chosen as a study region for this research (Fig 1).

Figure 1. Study area map

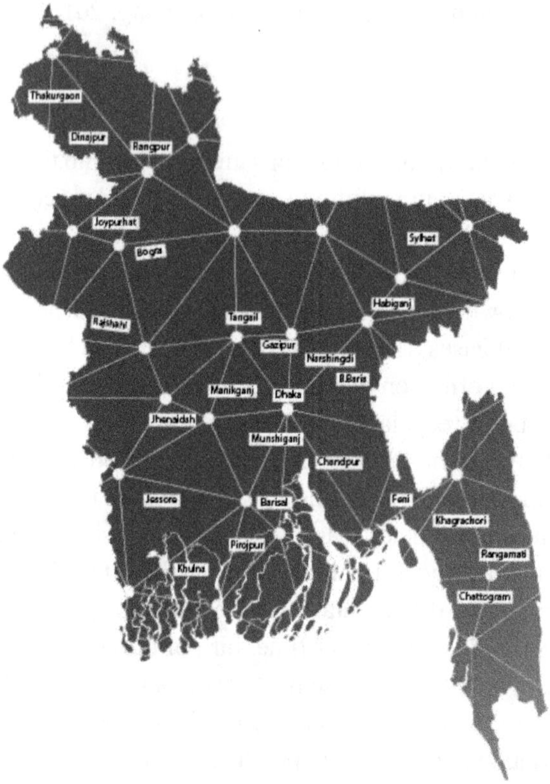

Primary data are gathered through interviews using questionnaires, and other methods. It obtained directly from the source and is considered to be the most reliable source of information on research. Determine the research's objective and target population before deciding on a data gathering sources. The study used the survey data and qualitative information for this study. The qualitative data were collected through Focus Group Discussion (FGD) and Key Informants' Interview (KIIs). Secondary data were collected through different sources including project documents (confidential), published journal articles for literature review. The Statistical Program of Social Sciences (SPSS) was used to enter the data; analysis and interpretation. After analysis, results were presented in the tabular and graphical forms to ease the reading of the audience. This study utilized an empirical method that combines qualitative and quantitative. The data was collected in accordance with the various policies of Bangladesh Bank and other banks and their confidentiality provisions. The Focus Group Discussion and Key Informants' Interviews (KIIs) were conducted for this study. Key Informant Interviews were carried out with the Bank professionals and environment specialists. The study used genuine and reliable data, which processed and managed using statistical software. Primary data is derived from its own observation and has never been used previously. First Security Islami Bank Limited data for the years 2019 and 2020 were analysed. The bank has 197 branches, and using FSIBL as a baseline, the approximate e-waste generated has been calculated using simple arithmetic deduction, i.e. dividing the total e-waste in Kg/year by the number of FSIBL branches yields a value per branch (Y), and because all commercial banks' branches are homogeneous in nature, the other bank's values were derived by multiplying the respective number of branches by Y. Thus, we estimated the total amount of e-waste generated by 41 commercial banks in Bangladesh in 2020. Secondary data were gathered from relevant sources. This is exploratory study that demonstrated for finding the gaps between e-waste management policy and practices in the private commercial Banks in Bangladesh. .Secondary data sources were from government censuses, journals, articles, magazines, newspapers, libraries, and the internet. SPSS (Statistical Package for the Social Sciences) used to analyze data, and Microsoft Office Excel used to exhibit data in the form of table, bar charts, pie charts, histograms, and line graphs. Pearson Correlations test and coefficient test has been carried out to look at the significance of growth rate of 41 commercial banks and First Security Islami Bank Limited.

4. RESULTS

4.1 E-Waste Production of 41 Commercial Banks (2016–2020)

The study found that 41 private commercial banks from Bangladesh discard PCs, monitors, UPS, IPS etc. as e-waste during a period of 2016 to 2020 (Table 1). These out-dated items pose a major threat to the environment. The study also found that banks release approximately 2176.30 tons of e-waste with average 400 tons during the period. The huge volume of e-waste (490.47 tons) disposed in 2021. Contrarily, this figure increased to 100 percent from 2016 to 2021. The bank used 385.26 tons of discarded devices in 2019, which is around 22% increased than the year 2018 (Figure 2). Additionally, the amount of e-waste generated in 2020 has increased. According to the study, a single bank generated an average of 5.84 tons of e-waste in 2016. In the recent five years, the single bank's maximum e-waste generation was 33.83 tons in 2019 and the lowest was 19.48 tons in 2016.

4.2 E-Waste Production Growth Rate

Figure 2 and Table 2 represent, an analysis of growth rate from 41 banks in Bangladesh reveals that growth rate percentage with the moderate growth 21.97 tons. The Pearson correlation of e-waste data from 2016 to 2020 demonstrates that the growth rate has changed significantly at the rate each year and that the correlation is significant at the 0.05 level for each year. This is highly significant with the p-value of 0.000 for all years.

Table 1. E-waste production (tons) from 2016 to 2020 of 41 commercial banks

Year	Minimum	Maximum	Mean	Std. Deviation
2016	0.19	19.48	5.84	3.74
2017	0.24	23.65	7.01	4.98
2018	0.29	26.26	8.19	5.51
2019	0.34	30.40	9.40	6.26
2020	0.39	33.83	10.67	6.93

Source: Authors, 2021

Figure 2. Yearly e-waste growth rate of 41 PCB

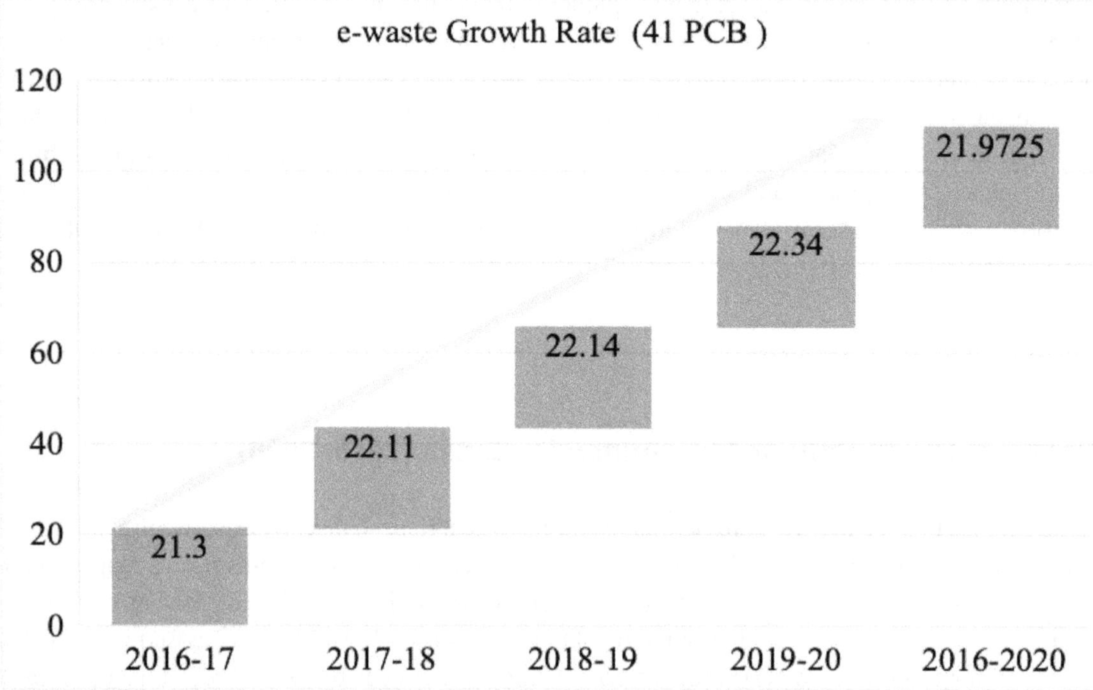

Table 2. E-waste production growth rate depends on branch count of 41 PCBs

Pearson Correlations	2016	2017	2018	2019	2020
Growth rate	-	.000	.000	.000	.000

**. Correlation is significant at the 0.01 level (2-tailed).

4.4 Product (E-Waste) Purchase Value of FSIBL

It was found that, FSIBL is one of the most popular private banks in Bangladesh. According to the bank's data from 2016 to 2020, the bank has discharged 24 different electronics as the following e-waste, with total purchase worth of BDT 100451196 in just four years. Electronics was purchased at maximum cost of BDT 5191393 in 2020 and minimum cost of BDT 1210691 in 2016 (Table 3) respectively.

4.5 Growth Rate (FSIBL)

The figure 4 describes that the e-waste production rate is in an upward trend. Where, it was 9.90 tons in the year 2016, and it was increased to 18.21 tons in 2020. The growth rate of the last 5 years is almost double and yearly almost 20 percent. Also, results from One Way ANOVA revealed that there is no statistically significant relationship between the year and the generation of e-waste, in the FSIBL with the P value of 0.132 (Table 4).

Table 3. E-waste purchase value from 2017 to 2020 of FSIBL

Year	2017(bdt)	2018 (bdt)	2019 (bdt)	2020 (bdt)
Laptop	340000	65934	317997	680513.58
Dot Matrix Printer	1624000	2706000	738000	3237606
Network Printer	1680000	1215410	640000	4793600
Scanner A4	319500	319000	275000	585000
Scanner Legal	0	0	45000	221490
Server	197500	6000000	1245000	1332150
Router	0	0	0	0
Switch	60000	0	0	0
Monitor	3780000	2321000	2002000	5140200
Offline UPS 650VA	1560000	1223680	0	0
Offline UPS 1200 VA	220000	118800	215000	2875625
Offline UPS 1800 VA	0	0	1150000	7012425
Online UPS	0	2500	35000	1048600
CPU	19050000	13860000	10255000	10972850
Laser Printer	228000	399349	752000	6887718.4
Hard Drive	0	24450	0	0
Projector	0	135000	0	0
External Hard Drive	0	28600	14300	10710.7
All in One Color printer	0	6650	6650	21346.5
SSD Hard Disk	0	5300	0	0
MICR Cheque Scanner	0	222600	0	0
BRTA Money Reciept Dot Printer	0	48900	544000	2095488
POS Printer	0	0	66400	106572
Apple MAC PC	0	0	254000	4892040
Total	1210791.67	28771125	18555347	51913935.2

*Total purchase value100451198.87
Source: Authors, 2021

Table 4. Growth rate 2017-2020 considering FSIBL

Coefficients				
Standardized Coefficients	t	Sig.	95.0% Confidence Interval for B	
Beta			Lower Bound	Upper Bound
-.867	-2.460	.133	-2.803	.764
Dependent Variable: e-waste/ Independent Variable: Year				

4.6 New Rules of Department of Environment (DOE) and Remedial measure

Bangladesh's Department of Environment (DOE) released the Hazardous Waste (e-waste) Management Rules, 2021 on June 10, 2021, in accordance with the 1995 Bangladesh Environmental Protection Act. The e-waste rule applies to the products listed in the Schedule (appliances, monitoring and control equipment, medical equipment, automatic machines, and information and communication technology) and imposes restrictions on the products producers, assemblers, collectors, sellers, and users. The respondents from Key Informants' Interviews mentioned that this regulation becoming effective immediately upon publication. In disobeying these rules, the offender shall be punished with imprisonment for a maximum period of two years or to a fine of up to BDT 2000, or to both, in accordance with Section 15(1) of the Bangladesh Environmental Protection Act, 1995. In case of further repeating, they shall be punished with imprisonment for a term ranging from two to ten years or a fine ranging from BDT 200,000 to BDT 1,000,000 or both. Now it will help in e-waste management in the banking sectors as well as for other sectors.

4.6.1 Remedial Public Measure and Law

In accordance to the e-waste regulation, Banks has given guarantee for the appropriate handling of e-waste in order to avoid negative health and environmental consequences and to reduce e-waste. It was mentioned by the respondents that formal recyclers are trying to manage electronic trash in accordance with the 3R (Reduce, Reuse, and Recycling) principles. Department of Environment (DOE) announced the Hazardous E-waste management Rules, 2021. According to this rule, Banks and financial organizations are thinking about their waste disposal strategies and the responsibility.

4.6.2 Present Scenarios of E-Waste Management by the PCBs

Private commercial banks in Bangladesh have some broad regulations regarding e-waste management but, they are not following appropriate management practices of such hazardous items. It was found that the banks dispose of various e-wastes in accordance with certain basic disposal regulations. It is not causing harm to the bank or any of its authorized personnel. As a consequence, collectors separate broken electrical items' batteries, circuit boards, and plastic casings. Certain pieces are melting in the fire. The workers do not utilize any safety equipment, which leads in a variety of physical difficulties. The country loses millions of mobile phones, computers, laptops, televisions, refrigerators, washing machines, air conditioners, and copying machines every year. The issue is that these electronic gadgets lack from formal environmental management system. They discarded with other household rubbish in the dustbin. Electronic equipment mostly includes lead, silicon, tin, cadmium, mercury, zinc, chromium, and nitrous oxide. These poisons are poisoning the country's soil and water. It enters the human body in a variety of ways, increasing the chance of developing numerous ailments. Pregnant women and children are particularly at danger. Premature delivery, poor birth weight, and even still birth are all caused by e-waste. Lead from e-waste causes serious harm to the neural system of newborns. Children can acquire complications of various diseases as they age, including lung, pulmonary, thyroid, cancer, and heart disease. Above all, while the bank has a disposition policy, it is limited to the bank's security. Banks must constantly assess their data policies as long as they have e-waste. According to the practices, the different types of e-waste crushed by any means that are causing major harm to the environment and human health. In Bangladesh, there is no facility or framework for the disposal or reuse of these obsolete devices.

5. DISCUSSION

The study found that Bangladesh produces around 2.7 million metric tons of e-waste per year. Safe disposal of these products is not being undertaken, and without an understanding of the harmful effects of improper e-waste disposal, this has been dumped in open landfills, agricultural land, and bodies of water. Regrettably, the fate of millions of tons of e-waste generated in Bangladesh each year is largely unknown. A study mentioned that the amount of e-waste such as computers, phones, and televisions - discarded each year, is increasing at an alarming rate (Hossain *et al.,* 2010). Banks wasted PCs, monitors, UPS, IPS, etc as e-waste. Result found that banks generated about 1200 tons of e-waste during 2016-2020. 365 tons of e-waste disposed in 2016 and this proportion jumped to 80 percent. In 2019, the bank used 246.18 tons of discarded devices, about 15% less than in 2018. Also, e-waste production in 2020 increased slightly from 2018 and 2019, reaching approximately 262 tons. In 2016, a single bank generated an average of 8.60 tons of e-waste. The volume decreased from 2018 to 2019, but rebounded to 4.59 tons in 2020. In the last four years, the single bank generated the most e-waste (52.41) and the least (34.42). It was 7.93 from 2016 to 2020 and 6.15 from 2016 to 2020. A study is also mentioned that the generation of e-waste near to 3,1 MMT/year (Roy et al., 2022). Certain banks have made a point of boosting their usage of environmentally friendly technologies. E-waste has a recycling capacity of up to 79 percent of its entire weight, but a dangerous content of up to 63 percent. WB must have a distinct sales strategy in order to achieve success (Raharjo *et al.,* 2021). An analysis of data from banks in Bangladesh reveals that the Pearson correlation of e-waste's data from 2016 to 2020 demonstrates that the growth rate has changed significantly every year. It was reported that the world e-waste growth rate is increasing (Jain, 2008). Department of Environment (DOE) has released the hazardous e-waste management Rules, 2021, which urge banks and financial institutions to reconsider their waste management strategy. Otherwise, environmental pollution and health risks will escalate substantially. The Bangladesh Environment Conservation Act of 1995 gives authority to the Director General to undertake any activity necessary to conserve and enhance the quality of the environment and to control, prevent and assuage pollution. E-waste management policy can help achieving targets regarding sustainable production and consumption (Almulhim, 2022). In the fiscal year 2013-2014, Bangladesh created around 10 million tons of e-waste. The volume of e-waste is increasing; it is difficult to identify methods and solutions for managing the volume of e-waste (ESDO, 2010). Suitable safeguards for e-waste disposal are important, rather than it can result in harmful environmental damage. A lack of knowledge caused by a scarcity of information on the proper management and recycling of these expired items might expose people to health risks. E-waste degrades the soil composition of our farming area and reduces crop yield. The issue arises when this e-waste is illegally dumped in landfills. Three types of hazardous discharge such as i) Primary emission: hazardous chemicals found in e-waste (i.e. lead, mercury, arsenic, Poly Chlorinated Biphenyl etc.); ii) Secondary emissions: hazardous reaction products produced by improper treatment of e-waste (e.g. dioxins or furans formed during open burring or improper smelting of plastics with halogenated flame retardants; and iii) Tertiary emissions: toxic compounds or reagents discharged into the environment as a result of improper handling and treatment (e.g. cyanide or other leaching agents, mercury for gold amalgamation). In Bangladesh, electronic waste management, disassembly, and recycling are carried out in an extremely rudimentary way in the informal sector; no safety precautions are used. As a result, employees are directly exposed to health risks, while the general public is indirectly exposed to the danger of environmental (air, water, and soil) contamination as a result of inefficient waste management and disposal. Hazardous and toxic elements found in e-wastes due to improper handling, processing, recycling, and disposal (CERM, 2018 and Grant *et al.,* 2013). Another study reported that Bangladeshi informal e-waste recycling sectors performing all activities without protections (Parvez et al., 2022)

6. CONCLUSION AND RECOMMENDATION

6.1 Conclusion

The data and analyses in this study demonstrate unequivocally that Bangladesh lacks regulations and/or policies governing e-waste disposal. Bangladesh clearly has policies and regulations in place to safeguard the environment from pollution. However, no law particularly, addresses e-waste management, and as a result, the country is advancing without e-waste management laws. On the other hand, the growing trend of e-waste creation has emerged as one of Bangladesh's most serious environmental concerns and difficulties. Bangladesh is required to create a legal framework for the proper management of e-waste as a signatory to the Basel Convention 64. It goes without saying that Bangladesh is committed to working with the international community to implement steps to combat electronic waste within its borders. As a result, the Department of Environment under the Ministry of Environment and Forest has drafted laws such as the 2011 e-waste Management Rules and the 2009 Bangladesh Hazardous Waste Management Policy. These rules include provisions governing the handling of e-waste in Bangladesh. However, these bills have not yet been approved and enacted. Additionally, this study discovered that e-waste might be detrimental to human health and the environment if not properly managed or disposed of. Numerous nations have already implemented new policies for the handling of electronic trash in light of the destruction it has wreaked. As such, the author advises that Bangladesh should either finalize existing laws or develop a new policy on e-waste management in order to effectively manage the country's growing electronic waste while also protecting the environment.

6.2 Recommendations

At the workplace, the BOYD (Bring your own device) principle may be adopted. ICT inventory disposal policies, which are largely absent in commercial banks, should be widely implemented. Also, inventory and database of e-waste (at least for chosen goods) should be developed. The Extended Producer Responsibility (EPR) principle should be used to facilitate the manufacturer and seller's procedure. Workers' health and safety should be prioritized throughout the recycling process. Bangladesh Bank, as regulator, may provide incentives to commercial banks to develop and introduce banking products that will facilitate the establishment of recycling and recovery facilities for electronic waste, thereby promoting sustainable development. If you don't need anything, donate it so that it can be used by someone else. Donations are excellent tax deductions since the amount is generally close to the worth of the item if sold. Alternatively, a few stores offer a buyback option. Before you go out and buy a new gadget, see whether the store will buy your old camera, laptop, or other item. Best Buy offers a trade-in program where you may bring in your old electronics for Best Buy gift cards.

6.3 Implication and Future Research

The e-waste of 41 private commercial banks in Bangladesh was the subject of the study. The research findings include thorough information on the creation of e-waste by Private Commercial Banks in Bangladesh, as well as its rise over time. This research provides a comprehensive picture of banks' and government's new e-waste policy, as well as what has to be done in the future. Finally, using various methods to control e-waste might help you make the best selection.

REFERENCES

Agbeshie, A. A., Adjei, R., Anokye, J., & Banunle, A. (2020). Municipal waste dumpsite: Impact on soil properties and heavy metal concentrations, Sunyani, Ghana. *Scientific African*, *8*, e00390. doi:10.1016/j.sciaf.2020.e00390

Almulhim, A. I. (2022). Household's awareness and participation in sustainable electronic waste management practices in Saudi Arabia. *Ain Shams Engineering Journal*, *13*(4), 101729. doi:10.1016/j.asej.2022.101729

Buet, C. (2018). *Assessment of Generation of E-Waste, its Impacts on Environment and Resource Recovery Potential in Bangladesh*. Academic Press.

ESDO. (2010). *Study on E-waste: Bangladesh Situation*. Environment and Social DevelopmentOrganization.

Fu, J., Zhang, H., Zhang, A., & Jiang, G. (2018). *E-waste recycling in China: a challenging field*. Academic Press.

Gaidajis, G., Angelakoglou, K., & Aktsoglou, D. (2010). E-waste: Environmental problems and current management. *Journal of Engineering Science and Technology Review*, *3*(1), 193–199. doi:10.25103/jestr.031.32

Grant, K., Goldizen, F. C., Sly, P. D., Brune, M. N., Neira, M., van den Berg, M., & Norman, R. E. (2013). Health consequences of exposure to e-waste: A systematic review. *The Lancet. Global Health*, *1*(6), e350–e361. doi:10.1016/S2214-109X(13)70101-3 PMID:25104600

Hossain, S., Sulatan, S., Shahnaz, F., Akram, A. B., Nesa, M., & Happell, J. (2010). E-waste: Bangladesh situation. Study Report Environment and Social Development Organization-Environmental and Social Development Organisation Dhaka. Bangladesh.

Huang, J., Nkrumah, P. N., Anim, D. O., & Mensah, E. (2014). E-waste disposal effects on the aquatic environment: Accra, Ghana. *Reviews of Environmental Contamination and Toxicology*, 19–34. PMID:24515808

Islam, M. N. (2016). E-waste management of Bangladesh. *International Journal of Innovative Human Ecology and Nature Studies*, *4*(2), 1–12.

Jain, A. (2008). Global e-waste growth. *E-waste: Implication, Regulations, and Management in India and Current Global Best Practices*, 3-22.

Karim, R., Bari, N., & Amin, M. (2014). E-waste management in Bangladesh. *2nd International Conference on Green Energy and Technology*.

Khaliq, A., Rhamdhani, M., Brooks, G., & Masood, S. (2014). *Metal Extraction Processes for Electronic Waste and Existing Industrial Routes: A Review and Australian Perspective*. Academic Press.

Kumar, S., & Rawat, S. (2018). Future e-Waste: Standardisation for reliable assessment. *Government Information Quarterly*, *35*(4), S33–S42. doi:10.1016/j.giq.2015.11.006

MoEF. (2011). *Implementation of e-waste rules in India*. CPCB.

Nnorom, I. C., & Osibanjo, O. (2008). Overview of electronic waste (e-waste) management practices and legislations, and their poor applications in the developing countries. *Resources, Conservation and Recycling*, *52*(6), 843–858. doi:10.1016/j.resconrec.2008.01.004

Noel-Brune, M., Goldizen, F. C., Neira, M., van den Berg, M., Lewis, N., King, M., Suk, W. A., Carpenter, D. O., Arnold, R. G., & Sly, P. D. (2013). Health effects of exposure to e-waste. *The Lancet. Global Health*, *1*(2), e70. doi:10.1016/S2214-109X(13)70020-2 PMID:25104155

Orisakwe, O. E., Frazzoli, C., Ilo, C. E., & Oritsemuelebi, B. (2019). Public health burden of e-waste in Africa. *Journal of Health & Pollution*, *9*(22), 190610. doi:10.5696/2156-9614-9.22.190610 PMID:31259086

Pariatamby, A., & Victor, D. (2013). Policy trends of e-waste management in Asia. *Journal of Material Cycles and Waste Management*, *15*(4), 411–419. doi:10.100710163-013-0136-7

Parvez, S. M., Hasan, S. S., Knibbs, L. D., Jahan, F., Rahman, M., Raqib, R., Islam, N., Aich, N., Moniruzzaman, M., Islam, Z., Fujimura, M., & Sly, P. D. (2022). Ecological Burden of e-Waste in Bangladesh—an Assessment to Measure the Exposure to e-Waste and Associated Health Outcomes: Protocol for a Cross-sectional Study. *JMIR Research Protocols*, *11*(8), e38201. doi:10.2196/38201 PMID:35972788

Quinn, M., Jordan, H., & Lacy-Nichols, J. (2021). Upstream and downstream explanations of the harms of ultra-processed foods in national dietary guidelines. *Public Health Nutrition*, *24*(16), 5426–5435. doi:10.1017/S1368980021003505 PMID:34392856

Raharjo, S., Wulandari, S., & Fitriani, S. (2021, June). Waste bank system improvement for electronic waste recycling in Indonesia: A case study of Padang City. *IOP Conference Series. Earth and Environmental Science*, *802*(1), 012053. doi:10.1088/1755-1315/802/1/012053

Roy, H., Rahman, T. U., Suhan, M. B. K., Al-Mamun, M. R., Haque, S., & Islam, M. S. (2022). A comprehensive review on hazardous aspects and management strategies of electronic waste: Bangladesh perspectives. *Heliyon*, *8*(7), e09802. doi:10.1016/j.heliyon.2022.e09802 PMID:35815143

Schwarzer, R., Rochau, U., Saverno, K., Jahn, B., Bornschein, B., Muehlberger, N., Flatscher-Thoeni, M., Schnell-Inderst, P., Sroczynski, G., Lackner, M., Schall, I., Hebborn, A., Pugner, K., Fehervary, A., Brixner, D., & Siebert, U. (2015). Systematic overview of cost–effectiveness thresholds in ten countries across four continents. *Journal of Comparative Effectiveness Research*, *4*(5), 485–504. doi:10.2217/cer.15.38 PMID:26490020

Sivaramanan, S. (2013). E-Waste Management, Disposal and Its Impacts on the Environment. *Universal Journal of Environmental Research and Technology*, *3*(5).

Sthiannopkao, S., & Wong, M. H. (2013). Handling e-waste in developed and developing countries: Initiatives, practices, and consequences. *The Science of the Total Environment*, *463*, 1147–1153. doi:10.1016/j.scitotenv.2012.06.088 PMID:22858354

Vasanthi, P., Kaliappan, S., & Srinivasaraghavan, R. (2008). Impact of poor solid waste management on ground water. *Environmental Monitoring and Assessment*, *143*(1), 227–238. doi:10.100710661-007-9971-0 PMID:17999155

Vats, M. C., & Singh, S. K. (2014). E-Waste characteristic and its disposal. *International Journal of Ecological Science and Environmental Engineering, 1*(2), 49-61.

Wang, Y., Peris, A., Rifat, M. R., Ahmed, S. I., Aich, N., Nguyen, L. V., Urík, J., Eljarrat, E., Vrana, B., Jantunen, L. M., & Diamond, M. L. (2020). Measuring exposure of e-waste dismantlers in Dhaka Bangladesh to organophosphate esters and halogenated flame retardants using silicone wristbands and T-shirts. *The Science of the Total Environment*, *720*, 137480. doi:10.1016/j.scitotenv.2020.137480 PMID:32146393

Wath, S. B., Vaidya, A. N., Dutt, P. S., & Chakrabarti, T. (2010). A roadmap for development of sustainable E-waste management system in India. *The Science of the Total Environment*, *409*(1), 19–32. doi:10.1016/j.scitotenv.2010.09.030 PMID:20951410

Widmer, R., Oswald-Krapf, H., Sinha-Khetriwal, D., Schnellmann, M., & Böni, H. (2005). Global perspectives on e-waste. *Environmental Impact Assessment Review*, *25*(5), 436–458. doi:10.1016/j.eiar.2005.04.001

Williams, E., Kahhat, R., Allenby, B., Kavazanjian, E., Kim, J., & Xu, M. (2008). Environmental, social, and economic implications of global reuse and recycling of personal computers. *Environmental Science & Technology*, *42*(17), 6446–6454. doi:10.1021/es702255z PMID:18800513

Zhang, Q., Zhou, T., Xu, X., Guo, Y., Zhao, Z., Zhu, M., Li, W., Yi, D., & Huo, X. (2011). Downregulation of placental S100P is associated with cadmium exposure in Guiyu, an e-waste recycling town in China. *The Science of the Total Environment*, *410*, 53–58. doi:10.1016/j.scitotenv.2011.09.032 PMID:21992841

Chapter 7

E-Waste Management in Developing Countries:
Current Practices, Challenges, Disposal, and Impact on Human Health and Environment

Laxmi Kant Bhardwaj
https://orcid.org/0000-0001-7518-4199
Amity University, Noida, India

Prangya Rath
https://orcid.org/0000-0002-6764-1625
Amity University, Noida, India

Rajat Tokas
Amity University, Noida, India

ABSTRACT

Information and telecommunications technology (ITT) has expanded into many aspects of modern life and has positively affected human life even in the most remote areas of developing countries. The rapid growth in ITT has led to an improvement in the capacity of computers but simultaneously to a decrease in the product's lifetime as a result of which increasingly large quantities of e-waste are generated annually. It contains hazardous components that if it is not properly managed can have adverse environmental and health effects. The management of e-waste poses significant challenges in developing countries. Due to the lack of adequate infrastructure, e-waste is burned or dumped in open areas. Informal and inefficient recycling practices are generally employed which further exacerbate pollution and health risks. So, the implementation of e-waste management, adequate recycling, and waste disposal facilities have become crucial concerns. This chapter reviews the concept of e-waste generation, challenges, disposal, and its impact on human health and environment in detail.

DOI: 10.4018/979-8-3693-1018-2.ch007

Figure 1. Graphical abstract

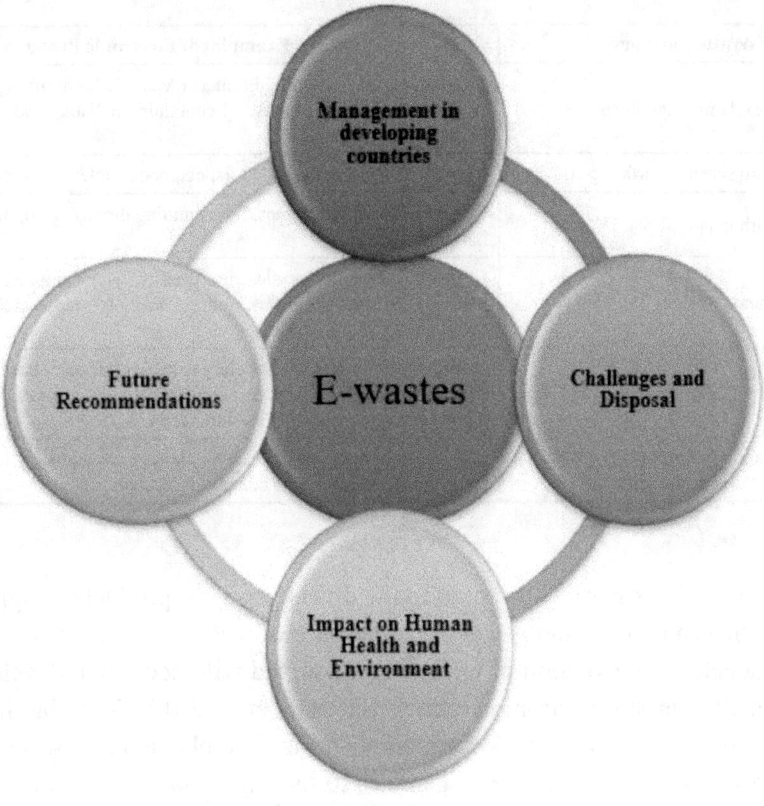

1. INTRODUCTION

Electrical waste (e-waste) constitutes one of the most rapidly expanding categories of solid waste globally (Bhardwaj et al. 2023a). Its growth rate stands at 3-5% per year, ~ three times quicker than other distinct waste streams (Schwarzer et al. 2005). This category encompasses ubiquitous items like mobile phones, computers, toys, televisions, iPods, printers, power tools, fluorescent lamps and various other small to large appliances prevalent in modern households and industries. E-waste, unfortunately, includes a myriad of noxious substances, containing plastics and heavy metals such as lead (Pb), nickel (Ni), chromium (Cr), cadmium (Cd), arsenic (As) and mercury (Hg). Additionally, it includes polybrominated diphenyl ethers (PBDEs) and polybrominated biphenyls (PBB), employed in cables, connectors and covers. The central processing units (CPUs) contain Cd, Pb and Hg, while printed circuit boards (PCBs) consist of antimony (Sb), silver (Ag), Cr, zinc (Zn), Pb, tin (Sn) and copper (Cu) (Herat and Agamuthu 2012). Notably, Pb finds common usage in computers, cabling, fluorescent tubes and tin-lead solders (Herat 2008). The presence of these hazardous constituents raises concerns about potential negative impacts on both the environment and human health if e-waste is not appropriately managed. Numerous studies have reported hazardous pollutants in samples collected from diverse sites of e-waste recycling, as detailed in Table 1.

Table 1. The occurrence of hazardous pollutants in the different samples collected from different e-waste recycling sites

S. No.	E-Waste Category	Examples of Electronic Products
1	Temperature exchange equipment	Appliances such as air conditioners (ACs), de-humidifying equipment, freezers, pumps with heat controllers, oil-containing radiators and other equipment with temperature exchange
2	Equipment with screens (surface >100 cm^2)	Laptops, photo frames, monitors, notebooks, televisions
3	Equipment with lamps	Compact fluorescent lamps, light emitting diode (LED), high-density discharge lamps
4	Large equipment (>50 cm)	Cookers, clothes dryers, electric stoves, copying equipment, dishwashing machines, electric hot plates, printing machines, medical devices, musical equipment, washing machines, etc.
5	Small equipment (<50 cm)	calculators, carpet sweepers, electric kettles, electric shavers, electric knives, hi-fi equipment, irons, microwaves, radio sets, scales, smoke detectors, toasters, toys, video cameras/ recorders, Vacuum cleaners, etc.
6	Small information technology as well as telecommunication equipment	Global positioning systems (GPSs), pocket calculators, telephones, mobile phones, printers, personal computers, routers

The surge in e-waste is primarily attributed to the diverse array of products, ranging from mechanical tools to highly incorporated systems and the rapid pace of technological advancements (Yla-Mella et al. 2015). The accelerated evolution of technology, coupled with economic development, results in a relatively shorter lifespan for electronic products (Hossain et al. 2015). The classification of electrical equipment into brown goods (such as computers and mobile phones) and white goods (including refrigerators and air conditioners) further contributes to the expanding volume of e-waste (Widmer et al. 2005). The escalating quantities of e-waste are closely tied to the disposal of electrical and electronic equipment (EEE). In developing countries, challenges intensify due to factors like population growth, rapid urbanization and industrial expansion, exacerbating e-waste issues. The lack of adequate infrastructure for safe waste management leads to prevalent practices of burying, open burning, or dumping e-waste into surface water bodies (Gupta et al. 2023). To categorize e-waste comprehensively, it is divided into six distinct categories, as outlined in Table 2.

2. REVIEW OF LITERATURE

The existing body of literature on e-waste management in developing nations such as India provides a comprehensive overview of the challenges, initiatives and potential solutions in handling the escalating issue of electronic waste. India, with its rapid technological advancements and increasing consumer electronics usage, has witnessed a surge in e-waste generation, necessitating effective management strategies. Studies highlight the substantial challenges faced by India in e-waste management, including inadequate infrastructure, informal recycling practices and insufficient awareness among the populace. The lack of proper disposal facilities and regulatory frameworks exacerbates the health and environmental risks associated with e-waste. Several research works delve into the environmental impact of improper e-waste disposal, emphasizing the presence of dangerous substances such as heavy metals and toxic chemicals. These studies underscore the urgent need for stringent regulations and improved recycling practices to

Table 2. Six different categories of e-wastes

S. No.	Type of Samples	Hazardous Pollutants	E-Waste Recycling Site	References
1	Sediment	PCBs, PAHs and heavy metals	Nanguan River, East China	Chen et al. 2010
			Wenling, Taizhou, China	Tang et al. 2010a
		Heavy metals	Bangalore, India	Ha et al. 2009
		PCDD/Fs	China	Ren et al. 2010
		Dechlorane plus (DP)	South China	Yu et al. 2010
2	Soil	PCBs, PAHs and heavy metals	Taizhou, East China	Tang et al. 2010b
		Heavy metals	Guiyu, Guangdong Province, China	Li et al. 2011
3	Soils and vegetables		Longtang, Guangdong province, China	Luo et al. 2011
		PBDEs	South China	Wang et al. 2011a
4	Air	PCDD/Fs	China	Sheng et al. 2011
		Dechlorane plus (DP)	China	Chen et al. 2011
5	Surface Dust	PBDEs and PCDD/Fs	South China	Leung et al. 2011
6	Mangrove wetland	PBDE	India	Binelli et al. 2007
7	Hair	Heavy metals	Longtang, South China	Zheng et al. 2011
		PBDEs and PCDD/Fs	Eastern China	Ma et al. 2011
8	Placenta	Heavy metals	Guiya, China	Guo et al. 2010
9	Urine		Taizhou, Zhejiang province, Southeast China	Wang et al. 2011b
			Taizhou, China	Wang et al. 2010
10	Human	PCBs, BFRs, HBCDs and PBDEs	Ghana, West Africa	Asante et al. 2011
		PBBs, PBDEs and PCBs	Zhejiang Province, China	Zhao et al. 2008
11	Children's blood	PCBs, PCDD/Fs, PBDEs and organochlorine pesticides (OCPs)	Zhejiang, China	Shen et al. 2010
12	Breast milk	BFRs and PCBs	Vietnam	Tue et al. 2010
13	Umbilical cord blood	PBDEs	Guiyu, China	Wu et al. 2010
14	Thyroid gland	PBDEs, dioxins and PCBs	China	Zhang et al. 2010
15	Skin	PCBs	Taizhou, China	Xing et al. 2011
16	Chicken tissue and eggs	PBDEs	Southeast China	Qin et al. 2011
17	Fish		Guiya, China	Luo et al. 2007

mitigate the adverse effects on ecosystems and public health. The literature also explores various initiatives and policies implemented by the Indian government to address the e-waste challenge. Researchers analyze the effectiveness of these policies, identifying areas for improvement and suggesting potential interventions. Case studies and empirical research contribute valuable insights into the practical aspects of e-waste management, shedding light on successful models and best practices.

Darby and Obara (2005) examined the extent of e-waste in the United Kingdom (UK), revealing that over 90% of Electrical and Electronic Equipment (EEE) is either incinerated or landfilled. The United States of America (USA) is recognized as the leading global producer of e-waste. In 2009, approximately

5 million tonnes of e-waste were stockpiled in the USA, with an additional 2.37 million tonnes slated for disposal (United Nations Environment Programme 2007). Within the European Union (EU), the overall production of e-waste was estimated at 9.3 million tonnes in 2005, encompassing 40 million computers and 32 million televisions (Fadaei 2022). A forecast by Computer Aid International (2010) anticipated a surge to 12 million tonnes of e-waste by 2020. In China, Veenstra et al. (2010) reported that about 83 million units of EEE were categorized as scrap in 2007, a figure expected to escalate to 227 million units by 2012. Meanwhile, in Japan, Moriguchi (2006) identified approximately 12.9 million units of EEE as scrap in 2008.

In Canada, approximately 5 million units of EEE are discarded annually (Deathe et al. 2008). Korea witnessed the end-of-life disposal of over 15 million mobile phones and 3 million computers in 2004 (Yoon and Jang 2006). Canning's study in 2006 revealed that more than 130 million phones in the USA and 105 million phones in Europe were discarded each year. The United Nations (UN) projected a substantial increase in e-waste of 200–400% in South Africa and China and a staggering 500% in India by 2020 (Secretariat 2011). Additionally, the UN anticipated a sevenfold increase in e-waste from discarded mobile phones in China and an eighteenfold increase in India. Furthermore, the UN estimates a 4–8 fold rise in e-waste from discarded personal computers by 2020 in developing countries such as Senegal and Uganda. The objective of this study is to provide a comprehensive review of e-waste, covering aspects such as its management, challenges, disposal methods and its impact on the environment and human health in developing countries.

3. E-WASTE MANAGEMENT

Addressing the challenges associated with e-waste management has become a top priority. The rising volume of e-waste generated globally is currently a significant fear for researchers. The surge in population and rapid development exacerbates the demand for efficient e-waste management in developing nations. Conventional waste management practices not only impact public health and the local environment but also extend their repercussions to neighboring countries (Murad and Siwar 2007). The substantial measures of e-waste generated, coupled with the availability of inexpensive land, further encourage the adoption of traditional approaches to e-waste disposal. Growing apprehensions revolve around the notion that a considerable portion of e-waste from developed countries is finding its way into economically challenged developing nations (Hossain et al. 2011).

Several developed nations have enacted legislation that enforces the concept of extended producer responsibility (EPR), which involves manufacturers taking back electronic products that have reached the end of their life cycle. EPR represents a strategy to incorporate principles of sustainable development into global trade, grounded in the law principle known as the "Polluter Pays Principle" (Kibert, 2004). Critical elements for the successful management of e-waste in developing nations include shifts in governmental perspectives, the enactment of specific laws addressing e-waste, regulation of e-waste disposal, the adoption of EPR measures and the handover of technology to facilitate environmentally sound e-waste recycling.

4. CHALLENGES AND DISPOSAL OF E-WASTE

There are two recycling processes: formal and informal. The informal process is widely employed, involving the burning and disassembly of e-waste in open areas to retrieve valuable components like steel, aluminum and copper (Leung et al. 2006). The recycling of e-waste is an intricate procedure, with the informal recycling sector playing a prominent role in Asian countries such as the Philippines, Pakistan, China, Vietnam and India, as well as in certain African nations like Ghana and Nigeria where e-waste disposal is prevalent (Nnorom and Osibanjo 2008). Osibanjo and Nnorom (2007) highlighted that the primary challenges faced by developing countries are centered around the increasing rate of e-waste production and its insufficient management. There are several challenges which are as follows:

(i) Greater threats are seen from inadequate handling
(ii) Lack of formal recycling systems
(iii) Weakly regulated or absence of recycling legislation
(iv) Poor or absence of inventory assessment of e-waste

Brigden et al. (2005) undertook a study in the vicinity of the sites of e-waste recycling in India and China, revealing elevated concentrations of Pb, polychlorinated dibenzodioxins (PCDDs), polychlorinated dibenzofurans (PCDFs) and polybrominated diphenyl ethers (PBDEs), as well as polybrominated dibenzodioxins (PBDDs) and polybrominated dibenzofurans (PBDFs) in freshwater, sediments, air and dust samples. A parallel investigation overseen by Brigden et al. (2008) in Ghana identified heightened levels of dibutyl phthalate (DBP), Pb and bis(2-ethylhexyl) phthalate (DEHP) in the samples collected from the sites of e-waste recycling. Several pollutants such as pesticides, heavy metals and DEHP were reported in the water/soil and sediment samples by the researchers in different studies (Bhardwaj and Jindal 2020; Bhardwaj et al. 2021; Bhardwaj et al. 2023b).

5. IMPACT ON HUMAN HEALTH AND ENVIRONMENT

The inadequate handling of e-waste poses substantial risks to both human and animal health while also negatively impacting the environment. Mismanagement of e-waste exposes the environment and human health to harmful toxic components, leading to potential harm through various routes such as skin contact, inhalation and ingestion. The harmful nature of e-waste disposal becomes a significant alarm, as improper methods elevate the risk of transferable diseases (Murad and Siwar 2007). The dumping of e-waste in open sites further exacerbates the issue, allowing harmful substances to enter the food chain through bioaccumulation, consequently contributing to the onset of various diseases (Mairaj et al. 2022; Alam et al. 2023). Researchers have extensively examined the contamination levels in PET bottled water samples, providing a detailed account of the migration of contaminants from plastic items into water bodies (Bhardwaj and Sharma 2021; Bhardwaj 2022). Wang and Xu (2015) assessed the environmental effect of handling PCBs and Song et al. (2018) investigated the environmental impact of cathode ray tube televisions in China. The toxic substances can leach into the soil, contaminating it and affecting plant growth (Bhardwaj et al. 2023c). These chemicals and heavy metals can seep into water bodies, causing water pollution. This affects aquatic life and various habitats that can ultimately impact human communities that rely on these resources. Burning or dismantling e-waste releases toxic fumes and particles into the environment, which affects flora and fauna. People residing in the vicinity have been observed to be affected with various problems like respiratory problems (such as lung irritation, and asthma) neurological disorders, developmental delays, organ damage, skin rashes, burns, eye irritation, etc. (Bhardwaj and Vikram 2023; Bhardwaj et al. 2023d).

6. CONCLUSION AND RECOMMENDATIONS

E-waste has evolved into a significant global concern, posing both social and environmental threats to numerous developing countries. The proper handling and disposal of e-waste have emerged as major challenges for many nations worldwide. Developing nations, in particular, face multiple challenges concerning the production, cross-border transport and efficient handling of e-waste. Urgent action is required to establish comprehensive legislation aimed at efficient e-waste management in these nations. Such legislation should also stipulate that the export of e-waste is permissible only when the receiving country demonstrates sufficient capacity to govern e-waste in a manner that safeguards both health and the environment.

As per the critical review, the following recommendations might be beneficial to minimize e-waste:

- The inclusion of informal recycling practices into the official municipal waste management system should be formally endorsed by the government.
- The government should allocate adequate funds to support the collection, storage, recycling and/or disposal of e-waste.
- The authority should provide training to recycling participants as well as enhance existing recycling facilities for e-waste.
- Authorities should initiate an awareness program for workers to educate them about the effects of e-waste on both the environment and human health.
- Developing countries should create and maintain an inventory assessment of e-waste.
- Robust legislation must be established in developing countries to standardize and control the impact of e-waste.
- Alternative or new techniques should replace traditional techniques to minimize e-waste.
- EPR principles should be adopted and included in the legislation of developing countries.
- Producers, importers and retailers ought to be obligated to cover the expenses associated with the recycling, collection and disposal of e-waste.
- Establishing standards and implementing a certification system is crucial for second-hand electrical appliances, as well as for recycling and disposal enterprises, to guarantee the adoption of safe and environmentally sound practices.

7. ACKNOWLEDGMENT

The authors thank Amity University for providing the platform to do this study.

8. STATEMENT AND DECLARATIONS

8.1. Funding

This study was not supported by any funding agency.

8.2. Conflicts of Interest/Competing Interest

The authors declare that they have no competing financial interest or personal relationship that could have appeared to influence the work reported in this chapter.

8.3. Availability of Data and Materials

Not applicable

8.4. Code Availability

Not applicable

8.5. Author's Contributions

All authors have equal contributions.

REFERENCES

Alam, S., Bhardwaj, L. K., Mallick, R., & Rai, S. (2023). Estimation of Heavy Metals and Fluoride Ion in Vegetables Grown Nearby the Stretch of River Yamuna, Delhi (NCR), India. *Indian Journal of Environmental Protection*, *43*(1), 64–73.

Asante, K. A., Adu-Kumi, S., Nakahiro, K., Takahashi, S., Isobe, T., Sudaryanto, A., Devanathan, G., Clarke, E., Ansa-Asare, O. D., Dapaah-Siakwan, S., & Tanabe, S. (2011). Human exposure to PCBs, PBDEs and HBCDs in Ghana: Temporal variation, sources of exposure and estimation of daily intakes by infants. *Environment International*, *37*(5), 921–928. doi:10.1016/j.envint.2011.03.011 PMID:21470682

Bhardwaj, L. K. (2022). Evaluation of bis (2-ethylhexyl) phthalate (DEHP) in the PET bottled mineral water of different brands and impact of heat by GC–MS/MS. *Chemistry Africa*, *5*(4), 929–942. doi:10.100742250-022-00385-6

Bhardwaj, L. K., & Jindal, T. (2020). Persistent organic pollutants in lakes of Grovnes Peninsula at Larsemann Hill area, East Antarctica. *Earth Systems and Environment*, *4*(2), 349–358. doi:10.100741748-020-00154-w

BhardwajL. K.KumarD.KumarA. (2023c). Phytoremediation Potential of Ocimum Sanctum: A Sustainable Approach for Remediation of Heavy Metals. doi:10.20944/preprints202308.0593.v1

BhardwajL. K.RathP.ChoudhuryM. (2023a). Assessment of E-Waste Pollution Level in Soil and Water and Its Impact on Human Health: A Review. doi:10.20944/preprints202311.1714.v1

Bhardwaj, L. K., & Sharma, A. (2021). Estimation of physico-chemical, trace metals, microbiological and phthalate in PET bottled water. *Chemistry Africa*, *4*(4), 981–991. doi:10.100742250-021-00267-3

Bhardwaj, L. K., Sharma, S., & Jindal, T. (2021). Occurrence of polycyclic aromatic hydrocarbons (PAHs) in the Lake water at Grovnes Peninsula Over East Antarctica. *Chemistry Africa*, *4*(4), 965–980. doi:10.100742250-021-00278-0

Bhardwaj, L. K., Sharma, S., & Jindal, T. (2023b). Estimation of Physico-Chemical and Heavy Metals in the Lakes of Grovnes & Broknes Peninsula, Larsemann Hill, East Antarctica. *Chemistry Africa*, 1-18.

BhardwajL. K.SinghV. V.DwivediK.RaiS. A. (2023d). Comprehensive Review on Sources, Types, Impact and Challenges of Air Pollution. https://doi.org/ doi:10.20944/preprints202311.1245.v1

BhardwajL. K.VikramV. (2023). Air Pollution and Its Effect on Human Health. https://doi.org/ doi:10.20944/preprints202307.1691.v1

Binelli, A., Sarkar, S. K., Chatterjee, M., Riva, C., Parolini, M., Bhattacharya, B., Bhattacharya, A. K., & Satpathy, K. K. (2007). Concentration of polybrominated diphenyl ethers (PBDEs) in sediment cores of Sundarban mangrove wetland, northeastern part of Bay of Bengal (India). *Marine Pollution Bulletin*, *54*(8), 1220–1229. doi:10.1016/j.marpolbul.2007.03.021 PMID:17512955

Brigden, K. (2005). *Recycling of electronic wastes in China and India: workplace and environmental contamination*. http://www. greenpeace. org/raw/content/china/en/press/reports/recycling-of-electronic-wastes. pdf

Brigden, K., Labunska, I., Santillo, D., & Johnston, P. (2008). Chemical contamination at e-waste recycling and disposal sites in Accra and Korforidua. Academic Press.

Canning, L. (2006). Rethinking market connections: Mobile phone recovery, reuse and recycling in the UK. *Journal of Business and Industrial Marketing*, *21*(5), 320–329. doi:10.1108/08858620610681623

Chen, L., Yu, C., Shen, C., Zhang, C., Liu, L., Shen, K., Tang, X., & Chen, Y. (2010). Study on adverse impact of e-waste disassembly on surface sediment in East China by chemical analysis and bioassays. *Journal of Soils and Sediments*, *10*(3), 359–367. doi:10.100711368-009-0176-8

Chen, S. J., Tian, M., Wang, J., Shi, T., Luo, Y., Luo, X. J., & Mai, B. X. (2011). Dechlorane Plus (DP) in air and plants at an electronic waste (e-waste) site in South China. *Environmental Pollution*, *159*(5), 1290–1296. doi:10.1016/j.envpol.2011.01.026 PMID:21324409

Darby, L., & Obara, L. (2005). Household recycling behaviour and attitudes towards the disposal of small electrical and electronic equipment. *Resources, Conservation and Recycling*, *44*(1), 17–35. doi:10.1016/j.resconrec.2004.09.002

Deathe, A. L., MacDonald, E., & Amos, W. (2008). E-waste Management Programmes and the Promotion of Design for the Environment: Assessing Canada's Contributions. *Review of European Community & International Environmental Law*, *17*(3), 319–334. doi:10.1111/j.1467-9388.2008.00610.x

Fadaei, A. (2022). E-waste management status worldwide: Major challenges and solutions. *Proceedings of the International Academy of Ecology and Environmental Sciences*, *12*(4), 281.

Guo, Y., Huo, X., Li, Y., Wu, K., Liu, J., Huang, J., ... Xu, X. (2010). Monitoring of lead, cadmium, chromium and nickel in placenta from an e-waste recycling town in China. *The Science of the Total Environment*, *408*(16), 3113–3117. doi:10.1016/j.scitotenv.2010.04.018 PMID:20451954

GuptaP.SharmaA.BhardwajL. K. (2023). Solid Waste Management (SWM) and Its Effect on Environment & Human Health. doi:10.20944/preprints202309.0384.v1

Ha, N. N., Agusa, T., Ramu, K., Tu, N. P. C., Murata, S., Bulbule, K. A., Parthasaraty, P., Takahashi, S., Subramanian, A., & Tanabe, S. (2009). Contamination by trace elements at e-waste recycling sites in Bangalore, India. *Chemosphere*, *76*(1), 9–15. doi:10.1016/j.chemosphere.2009.02.056 PMID:19345395

Herat, S. (2008). Contamination of solid waste from toxic materials in electronic waste (E-waste). *Journal of Solid Waste Technology Management*, *34*(4), 1–18.

Herat, S., & Agamuthu, P. (2012). E-waste: A problem or an opportunity? Review of issues, challenges and solutions in Asian countries. *Waste Management & Research*, *30*(11), 1113–1129. doi:10.1177/0734242X12453378 PMID:22851536

Hossain, M. S., Al-Hamadani, S. M., & Rahman, M. T. (2015). E-waste: A challenge for sustainable development. *Journal of Health & Pollution*, *5*(9), 3–11. doi:10.5696/2156-9614-5-9.3 PMID:30524771

Hossain, S., Sulatan, S., Shahnaz, F., & Hossain, M. L. (2011). *Illegal import and trade off of E-waste in Bangladesh. Environment and Social Development Organization*. ESDO.

Kibert, C. J. (2004). Green buildings: An overview of progress. *Journal of Land Use & Environmental Law*, *19*(2), 491–502.

Leung, A., Cai, Z. W., & Wong, M. H. (2006). Environmental contamination from electronic waste recycling at Guiyu, southeast China. *Journal of Material Cycles and Waste Management*, *8*(1), 21–33. doi:10.100710163-005-0141-6

Leung, A. O., Zheng, J., Yu, C. K., Liu, W. K., Wong, C. K., Cai, Z., & Wong, M. H. (2011). Polybrominated diphenyl ethers and polychlorinated dibenzo-p-dioxins and dibenzofurans in surface dust at an E-waste processing site in Southeast China. *Environmental Science & Technology*, *45*(13), 5775–5782. doi:10.1021/es103915w PMID:21639085

Li, J., Duan, H., & Shi, P. (2011). Heavy metal contamination of surface soil in electronic waste dismantling area: Site investigation and source-apportionment analysis. *Waste Management & Research*, *29*(7), 727–738. doi:10.1177/0734242X10397580 PMID:21406506

Luo, C., Liu, C., Wang, Y., Liu, X., Li, F., Zhang, G., & Li, X. (2011). Heavy metal contamination in soils and vegetables near an e-waste processing site, south China. *Journal of Hazardous Materials*, *186*(1), 481–490. doi:10.1016/j.jhazmat.2010.11.024 PMID:21144651

Luo, Q., Cai, Z. W., & Wong, M. H. (2007). Polybrominated diphenyl ethers in fish and sediment from river polluted by electronic waste. *The Science of the Total Environment*, *383*(1-3), 115–127. doi:10.1016/j.scitotenv.2007.05.009 PMID:17560632

Ma, J., Cheng, J., Wang, W., Kunisue, T., Wu, M., & Kannan, K. (2011). Elevated concentrations of polychlorinated dibenzo-p-dioxins and polychlorinated dibenzofurans and polybrominated diphenyl ethers in hair from workers at an electronic waste recycling facility in Eastern China. *Journal of Hazardous Materials, 186*(2-3), 1966–1971. doi:10.1016/j.jhazmat.2010.12.091 PMID:21255916

Mairaj, S., Nagar, R. D., Bhardwaj, L. K., Rehman, F., Punnakal, A., Chauhan, A., & Jindal, T. (2022). Antioxidant responses of arsenite-induced oxidative stress in rice (Oryza sativa L.) and its modulation by eugenol (extracted from Ocimum sanctum). *Indian Journal of Agricultural Research, 56*(4), 396–400.

Moriguchi, Y. (2006). *Establishing a sound material cycle society in Asia*. In The Asia 3R Conference, Tokyo, Japan.

Murad, W., & Siwar, C. (2007). Waste management and recycling practices of the urban poor: A case study in Kuala Lumpur city, Malaysia. *Waste Management & Research, 25*(1), 3–13. doi:10.1177/0734242X07070766 PMID:17346002

Nnorom, I. C., & Osibanjo, O. (2008). Overview of electronic waste (e-waste) management practices and legislations, and their poor applications in the developing countries. *Resources, Conservation and Recycling, 52*(6), 843–858. doi:10.1016/j.resconrec.2008.01.004

Osibanjo, O., & Nnorom, I. C. (2007). The challenge of electronic waste (e-waste) management in developing countries. *Waste Management & Research, 25*(6), 489–501. doi:10.1177/0734242X07082028 PMID:18229743

Qin, X., Qin, Z., Li, Y., Zhao, Y., Xia, X., Yan, S., Tian, M., Zhao, X., Xu, X., & Yang, Y. (2011). Polybrominated diphenyl ethers in chicken tissues and eggs from an electronic waste recycling area in southeast China. *Journal of Environmental Sciences (China), 23*(1), 133–138. doi:10.1016/S1001-0742(10)60384-1 PMID:21476352

Ren, M., Zhang, S., Liu, X., Feng, G., Xu, Z., & Fang, J. (2010). Concentrations and profiles of polychlorinated dioxins and furans in a discarded electronic waste open burning site. In *2010 4th International Conference on Bioinformatics and Biomedical Engineering* (pp. 1-4). IEEE.

Schwarzer, S., De Bono, A., Giuliani, G., Kluser, S., & Peduzzi, P. (2005). E-waste, the hidden side of IT equipment's manufacturing and use. *Environment Alert Bulletin*, 5.

Secretariat, R. S. (2011). *E-waste in India. India Research Unit (Larrdis)*. Rajya Sabha Secretariat.

Shen, H., Ding, G., Han, G., Wang, X., Xu, X., Han, J., Lou, X., Xu, C., Cai, D., Song, Y., & Lu, W. (2010). Distribution of PCDD/Fs, PCBs, PBDEs and organochlorine residues in children's blood from Zhejiang, China. *Chemosphere, 80*(2), 170–175. doi:10.1016/j.chemosphere.2010.03.012 PMID:20378148

Sheng, W., & Yan, G., JingGuang, L., TingMing, S., YunFeng, Z., & YongNing, W. (2011). Particle-bound PCDD/Fs in the atmosphere of an electronic waste dismantling area in China. *Biomedical and Environmental Sciences, 24*(2), 102–111. PMID:21565680

Song, X., Zhang, C., Yuan, W., & Yang, D. (2018). Life-cycle energy use and GHG emissions of waste television treatment system in China. *Resources, Conservation and Recycling, 128*, 470–478. doi:10.1016/j.resconrec.2016.09.004

Tang, X., Shen, C., Chen, L., Xiao, X., Wu, J., Khan, M. I., Dou, C., & Chen, Y. (2010). Inorganic and organic pollution in agricultural soil from an emerging e-waste recycling town in Taizhou area, China. *Journal of Soils and Sediments*, *10*(5), 895–906. doi:10.100711368-010-0252-0

Tang, X., Shen, C., Shi, D., Cheema, S. A., Khan, M. I., Zhang, C., & Chen, Y. (2010). Heavy metal and persistent organic compound contamination in soil from Wenling: An emerging e-waste recycling city in Taizhou area, China. *Journal of Hazardous Materials*, *173*(1-3), 653–660. doi:10.1016/j.jhazmat.2009.08.134 PMID:19783093

Tue, N. M., Sudaryanto, A., Minh, T. B., Isobe, T., Takahashi, S., Viet, P. H., & Tanabe, S. (2010). Accumulation of polychlorinated biphenyls and brominated flame retardants in breast milk from women living in Vietnamese e-waste recycling sites. *The Science of the Total Environment*, *408*(9), 2155–2162. doi:10.1016/j.scitotenv.2010.01.012 PMID:20116831

UNEP. (2007). *E-waste-Volume I: Inventory assessment manual*. United Nations Environmental Programme.

Veenstra, A., Wang, C., Fan, W., & Ru, Y. (2010). An analysis of E-waste flows in China. *International Journal of Advanced Manufacturing Technology*, *47*(5-8), 449–459. doi:10.100700170-009-2356-5

Wang, H., Han, M., Yang, S., Chen, Y., Liu, Q., & Ke, S. (2011b). Urinary heavy metal levels and relevant factors among people exposed to e-waste dismantling. *Environment International*, *37*(1), 80–85. doi:10.1016/j.envint.2010.07.005 PMID:20678797

Wang, H., Lv, S., Li, F., Liu, Q., & Ke, S. (2010). Study on the changes of urinary 8-hydroxydeoxyguanosine levels and burden of heavy metal around e-waste dismantling site. *The Science of the Total Environment*, *408*(24), 6092–6099. doi:10.1016/j.scitotenv.2010.09.019 PMID:20889187

Wang, J., & Xu, Z. (2015). Disposing and recycling waste printed circuit boards: Disconnecting, resource recovery, and pollution control. *Environmental Science & Technology*, *49*(2), 721–733. doi:10.1021/es504833y PMID:25525865

Wang, Y., Luo, C., Li, J., Yin, H., Li, X., & Zhang, G. (2011a). Characterization of PBDEs in soils and vegetations near an e-waste recycling site in South China. *Environmental Pollution*, *159*(10), 2443–2448. doi:10.1016/j.envpol.2011.06.030 PMID:21763040

Widmer, R., Oswald-Krapf, H., Sinha-Khetriwal, D., Schnellmann, M., & Böni, H. (2005). Global perspectives on e-waste. *Environmental Impact Assessment Review*, *25*(5), 436–458. doi:10.1016/j.eiar.2005.04.001

Wu, K., Xu, X., Liu, J., Guo, Y., Li, Y., & Huo, X. (2010). Polybrominated diphenyl ethers in umbilical cord blood and relevant factors in neonates from Guiyu, China. *Environmental Science & Technology*, *44*(2), 813–819. doi:10.1021/es9024518 PMID:20000818

Xing, G. H., Liang, Y., Chen, L. X., Wu, S. C., & Wong, M. H. (2011). Exposure to PCBs, through inhalation, dermal contact and dust ingestion at Taizhou, China–A major site for recycling transformers. *Chemosphere*, *83*(4), 605–611. doi:10.1016/j.chemosphere.2010.12.018 PMID:21295325

Ylä-Mella, J., Keiski, R. L., & Pongrácz, E. (2015). Electronic waste recovery in Finland: Consumers' perceptions towards recycling and re-use of mobile phones. *Waste Management (New York, N.Y.)*, *45*, 374–384. doi:10.1016/j.wasman.2015.02.031 PMID:25797074

Yoon, H., & Jang, Y. C. (2006, May). The practice and challenges of electronic waste recycling in Korea with emphasis on extended producer responsibility (EPR). In *Proceedings of the 2006 IEEE International Symposium on Electronics and the Environment* (pp. 326-330). IEEE. 10.1109/ISEE.2006.1650085

Yu, Z., Lu, S., Gao, S., Wang, J., Li, H., Zeng, X., Sheng, G., & Fu, J. (2010). Levels and isomer profiles of Dechlorane Plus in the surface soils from e-waste recycling areas and industrial areas in South China. *Environmental Pollution*, *158*(9), 2920–2925. doi:10.1016/j.envpol.2010.06.003 PMID:20630636

Zhang, J., Jiang, Y., Zhou, J., Wu, B., Liang, Y., Peng, Z., Fang, D., Liu, B., Huang, H., He, C., Wang, C., & Lu, F. (2010). Elevated body burdens of PBDEs, dioxins, and PCBs on thyroid hormone homeostasis at an electronic waste recycling site in China. *Environmental Science & Technology*, *44*(10), 3956–3962. doi:10.1021/es902883a PMID:20408536

Zhao, G., Wang, Z., Dong, M. H., Rao, K., Luo, J., Wang, D., Zha, J., Huang, S., Xu, Y., & Ma, M. (2008). PBBs, PBDEs, and PCBs levels in hair of residents around e-waste disassembly sites in Zhejiang Province, China, and their potential sources. *The Science of the Total Environment*, *397*(1-3), 46–57. doi:10.1016/j.scitotenv.2008.03.010 PMID:18439655

Zheng, J., Luo, X. J., Yuan, J. G., He, L. Y., Zhou, Y. H., Luo, Y., Chen, S.-J., Mai, B.-X., & Yang, Z. Y. (2011). Heavy metals in hair of residents in an e-waste recycling area, South China: Contents and assessment of bodily state. *Archives of Environmental Contamination and Toxicology*, *61*(4), 696–703. doi:10.100700244-011-9650-6 PMID:21360078

Chapter 8
Economic Perspectives: Opportunity and Challenges in E-Waste Management

Avula Jhansy
Woxsen University, India

Renuka Oladri
Woxsen University, India

Bodapatla Sindhu Priya
Woxsen University, India

Bhanu Prakash Saripalli
Woxsen University, India

ABSTRACT

Managing electronic trash (or "e-waste") offers both opportunities and challenges in the modern era. There are excellent opportunities for recycling and recovering valuable resources via good management as electronic devices and technology use grow. Lowering the demand for new materials and increasing employment possibilities in the recycling and waste management sectors can result in cost savings. But controlling e-waste can be difficult. Due to the toxic compounds, including lead and mercury, that are present in electronic gadgets, the immense amount of electronic consumption has the potential to overwhelm the waste management infrastructure that is already in place, causing environmental degradation. Adequate regulatory frameworks and efficient e-waste policy enforcement are essential to addressing the issues related to e-waste management. This study explored the potential and barriers to promoting sustainable practices to analyze the economic aspects of managing e-waste.

1. INTRODUCTION

Electronic waste (e-waste), which results from the use of electronic and electrical gadgets, has become a major global issue. Due mostly to the poisonous chemicals and hazardous materials found in e-waste,

DOI: 10.4018/979-8-3693-1018-2.ch008

this issue poses a risk to both human health and the environment. Additionally, an excellent motivation for recovery initiatives is the possibility of financial benefit from the recovery of valuable materials, particularly precious metals inherent in e-waste. Even with environmental regulations in place, a simulation predicts that industrialized countries would see the greatest growth in the manufacturing of electrical and electronic equipment. The entire mass flow of Waste Electrical and Electronic Equipment (WEEE) (Azevedo, L. P. et al. 2019) increases by a factor of 2.7–4.0 even within an environmental regulatory framework, but a factor of 3.1–7.0 is projected in the absence of regulation. The quick economic growth of the sector, the short product lifecycle of electronic equipment, and the recent development of a wide range of electronics-related items are all significant causes (Bazargan, A., et al. 2012). There is a lot of growth in the production of electronic and electrical devices and there are many industries set up globally to produce these devices. E-waste is classified based on the origin and intended use of electronic items. E-waste includes a wide range of electronic devices, including telecommunications and information technology equipment, sizable home appliances, automated dispensers, lighting systems, medical equipment, monitoring tools, consumer electronics, and even items like electronic tools, electrical tools, sports equipment, toys, mobile phones, and computers. New models are frequently developed with faster mechanical speeds, simplifying duties even further. However, the practice of replacing older versions with newer ones generates a substantial amount of e-waste, which causes serious environmental and health issues. For this reason, e-waste management is required (Shahabuddin, M., et al.2023). With an annual e-waste production of around two million tonnes that is increasing at an alarming pace of 30%, India faces a huge challenge. Unfortunately, due to insufficient infrastructure management and a lack of collection centers, along with the absence of a structured incentive system for both makers and recyclers, a startling 95% of e-waste ends up in the unregulated sector for disposal. Consumer awareness of the critical importance of safe e-waste disposal remains depressingly low, and the lack of appealing incentives restricts involvement in recycling efforts (Sheoran, M., & Das Gupta, D. 2023). On the other hand, several studies have evaluated how much e-waste is produced in the Netherlands according to ten categories included in the European WEEE Directive. These evaluations have been based on certain technologies, such as but not limited to liquid crystal display (LCD) and cathode ray tube (CRT) devices, as well as several other electrical products, including monitoring equipment, televisions (TVs), laptops, and mobile phones. Several other studies have concentrated on analyzing e-waste generation specifically from electrical equipment, such as TVs, refrigerators, and personal computers (Ismail, H., and Hanafiah, M. M. 2020). The negative consequences of e-waste on people and ecosystems are of the utmost importance in South Asian countries that are undergoing rapid economic development, lifestyle changes, and sociotechnical transformations. Contrary to popular belief, these nations frequently lack effective waste management systems. Consider Bangladesh, where only 20% to 30% of the 3.2 million tonnes of e-waste produced each year is recycled, with the balance being dumped in landfills. This huge variation emphasizes how urgently these areas require comprehensive e-waste management solutions (Hossain, M. S. et al., 2015).

India ranks third worldwide in e-waste production, following China and the United States. The volume of e-waste has surged dramatically, escalating from 700,000 tonnes in FY18 to 1.6 million tonnes in FY22. The e-waste management sector in India is predominantly characterized by unregulated and informal entities (https://www.businesstoday.in/magazine/deep-dive/story/will-indias-bet-on-new-policy-to-tackle-e-waste-pay-off-385530-2023-06-14). Increased production of electronic waste is closely related to increased demand for electronic products in India. Experts in the field estimate that the nation sells 6.5 million washing machines, 148 million cell phones, 14 million refrigerators, 19 million audio gadgets, and over 17 million TVs each year. Additionally, these figures keep rising every year.

The chapter's first section examines the potential financial rewards of using sustainable e-waste management strategies, including job creation, resource recovery, and reduced environmental impact. By comprehending these economic variables, policymakers and businesses may create efficient plans to lessen the negative effects of e-waste and make use of its potential to stimulate economic growth. The benefits of applying circular economy principles to e-waste management are covered in this chapter. By converting from a linear "take-make-dispose" strategy to a circular one, which encourages product repair, refurbishing, and recycling, a framework that is sustainable and profitable can be built. It also covered the challenges associated with e-waste management. These include a lack of infrastructure and technology to handle e-waste efficiently, a lack of education among consumers and businesses about ethical e-waste disposal, and the prevalence of illegal e-waste dumping in underdeveloped countries. To address these difficulties, governments, corporations, and civil society organizations must collaborate, and money must be set aside for research and development. To address India's e-waste management challenges, a systematic approach is required. E-waste management and disposal must be done with attention and responsibility since they may contain potentially hazardous components that, if handled improperly, are bad for the environment and human health.

1.1. What Happens If We Ignore E-Waste?

More than a thousand different substances make up e-waste, which can also be divided into metals and non-metals. For instance, if we consider a single laptop, it is a composition of plastic, glass, non-metals, and metals and it holds for a very negligible quantity and we can easily ignore them but from the period 1994 to 2003, 500 million PC disposals were weighing about 2,872,000 tonnes of plastic, 718, 000 tonnes of lead, 1363 tonnes of cadmium, and 287 tonnes of mercury. Table 1 illustrates the metal content in a typical desktop computer. Mobile phone disposal is another illustration; over a billion of them are created worldwide each year. These cell phones and computers typically hold 3% of the world's gold, 2.5% of its silver, and 12% of its palladium. Additionally, these gadgets use 15% of the annual cobalt produced, mostly as a component of Li-ion batteries. So, we cannot ignore E-waste because it contains many important components in great quantities which will help us to regenerate the products again and again (Bazargan, A., et al.2012).

The recycling of e-waste is a valuable endeavor as it encompasses a diverse array of materials such as metals, glass, and plastics. Referencing Table 2 provides a comprehensive breakdown of the constituent components of e-waste. Nevertheless, only an inadequate 17.4% of the global e-waste is successfully collected and recycled, primarily owing to the intricate nature of the collection and recycling processes. E-waste recycling poses a formidable challenge due to the heterogeneous composition of metals with varying physicochemical properties, including the presence of hazardous halogen compounds. Inadequate disposal of substances containing these perilous compounds can instigate environmental peril. Consequently, implementing proper waste management techniques and ensuring the safe disposal of e-waste has assumed global significance. This is essential for mitigating human health risks and preventing environmental degradation, which can transpire because of indiscriminate landfilling practices.

Table 1. Metal content in a typical desktop computer

Metal	Content%
Aluminium	14.17
Cadmium	0.0094
Copper	7
Chromium	0.0063
Gold	0.0016
Iron	20.47
Lead	6
Nickel	0.8503
Mercury	0.0022
Silver	0.0189
Zinc	2.024
Others	52.12

Table 2. A typical description of the components that compose e-waste (Source: Shahabuddin, M., et al.2023)

Materials	Composition [%]
Printed circuit boards	1.71
Cables	1.97
Metals	60.19
Screens	11.87
Metal-plastic mixture	4.97
Pollutants	2.7
plastics	15.21
Others	1.38

1.2. Impacts of E-Waste on Human Health and Environment

One of the greatest difficulties facing the world in the twenty-first century is the management of e-waste, which poses major risks to both the environment and human health. The production of electronic goods involves the usage of heavy metals, such as polycyclic aromatic hydrocarbons (PAHs), which are extremely hazardous to the environment and are created while burning e-waste at low temperatures. In addition, burning plastic in e-waste releases 100 times more dioxins than burning residential waste does. Aquatic organisms are poisoned by the chemical cocktail produced as leachate by the Toxicity Characteristic Leaching Procedure (TCLP) from various electrical products. Even if we can recover 95% of the useful materials from computers through recycling, recycling processes have a significant negative impact on the environment, necessitating the development of new technology. (Gaidajis, G., et al., 2010).

Exposure to e-waste can have several extreme effects on human health, so it needs to be adequately managed. Back discomfort, red, itchy eyes, and other work-related ailments were more frequent, according to comparative cross-sectional research of e-waste workers and onlookers. Residents of e-waste locations are exposed to large levels of persistent pollutants, and it has been discovered that these hazardous substances significantly affect the hormonal and neonatal growth of such populations. E-waste burning generates toxins that can irritate the eyes and cause asthma and respiratory problems. Lead exposure increases the probability of periodontitis and other mouth problems in preschool children living in e-waste zones. Other possible health effects of informal recycling and disposal include harm to organs, aberrant reproductive growth, immunotoxicity, and genotoxicity (Ghulam, S. T., & Abushammala, H. 2023)

1.3. Contribution of Developed Countries Towards E-Waste

Governments of industrialized nations have set up new environmental policies and guidelines to address the issue of excessive e-waste generation to reduce the amount of WEEE produced. The extended producer responsibility (EPR) practice (Turaga, R. M. R. et al. 2019), supported by the Organisation for Economic Co-operation and Development (OECD), is one strategy for getting manufacturers to recycle their products when they reach the end of their useful lives. The Restrictions of Hazardous Compounds Directive (RoHS), which limits the use of specific hazardous compounds in electrical and electronic equipment, was recently adopted by the European Union. The six substances are polybrominated biphenyls (PBB), polybrominated diphenyl ethers (PBDEs), hexavalent chromium, lead, mercury, and cadmium (Bazargan, A., et al.2012). The business sector, which includes government agencies, businesses both large and small, offices of multinational corporations, and other similar organizations, is where the majority of computer usage and, consequently, the creation of e-waste originate. This sector is responsible for 78% of the current installed PCs. For the remaining 22%, additional sources include individual houses, foreign embassies, PC manufacturing plants, PC stores, secondary markets for used PCs, and the importation of electronic garbage from other countries (Pinto, V. N. 2008).

Based on extensive data gathered from nationwide sales and the prescribed average lifespan of electrical and electronic equipment (EEE) as stipulated by the E-waste Management Rules, 2016, India recorded an estimated generation of approximately 13,46,496.31 tonnes and 16,01,155.36 tonnes of e-waste during the financial years 2020-21 and 2021-22, respectively (https://legalitysimplified.com/2023/07/24/cpcb-estimates-e-waste-generation-in-india-and-ministry-introduces-new-rules-for-better-management/#:~:text=According%20to%20the%20data%20collected,and%202021%2D22%2C%20respectively). The Ministry has revised the previous regulatory framework in an innovative effort to raise the bar for e-waste management standards and promote environmental sustainability. As a result, a wide range of actions have been implemented. The newly passed E-Waste (Management) Rules, 2022 took effect in November 2022 and began to be fully implemented on April 1, 2023. The EPR framework for recycling e-waste is introduced by these standards, which were carefully constructed to ensure the ecologically responsible management of e-waste. All makers, producers, refurbishers, and recyclers must register on a specific website created by the Central Pollution Control Board (CPCB) by the EPR regulations. Emphasizing the promotion of a Circular Economy, the new rules underscore the importance of scientific recycling and the proper disposal of e-waste as fundamental principles guiding responsible e-waste management.

The Union Ministry of Environment, Forest and Climate Change has recently unveiled the preliminary notification for forthcoming electronic waste management regulations (https://www.dhyeyaias.

com/current-affairs/daily-current-affairs/e-waste-roadmap-2023-for-india). These regulations substantially impact consumer goods manufacturers and electronic goods producers. They are now obliged to ensure that a minimum of 60% of their electronic waste is gathered and processed by the year 2023, with progressive targets of 70% and 80% for 2024 and 2025, respectively. To validate their compliance, they will be issued EPR certificates, which document the quantity of e-waste a company has responsibly collected and processed each year. This system permits organizations to trade surplus EPR certificates, aiding one another in fulfilling their obligations. Under these regulations, companies will be required to register on an online platform, where they must delineate their yearly production and e-waste collection objectives. Oversight and coordination of EPR certificate trading will be managed by the CPCB, which will also monitor compliance with the stipulated targets. Failure to meet these annual targets will entail financial penalties or what is referred to as "environmental compensation." However, the precise quantum of these penalties is not detailed in the current draft. Furthermore, these regulations include provisions for safeguarding the health and well-being of laborers employed in the dismantling and recycling operations associated with e-waste, recognizing the critical importance of worker safety in this sector.

In February 2003, the European Union (EU) implemented legislation aimed at limiting the use of hazardous substances in electrical and electronic equipment (Directive 2002/95/EC) and encouraging the collection and recycling of these devices (Directive 2002/96/EC). These regulations outline the establishment of collection programs where consumers can return their used electronic equipment without incurring any fees. The primary goal of these programs is to promote increased recycling and the potential for reuse of these products. Significant efforts to address similar issues have been undertaken in various regions worldwide, including California, Norway, China, South Korea, and Japan, with substantial initiatives on the RoHS. Numerous other countries, such as Australia, New Zealand, Thailand, Malaysia, Taiwan, Canada, and Brazil, are actively contemplating similar measures.

1.4. Methods of Disposal

Disposal of e-waste should be done appropriately, or else it results in hazardous effects on the environment and human health. Figure 1 illustrates the methods of disposal of E-waste. Different methods of disposal are in use (Annamalai, J. 2015).

1.4.1. Land Fill

This method is bad for the environment because it can contaminate the soil, and underground water, or release toxic acids that are hazardous to the environment. The waste is piled up in a heap and then covered with other domestic waste, including solid and other degradable waste. And rivers, animals, and people can all be affected by the contaminated water. E-waste is the source of almost 70% of the heavy metals found in landfills. Additionally, electronic garbage should not be disposed of using this approach.

1.4.2. Incineration

This technique involves burning the produced e-waste in a controlled setting and an open area. Another technique employed occasionally is pyrolysis, which involves heating the substance in the absence of oxygen to separate the constituent parts into fumes, oils, and salts. One benefit of this process is that it decreases the amount of waste produced and makes use of the energy contained in the combustible materials found therein. It also contributes to global warming and emits toxic chemicals into the environment, which are just a few of its many drawbacks. It is also not a proper method of trash disposal, much like landfills.

1.4.3. Reuse

Using the product in its entirety or the spare parts, with or without modifications, is the goal of this strategy. This could be used by desktop computers, mobile phones, and other devices. This method reduces the quantity of e-waste created even if it is not a direct disposal option. The things should preferably be reused compared to the previous two, as it will ultimately lower the demand for electronic devices and the generation of electronic waste by a certain amount.

1.4.4. Recycle

Recycling electronic waste, or the management of e-waste, allows for the recovery of all the components, their regeneration, and reuse, eliminating the need for additional mining to obtain metals and non-metals. E-waste recycling Despite the ideology's seeming intelligence, recycling e-waste is a highly challenging operation because it contains numerous harmful components. If e-waste recycling techniques including open burning, metal acid stripping, acid baths (Naik, S., & Eswari, J. S. 2022), and incineration are not properly disposed of, they can yield compounds like heavy metals, furans, and dioxins that constitute harm to the environment.

Figure 1. Methods of disposal of e-waste

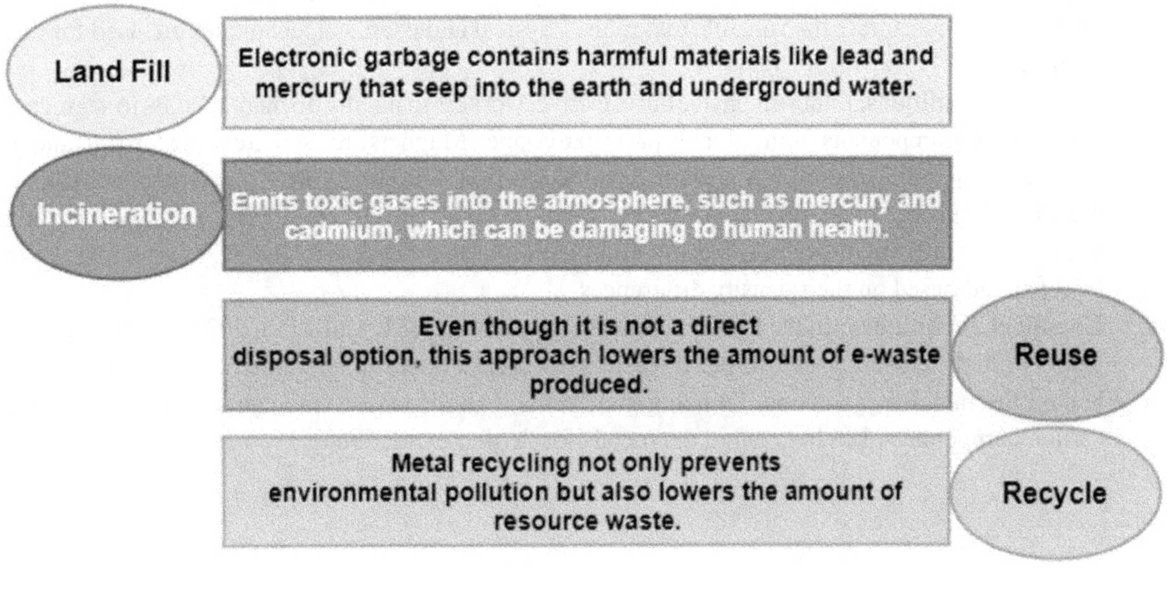

2. OPPORTUNITIES

As the volume of electronic waste (e-waste) continues to increase across the world, there are significant opportunities for businesses, organizations, and individuals to contribute to sustainable e-waste management. Let's explore some of these opportunities:

2.1. Resource Recovery

E-waste contains valuable components that can be recovered and reused, including metals, polymers, and precious metals. We can recover these resources by putting in place effective recycling procedures, which will cut down on the demand for new materials and encourage a circular economy.

2.1.1. Prospects Are Essential to Resource Recovery

Interwoven into the very fabric of e-waste management lies the core potential epitomized by resource recovery. Electronic devices, encompassing smartphones to laptops, encapsulate a complex fusion of precious metals like gold, silver, and palladium, intermingled with copper, plastics, and the elusive rare earth elements. When these devices reach the end of their useful life, they frequently become useless, leading to the growing e-waste problem. However, strategic management methodologies can unearth, rejuvenate, and repurpose these discarded electronic fragments, effectively reducing the demand for fresh raw materials.

The following are the resource recovery methods for the extraction of copper from printed circuit boards (Hsu E. et al., 2019):

i. **Pre-treatment: Size Reduction and Separation:** Before extracting valuable metals like copper from printed circuit boards (PCBs), the e-waste is subjected to pre-treatment processes. Size reduction involves shredding the PCBs into smaller pieces, making them easier to handle and process. Separation techniques are then employed to isolate the different factions within the PCBs.

ii. **Separation of Non-Metallic Fractions:** In this stage, the non-metallic fractions are separated from the PCBs to concentrate the valuable materials. Physical and chemical techniques are used for this purpose.

iii. **Physical Techniques:** Labor-intensive but effective, workers manually dismantle PCBs to segregate non-metallic components from valuable parts like copper. Magnets are used to extract ferromagnetic materials, separating them from the non-metallic fraction. Shredding: PCBs are shredded further, breaking them into even smaller pieces for better separation. Different materials have varying densities. By utilizing processes like air classification or hydrocyclones, the PCB components can be separated based on their density differences.

iv. **Chemical Techniques:** In the Pyrolysis technique (Naik, S., & Eswari, J. S. 2022), PCBs are subjected to high temperatures without oxygen. This breaks down the organic materials and leaves behind the metal-rich fraction. Utilizing supercritical fluids or solvents can selectively dissolve certain components, leaving behind the metals for further extraction.

2.1.2. Separation and Recovery of Metals

Once the non-metallic fractions are separated, the focus shifts to extracting valuable metals like copper from the PCBs. Pyrometallurgical methods involve high-temperature processes to extract metals. In the case of copper extraction, smelting the metal-rich fraction at high temperatures can separate copper from other materials. Hydrometallurgical methods use liquid solutions to leach out metals. In copper extraction, the metal-rich fraction is subjected to chemical leaching, where acidic or alkaline solutions are used to dissolve the copper. The resulting solution, called a leachate, contains copper ions that can be further processed to recover the copper metal. In the context of e-waste management, these resource recovery methods play a vital role in extracting valuable materials like copper from printed circuit boards. These techniques not only help mitigate the environmental impact of e-waste but also contribute to conserving valuable resources and reducing the need for mining new raw materials.

2.1.3. Mitigating Ecological Footprints

E-waste, often inappropriately disposed of, disseminates a cocktail of hazardous substances, including lead, mercury, and flame-retardant compounds, perpetuating contamination of soils, water bodies, and the very air we breathe. The astute retrieval of valuable resources from forsaken devices assuages the urgency for fresh material extraction, inevitably diminishing the call for ecologically detrimental extraction methodologies. In this vein, ecosystems experience respite, and the carbon footprint typically associated with mining and manufacturing witnesses a notable decline.

2.2. Job Creation

To collect, disassemble, and recycle e-waste, professional labor is required. As a result, it can create job opportunities and aid in the expansion of the recycling and waste management industries' economies. The mounting tide of e-waste offers a unique avenue for job creation, spanning various stages of the waste management process. From collection and sorting to refurbishment and recycling, e-waste management demands a diverse range of skills and expertise. The proper dismantling, sorting, and processing of electronic components require technicians with specialized knowledge, creating a demand for skilled labor (So, W. W. M., et al. 2019).

2.2.1. Collection and Sorting

The initial step in effective e-waste management involves collecting discarded electronic devices from consumers, businesses, and disposal centers. This process necessitates a workforce skilled in logistics, transportation, and inventory management. Additionally, the sorting of e-waste into categories for recycling, repair, or proper disposal requires manual dexterity and awareness of different electronic components (Awasthi, A. K. et al. 2023).

2.2.2. Refurbishment and Repair

Many discarded electronic devices still possess functional components that can be refurbished and repurposed. Skilled technicians proficient in diagnosing and repairing electronic malfunctions are integral to this process. By refurbishing and extending the lifespan of devices, these technicians contribute to reducing electronic waste while also making technology more accessible to individuals with limited resources.

2.2.3. Recycling and Recovery

Recovering valuable materials from e-waste, such as precious metals and rare earth elements, requires specialized expertise. Technicians skilled in advanced recycling techniques are needed to extract these valuable resources without causing harm to the environment. These professionals play a vital role in reducing the need for environmentally damaging raw material extraction.

2.2.4. Data Cleaning and Security

With the proliferation of personal and sensitive data on electronic devices, ensuring data privacy and security during the recycling process is of paramount importance. Experts in data sanitization and cybersecurity are essential to prevent identity theft and privacy breaches.

2.3. Environmental Protection

By managing e-waste properly, dangerous components are kept out of the soil and water, protecting the environment and fragile ecosystems from harm. One of the most prominent opportunities in e-waste management is the preservation and restoration of fragile ecosystems. The improper disposal of electronic devices, laden with hazardous substances like lead, mercury, and flame retardants, poses a grave threat to soil, water, and air quality. The sheer volume of e-waste generated exacerbates this challenge. However, through effective e-waste management practices, these toxins can be safely contained and extracted, preventing their seepage into the environment, and protecting both human health and biodiversity.

2.3.1. Reducing Resource Depletion

E-waste management offers a tangible opportunity to curb the rampant depletion of natural resources. Electronic devices rely on a myriad of materials, many of which are non-renewable and extracted through environmentally damaging processes. By recycling and recovering valuable components from discarded devices, e-waste management directly reduces the demand for new raw materials, mitigating the ecological impact of mining, refining, and processing. This, in turn, contributes to the conservation of ecosystems and promotes a more sustainable approach to resource utilization.

2.3.2. Mitigating Carbon Footprints

The environmentally conscious management of e-waste aligns with global efforts to mitigate carbon footprints and combat climate change. Traditional methods of electronic waste disposal, such as landfilling or incineration, release greenhouse gases and contribute to global warming. Opting for sustainable practices, such as recycling and refurbishing, reduces the release of harmful emissions and conserves energy that would otherwise be expended in the manufacturing of new electronic devices. This process is inherently linked to the reduction of carbon dioxide emissions, contributing to international climate goals.

2.3.3. Promoting Circular Economy

Embracing a circular economy model is a powerful opportunity embedded within e-waste management. Unlike the linear "take-make-dispose" approach, a circular economy emphasizes the continual use, repair, and recycling of products. E-waste management can facilitate this transition by promoting the repair and refurbishment of electronic devices, extending their lifespan, and reducing the need for constant replacements. This not only conserves resources but also fosters a culture of sustainable consumption, wherein products are valued for their longevity rather than their disposability (Cesaro et al. 2018).

2.3.4. Creating an Ecologically Conscious Society

Perhaps one of the most profound opportunities lies in the transformation of societal attitudes towards consumption and waste. E-waste management offers a platform to educate consumers about the environmental impact of their choices. As individuals become more informed about responsible disposal methods and the significance of recycling, they are empowered to make greener decisions. This shift in consumer behavior can permeate into other aspects of their lives, contributing to a broader culture of sustainability and environmental stewardship. (So, W. W. M., et al.2019).

2.4. Technological Advancements

E-waste treatment is in greater demand, which has led to innovation in recycling and recovery methods, resulting in more effective and sustainable practices. In the realm of waste management, traditional methods have often proven inadequate in addressing the mounting challenges posed by electronic waste (e-waste). However, recent technological advancements offer promising opportunities to revolutionize e-waste management practices.

The implementation of a deposit return system (Sheoran, M., & Das Gupta, D. 2023), in which customers are responsible for paying a refundable charge at the time of product purchase, constitutes a proactive strategy. The fee should be calculated to consider the costs associated with managing e-waste while giving customers an incentive to return used goods. The incorporation of Radio Frequency Identification (RFID) tags is essential for ensuring thorough product tracking. When a product is safely delivered to designated recycling facilities or recyclers, the tags activate at the point of sale and remain active until then. At that point, the consumer can get their deposit fee back. Government subsidies for producers and recyclers might further energize the entire process. We may use mobile phones as a model to illustrate this strategy, including suggested deposit fees and the accompanying cost structure.

3. CHALLENGES

The E-waste problem is of global concern because of the nature of the production and disposal of waste in a globalized world. To detoxify contaminated environments, make them safe for human occupancy and consumption, and safeguard the health of the ecosystems that sustain life, requires the development of creative and affordable solutions.

3.1. Rapid E-Waste Growth

The increasing use of electronic devices results in a significant surge in e-waste generation. This overwhelming increase strains existing waste management infrastructure and poses challenges for appropriate disposal and recycling.

3.1.1. Inadequate Global E-Waste Infrastructure

There has been significant growth in the production of electronic garbage around the world, yet there is a lack of infrastructure, such as recycling facilities, to handle these massive volumes of e-waste. The yearly e-waste created is very minimally handled by these recycling facilities. Therefore, the government should prioritize financing for recycling facilities to reduce e-waste. Additionally, because of ineffective cooperation with e-waste collectors, the current recycling centers manage e-waste below their capacity.

For instance, just 17.4% of the estimated 53.6 million metric tonnes of e-waste produced worldwide in 2020 was formally collected and recycled. Governments must fund the expansion of recycling facilities and provide incentives for private sector involvement. For instance, the E-Waste Recycling Act of Japan created a framework for shared responsibility involving producers, merchants, and consumers, which improved recycling rates to over 40% (Pramila, S., et. al., 2012).

3.1.2. Constraints in Formal E-Waste Recycling

Due to a lack of appropriately qualified e-waste managers or workers who can apply cutting-edge recycling procedures, most methods used are mechanical disassembly and manual sorting. These provide challenges for recycling e-waste. While a few emerging companies are engaged in metal extraction from e-waste, their processing capacity is restricted. As a result, a notable portion of formal sector e-waste is exported to countries with advanced metal extraction infrastructure. Conversely, informal practices such as open-air incineration and acid leaching are rampant, causing environmental contamination and health risks (Rautela, R. et al. 2021).

3.2. Hazardous Components

If not properly handled and disposed of, dangerous compounds included in e-waste, such as lead, mercury, and flame retardants, can pose health concerns.

3.2.1. Environmental Contamination

Environmental contamination emphasizes different potential pollutants of air, water, and soil because of inefficient disposal behaviors of e-waste and due to the presence of different chemical compounds like mercury, sulfur, cadmium, and lead. The dangerous nature of these chemical compounds causes environmental effects that include the death of animals, and slower growth, and development in animals because of improper e-waste management. These toxic chemical substances and persistent organic pollutants (POP) will contaminate the environment through the ecological food chain. The Agbogbloshie e-waste dumpsite in Ghana has gained notoriety as one of the world's largest e-waste processing areas, with reports of severe environmental contamination and health issues among the local population.

3.2.2. Health Risks

Different health risks can be encountered due to the use of different Americium, sulfur, and Brominated Flame Retardants (BFRs) (Avis, W. 2021). Some hazardous components, such as heavy metals, can persist in the environment for a long time. Exposure to these hazardous materials can lead to various health problems, such as respiratory issues and cancers, E-waste contains nickel (Ni), which is a carcinogen, an asthma trigger, a lung irritant, and a skin irritant. Through the air, it penetrates the surroundings. Skin irritation, hair loss, heart and lung damage, and fertility issues are all brought on by antimony (Sb). Arsenic is carcinogenic and causes skin and lung cancer in humans, while polyvinyl chloride (PVC) harms animal kidneys and is soluble in water. The effects of bioaccumulation on human health, liver damage, particularly in pregnant and nursing women, are caused by Polychlorinated Biphenyls (Leung, A. O. W. 2019), BFRs, and other chemical components that are present in the food chain.

A notable example is Guiyu in China, where extensive e-waste processing has led to alarming rates of lead poisoning among residents. This underscores the urgent need to address the health impacts on vulnerable populations (Pramila, S., et. al., 2012).

3.2.3. Air and Water Pollution

Improper disposal methods such as incineration or dumping in landfills can release hazardous chemicals into the air and water. Burning e-waste can release toxic fumes and pollutants, contributing to air pollution. When hazardous materials reach water bodies, they can contaminate water sources, affecting aquatic life and potentially entering the human water supply.

Landfills can lead to the leaching into the groundwater, contaminating drinking water. The cadmium from one mobile phone battery is enough to pollute 600 m^3 of water.

Most of the e-waste is led to (sanitary) landfill sites. E-waste can liberate heavy metals that are enclosed in plastic-covered electronic circuits if it is burned before being disposed of in landfills. Even though these heavy metals may not be accessible after washing out, burning can release them into the air, which contributes to air pollution and could have an impact on human health (Gaidajis, G. et al., 2010).

3.3. Data Security

Sensitive commercial or private data are frequently found in e-waste. To avoid breaches of data and identity theft, it is essential to ensure data sanitization and destruction during recycling. As the digital era advances, the issue of data security in e-waste management becomes increasingly critical. The challenges outlined earlier highlight the intricate web of concerns that must be navigated to ensure the responsible disposal of electronic devices without compromising sensitive information.

3.3.1. Data Breaches

When electronic gadgets like laptops, mobile phones, and other computing devices are disposed of improperly, data breaches occur. When we fail to properly delete the data from the device, insecure data erasing also results in data breaches, which cause financial theft.

For instance, a 2019 study by the University of Hertfordshire found that almost 40% of used electronic equipment included personally identifiable information (PII) from the owners before them. This worrying statistic emphasizes how exposed abandoned devices are to data leakage risks.

3.3.2. Identity Theft

Electronic devices like smartphones, laptops, and tablets may contain personal identification information. When these devices are not properly disposed of, the data or sensitive information, such as banking information, and passwords, might be accessed by unauthorized individuals who get into possession of discarded devices. Most people use these devices to store sensitive data, including passwords and bank credentials.

Credit card numbers, names, addresses, usernames, passwords, business and personal information, tax information, and more were among the PII that were recovered. Tablets were alarmingly discovered to have the largest quantity of recoverable PII, at 50%, while mobile phones had the least, at 13%. Additionally, 44% of hard discs had PII. 40% of the devices in total produced PII.

3.3.3. Data Resale and Black Market

Discarded electronics with intact data can end up in the second-hand market, The existence of a second-hand market for electronics introduces another layer of risk. Devices with intact data can be sold, potentially exposing sensitive information to malicious actors. Recent cases have demonstrated the presence of personal data on devices purchased from online marketplaces, highlighting the pressing need for improved data sanitization practices.

3.4. Global Trade Dynamics

E-waste is constantly exported from developed countries to developing ones, posing risks to the environment and public health in such countries. Collaboration across borders and strict rules are needed to solve this problem. E-waste generated by several nations is transferred to other locations for "disposal" in massive quantities. An abrupt rise in illegal trades has been brought on by the rising transboundary movements of e-waste between industrialized and poor countries. E-waste imports and exports are governed by laws and regulations in a variety of nations across the world.

The trading of e-waste is prohibited if these laws are violated with the intent to commit a crime or for purposes of economic, political, or social gain. The necessity to lower service import and export fees on e-waste or to obtain financial advantage may be the driving force behind this illegal trade in e-waste. Trash producers, trash gatherers, waste management firms, waste treatment specialists, shipping firms, and shipping agents are among the groups involved in such transactions.

These trades are conducted between various nations, either interregional or intraregional, and both types of trade are illegal. One of the nations producing a significant amount of e-waste is Bangladesh, which also buys wealthy nations' e-waste for disposal. The traders effectively employ the garbage's precious components to create other useful products while extracting high prices from both developed and developing nations to dispose of their electronic waste. In contrast to this trade, Bangladesh neither lacks the necessary legal authorizations to exchange and dispose of this e-waste nor the proper locations for doing so. As a result, the transaction is prohibited. Similar unlawful trade is visible in China, which imports e-waste from the European Union (EU) and uses this garbage to regenerate useful electronic products. It has been shown that Italy participates in these criminal activities by smuggling e-waste over international borders. Amounts of illegal trash were seen being transported from Germany to the Czech Republic, according to the authorities.

Poor e-waste management leads to these unlawful trades, which allow smugglers, brokers, and criminals a method to satisfy consumer expectations for inexpensive e-waste disposal. In addition, they remove parts from waste and sell them to other businesses so they can reuse them, which is a big source of income for them. When such unlawful trades become obvious, it is necessary to thoroughly analyze these trade lines and determine how they are moving throughout the world. The detection of such illegal trades, the marking of the trade route, and the investigation of what, how, how much, and where the transaction is being conducted all fall under the purview of forensics. Additionally, stringent legislation must be put in place, and proper import and export customs checks must be carried out. (Bisschop, L. 2012).

3.5. Lack of Awareness

Inappropriate e-waste disposal has negative effects on the environment and human health, which many people are unaware of. It is crucial to increase the public's awareness of ethical e-waste management. The lack of awareness surrounding data security risks in e-waste disposal exacerbates the problem. Many individuals and organizations remain unaware of the potential consequences of improper disposal. Raising awareness through educational campaigns for consumers and businesses can play a pivotal role in mitigating this challenge.

In GCC nations, there is an absence of knowledge on data security and privacy in e-waste products. Most people think that once data is deleted, it cannot be recovered. Because disclosing personal information has a cultural influence, consumers of the GCC are especially sensitive to privacy issues. The authors contend that adequate steps will be taken by homes to properly dispose of data if the public is aware of privacy issues. Institutions in the GCC region frequently give or sell outdated computers without properly removing their hard discs and memory components. To determine the awareness of and disposal practices for electronic devices carrying sensitive personal data, more research is required. To preserve data security and prevent illegal recovery during recycling, certain steps should be performed (Alghazo, J., & Ouda, O. K., 2016).

There are two ways to comprehend a lack of awareness.

i. While Buying Products
ii. While E-waste management
    i. While Buying Electronic Products:

Customers for their benefit do buy these products more and more. Customers are just told the speed and battery power and all about the product, but no one insists them about what they are made how they will affect the environment, whether it is good to use them, whether recycling is possible, what happens to human health and environment if we rely on them. So, no customer cannot get to know the consequences and no industry takes care to explain to people the consequences. It's everyone with their benefit and people should also be known about these so that people can get to know proper methods of disposal after using them.

3.6 While E-Waste Management

The people who are present at the sites of recycling industries and the people who are working at the recycling sites do not know the consequences of toxic elements released from the process of e-waste recycling, they work for their offspring without any proper guidance on safety measures. This promotes a negative way with a huge impact on workers' health and their surrounding environment, and they also don't even know that they are affecting their children indirectly these toxic elements enter their bodies and their food (Heacock, M. et al. 2016). This awareness should be provided by their respective recycling industries, and they must take care of them by suggesting all the safety precautions and creating awareness of the toxic elements released how they enter their bodies, and which element harms and which elements do not harm them. Also, they should provide clear hygienic suggestions like taking a bath and washing their hands, required after going home and meeting their children.

Even though the recycling methods are beneficial they harm us. Industries should be aware of new methods which cause less environmental harm. These two examples demonstrate the need for increased education on these topics. They will be able to make wiser purchases and understand how to properly dispose of outdated technology as a result. Information must be shared widely for people to act more responsibly and effectively. (Bazargan, A., et al., 2012).

4. SWOT

SWOT analysis is a well-known technique for assessing and managing strategic goals. SWOT stands for Strengths, Weaknesses, Opportunities, and Threats. Opportunities and risks are external, whereas advantages and drawbacks are internal. SWOT analysis is a strategic planning tool that is used to examine both internal and external aspects that may have an impact on an initiative, organization, or effort. A SWOT analysis (Sheoran, M., & Das Gupta, D. 2023) of the electronic waste (e-waste) issue in India demonstrates several essential aspects as described in Table 3.

Table 3. E-waste in India: SWOT analysis

S	**STRENGTH**	i. The rapidly developing consumer electronics market in India represents economic prosperity and technological advancement. ii. For effective e-waste treatment and recycling, India has a significant labor force that is technically skilled. iii. By adopting laws and policies like the E-Waste (Management and Handling) Rules of 2011, India has made progress.
W	**WEAKNESS**	i. The risks to the environment and public health associated with incorrect e-waste disposal are still unknown to customers as well as companies. ii. Efficient e-waste management in India is hindered by problems with insufficient recycling facilities and waste collection methods.
O	**OPPORTUNITIES**	i. India has the chance to create a sustainable e-waste management ecosystem because it transitioned to a circular economy that places a strong emphasis on recycling and reuse. ii. Technologies for recycling e-waste have the potential to be innovative, attract investment, and produce green employment possibilities.
T	**THREATS**	i. E-waste poses serious environmental hazards, such as soil and water contamination and harm to ecosystems, when it is improperly disposed of or recycled in an unregulated manner. ii. For people involved in informal recycling as well as communities living close to disposal sites, exposure to dangerous elements in e-waste can have serious health consequences.

5. CONCLUSION

A systematic strategy is necessary to address India's e-waste management issues. The goal of this approach should be to combine both large and small organized and unorganized sector units into a single value chain. The fact that e-waste contains potentially dangerous materials that are harmful to the environment and human health when handled improperly, it is crucial to use caution and take responsibility in its management and disposal. Significant opportunities for resource recovery, job generation, and environmental protection have been provided by e-waste management. To achieve sustainable and effective e-waste management practices, it is essential to address the issues linked to rapid e-waste growth, hazardous components, data security, global trade dynamics, and lack of knowledge. We can make use of the opportunities and overcome the obstacles to build a greener, more sustainable future by using responsible and innovative strategies.

A circular economy, a framework designed to reduce negative environmental effects and maximize resource use, might be developed if elements from e-waste are effectively recovered. The importance of recycling and properly disposing of e-waste can be greatly increased by utilizing artificial intelligence-driven promotional efforts, which will also encourage responsible consumer behavior. Artificial intelligence has the potential to greatly improve productivity, mitigate negative environmental effects, and promote sustainable practices when it comes to managing e-waste. Critical difficulties related to e-waste and its management are highlighted by the current review of the literature. It is a useful tool for decision-makers and stakeholders involved in recycling since it provides insightful information, a thorough SWOT analysis, and a summary of India's regulatory procedures. Stakeholders may skilfully negotiate problems, exploit opportunities, and create strategies that are in line with the goals of sustainable waste management by assessing their strengths, weaknesses, opportunities, and threats. In summary, the proper management of e-waste has the potential to lead to positive effects including the recovery of resources, the creation of jobs, and the development of a circular economy. However, there are many obstacles to overcome in the development of infrastructure, the management of hazardous substances, the creation of effective collection systems, and the adoption of strong regulatory frameworks. E-waste management must address these concerns through investments, innovation, and cooperative efforts among stakeholders if the environmental and financial benefits are to be attained.

REFERENCES

Alghazo, J., & Ouda, O. K. (2016, October). Electronic waste management and security in GCC countries: A growing challenge. In *Proceedings of the ICIEM International Conference, Sousse, Tunisia* (pp. 27-30). Academic Press.

Annamalai, J. (2015). Occupational health hazards related to informal recycling of E-waste in India: An overview. *Indian Journal of Occupational and Environmental Medicine*, *19*(1), 61. doi:10.4103/0019-5278.157013 PMID:26023273

Avis, W. (2021). *Drivers, barriers, and opportunities of e-waste management in Africa*. Academic Press.

Awasthi, A. K., Iacovidou, E., Awasthi, M. K., Johnson, M., Parajuly, K., Zhao, M., Mishra, S., & Pandey, A. K. (2023). Assessing Strategic Management of E-Waste in Developing Countries. *Sustainability (Basel), 15*(9), 7263. doi:10.3390u15097263

Azevedo, L. P., Araújo, F. G. D. S., Lagarinhos, C. A. F., Tenório, J. A. S., & Espinosa, D. C. (2019). Resource recovery from e-waste for environmental sustainability: a case study in Brazil. In *Electronic waste management and treatment technology* (pp. 175–200). Butterworth-Heinemann. doi:10.1016/B978-0-12-816190-6.00008-X

Bazargan, A., Lam, K. F., & McKay, G. (2012). Challenges and opportunities of e-waste management. *Nova Scientia*.

Bisschop, L. (2012). Is it all going to waste? Illegal transport of e-waste in a European trade hub. *Crime, Law, and Social Change, 58*(3), 221–249. doi:10.100710611-012-9383-0

Cesaro, A., Marra, A., Kuchta, K., Belgıorno, V., & Van Hullebusch, E. D. (2018). WEEE management in a circular economy perspective: An overview. *Global NEST Journal, 20*(4), 743–750. doi:10.30955/gnj.002623

CPCB Estimates E-Waste Generation in India and Ministry Introduces New Rules for Better Management. (n.d.). https://legalitysimplified.com/2023/07/24/cpcb-estimates-e-waste-generation-in-india-and-ministry-introduces-new-rules-for-better-management/#:~:text=According%20to%20the%20data%20collected,and%202021%2D22%2C%20respectively

E-waste roadmap 2023 for India: Daily Current Affairs. (n.d.). https://www.dhyeyaias.com/current-affairs/daily-current-affairs/e-waste-roadmap-2023-for-india

Gaidajis, G., Angelakoglou, K., & Aktsoglou, D. (2010). E-waste: Environmental problems and current management. *Journal of Engineering Science and Technology Review, 3*(1), 193–199. doi:10.25103/jestr.031.32

Ghulam, S. T., & Abushammala, H. (2023). Challenges and Opportunities in the Management of Electronic Waste and Its Impact on Human Health and Environment. *Sustainability (Basel), 15*(3), 1837. doi:10.3390u15031837

Heacock, M., Kelly, C. B., Asante, K. A., Birnbaum, L. S., Bergman, Å. L., Bruné, M. N., Buka, I., Carpenter, D. O., Chen, A., Huo, X., Kamel, M., Landrigan, P. J., Magalini, F., Diaz-Barriga, F., Neira, M., Omar, M., Pascale, A., Ruchirawat, M., Sly, L., ... Suk, W. A. (2016). E-waste and harm to vulnerable populations: A growing global problem. *Environmental Health Perspectives, 124*(5), 550–555. doi:10.1289/ehp.1509699 PMID:26418733

Hossain, M. S., Al-Hamadani, S. M., & Rahman, M. T. (2015). E-waste: A challenge for sustainable development. *Journal of Health & Pollution, 5*(9), 3–11. doi:10.5696/2156-9614-5-9.3 PMID:30524771

Hsu, E., Barmak, K., West, A. C., & Park, A. H. A. (2019). Advancements in the treatment and processing of electronic waste with sustainability: A review of metal extraction and recovery technologies. *Green Chemistry, 21*(5), 919–936. doi:10.1039/C8GC03688H

India e-Waste Management Market Report 2023: India has Become the Largest Producer of e-Waste after China and the United States - Analysis & Forecasts 2018-2022 & 2023-2027. (n.d.). https://www.globenewswire.com/en/news-release/2023/03/01/2618469/28124/en/India-e-Waste-Management-Market-Report-2023-India-has-Become-the-Largest-Producer-of-e-Waste-after-China-and-United-States-Analysis-Forecasts-2018-2022-2023-2027.html

Ismail, H., & Hanafiah, M. M. (2020). A review of sustainable e-waste generation and management: Present and future perspectives. *Journal of Environmental Management*, *264*, 110495. doi:10.1016/j.jenvman.2020.110495 PMID:32250915

Leung, A. O. W. (2019). Environmental contamination and health effects due to E-waste recycling. *Electronic waste management and treatment technology*, 335-362.

Naik, S., & Eswari, J. S. (2022). Electrical waste management: Recent advances challenges and outlook. *Total Environment Research Themes*, 100002.

Pinto, V. N. (2008). E-waste hazard: The impending challenge. *Indian Journal of Occupational and Environmental Medicine*, *12*(2), 65. doi:10.4103/0019-5278.43263 PMID:20040981

Pramila, S., Fulekar, M., & Bhawana, P. (2012). E-waste- A challenge for tomorrow. *Research Journal of Recent Sciences ISSN*, *2277*, 2502.

Rautela, R., Arya, S., Vishwakarma, S., Lee, J., Kim, K. H., & Kumar, S. (2021). E-waste management and its effects on the environment and human health. *The Science of the Total Environment*, *773*, 145623. doi:10.1016/j.scitotenv.2021.145623 PMID:33592459

Shahabuddin, M., Uddin, M. N., Chowdhury, J. I., Ahmed, S. F., Uddin, M. N., Mofijur, M., & Uddin, M. A. (2023). A review of the recent developments, challenges, and opportunities of electronic waste (e-waste). *International Journal of Environmental Science and Technology*, *20*(4), 4513–4520. doi:10.100713762-022-04274-w

Sheoran, M., & Das Gupta, D. (2023). *International best practices for e-waste take back and policy interventions for India*. Facilities. doi:10.1108/F-03-2023-0027

So, W. W. M., Lee, J. C. K., & Chow, C. F. (2019). *Environmental sustainability and education for waste management*. Springer Singapore. doi:10.1007/978-981-13-9173-6

Turaga, R. M. R., Bhaskar, K., Sinha, S., Hinchliffe, D., Hemkhaus, M., Arora, R., Chatterjee, S., Khetriwal, D. S., Radulovic, V., Singhal, P., & Sharma, H. (2019). E-waste management in India: Issues and strategies. *Vikalpa*, *44*(3), 127–162. doi:10.1177/0256090919880655

Chapter 9
E-Waste Reduction Methods in Sustainable Ways

Deekshitha Lingareddy
Woxsen University, India

Thanmayee Isanaka
Woxsen University, India

C. Dinesh
Woxsen University, India

K. V. Rajesh Kumar
https://orcid.org/0000-0002-7989-1824
Woxsen University, India

ABSTRACT

The problem of e-waste generation is increasing day by day in India and worldwide. This rise in e-waste poses damage to public health and the environment as a whole. Toxic substances and poor disposal of e-waste significantly affects plant and animal lives. When it comes to e-waste generation, India is the fastest growing country when it comes to electronics use. Various policy measures and initiatives have been introduced over the past decades in India related to e-waste management. It is still a source of distress and unsolved matter. Sustainable development needs a lot of effort and well-organized e-waste management can be possible only with a systematic waste collection system, accurate estimations, and early forecasting. This chapter presents existing e-waste scenarios in India that are posing significant challenges to meet sustainable development goals and highlights some of the sustainable ways to reduce e-waste. This chapter serves as a basis for awareness and knowledge about the potential harms of e-waste to human health and the environment to promote glsustainabilitynable.

DOI: 10.4018/979-8-3693-1018-2.ch009

INTRODUCTION

The "electrical or electronic equipment (EEE)" turns "Waste electrical or electronic equipment (WEEE)" when it is thrown or discarded inappropriately or without any recycling approach (Borthakur & Singh, 2012). There has been a steep rise in demand and manufacturing of electronics items due to the huge consumption pattern (Borthakur & Singh, 2012). Hence, it becomes important to promote sustainable consumption of those goods. It goes without saying that e-waste consists of toxic elements which may pose an ill effect on both environment and public health (Chandra, 2015). It becomes vital to avoid the unwanted use of EEE to make lives easier for current and next generations and protect the planet as a whole.

There is a need to switch to the modern approach of a circular economy from a typical linear economy in order to protect the environment (Stefansdotter et al., 2016). In a linear economy, a product is produced and thrown away after its final use. In a circular economy, once a product is used, it is recycled. There are significant economic benefits of a circular economy for electrical and electronic industries. It is possible to reduce consumer cost by 7% to 14% by 2030 and 2040, respectively, with a circular business model (Morlet, 2018). It is also seen that non-functional items are wrongly classified as "used items" and sent to third world countries of Asia and Africa with a license to export surplus items for refurbishment or reuse (Ananno et al., 2021).

There is a need to implement 4 Rs of recycling – (1) Reuse – Reusing items to maximum possible extent; (2) Reduce – Cutting down waste by avoiding excessive use of equipment; (3) Recycle – New products can be made of e-waste; and (4) Recovery – Valuable resources can be restored from e-waste (Stone, 2020). With technological advancements and changing scenarios, recovery is very important. There are three design-specific principles which reportedly serve as a basis of the circular economy - (1) Circulation of supplies and goods at highest value; (2) Reducing waste to control pollution; and (3) Recovery (Morlet, 2018).

Because of constant technological improvements and upgrades, new and latest devices and electronics have become the center of attraction for young consumers. This urge of being relevant to latest technological advancements and social validation is creating a mountain of e-waste. The 4Rs are much needed to reduce the impact of e-waste and manage the same to save the environment. Proper e-waste reuse and recycling can save the environment by keeping a lot of valuable metals like Gold, Silver, Cerium, Gadolinium, and Palladium into landfill as they can be reused for making other electronic items (Sethurajan et al., 2019; Anand, 2022).

If everyone switches to energy-saving lights, it is possible to save up to US$120 billion every year and it can take the natural resources of over 3 planets to maintain the existing lifestyle of 9.6 billion people on earth by 2050 (UNEP, 2022). There is a 38% rise in e-waste from 2010 to 2019, according to Goal-12 of "Sustainable Development Goals (SDG) Report 2020", which titles "sustainable consumption and production patterns" (Guterres, 2020). It is among the most important reasons to look closely at sustainable development goals for governments in both developing and developed countries (CPCB, 2021). However, sustainability is more than the responsibility of any organization or the government as everyone should play their role in adopting sustainable activities in daily life.

Every year, the "weight of e-waste generated every year in the world is similar to the weight of five thousand Eiffel Towers (Parajuly et al., 2019). In 2019-20 alone, 1,014,961 tons of e-waste will be generated in India, i.e., 32% more than the amount of e-waste generated in 2018-19. In 2018 and 2019, only 3.6 percent and 10 percent of wastes were collected, respectively (CPCB, 2021). These numbers are so huge to create concern for landfill for the planet. There have been significant advancements of e-products and technological improvements over the past century. Over 50,000 tons of e-waste is generated every month in India alone. It is not shocking that e-waste is being generated at a rapid rise of 23.7% every year since the past decade.

Being the third largest producer of e-waste in the world, India needs to be responsible to come up with proper strategies for e-waste management (Namo eWaste, 2022). Delhi is the second state after Maharashtra in e-waste generation (Pendharkar, 2018). With 3230 kilotons of e-waste generated every year, India is one of top 3 countries following China and the US with 10,129 and 6918 kilotons of e-waste, respectively. However, China and the US also have significantly higher recycling rates, i.e., 16% and 15%, respectively, while India recycles only 1% of e-waste every year which is alarming (Ruiz, 2022). When it comes to ranking of 180 countries at the "Environmental Performance Index (EPI)", India stands at the bottom.

In India, consumers aggressively replace their digital devices with every new launch due to shortened life cycle of products and latest innovations, causing a pile of e-waste (Jayaraman et al., 2019). In 2019, 2.4 kg of e-waste per capita will be generated in India (UNITAR, 2020) (Figure 1). In 20 states, there are only 51 e-waste collection points in India. In Assam, Andhra Pradesh, Gujarat, Chhattisgarh, and other 20 states in India, there are only 400 e-waste dismantlers and recyclers. However, these licensed recyclers and dismantlers can process 1,068,542.72 tons of e-waste annually (Ministry of Environment, Forest, and Climate Change, 2022). This recycling capacity is still underused as the informal sector still handles most of scrap and waste materials (Forti et al., 2020). There are Non-Governmental Organizations (NGOs) which act as a channel among waste collectors but "original equipment manufacturers (OEMs)" are still pressured to pick and recycle their products. Informal sector recycles 95% of consumer electronics in India (Ruiz, 2020; Singhal et al., 2019).

Figure 1. Global stats for e-waste generation
Source: UNITAR (2020)

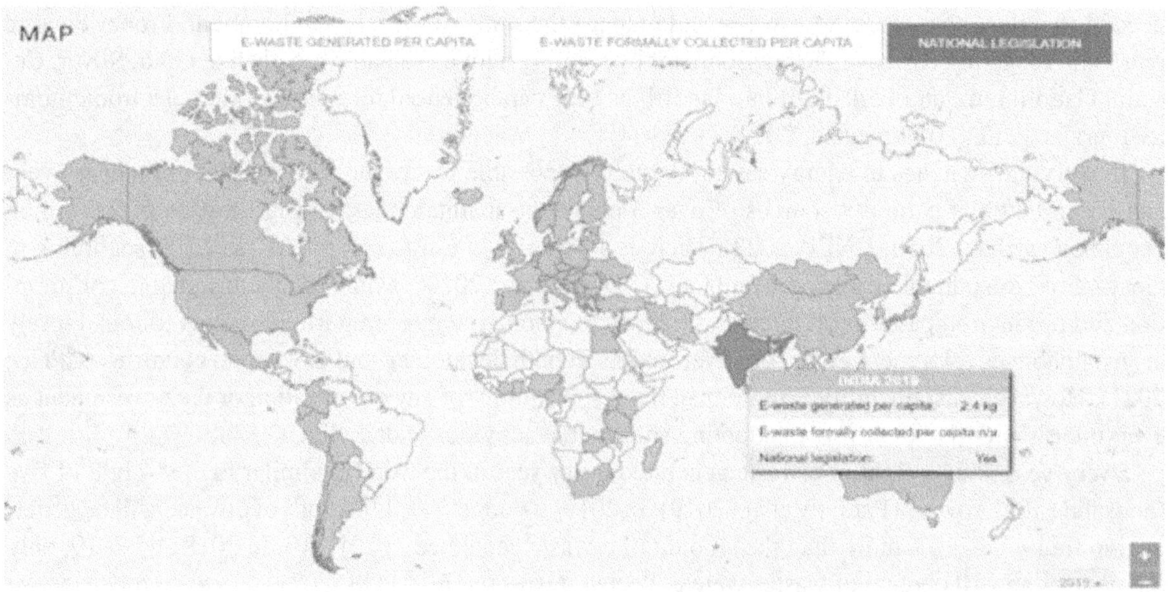

E-waste management has become a global concern which could jeopardize the environment in different ways. India has a rise in consumer durables and their demands being one of the fastest growing countries worldwide (Reynolds, 2021). India is one of the top countries in e-waste generation with a huge growth rate. However, South Asian countries have finally started to recognize the value of proper e-waste management. India is the only country with e-waste laws in Southern Asia. Only licensed recyclers and dismantlers can collect e-waste because of a law which was enacted in 2011. According to a survey report by Kandel (2022), several heavy metals are found in e-waste like Zn, Pb, Hg, Cr, Cd, etc. Bagwan (2023) predicted that the average capacity of dismantling/recycling e-waste is going to be 163563.15 MT by 2030, reaching over 248 recyclers. It highlights the scope for entrepreneurship in e-waste recycling with expansion of 7.23% a year on average (Bagwan, 2023). According to a survey conducted by Ananno et al. (2021), factors like monthly income, environmental awareness, and education level of consumers affect their willingness to process e-waste. According to them, e-waste recycling can be improved by environmental awareness and educating public on the benefits of e-waste recycling and making the process more convenient.

Types and Components of E-Waste

There are thousands of hazardous and non-hazardous components used in e-waste (UNEP, 2006). Technological advancement is so rapid that every new electrical and electronic item replaces current models or makes some electronic items obsolete, non-functioning, and useless soon enough to generate a lot of e-waste. There are generally two types of e-waste as per their physical composition (Hossain et al., 2015) (Figure 2).

Figure 2. Types of e-wastes as per their physical composition
Source: Hossain et al. (2015)

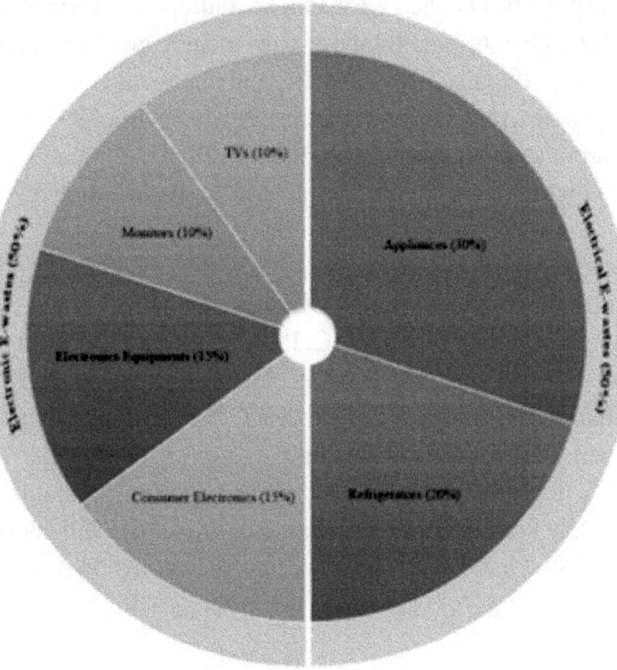

Figure 3. Chemical components in e-waste
Source: Sodhi & Reimer (2001)

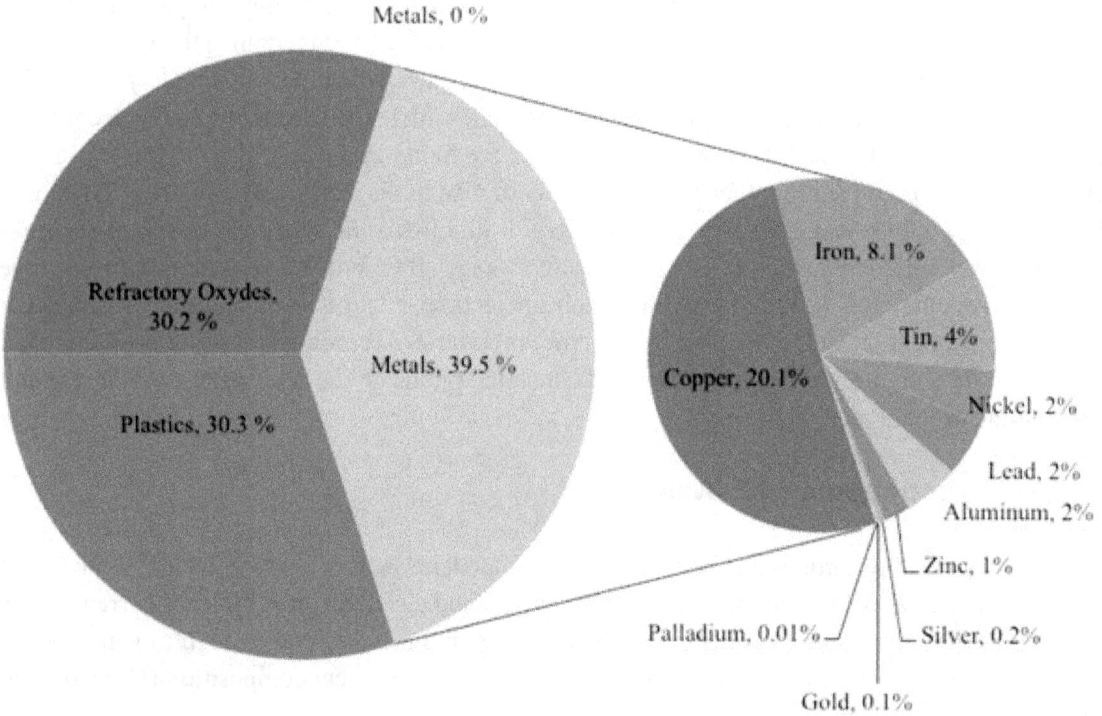

Electrical wastes consist of household and refrigeration appliances like bulbs, cables, dryers, washing machines, vacuum cleaners, AC, water heaters, coffee machines, irons, and toasters. Electronic wastes include monitors, TVs, CD players, DVDs, VCRs, routers, radio, GPS devices, calculators, smartphones, cameras, tablets, laptops, etc. Over 60% of total e-wastes include gold, iron, copper, and aluminum and out of which, 2.7% are toxic pollutants, heavy metals, and chemical compounds (Widmer et al, 2005). E-wastes usually include plastics, metals, and refractory oxides (Sum, 1991) (Figure 3). Plastic e-wastes include polyesters, polycarbonates, polypropylene, and polyethylene; metal scrap includes iron, tin, and copper; and major oxides include silica, alumina, potassium, barium titanate, aluminum silicates, and magnesium.

Some of the components can be reused when it comes to assemble new products while others are cut to metals to be reused in different applications like flatware, construction, etc. Some of the wide range of components used are fiberglass, epoxy resins, PVC, PCBs, lead, thermosetting plastics, silicon, copper, carbon, beryllium, aluminum, and iron. Some of the small amounts of elements found in e-waste are mercury, cadmium, and thallium (Muszyńska et al., 2015). Some of the components found in traces are americium, arsenic, antimony, barium, boron, bismuth, europium, cobalt, germanium, gallium, indium, gold, manganese, lithium, niobium, platinum, palladium, selenium, tantalum, etc. (Hieronymi et al., 2012). Tin and lead are widely used in almost all electronics as solder and copper as circuit board tracks and as wire. Table 1 and 2 list some of the toxic and non-toxic components of e-waste and their use.

Table 1. Toxic components of e-waste and their use

Components	Electric Appliances for Use	Harmful Effects	Source
Lead	Solder, batteries, CRT monitor, and some PVC formulations.	Behavioral disturbance, impaired cognitive function, ADHD, and lower IQ	Chen et al. (2011)
Americium	Smoke alarms as radioactive source	Carcinogenic	Silva et al. (2012)
Mercury	Various applications including fluorescent tubes, backlights in monitors, tilt switches (thermostats and doorbells)	Dermatitis, sensory impairment, muscle weakness, and memory loss. In-utero exposure leads to fetal deficits in attention, motor function, and verbal areas	Chen et al. (2011)
Sulfur	Lead-acid batteries	Kidney damage, liver damage, throat and eye irritation. It can release sulfur dioxide and form sulfuric acid when open in the environment.	Chen et al. (2011)
Cadmium	Light-sensitive resistors, nickel-cadmium rechargeable batteries, and aviation and marine alloys.	Inhalation can cause severe lung and kidney damage.	Sharma et al. (2015)
PVC	Found widely in electronics as insulator in cables	Over time, it can become pollutant and pose health problems for humans and animals. Its exposure can cause developmental and reproductive health effects	Nnorom & Osibanjo (2008); Jonkers et al. (2016)

Table 2. Non-toxic components used in e-waste

Component	Usage
Copper	ICs, component leads, and copper wire
Aluminum	ICs, capacitors, and almost all electronic items consuming at least few watts of energy
Gold	Widely used in digital devices as connector plating
Germanium	Transistors used in 1950s or 60s
Lithium	Rechargeable lithium-ion batteries used in smartphones, tablets, wireless headphones, EVs, pacemakers, digital cameras, laptops, smart watches, power banks, UPS, alarm systems, and any other digital device used today.
Silicon	ICs, glass, transistors, and circuit boards
Nickel	Nickel-cadmium batteries
Tin	Solder
Zinc	Steel part plating

Source - US EPA (2014)

Sustainable E-Waste Reduction

There is a significant impact of e-waste on the environment, especially when there is a rise in use of digital devices. According to research conducted by a top IoT and mobile device management provider SOTI (2022), at least 7 out of 10 IT leaders are replacing devices hastily and 41.6% of them replaced wearables, mobile phones, and handheld devices for a newer model. Companies are making changes in their IT departments to control corporate e-waste and bring green digital solutions to reduce the ill effects of IT operations on the environment. According to Shash Anand, VP of SOTI (Product Strategy), the lifespan of a device starts with knowledge of leadership. The longer the lifespan of a device, the fewer hardware replacements would be required.

Extreme environmental impact, toxicity, and financial recovery from e-waste have made it important to recover metal from e-waste. Using e-waste can possibly be a secondary source of base and valuable metals (Otto et al., 2018). Precious metals which are over 10 times purer than rich ore content have been the most important reason to extract valuable metal from e-waste. Table 3 lists the existing international prices of valuable metals. Hence, in order to reduce e-waste, it becomes highly important to extract valuable metals like Copper, Aluminum, Gold, Lead, Germanium, Palladium, Platinum, etc. from e-waste. Recycling can ensure utmost recovery and profitability and least environmental impact.

Various developed and developing countries have laws and regulations for e-waste management as described in (Figure 4). Each country has their e-waste management process which is described in Table 4). Developed countries have taken various initiatives like extended producer responsibility (EPR)" to deal with the issue of e-waste. According to EPR, the manufacturer needs to take the product back after its final use. This policy is mainly aimed to reduce over-exploitation of natural resources and waste. Figure 5 illustrates the process of converting e-waste into energy through a general e-waste management system.

It is observed that e-waste is separated into plastic and metal parts with mechanism/gravity separation method. Various noble metals like aluminum, iron, copper, gold, silver, etc. can be recovered by disposing of other e-waste parts. It is possible to recycle plastic parts and transform them into plastic bottles or energy through gasification, pyrolysis, and incineration (Figure 5). Plastic part of e-waste can be fueled with pyrolysis (Jayaraman et al., 2019). Pyrolysis of e-waste plastic can produce up to 51 to 71% of oil at the temperature of 430 to 460 degrees Celsius. In addition, incineration can also help generate energy or heat.

E-waste management should also be accompanied with deposit-refund schemes like "advance disposal fee (ADF)" and "advance recycling fee (ARF)". The sale of e-waste is usually taxed by ARF to cover

Table 3. Existing international prices of valuable metals (in USD)

Valuable Metals in E-Waste	Prices (in USD)
Gold	1306.10
Palladium	989.80
Platinum	916.65
Silver	16.59
Copper	3.06
Rhodium	2240

Source: Islam et al. (2020)

Figure 4. Laws and regulations in different countries regarding e-waste management
Source: Tansel, 2017; Themelis and Bourtsalas, 2019

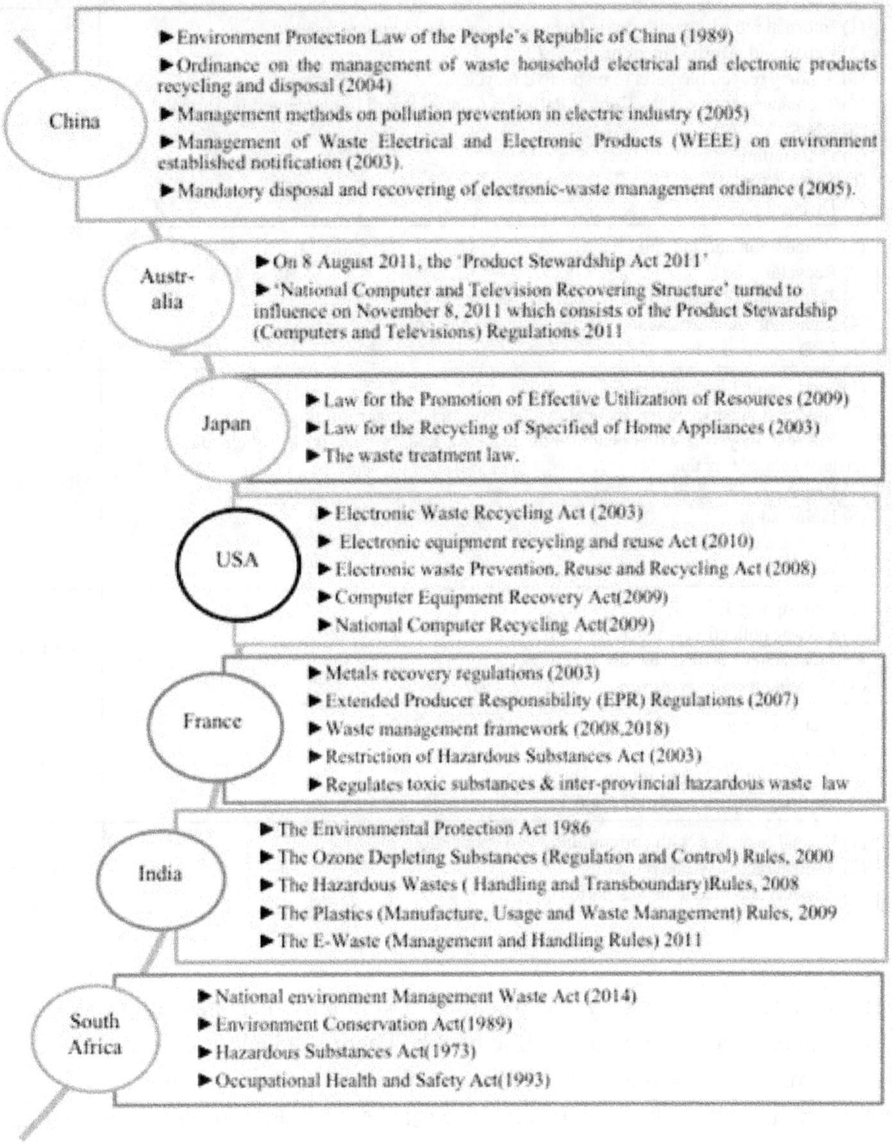

recycling costs and ADF consists of funds to bear the cost of disposal. ARF can be calculated on per unit of item sold on the basis of weight (Zarei et al., 2018).

CONCLUSION

In order to keep electronic and electrical equipment at their highest value all the time in the long term, it is high time to switch to a circular economy for all developed and developing countries. The problem of e-waste is rising significantly and e-waste management is getting even worse due to poor strategies

Table 4. E-waste management procedures followed in various countries

Countries	E-Waste Management Process	References
India	(1) Informal recycling with crude techniques for valuable metal recovery (2) Repair and refurbishment of used EEE (3) Sending recyclable parts to respective recyclers (4) Techniques approved by Central Pollution Control Board implemented by formal recyclers (5) Dismantling (6) Segregation (7) Disposal	Ghosh et al. (2016)
US	(1) Incineration and landfill (2) Recycling (3) Transport, collection, sorting, and dismantling (4) separation of chemical and mechanical materials (5) EPR	Lee et al. (2018)
China	(1) Waste reduction and prevention (2) Association between mechanical dismantling and formal recycling (3) Informal recycling with open burning, leaching acid, or other backyard techniques to retrieve valuable metals (4) Reuse (5) Incineration (6) Selling recyclable parts to recyclers (7) Repairing and refurbishment of used EEE	Ghosh et al. (2016)
EU	(1) Redesigning EEE (2) E-waste collection (3) Treatment of e-waste metals (4) Valuable metals recovery	Ghosh et al. (2016)
UK	(1) Metal recovery (2) Reuse (3) Segregation, dismantling, and shredding (4) "Life Cycle Assessment (LCA)"	Ongondo et al. (2011)
Brazil	(1) Formal recycling with comminuting and dismantling (2) Targeted refining and recycling (3) Physical and mechanical isolation	Mourshed et al. (2017)

without formal collection, recycling, and segregation. Eventually, it is affecting the environment and posing harm to the natural ecosystem and public health. There is a lack of efficiency and segmentation in existing e-waste management practices in India. A proper regulatory framework is needed to take concrete actions to reduce the harmful effects of e-waste. At the same time, social awareness is really important about the value of recycling and saving the environment. These practices can secure the environment in a sustainable way and curb the ill effects of increasing e-waste.

Several studies have observed an extreme lack of waste which is managed out of a huge pile of e-waste. Hence, all stakeholders should collaborate to spread awareness about the perils of e-waste and keep them together to deal with this environmental concern. It can help countries worldwide to manage e-waste in a sustainable manner in the long term. This chapter would help policymakers, decision-makers, and corporate leaders to play a vital role in building sustainable behavior among young consumers. Marketers can consider this chapter and make strategies with both environment and profitability in mind to make a win-win for both consumers and companies. Government can formulate plans with environmental sustainability in mind and readers can understand the value of e-waste management in daily lives. All in all, awareness is important to address the problem of e-waste as most people don't have any idea about this burning issue in developing countries.

Figure 5. Standard e-waste management procedure
Source: Ikhlayel, 2018; Mourshed et al., 2017

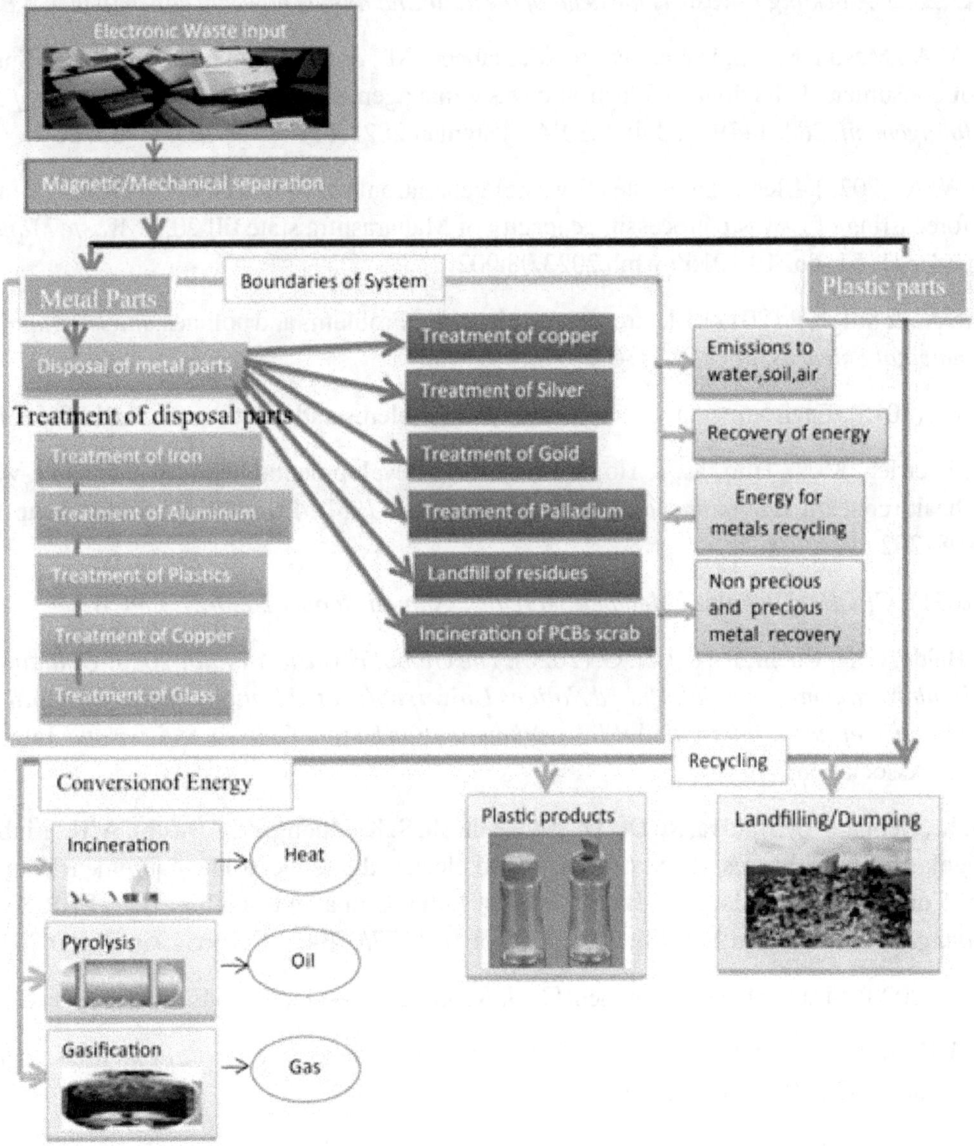

REFERENCES

Anand, A. (2022). *Tackling e-waste is the need of the hour and here is how you can chip in.* CNBCTV18.

Ananno, A. A., Masud, M. H., Dabnichki, P., Mahjabeen, M., & Chowdhury, S. A. (2021). Survey and analysis of consumers' behaviour for electronic waste management in Bangladesh. *Journal of Environmental Management, 282*, 111943. doi:10.1016/j.jenvman.2021.111943 PMID:33465720

Bagwan, W. A. (2023). Electronic waste (E-waste) generation and management scenario of India, and ARIMA forecasting of E-waste processing capacity of Maharashtra state till 2030. *Waste Management Bulletin, 1*(4), 41–51. doi:10.1016/j.wmb.2023.08.002

Borthakur, A., & Singh, P. (2012). Electronic waste in India: Problems and policies. *International Journal of Environmental Sciences, 3*(1), 353–362.

Chandra, M. (2015). Environmental concerns in India: Problems and solutions. *J. Int'l Bus. & L., 15*, 1.

Chen, A., Dietrich, K. N., Huo, X., & Ho, S. M. (2011). Developmental neurotoxicants in e-waste: An emerging health concern. *Environmental Health Perspectives, 119*(4), 431–438. doi:10.1289/ehp.1002452 PMID:21081302

CPCB. (2021). *CPCB on behalf of Member Secretary. Annual Report 2020-21.* CPCB.

Forti, V., Balde, C. P., Kuehr, R., & Bel, G. (2020). *The Global E-waste Monitor 2020: Quantities, flows and the circular economy potential. United Nations University/United Nations Institute for Training and Research: Bonn, Germany; International Telecommunication Union: Geneva, Switzerland.* International Solid Waste Association.

Ghosh, S. K., Debnath, B., Baidya, R., De, D., Li, J., Ghosh, S. K., Zheng, L., Awasthi, A. K., Liubarskaia, M. A., Ogola, J. S., & Tavares, A. N. (2016). Waste electrical and electronic equipment management and Basel Convention compliance in Brazil, Russia, India, China and South Africa (BRICS) nations. *Waste Management & Research, 34*(8), 693–707. doi:10.1177/0734242X16652956 PMID:27422616

Guterres, A. (2020). Sustainable Development Goals Report 2020—Goal 12—Electronic Waste. UNStats.

Hieronymi, K., Kahhat, R., & Williams, E. (Eds.). (2012). *E-waste management: From waste to resource.* Routledge. doi:10.4324/9780203116456

Hossain, M. S., Al-Hamadani, S. M., & Rahman, M. T. (2015). E-waste: A challenge for sustainable development. *Journal of Health & Pollution, 5*(9), 3–11. doi:10.5696/2156-9614-5-9.3 PMID:30524771

Ikhlayel, M. (2018). An integrated approach to establish e-waste management systems for developing countries. *Journal of Cleaner Production, 170*, 119–130. doi:10.1016/j.jclepro.2017.09.137

Islam, A., Ahmed, T., Awual, M. R., Rahman, A., Sultana, M., Abd Aziz, A., ... Hasan, M. (2020). Advances in sustainable approaches to recover metals from e-waste-A review. *Journal of Cleaner Production, 244*, 118815. doi:10.1016/j.jclepro.2019.118815

Jayaraman, K., Vejayon, S., Raman, S., & Mostafiz, I. (2019). The proposed e-waste management model from the conviction of individual laptop disposal practices-An empirical study in Malaysia. *Journal of Cleaner Production*, *208*, 688–696. doi:10.1016/j.jclepro.2018.10.125

Jonkers, N., Krop, H., van Ewijk, H., & Leonards, P. E. (2016). Life cycle assessment of flame retardants in an electronics application. *The International Journal of Life Cycle Assessment*, *21*(2), 146–161. doi:10.100711367-015-0999-z

Kandel, A. (2022). *A REPORT On E-waste Management E-waste Management*. Sharda University.

Lee, D., Offenhuber, D., Duarte, F., Biderman, A., & Ratti, C. (2018). Monitour: Tracking global routes of electronic waste. *Waste Management, 72*, 362-370.

Ministry of Environment, Forest and Climate Change. (2022). Annual Report of 2021-22. Government of India.

Morlet, A. (2018). *The Circular Economy Opportunity for Urban and Industrial Innovation in China*. Ellen MacArthur Foundation.

Mourshed, M., Masud, M. H., Rashid, F., & Joardder, M. U. H. (2017). Towards the effective plastic waste management in Bangladesh: A review. *Environmental Science and Pollution Research International*, *24*(35), 27021–27046. doi:10.100711356-017-0429-9 PMID:29079979

Muszyńska, B., Rojowski, J., Dobosz, K., & Opoka, W. (2015). Biological and physico-chemical properties of thallium. *Medicina Internacia Revuo*, (105), 180–185.

Namo eWaste. (2022). *Namo eWaste—E-Waste Recycling & Management Solution in India*. Namo eWaste.

Nnorom, I. C., & Osibanjo, O. (2008). Sound management of brominated flame retarded (BFR) plastics from electronic wastes: State of the art and options in Nigeria. *Resources, Conservation and Recycling*, *52*(12), 1362–1372. doi:10.1016/j.resconrec.2008.08.001

Ongondo, F. O., Williams, I. D., & Cherrett, T. J. (2011). How are WEEE doing? A global review of the management of electrical and electronic wastes. *Waste Management, 31*(4), 714-730.

Otto, S., Kibbe, A., Henn, L., Hentschke, L., & Kaiser, F. G. (2018). The economy of E-waste collection at the individual level: A practice oriented approach of categorizing determinants of E-waste collection into behavioral costs and motivation. *Journal of Cleaner Production*, *204*, 33–40. doi:10.1016/j.jclepro.2018.08.293

Parajuly, K., Kuehr, R., Awasthi, A. K., Fitzpatrick, C., Lepawsky, J., Smith, E., & Zeng, X. (2019). *Future e-waste scenarios. StEP Initiative*. UNU ViE-SCYCLE, and UNEP IETC.

Pendharkar, V. (2018). *Indian cities stare at a mountain of e-waste, with little idea of how to manage it*. Citizen Matters.

Reynolds, O. (2021). The World's Fastest Growing Economies. *Focus Economics, 16*.

Ruiz, A. (2022). *Latest global E-waste statistics and what they tell us*. TheRoundup.

Sethurajan, M., van Hullebusch, E. D., Fontana, D., Akcil, A., Deveci, H., Batinic, B., Leal, J. P., Gasche, T. A., Ali Kucuker, M., Kuchta, K., Neto, I. F. F., Soares, H. M. V. M., & Chmielarz, A. (2019). Recent advances on hydrometallurgical recovery of critical and precious elements from end of life electronic wastes-a review. *Critical Reviews in Environmental Science and Technology*, *49*(3), 212–275. doi:10.1080/10643389.2018.1540760

Sharma, H., Rawal, N., & Mathew, B. B. (2015). The characteristics, toxicity and effects of cadmium. *International Journal of Nanotechnology and Nanoscience, 3*(10).

Silva, R. J., Bidoglio, G., Robouch, P. B., Puigdomenech, I., Wanner, H., & Rand, M. H. (2012). *Chemical thermodynamics of americium*. Newnes.

Singhal, D., Tripathy, S., & Jena, S. K. (2019). Sustainability through remanufacturing of e-waste: Examination of critical factors in the Indian context. *Sustainable Production and Consumption*, *20*, 128–139. doi:10.1016/j.spc.2019.06.001

Sodhi, M. S., & Reimer, B. (2001). Models for recycling electronics end-of-life products: Modelle zum Recycling von Altprodukten aus dem Elektronikbereich. *OR-Spektrum*, *23*(1), 97–115. doi:10.1007/PL00013347

SOTI. (2022). *Reduce, reuse, rethink: From discard mentality to tech sustainability*. Available at https://soti.net/media/6647/industry-report-reduce-reuse-rethink-from-discard-mentality-to-tech-sustainability-letter.pdf

Stefansdotter, A., Knusden, J. S., Flack, M., & Hansen, P. G. (2016). *Nudging for sustainable consumption of electronics. Summary Report*. Denmark: Nordic Council of Ministers. Available online at: https://norden.diva-portal.org/smash/get/diva

Stone, S. (2020). *The 4Rs: Reduce, Reuse, Recycle and Recover with the Gloucestershire Energy from Waste Facility*. Gloucester Energy from Waste Facility.

Sum, E. Y. (1991). The recovery of metals from electronic scrap. *Journal of the Minerals Metals & Materials Society*, *43*(4), 53–61. doi:10.1007/BF03220549

Tansel, B. (2017). From electronic consumer products to e-wastes: Global outlook, waste quantities, recycling challenges. *Environment International*, *98*, 35–45. doi:10.1016/j.envint.2016.10.002 PMID:27726897

Themelis, N. J. (2019). *Recovery of materials and energy from urban wastes*. Springer Berlin Heidelberg. doi:10.1007/978-1-4939-7850-2

UNEP. (2006). *Basel conference addresses electronic wastes challenge*. United Nations Environment Programme.

UNEP. (2022). *Sustainable Consumption and Production Policies*. United Nations Environment Programme.

UNITAR. (2020). *Global E-waste Statistics (2020)—Sustainable Cycles (SCYCLE) Programme Map*. The United Nations Institute for Training and Research. Available online: https://globalewaste.org/map/

US EPA. (2014). *Regulatory and Guidance Information by Topic: Waste – Guide for Industrial Waste Management.* Retrieved from https://www.epa.gov/regulatory-information-topic/regulatory-and-guidance-information-topic-waste#solid

Widmer, R., Oswald-Krapf, H., Sinha-Khetriwal, D., Schnellmann, M., & Böni, H. (2005). Global perspectives on e-waste. *Environmental Impact Assessment Review*, *25*(5), 436–458. doi:10.1016/j.eiar.2005.04.001

Zarei, M., Taghipour, H., & Hassanzadeh, Y. (2018). Survey of quantity and management condition of end-of-life tires in Iran: A case study in Tabriz. *Journal of Material Cycles and Waste Management*, *20*(2), 1099–1105. doi:10.100710163-017-0674-5

Chapter 10
Extended Producer Responsibility (EPR) and E-Waste Management:
A Roadmap for Entrepreneurs

Bhairab Chandra Patra
Indian Institute of Management, Sambalpur (IIM Sambalpur), India

Dibya Nandan Mishra
https://orcid.org/0000-0002-9918-115X
Department of Management Studies, National Institute of Technology, Silchar, India

ABSTRACT

Electronic trash (e-waste) creation has increased exponentially as a result of the widespread use of electronic devices and the quick rate of technical development, posing a serious environmental concern. Extended producer responsibility (EPR) has become a crucial tactic for addressing the negative social and environmental effects of e-waste. This chapter provides entrepreneurs with a detailed roadmap for navigating the challenges of EPR adoption in e-waste management. For entrepreneurs, the application of extended producer responsibility in the management of e-waste poses both a difficulty and an opportunity. By embracing the EPR precepts and following the recommended roadmap, entrepreneurs may significantly minimize the environmental impact of e-waste, promote resource conservation, and create a more sustainable future.

1. INTRODUCTION

Ecosystems, human health, and resource sustainability are seriously threatened by the incorrect management and disposal of e-waste. Extended producer responsibility (EPR) is a political strategy that places the onus of responsible e-waste disposal on manufacturers and producers rather than on consumers and local governments (Khetriwal, Kraeuchi, & Widmer, 2009). This article offers a succinct summary

DOI: 10.4018/979-8-3693-1018-2.ch010

of the essential elements of EPR and proposes a roadmap that gives entrepreneurs the knowledge and techniques they need to put good e-waste management practices into reality.

EPR requires producers to take responsibility for the whole lifespan of their products, including disposal. The management of e-waste is a crucial topic that has attracted a lot of attention recently. E-waste creation has increased dramatically as a result of the expanding usage of electronic gadgets and the quickening rate of technical innovation. E-waste improperly disposed of has the potential to seriously harm the environment and human health. In this situation, EPR has shown to be a useful tool for managing e-waste by directing it to approved recyclers and dismantlers for safe handling and disposal (Arya & Kumar, 2020). This article provides a comprehensive guide that gives insights into the concept of EPR and its implementation in e-waste management. The article gives company owners a road map for implementing EPR ideas into their daily operations in order to embrace sustainable business practices. It covers a range of EPR-related topics, including as its history, the legal system, and difficulties with implementation. The article offers examples of effective EPR implementation in many nations, making it a useful tool for businesspeople, legislators, and scholars.

EPR may help enterprises in a number of ways in addition to the advantages already discussed. Entrepreneurs can use it to comply with environmental laws and lessen their environmental impact. Entrepreneurs may improve their brand reputation and image among customers who are becoming more environmentally sensitive by using sustainable business practices (Khan, Hussain & Sampene, 2023). EPR can also open up new economic avenues by encouraging the development of cutting-edge recycling technology and establishing new markets for recovered goods. EPR may help brands by facilitating the development of a circular economy, creating additional money, increasing your customer base, promoting sustainability or environmental friendliness, and lowering the amount of garbage that ends up in landfills (Marks, Miller & Vassanadumrongdee, 2023).

2. EXTENDED PRODUCER RESPONSIBILITY (EPR)

Extended Producer Responsibility (EPR) is a political strategy intended to encourage the responsible handling and disposal of goods and the trash they generate. It transfers control of a product's lifetime, including its end-of-life disposal, from the consumer or municipality to the manufacturer and producer of that product (Mayanti & Helo, 2023). E-waste management, or e-waste, is the process of applying EPR concepts to electronic trash, which includes discarded electronic equipment and gadgets (Kaur, Kaur & Kaur, 2023). The key components of EPR are as elaborated in the following sub-sections.

2.1. Legal Framework and Regulations

Entrepreneurs need to be aware of their jurisdictions' e-waste management legal requirements and regulatory frameworks. Avoiding fines and promoting sustainable activities depend on adherence to these requirements. The framework mandates that producers, manufacturers, and brand owners recover the plastic trash produced as a result of their products under extended producer responsibility (Gupta & Dash, 2023). The garbage producers are required to reduce the production of plastic trash by source segregation and to turn it over to local organizations. The manufacturer, importer, or brand owner must participate in EPR framework implementation initiatives other than data collecting (Pani & Pathak, 2021).

2.2. Product Design and Lifecycle

EPR can be improved if eco-design and circular economy ideas are applied from the beginning of the product's lifecycle. Entrepreneurs should use tactics that make it easier to disassemble a product and retrieve its parts once it has reached the end of its useful life (Dissanayake & Weerasinghe, 2021). The lifespan and environmental effect of a product are greatly influenced by its design. The design of the product should make it simple to disassemble, repair, and recycle.

The Extended Producer Responsibility (EPR) framework requires manufacturers to provide items that are easily recyclable and favorable to the environment. In accordance with the EPR framework, producers must also collect the plastic trash that their goods create. The garbage producers are required to reduce the production of plastic trash by source segregation and to turn it over to local organizations (Fiksel, Sanjay & Raman, 2021). The manufacturer, importer, or brand owner must participate in EPR framework implementation initiatives other than data collecting.

2.3. Collection Infrastructure

For electronic waste to be collected, an effective collection network must be set up. To provide consumers with simple and ethical disposal options, business owners should work with approved collection stations, merchants, and trash management organizations (Kurniawan, Othman, Hwang & Gikas, 2022).

For Extended Producer Responsibility (EPR) to be successfully implemented, the collecting infrastructure is essential. It entails establishing efficient systems and infrastructure for gathering, classifying, and recycling product and packaging trash (Ramasubramanian, Tan, Chellappan, & Ramakrishna, 2023). Consumers should be able to easily access and use the infrastructure. It promotes customer engagement in recycling initiatives by facilitating access to collection programs.

2.4. Logistics and Transportation

For the safe movement of e-waste from collection locations to recycling facilities, effective transportation and logistics are essential. Entrepreneurs must streamline transportation routes, reduce pollutants, and guarantee adherence to safety regulations (Kumar & Jha, 2019).

Moving goods and packaging from consumers to recycling facilities or other suitable disposal locations is part of Extended Producer Responsibility's (EPR) logistics and transportation component (Ahlers, Hemkhaus, Hibler & Hannak, 2021). To collect, move, and deliver waste materials to the specified facilities, effective transportation networks and systems must be established. In order to reduce greenhouse gas emissions, maximize fuel economy, and minimize environmental effect, the transportation process should be planned (Kazancoglu, Ozbiltekin-Pala, & Ozkan-Ozen, 2021).

2.5. Recycling and Resource Recovery

To extract valuable resources from e-waste, entrepreneurs should collaborate with accredited recycling facilities that are outfitted with cutting-edge technology. By reusing recovered materials, proper recycling not only lessens its negative environmental effects but also promotes the circular economy (Purchase, Al Zulayq, O'Brien, Kowalewski, Berenjian, Tarighaleslami, & Seifan, 2021).

Extended Producer Responsibility (EPR)'s recycling and resource recovery component focuses on the effective management of waste materials produced by goods and packaging. It entails applying technology and methods to recycle, reuse, or repurpose waste products to recover valuable resources. Reduced trash creation, resource conservation, and environmental impact are the objectives (Pouikli, 2020).

2.6. Consumer Awareness and Education

It is crucial to inform customers on appropriate disposal practices, the value of recycling, and the risks associated with incorrect e-waste handling. In order to increase awareness and encourage ethical consumer behavior, entrepreneurs may play a critical role (Islam, Huda, Baumber, Shumon, Zaman, Ali & Sahajwalla, 2021).

With regards to Extended Producer Responsibility (EPR), a number of organizations and projects concentrate on raising consumer awareness and educating them. As an illustration, many industries have put into effect an EPR plan that addresses the collecting and reprocessing of plastic packaging waste on a brand-neutral basis throughout important markets (Kiehn & Weller Vojkovic, 2018). Additionally, they educate their workforce about trash management through volunteerism initiatives and training for sanitation and garbage staff. As part of its EPR activities, researchers place a strong emphasis on the necessity of educating consumers about responsible consumption and disposal.

3. ENTREPRENEURIAL ROADMAP

Outlining the procedures and methods that company owners or entrepreneurs might use to create and succeed in this industry is a necessary step in developing an entrepreneurial roadmap for Extended Producer Responsibility (EPR) in e-waste management. The handling of e-waste has both enormous commercial and environmental potential (Appolloni, D'Adamo, Gastaldi, Santibanez-Gonzalez, & Settembre-Blundo, 2021). The key components of the entrepreneurial roadmap for EPR are as follows:

3.1. Research and Regulatory Compliance

Entrepreneurs have to thoroughly investigate the EPR laws that apply to their sector and locality. Understanding statutory requirements, registration procedures, and reporting duties is necessary for this (Gui, Atasu, Ergun & Toktay, 2013).

EPR policies encourage companies to take environmental considerations into account when developing their goods with the goal of moving responsibility upstream and away from municipalities. Instead of focusing on a single link in the chain, this strategy aims to include environmental factors throughout the whole supply chain (Nnorom & Osibanjo, 2008).

3.2 Collaborative Partnerships

Entrepreneurs should establish partnerships with many stakeholders, including waste management firms, recycling facilities, governmental bodies, and consumer advocacy groups. These collaborations make e-waste collection, transportation, and recycling more effective (Camilleri, 2022).

In order to successfully execute Extended Producer Responsibility (EPR) initiatives, collaborative relationships are essential. EPR's objectives—reducing environmental impact, increasing recycling and responsible product design, and moving the financial burden of waste management from taxpayers to producers—must be accomplished via collaboration amongst a variety of stakeholders. Creating a sustainable and efficient framework for managing products throughout their lifecycles requires collaborative relationships in EPR. EPR initiatives can achieve their economic and environmental goals while putting as little stress as possible on the environment and taxpayers by bringing together producers, governmental organizations, consumers, and other stakeholders (Pouikli, 2020).

3.3 Product Design Innovation

Products with a less environmental impact may be produced by using sustainable design concepts that promote recyclability, simplicity of disassembly, and material efficiency. Innovation in product design is crucial to the execution of Extended Producer Responsibility (EPR) initiatives. Innovative product designs with a positive environmental impact can also increase recycling and promote EPR objectives (Maitre-Ekern, 2021). To achieve the goals of EPR, these design ideas and concepts must be incorporated into the product development process. Producers may reduce the environmental effect of their goods and enable their responsible management throughout their lifespan, eventually lessening the burden on society and the environment, by designing products with sustainability, recyclability, and resource efficiency in mind (Giovanardi, Konstantinou, Pollo, & Klein, 2023).

3.4 Collection Network Establishment

Consumers may conveniently dispose of their electronic trash thanks to the development of an accessible and extensive collection network. Entrepreneurs want also look into collaborations with merchants, community centers, and e-waste collecting activities. Producers are required to manage their goods throughout their full lifespan, including the collection and recycling of waste products, under the terms of the EPR. EPR goals, such as waste reduction, resource recovery, and environmental preservation, are efficiently realized with the help of a well-designed collection network. The success of EPR programs depends on a well-run collection network since it guarantees that items are handled correctly at the end of their lives. It advances sustainability, lowers waste, saves resources, and lessens the environmental impact of products—all key objectives of EPR. The collecting network needs cooperation from manufacturers, governments, waste management companies, and consumers to be efficient and successful (Rizzi, Bartolozzi, Borghini & Frey, 2013).

3.5. Technological Integration

Utilizing digital tools may simplify e-waste collection scheduling, tracking, and consumer communication. Examples of these tools include mobile applications and web platforms. The successful implementation of Extended Producer Responsibility (EPR) initiatives depends critically on technological integration. The efficacy, efficiency, and openness of EPR projects may all be considerably improved by technology. EPR initiatives may become more effective, transparent, and responsive to environmental and economic goals by adopting these technology advancements. The improved capabilities that technology offers in controlling the lifetime of items and lessening their environmental effect may benefit producers, consumers, governmental organizations, and waste management organizations (Zhao, Peng, Elahi & Wan, 2021).

3.6. Public Awareness Campaigns

Entrepreneurs may start educational campaigns to spread the word about ethical e-waste management and the value of EPR using social media, workshops, and neighbourhood activities. Campaigns to raise public awareness are crucial elements of Extended Producer Responsibility (EPR) initiatives. These initiatives aim to inform and involve the general public in the value of responsible consumption, recycling, and appropriate disposal of goods subject to EPR legislation. Public awareness initiatives that are successful may encourage a sense of shared responsibility among customers while also advancing the environmental and sustainability aims of EPR (Vishwakarma, Kumar, Arya, Tembhare, Dutta & Kumar, 2022). In order to garner public support, promote responsible consumption and disposal habits, and eventually accomplish the program's environmental and sustainability goals, public awareness campaigns are essential in EPR initiatives. EPR initiatives can be successful in lowering waste, preserving resources, and decreasing their environmental footprint by encouraging a culture of responsibility and environmental care among customers.

4. CONCLUSION

The implementation of Extended Producer Responsibility in e-waste management presents both a challenge and an opportunity for entrepreneurs. Entrepreneurs may considerably help reduce the environmental effect of e-waste, promote resource conservation, and develop a more sustainable future by adopting the EPR tenets and adhering to the suggested roadmap.

To sum up, Extended Producer Responsibility (EPR) is an essential and dynamic legislative framework created to handle the expanding issues surrounding the management of e-waste and advance environmentally friendly practices in the electronics sector. Manufacturers and producers now bear full responsibility for the whole lifespan of electronic items, from design and manufacturing through end-of-life disposal (Turaga, Bhaskar, Sinha, Hinchliffe, Hemkhaus, Arora & Sharma, 2019). This strategy strives to lessen the risks to the environment and human health brought on by electronic trash while promoting eco-friendly product design and ethical disposal techniques.

Our environment, health, and resources are seriously threatened by e-waste, a problem that is spreading quickly around the globe. Electronics are being thrown away at an alarming rate due to the prevalence of these gadgets and their short lifespans, and many of them are exported to nations that are ill-prepared to manage the toxic compounds they contain or are improperly disposed at the end (Kitila & Woldemikael, 2021).

EPR requires producers to bear the cost of handling the end-of-life of their products, which encourages the creation of more robust and readily recyclable electronic items. It promotes the development of effective systems for garbage collection and recycling, ensuring that e-waste is handled in an ecologically friendly way. EPR encourages producers to develop electronics with longer lifespans, simple-to-repair components, and eco-friendly materials. EPR also stimulates innovation in product design. Public awareness initiatives that inform customers about ethical e-waste disposal and the advantages of recycling are frequently included in EPR programs. The global adoption of EPR and the harmonization of standards can aid in addressing the global aspects of e-waste management and stop the export of e-waste to developing nations (Xavier, Ottoni & Lepawsky, 2021).

However, cooperation between governments, industry stakeholders, consumers, and the larger society is essential to the success of EPR and e-waste management. In order to guarantee that manufacturers uphold their duties and customers engage in recycling programs, effective legislation, monitoring, and enforcement are essential. Additionally, technical advancements that streamline collection and recycling procedures include product monitoring and internet platforms (Kumar, Singh, Gupta, & Madaan, 2020).

EPR is a glimmer of hope for a more sustainable and ethical way to handling e-waste in a society that is becoming more and more dependent on electronics (Dawson, 2019). EPR advocates a transition toward a circular economy, where resources are saved and waste is reduced, in addition to mitigating environmental harm by making manufacturers responsible for the items they release onto the market. In the area of managing e-waste, EPR provides a potential road towards a more sustainable future as we continue to watch the quick development of technology.

REFERENCES

Ahlers, J., Hemkhaus, M., Hibler, S., & Hannak, J. (2021). *Analysis of Extended Producer Responsibility Schemes*. Academic Press.

Appolloni, A., D'Adamo, I., Gastaldi, M., Santibanez-Gonzalez, E. D., & Settembre-Blundo, D. (2021). Growing e-waste management risk awareness points towards new recycling scenarios: The view of the Big Four's youngest consultants. *Environmental Technology & Innovation*, *23*, 101716. doi:10.1016/j.eti.2021.101716

Arya, S., & Kumar, S. (2020). E-waste in India at a glance: Current trends, regulations, challenges and management strategies. *Journal of Cleaner Production*, *271*, 122707. doi:10.1016/j.jclepro.2020.122707

Camilleri, M. A. (2022). Strategic attributions of corporate social responsibility and environmental management: The business case for doing well by doing good! *Sustainable Development (Bradford)*, *30*(3), 409–422. doi:10.1002d.2256

Dawson, L. (2019). 'Our Waste, our Resources; A Strategy for England'–Switching to a circular economy through the use of extended producer responsibility. *Environmental Law Review*, *21*(3), 210–218. doi:10.1177/1461452919851943

Dissanayake, D. G. K., & Weerasinghe, D. (2021). Towards circular economy in fashion: Review of strategies, barriers and enablers. *Circular Economy and Sustainability*, 1–21.

Fiksel, J., Sanjay, P., & Raman, K. (2021). Steps toward a resilient circular economy in India. *Clean Technologies and Environmental Policy*, *23*(1), 203–218. doi:10.100710098-020-01982-0 PMID:33199980

Giovanardi, M., Konstantinou, T., Pollo, R., & Klein, T. (2023). Internet of Things for building façade traceability: A theoretical framework to enable circular economy through life-cycle information flows. *Journal of Cleaner Production*, *382*, 135261. doi:10.1016/j.jclepro.2022.135261

Gui, L., Atasu, A., Ergun, Ö., & Toktay, L. B. (2013). Implementing extended producer responsibility legislation: A multi-stakeholder case analysis. *Journal of Industrial Ecology*, *17*(2), 262–276. doi:10.1111/j.1530-9290.2012.00574.x

Gupta, D., & Dash, S. (2023). Challenges of implementing extended producer responsibility for plastic-waste management: Lessons from India. *Social Responsibility Journal*, *19*(9), 1595–1612. doi:10.1108/SRJ-08-2022-0326

Islam, M. T., Huda, N., Baumber, A., Shumon, R., Zaman, A., Ali, F., Hossain, R., & Sahajwalla, V. (2021). A global review of consumer behavior towards e-waste and implications for the circular economy. *Journal of Cleaner Production*, *316*, 128297. doi:10.1016/j.jclepro.2021.128297

Kaur, B., Kaur, K., & Kaur, H. (2023). E-waste Management: A Transition Towards a Circular Economy. *The Impact of Climate Change and Sustainability Standards on the Insurance Market*, 403-416.

Kazancoglu, Y., Ozbiltekin-Pala, M., & Ozkan-Ozen, Y. D. (2021). Prediction and evaluation of greenhouse gas emissions for sustainable road transport within Europe. *Sustainable Cities and Society*, *70*, 102924. doi:10.1016/j.scs.2021.102924

Khan, A., Hussain, S., & Sampene, A. K. (2023). Investing in green intellectual capital to enhance green corporate image under the Influence of green innovation climate: A Case of Chinese Entrepreneurial SMEs. *Journal of Cleaner Production*, *418*, 138177. doi:10.1016/j.jclepro.2023.138177

Khetriwal, D. S., Kraeuchi, P., & Widmer, R. (2009). Producer responsibility for e-waste management: Key issues for consideration–learning from the Swiss experience. *Journal of Environmental Management*, *90*(1), 153–165. doi:10.1016/j.jenvman.2007.08.019 PMID:18162284

Kiehn, K., & Weller Vojkovic, A. (2018). *Millennials motivations for shopping second-hand clothing as part of a sustainable consumption practice*. Academic Press.

Kitila, A. W., & Woldemikael, S. M. (2021). Electronic waste management in Addis Ababa: The case of Bole and Nefas Silk Lafto sub-cities. *African Journal of Science, Technology, Innovation and Development*, *13*(2), 235–246. doi:10.1080/20421338.2020.1712014

Kumar, H., Singh, M. K., Gupta, M. P., & Madaan, J. (2020). Moving towards smart cities: Solutions that lead to the Smart City Transformation Framework. *Technological Forecasting and Social Change*, *153*, 119281. doi:10.1016/j.techfore.2018.04.024

Kumar, N., & Jha, A. (2019). Application of principles of supply chain management to the pharmaceutical good transportation practices. *International Journal of Pharmaceutical and Healthcare Marketing*, *13*(3), 306–330. doi:10.1108/IJPHM-09-2017-0048

Kumari, N., & Lenka, U. (2023). Employment and Retention of Differently-abled People in the Workplace Through Assistive Technologies. *International Journal of Digital Technologies*, *2*(1).

Kumari, N., & Lenka, U. (2023). Scientometric review of disability inclusion in the workplace.

Kurniawan, T. A., Othman, M. H. D., Hwang, G. H., & Gikas, P. (2022). Unlocking digital technologies for waste recycling in Industry 4.0 era: A transformation towards a digitalization-based circular economy in Indonesia. *Journal of Cleaner Production*, *357*, 131911. doi:10.1016/j.jclepro.2022.131911

Maitre-Ekern, E. (2021). Re-thinking producer responsibility for a sustainable circular economy from extended producer responsibility to pre-market producer responsibility. *Journal of Cleaner Production*, *286*, 125454. doi:10.1016/j.jclepro.2020.125454

Marks, D., Miller, M. A., & Vassanadumrongdee, S. (2023). Closing the loop or widening the gap? The unequal politics of Thailand's circular economy in addressing marine plastic pollution. *Journal of Cleaner Production*, *391*, 136218. doi:10.1016/j.jclepro.2023.136218

Mayanti, B., & Helo, P. (2023). Circular economy through waste reverse logistics under extended producer responsibility in Finland. *Waste Management & Research.*

Mishra, D. N., & Panda, R. K. (2021). How delightful is Indian wellness tourism? A netnographic study. [AHTR]. *Advances in Hospitality and Tourism Research*, *9*(1), 132–156. doi:10.30519/ahtr.784232

Mishra, D. N., & Panda, R. K. (2023). Evaluating visitor–therapist relationship in Indian spa and wellness resorts. *Journal of Hospitality and Tourism Insights*, *6*(5), 2433–2461. doi:10.1108/JHTI-06-2022-0246

Nnorom, I. C., & Osibanjo, O. (2008). Overview of electronic waste (e-waste) management practices and legislations, and their poor applications in the developing countries. *Resources, Conservation and Recycling*, *52*(6), 843–858. doi:10.1016/j.resconrec.2008.01.004

Pani, S. K., & Pathak, A. A. (2021). Managing plastic packaging waste in emerging economies: The case of EPR in India. *Journal of Environmental Management*, *288*, 112405. doi:10.1016/j.jenvman.2021.112405 PMID:33780821

Patra, B. C., & Lenka, U. (2021). An AHP analysis of scientometrically derived factors of entrepreneurial intentions of women and constructing a conceptual research framework. *Journal of Research in Marketing and Entrepreneurship*, *23*(1), 139–158. doi:10.1108/JRME-02-2020-0017

Patra, B. C., & Lenka, U. (2022). Barriers to entrepreneurial intentions of women: Nominal group technique, analytic hierarchy process, and scientometric approach instigating the necessity of policy intervention. *Journal of Public Affairs*, *22*(2), e2429.

Patra, B. C., & Lenka, U. (2023). Scientometric, fuzzy NGT and DEMATEL analysis for determining sustainable business practices for entrepreneurial firms. *Benchmarking.*

Patra, B. C., & Mishra, D. N. (2023). Boosting entrepreneurial intentions: A novel EEG-guided protocol. *MethodsX*, *11*, 102358. doi:10.1016/j.mex.2023.102358 PMID:37701733

Pouikli, K. (2020). Concretising the role of extended producer responsibility in European Union waste law and policy through the lens of the circular economy. Springer Berlin Heidelberg.

Purchase, C. K., Al Zulayq, D. M., O'Brien, B. T., Kowalewski, M. J., Berenjian, A., Tarighaleslami, A. H., & Seifan, M. (2021). Circular economy of construction and demolition waste: A literature review on lessons, challenges, and benefits. *Materials (Basel)*, *15*(1), 76. doi:10.3390/ma15010076 PMID:35009222

Ramasubramanian, B., Tan, J., Chellappan, V., & Ramakrishna, S. (2023). Recent Advances in Extended Producer Responsibility Initiatives for Plastic Waste Management in Germany and UK. *Materials Circular Economy*, *5*(1), 6. doi:10.100742824-023-00076-8

Rizzi, F., Bartolozzi, I., Borghini, A., & Frey, M. (2013). Environmental management of end-of-life products: Nine factors of sustainability in collaborative networks. *Business Strategy and the Environment, 22*(8), 561–572. doi:10.1002/bse.1766

Turaga, R. M. R., Bhaskar, K., Sinha, S., Hinchliffe, D., Hemkhaus, M., Arora, R., Chatterjee, S., Khetriwal, D. S., Radulovic, V., Singhal, P., & Sharma, H. (2019). E-waste management in India: Issues and strategies. *Vikalpa, 44*(3), 127–162. doi:10.1177/0256090919880655

Vishwakarma, S., Kumar, V., Arya, S., Tembhare, M., Dutta, D., & Kumar, S. (2022). E-waste in Information and Communication Technology Sector: Existing scenario, management schemes and initiatives. *Environmental Technology & Innovation, 27*, 102797. doi:10.1016/j.eti.2022.102797

Xavier, L. H., Ottoni, M., & Lepawsky, J. (2021). Circular economy and e-waste management in the Americas: Brazilian and Canadian frameworks. *Journal of Cleaner Production, 297*, 126570. doi:10.1016/j.jclepro.2021.126570

Zhao, Y., Peng, B., Elahi, E., & Wan, A. (2021). Does the extended producer responsibility system promote the green technological innovation of enterprises? An empirical study based on the difference-in-differences model. *Journal of Cleaner Production, 319*, 128631. doi:10.1016/j.jclepro.2021.128631

Chapter 11
Equity-Driven Solutions:
Harnessing IoT Technology to Mitigate Health Risks and Bridge Gaps in E-Waste Management

C. V. Suresh Babu
https://orcid.org/0000-0002-8474-2882
Hindustan Institute of Technolgy and Science, India

C. S. Akkash Anniyappa
Sri Sivasubramaniya Nadar College of Engineering, India

Anamanamudi Sai Karthik
Hindustan Institute of Technology and Science, India

Sekhar Babu
Hindustan Institute of Technology and Science, India

Yedlapalli Sai Durga Pavan
Hindustan Institute of Technology and Science, India

Ch Bala Brahanandam
Hindustan Institute of Technology and Science, India

ABSTRACT

This chapter delves into the global e-waste challenge, emphasizing health risks and equity issues. Positioned within the book's theme, it justifies the relevance of IoT-driven solutions for addressing e-waste's social impacts. The discussion navigates through health risks and disparities in developing nations. IoT-enabled e-waste monitoring is explored, detailing how sensors create real-time databases for precise tracking. Analysis of IoT data assesses health risks, identifies contamination hotspots, and facilitates equitable resource allocation. The chapter highlights IoT's role in tailored awareness campaigns, particularly for underserved populations, and underscores its impact on policy development and enforcement. The conclusion summarizes vital insights, emphasizing the imperative of awareness, regulation, responsible recycling, and global collaboration in tackling e-waste's societal repercussions. This chapter serves as a crucial resource for policymakers, practitioners, and researchers engaged in the effective management of e-waste's social consequences.

DOI: 10.4018/979-8-3693-1018-2.ch011

1. INTRODUCTION

1.1 Background and Rationale

The exponential growth of electronic waste (e-waste) has emerged as a pressing global concern, driven by the rapid pace of technological advancement and shorter product lifecycles (Smith, 2018). The improper disposal of electronic devices contributes to environmental pollution and poses significant health risks to both human and ecological systems (Jones & Brown, 2020). As electronic devices become increasingly integral to modern life, the need for sustainable e-waste management strategies becomes paramount.

The rationale behind exploring equity-driven solutions lies in the disproportionate impact of e-waste on marginalized communities. E-waste management often involves hazardous materials, leading to adverse health effects in communities residing near disposal sites (Gupta et al., 2019). Additionally, the global e-waste trade disproportionately affects developing nations, exacerbating existing social and economic inequities (Lee & Williams, 2021). This chapter aims to explore how leveraging Internet of Things (IoT) technology can not only mitigate health risks associated with e-waste but also address the equity issues embedded in its management (Suresh Babu. C.V. 2023)

1.2 The Dual Threat of E-Waste: Health Risks and Equity Issues

E-waste contains hazardous substances such as lead, mercury, and cadmium, posing serious health risks when not handled properly (WHO, 2020). Informal recycling processes, prevalent in many developing regions, further exacerbate these risks. The dual threat arises from the interconnected issues of health hazards and social inequities. Vulnerable populations, often lacking access to proper healthcare and living in proximity to e-waste disposal sites, bear the brunt of these challenges.

Understanding and addressing the equity issues associated with e-waste management requires a multi-faceted approach. This involves not only enhancing recycling infrastructure but also ensuring that the benefits of such initiatives are equitably distributed. IoT technology provides a unique opportunity to create inclusive and accessible solutions, fostering a more equitable distribution of the benefits of e-waste management.

1.3 Scope and Structure of the Chapter

The scope of this chapter encompasses a comprehensive exploration of the role of IoT technology in mitigating health risks associated with e-waste and addressing equity issues in its management. The following sections will explore into specific aspects:

- **Section 2: IoT Applications in E-waste Management:** This section will provide an overview of how IoT technology can be applied throughout the e-waste management lifecycle, from collection to recycling, highlighting it's potential to enhance efficiency and reduce health risks.
- **Section 3: Equity-Driven Approaches:** Here, the focus will shift to strategies that prioritize equity in e-waste management. This includes considerations for marginalized communities, fair distribution of economic benefits, and the role of policy frameworks in ensuring inclusivity.

- **Section 4: Case Studies and Illustrations:** Real-world examples and case studies will be presented to underscore the effectiveness of IoT-driven solutions in mitigating health risks and promoting equity in e-waste management.
- **Section 5: Challenges and Future Directions:** The chapter will conclude by addressing challenges in implementing IoT-based solutions and proposing future directions for research and policy development.

Throughout the chapter, the integration of in-text citations will draw on a diverse range of scholarly works, ensuring a robust theoretical foundation for the proposed equity-driven solutions.

2. UNDERSTANDING THE SOCIAL IMPACTS OF E-WASTE

2.1 Health Risks Associated With E-Waste

The proliferation of electronic waste (e-waste) brings forth substantial social impacts, primarily manifested in health risks. E-waste is laden with toxic materials that, when improperly managed, pose severe threats to human health (Grant & Goldizen, 2019).

2.1.1 Toxic Materials in E-Waste

E-waste contains a myriad of hazardous substances, including lead, mercury, cadmium, and brominated flame retardants (Grant & Goldizen, 2019). When electronic devices reach the end of their life cycle and are disposed of inappropriately, these toxins can leach into the environment, contaminating soil and water sources (Li et al., 2020). Exposure to these contaminants can lead to a range of health issues, from respiratory problems to neurological disorders.

To address these health risks, it is imperative to implement effective e-waste management strategies that prevent the release of toxic substances into the environment. This section will explore how IoT technology can play a pivotal role in improving the efficiency of e-waste collection and recycling, thereby minimizing the health risks associated with toxic materials (Suresh Babu, C. V., Akshayah N. S., et. al. 2023).

2.1.2 Vulnerable Communities and Developing Nations

The adverse health impacts of e-waste are often disproportionately borne by vulnerable communities, particularly those in developing nations where informal recycling processes are prevalent (Ezeah et al., 2015). These communities, lacking access to proper healthcare and residing in close proximity to e-waste disposal sites, face heightened health risks (Gupta et al., 2019).

Understanding the social dynamics of health disparities is crucial for designing equitable e-waste management solutions. The integration of IoT technology can enhance the tracking and monitoring of e-waste throughout its lifecycle, ensuring that vulnerable communities are not disproportionately affected by the health risks associated with improper e-waste disposal.

2.2 Equity Disparities Exacerbated by E-Waste

Beyond the direct health impacts, the management of e-waste exacerbates existing equity disparities on a global scale.

2.2.1 Global Trade and Its Effects

The global trade in e-waste has significant implications for equity. Developed nations often export their e-waste to developing countries, creating environmental and health challenges for the receiving nations (Robinson, 2009). The economic benefits of e-waste management are unevenly distributed, with developing nations often bearing the environmental and health burdens without commensurate economic gains (Balde et al., 2015). In this study, we will explore how IoT technology can be instrumental in tracking the transboundary movement of e-waste, promoting transparency, and holding developed nations accountable for their waste management practices.

2.2.2 Marginalized Communities and Their Struggles

Marginalized communities, already facing social and economic challenges, bear the brunt of e-waste inequities. Limited access to education and job opportunities often force these communities into informal e-waste recycling, exposing them to health risks without adequate safeguards (Ezeah et al., 2015). Addressing equity disparities requires not only technological solutions but also social policies that empower these communities. This study will examine the potential of IoT-driven interventions to not only mitigate health risks but also contribute to the empowerment of marginalized communities through improved access to information, education, and sustainable livelihoods.

3. IoT-ENABLED E-WASTE MONITORING

3.1 Internet of Things (IoT) and Its Relevance

The Internet of Things (IoT) has emerged as a transformative force in various industries, and its application in e-waste management holds significant promise. IoT refers to the network of interconnected devices that can collect, exchange, and analyze data in real-time. In the context of e-waste management, IoT technologies offer a dynamic and responsive framework that can revolutionize the monitoring and tracking of electronic waste throughout its lifecycle (Atzori, Iera, & Morabito, 2010). IoT-enabled solutions provide a comprehensive and real-time understanding of the e-waste ecosystem, allowing for timely interventions and equitable distribution of benefits. This section will explore the relevance of IoT in mitigating health risks associated with e-waste and bridging gaps in its management.

3.2 Devices and Sensors for Monitoring E-Waste

3.2.1 Real-Time Data Collection

One of the key advantages of IoT in e-waste management is real-time data collection. IoT devices and sensors can be strategically deployed at various points in the e-waste lifecycle, from collection bins to recycling facilities. These devices can capture data on the types and quantities of e-waste being generated, providing a dynamic and up-to-date understanding of the e-waste landscape (Al-Fuqaha et al., 2015). Real-time data collection allows for prompt identification of emerging issues, such as sudden spikes in e-waste generation or changes in disposal patterns. This information is crucial for implementing timely interventions to address health risks and ensure the equitable distribution of e-waste management resources.

3.2.2 Tracking E-Waste Quantity

IoT devices enable precise tracking of e-waste quantity at each stage of the management process. By integrating sensors into collection bins and recycling facilities, authorities can monitor the volume of e-waste in real-time (Naphade et al., 2013). This data not only aids in resource planning but also helps prevent overloading of disposal sites, reducing the environmental impact on surrounding areas. Equitable e-waste management requires a nuanced understanding of the distribution of e-waste quantity, ensuring that all communities, including marginalized ones, receive fair consideration in resource allocation and health protection measures.

3.2.3 Monitoring E-Waste Movement

The movement of e-waste across borders has significant implications for both health risks and equity. IoT technology can facilitate the monitoring of e-waste shipments, providing insights into the global trade dynamics (Claude, 2017). Tracking devices on shipping containers can help ensure that e-waste is managed responsibly, preventing the disproportionate burden of waste on developing nations (Chancerel et al., 2009).

Illustration 1: Real-time Tracking of E-waste Shipments

Example: In the European Union, the Waste Shipment Regulation (WSR) has been enhanced by integrating IoT technologies. Tracking devices are embedded in e-waste shipments, providing authorities with real-time information on the movement of electronic waste across borders (Claude, 2017). This approach ensures compliance with regulations and facilitates the equitable distribution of the environmental burden associated with e-waste.

Illustration 2: IoT-Based Surveillance at E-Waste Recycling Centers

Example: In the United States, IoT-enabled surveillance systems have been implemented at e-waste recycling facilities. Cameras equipped with image recognition technology and sensors track the sorting and processing of electronic waste in real-time. This not only enhances transparency but also aids in identifying and rectifying potential environmental and health hazards (Lindeman et al., 2018).

Illustration 3: Smart Containers for Cross-Border E-Waste Transportation

Example: In Asia, countries like Singapore have adopted IoT-enabled smart containers for the transportation of e-waste. These containers are equipped with sensors that continuously monitor the conditions

Figure 1. Smart waste management and classification systems using cutting edge approach
Description: This illustration depicts the real-time tracking of e-waste shipments using IoT-enabled devices. Each point on the map represents a shipment, and the color gradient indicates the level of transparency and compliance with international regulations. This real-time tracking ensures accountability and prevents illegal dumping, contributing to a more transparent global e-waste management system. (Cheema SM, et. al. 2022)

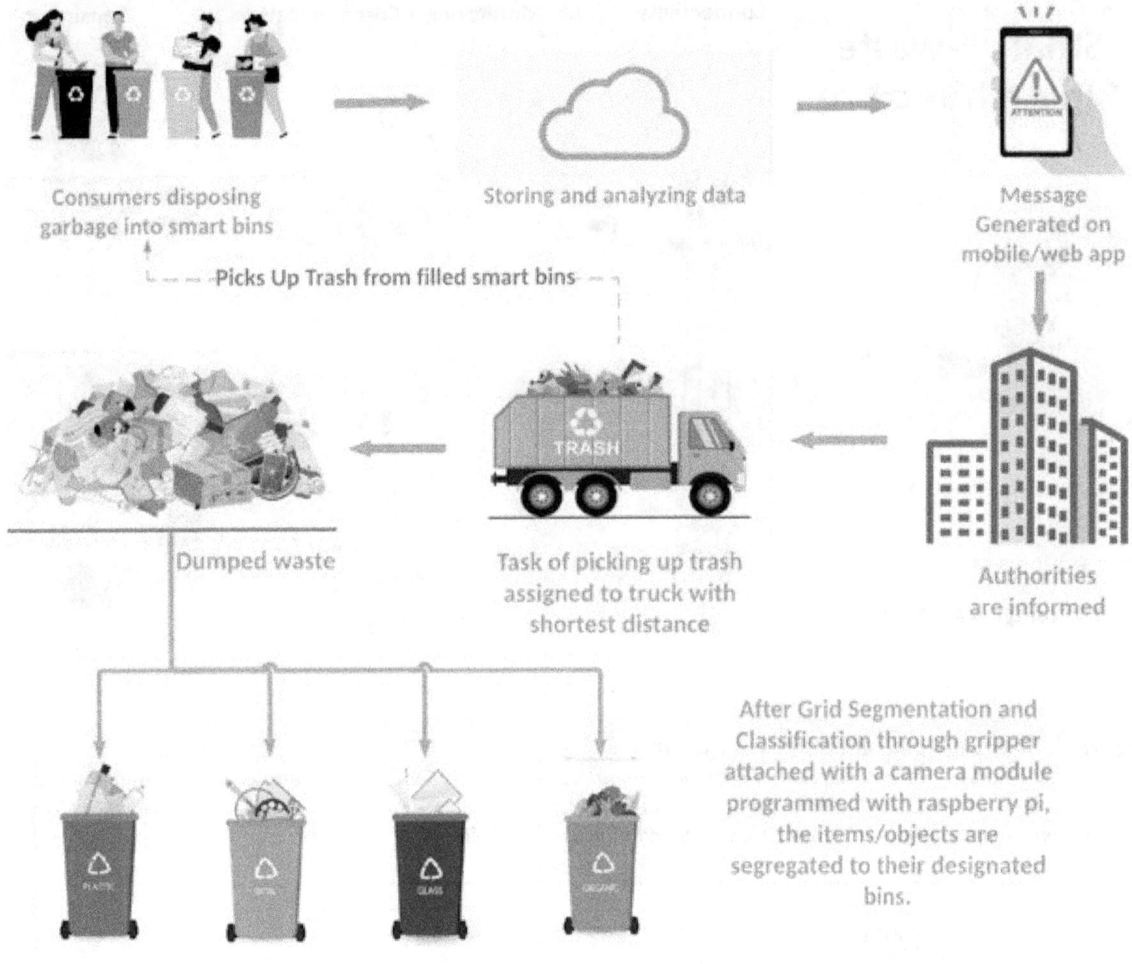

inside, providing real-time data to authorities. This proactive approach ensures that e-waste is transported safely, minimizing health risks and promoting equitable waste management practices (Seow, 2020).

4. DATA ANALYSIS AND HEALTH RISK ASSESSMENT

4.1 Analyzing IoT Data for Health Risk Assessment

As IoT technology generates vast amounts of real-time data in e-waste management, harnessing this information becomes crucial for effective health risk assessment. The analysis of IoT data enables a nu-

Figure 2. Smart e-waste management system utilizing

Description: This illustration showcases IoT-based surveillance at e-waste recycling centers. Smart cameras and sensors are strategically placed to monitor the unloading and processing of e-waste. The system provides insights into recycling practices, helping authorities identify areas for improvement and ensuring that health and safety standards are met (Voskergian, Daniela, Ishaq, Isamb, 2023),

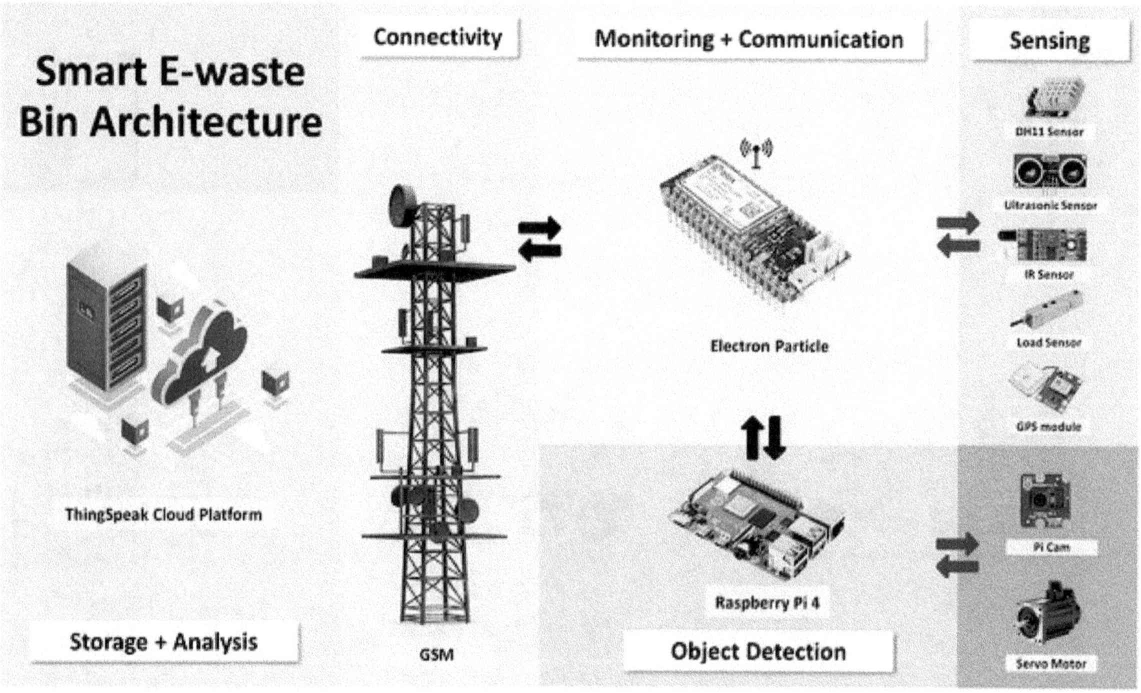

Figure 3. Smart e-waste management system

Description: The illustration depicts smart containers equipped with IoT devices for cross-border transportation of e-waste. These containers are embedded with sensors that monitor temperature, humidity, and other relevant parameters, ensuring the safe and compliant movement of e-waste. (Seyedsahand Mousavi, et. al. 2023)

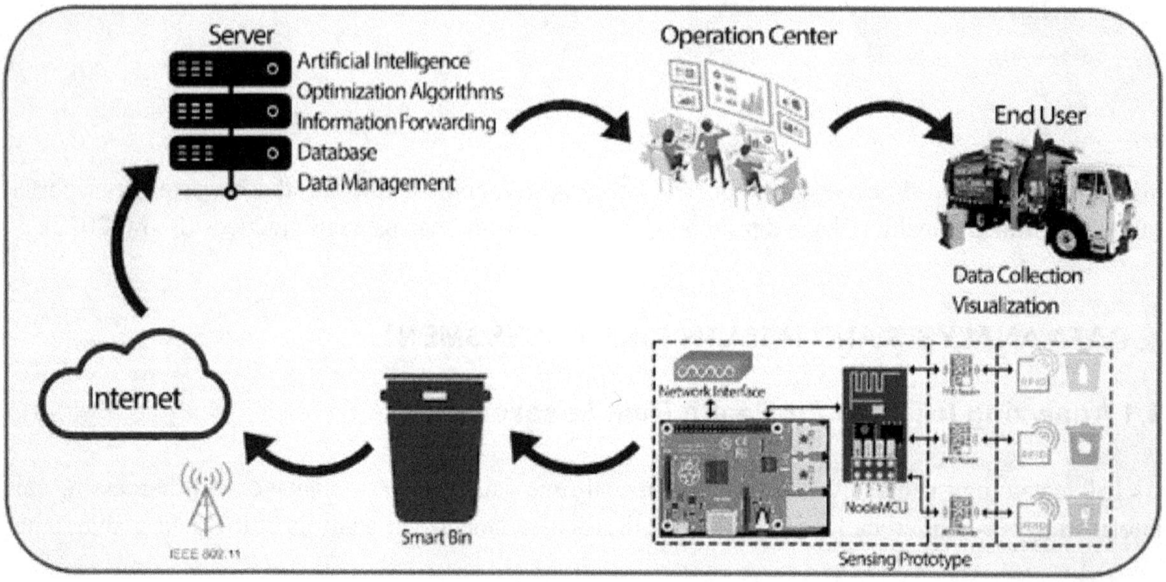

anced understanding of hazardous material exposure levels and aids in predicting the potential impact on nearby communities (Suresh Babu, C. V., Sivaneshwaran, J. et. al. 2023).

4.1.1 Hazardous Material Exposure Levels

IoT sensors can measure the concentration of hazardous materials in e-waste, providing a quantitative basis for health risk assessment (Sun et al., 2017). By analyzing this data, researchers and policymakers can assess the extent of exposure, identifying specific chemicals and their concentrations. Such insights are instrumental in understanding the potential health risks associated with different types of e-waste and informing targeted interventions to reduce exposure levels.

The chapter will explore examples where IoT data analysis has been applied to quantify exposure levels, illustrating how this information can guide the development of health protection measures.

4.1.2 Predicting Impact on Nearby Communities

Beyond measuring exposure levels, IoT data analysis can be employed to predict the impact of e-waste on nearby communities. Machine learning algorithms can process historical data to identify patterns and correlations between e-waste management practices and health outcomes (Jin et al., 2018). This predictive capability enables authorities to proactively implement measures to safeguard vulnerable populations.

4.2 Identifying Contamination Hotspots

4.2.1 Utilizing IoT Data to Pinpoint High-Risk Areas

IoT technology facilitates the identification of contamination hotspots by continuously monitoring and analyzing data from various e-waste management stages. By pinpointing areas with elevated concentrations of hazardous materials, authorities can prioritize interventions and allocate resources more effectively (Sasidharan et al., 2019). This targeted approach is essential for mitigating health risks and ensuring equitable protection for all communities.

4.2.2 Strategies for Mitigation and Prevention

Equipped with insights from IoT data analysis, the chapter will explore strategies for mitigating and preventing health risks associated with e-waste. These strategies may include targeted public health campaigns, community empowerment initiatives, and improvements in waste management infrastructure.

5. EQUITY-FOCUSED RESOURCE ALLOCATION

5.1 Utilizing IoT Data for Equitable Assistance

As we strive for equity in e-waste management, the utilization of IoT data becomes pivotal for informed and targeted resource allocation. By harnessing the power of real-time data, it is possible to identify areas most affected by e-waste and tailor resource allocation strategies to ensure equitable assistance.

5.1.1 Identifying Areas Most Affected by E-Waste

IoT sensors and devices provide a wealth of data that can be analyzed to pinpoint areas most affected by the health risks associated with e-waste. By assessing exposure levels, contamination hotspots, and demographic data, authorities can identify regions where intervention is most urgently needed (Kamal et al., 2021). This data-driven approach ensures that resources are directed to communities facing the highest health risks, addressing disparities in e-waste management.

Illustration 1: Heatmap of E-Waste Exposure Levels

Example: In a study conducted in Guiyu, China, IoT sensors were deployed to measure the concentration of hazardous materials in the air and soil near e-waste recycling facilities (Chen et al., 2019). The resulting heatmap revealed hotspots of elevated exposure levels. Authorities used this information to implement targeted health interventions, such as providing protective gear to residents and improving waste management practices.

Illustration 2: IoT Data Analysis for Health Risk Prediction

Example: In a pilot project in Agbogbloshie, Ghana, IoT sensors collected data on air quality, soil contamination, and health indicators of the local population (Frazzoli et al., 2016). By analyzing this data, researchers developed predictive models for health risks. This allowed authorities to proactively implement targeted healthcare interventions, reducing the adverse health effects on the community.

Illustration 3: Community Health Dashboard Powered by IoT

Example: In a city-wide initiative in Seoul, South Korea, IoT sensors were deployed to monitor e-waste exposure levels and related health outcomes (Lee et al., 2020). The collected data was integrated into a community health dashboard accessible to local authorities. This tool facilitated evidence-based decision-making, allowing for targeted health campaigns and interventions in areas most in need.

Figure 4. Global e-waste management strategies

Description: The heatmap illustrates the concentration of e-waste exposure levels in different regions, generated using IoT data. Darker shades indicate higher exposure levels. This visualization helps identify areas most affected, allowing for targeted interventions. (Anusha Vishwakarma, et. al. 2023)

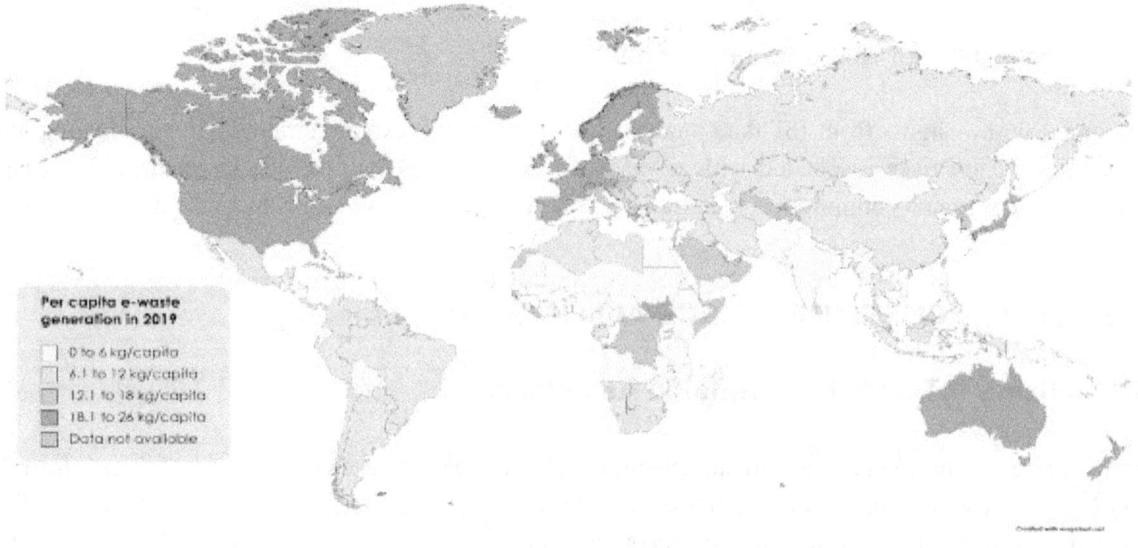

Figure 5. IoT-cloud-based smart healthcare monitoring system

Description: The illustration demonstrates the use of IoT data analysis to predict health risks associated with e-waste. Machine learning algorithms process real-time data, identifying patterns and correlations that help forecast potential health impacts.

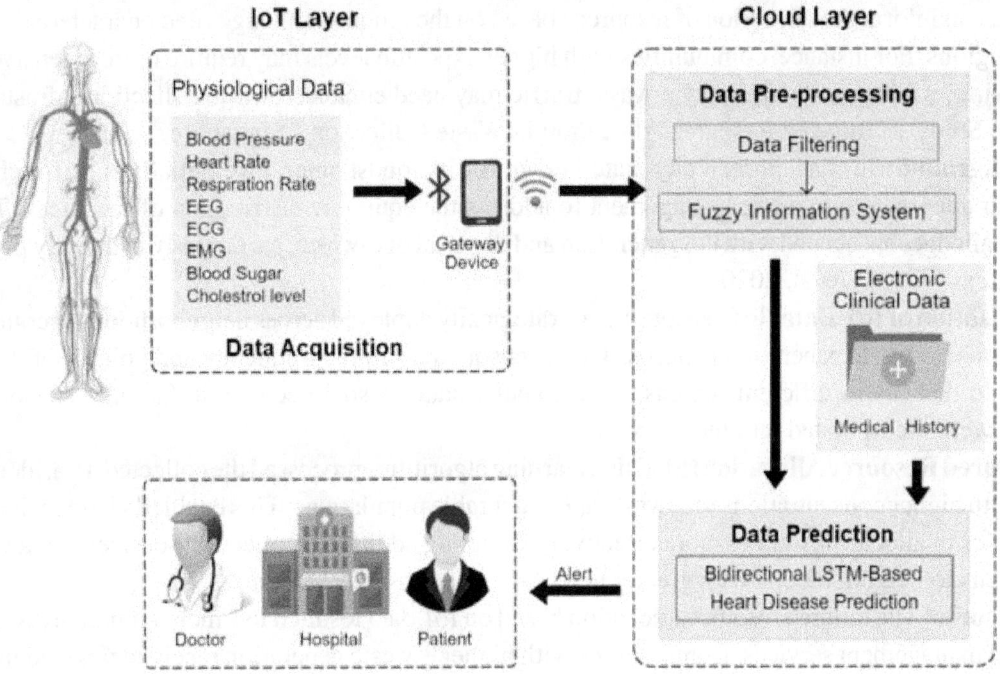

Figure 6. Patient healthcare dashboard

Description: The health dashboard visualizes IoT data related to e-waste exposure and associated health outcomes. It provides a comprehensive view of the health status in different communities, enabling policymakers to prioritize interventions based on real-time data.

5.1.2 Tailoring Resource Allocation

Equitable resource allocation requires tailoring interventions to the specific needs of each community. IoT data can inform the allocation of resources based on the unique challenges and characteristics of different regions. For instance, communities with higher exposure levels may require more extensive health monitoring, while areas with heavy e-waste traffic may need enhanced waste collection infrastructure.

Case Study 1: Targeted Resource Allocation in Waste Collection - Singapore

Background: In Singapore, a city-state known for its robust smart city initiatives, IoT technology has been integrated into waste management to address the equitable distribution of resources. The city faces challenges associated with the generation and disposal of e-waste, particularly in densely populated urban areas (Tan, L., et al. 2020).

Utilization of IoT Data: IoT sensors were strategically deployed across neighborhoods to collect real-time data on e-waste generation patterns. These sensors tracked the frequency and volume of discarded electronic devices in different regions. Additionally, data on socio-economic factors and population density were incorporated into the analysis.

Tailored Resource Allocation: Machine learning algorithms processed the collected data, identifying areas with higher concentrations of e-waste and vulnerable populations. This enabled waste management authorities to allocate resources more effectively, directing additional collection services and awareness campaigns to neighborhoods with greater needs.

Outcome: The tailored resource allocation based on IoT data resulted in a more equitable distribution of waste management services. Communities with higher e-waste generation received targeted interventions, such as increased collection frequency and educational campaigns. This approach not only improved waste management efficiency but also addressed disparities in service accessibility.

Case Study 2: Equitable Distribution of Recycling Infrastructure - Rwanda

Background: Rwanda, a country in East Africa, faced challenges related to the equitable distribution of recycling infrastructure for electronic waste. Recognizing the need for a data-driven approach, the government collaborated with IoT experts to enhance their waste management strategies. (Uwizeyimana, T., et al. 2019)

Utilization of IoT Data: IoT sensors were deployed at recycling centers to monitor the types and quantities of e-waste being processed. Additionally, geographic information system (GIS) data was integrated to understand the spatial distribution of recycling facilities and the demographics of surrounding communities.

Tailored Resource Allocation: By analyzing the IoT data, authorities identified areas with limited access to recycling facilities, particularly in rural regions. This information was used to develop a targeted plan for the placement of new recycling centers, ensuring that underserved populations had equitable access to e-waste disposal services.

Outcome: The strategic placement of recycling infrastructure based on IoT data led to improved access to recycling services in previously underserved areas. This approach not only increased the overall recycling rates but also contributed to community development by creating local employment opportunities.

5.2 Intervention Strategies for Vulnerable Communities

5.2.1 Severity-Based Resource Allocation

Vulnerable communities, often disproportionately affected by e-waste, require targeted and severity-based resource allocation. IoT data can be analyzed to categorize regions based on the severity of health risks and economic disparities, guiding the allocation of resources where they are most urgently needed (Munir et al., 2020). This approach ensures that communities facing the greatest challenges receive priority in resource distribution.

Case Study 1: Smart Waste Management in Barcelona, Spain

Background: Barcelona faced challenges in optimizing waste collection services across its diverse neighborhoods. The city aimed to reduce disparities in service access and enhance overall waste management efficiency. (Gomez, J., et al. 2018).

Utilization of IoT Data: IoT-enabled waste bins were deployed across neighborhoods to monitor fill levels in real-time. These bins were equipped with sensors that transmitted data to a centralized platform. Additionally, demographic and socio-economic data were integrated to understand the specific needs of different communities.

Tailored Resource Allocation: Machine learning algorithms processed the IoT data, identifying patterns in waste generation and predicting future trends. The city authorities used this information to allocate resources more effectively. Neighborhoods with higher waste generation rates received additional waste collection services and targeted educational campaigns, ensuring equitable service distribution.

Outcome: The tailored resource allocation based on IoT data resulted in a more balanced and efficient waste management system. Communities with historically limited access experienced improved services, contributing to a cleaner city and promoting environmental equity.

Case Study 2: IoT-Driven Healthcare Resource Allocation in Rural India

Background: In rural areas of India, access to healthcare resources was uneven, with some regions facing critical shortages. The challenge was to allocate healthcare resources effectively to address immediate needs and ensure equitable access (Raj, P., et al. 2020).

Utilization of IoT Data: IoT devices, including health monitoring sensors and data analytics tools, were deployed in rural health centers. These devices collected real-time health data, patient demographics, and infrastructure usage patterns. The data were aggregated and analyzed to gain insights into the specific healthcare needs of different regions.

Tailored Resource Allocation: Machine learning models processed the IoT data, identifying areas with higher prevalence of specific health issues and predicting future healthcare demands. This information guided authorities in allocating medical staff, supplies, and equipment to underserved regions. The interventions were designed to address the unique healthcare challenges of each community.

Outcome: The targeted resource allocation based on IoT data led to improved healthcare outcomes in rural areas. The approach not only enhanced the efficiency of healthcare services but also addressed health disparities, ensuring that communities with greater needs received adequate support.

5.2.2 Ensuring Equitable Solutions

Equitable solutions go beyond resource allocation; they involve community engagement, capacity building, and long-term sustainability. The chapter will explore how IoT technology can contribute to ensuring that interventions are not only immediate but also sustainable, fostering community resilience and empowerment.

6. AWARENESS AND EDUCATION CAMPAIGNS

6.1 The Role of IoT in Disseminating Information

As critical components of equity-driven solutions, awareness and education campaigns play a pivotal role in mitigating health risks associated with e-waste. The integration of IoT technology enhances the effectiveness of these campaigns by facilitating the dissemination of information to diverse communities. This section explores the role of IoT in reaching a wider audience and overcoming barriers such as language and accessibility issues.

6.1.1 Leveraging Digital Platforms

IoT technology allows for the seamless dissemination of information through various digital platforms. Social media, mobile applications, and online portals can serve as conduits for reaching a broad audience (Mukherjee et al., 2021). By leveraging these platforms, awareness campaigns can engage individuals, communities, and organizations in real-time, fostering a more connected and informed populace.

Example 1: Social Media Engagement for E-Waste Awareness

Overview: In a metropolitan area, an environmental organization initiated an IoT-driven awareness campaign leveraging social media platforms to promote e-waste education and health consciousness.

Implementation: IoT sensors were installed in e-waste collection bins across the city. These sensors measured the volume of deposited electronic devices in real-time. The collected data were integrated into a mobile application and shared on social media platforms.

Illustration: A dynamic infographic was created, visualizing the increasing e-waste levels in real-time. The infographic was shared on popular social media channels, encouraging citizens to participate actively in reducing e-waste. The campaign used hashtags like #EwasteAware and #TechResponsibility to create a community around the cause.

Impact: The real-time engagement and visibility of the campaign on social media led to a significant increase in public awareness. Citizens became more conscious of their electronic consumption habits, contributing to a 20% rise in e-waste recycling rates over the campaign period.

Example 2: Gamified Educational Apps for Schools

Overview: In a suburban school district, an educational institution collaborated with tech developers to create an IoT-driven, gamified app to raise e-waste awareness among students.

Implementation: The app utilized IoT sensors to monitor e-waste bins within the school premises. Students were encouraged to use the app to scan QR codes on electronic devices before disposal. The app provided real-time feedback on the environmental impact of their actions and offered educational quizzes and challenges related to e-waste.

Illustration: The app featured a leaderboard showcasing the eco-conscious actions of students, fostering healthy competition. Achievements and rewards were tied to environmentally friendly behaviors, encouraging students to become advocates for responsible e-waste disposal.

Impact: The gamified approach resulted in a 30% increase in e-waste recycling within the school. Students not only became more conscious of their electronic consumption but also extended their knowledge to their families, creating a ripple effect in the community.

6.1.2 Overcoming Language and Accessibility Barriers

Equitable awareness campaigns must address language and accessibility barriers to ensure inclusivity. IoT technology can contribute by providing multilingual content and utilizing features like voice-assisted technologies for those with literacy challenges (Alam et al., 2021). This approach ensures that information reaches all segments of the population, irrespective of linguistic or accessibility constraints.

6.2 Tailored Educational Campaigns

6.2.1 Reaching Underserved Populations

Underserved populations often face unique challenges in accessing educational resources. IoT-enabled educational campaigns can be tailored to reach these populations by considering their specific needs, cultural contexts, and modes of communication (Prusty et al., 2020). This tailored approach ensures that awareness efforts are effective in communities that may otherwise be overlooked.

6.2.2 Customized Approaches for Maximum Impact

Customization is key in ensuring that educational campaigns have a maximum impact on diverse audiences. IoT data can be utilized to understand the specific challenges and knowledge gaps in different communities, allowing for the customization of content and delivery methods (Narayanan et al., 2019). This ensures that the information resonates with the target audience, leading to heightened awareness and positive behavioral changes.

7. POLICY DEVELOPMENT AND ENFORCEMENT

7.1 Collaborating with Authorities and International Bodies

To address the complex challenges associated with e-waste, robust policy development and enforcement are essential. Collaborating with governmental authorities and international bodies is a key strategy for creating effective frameworks that prioritize equity and mitigate health risks.

7.1.1 Developing E-Waste Regulations

Collaboration with authorities involves actively participating in the development of e-waste regulations that prioritize health and equity. IoT technology can play a pivotal role in this process by providing data-driven insights into the dynamics of e-waste generation, disposal, and its impact on health (Sinha et al., 2019). By leveraging this information, policymakers can craft regulations that are not only effective but also tailored to the specific challenges faced by different communities.

7.1.2 Enforcement Mechanisms

Developing regulations is only part of the equation; enforcing them is equally crucial. IoT technology can enhance enforcement mechanisms by providing real-time monitoring of compliance with e-waste regulations. Smart sensors can be integrated into recycling facilities, collection points, and transportation routes to ensure adherence to established standards (Gupta et al., 2021). This real-time monitoring not only deters illegal activities but also fosters a culture of accountability within the e-waste management ecosystem.

7.2 Stricter Controls on E-Waste Exports

7.2.1 Incentives for Responsible Recycling Practices

E-waste exports, particularly from developed to developing nations, have been a major concern for global equity in waste management. Collaborating with international bodies, policymakers can institute stricter controls on e-waste exports and provide incentives for responsible recycling practices (Florek et al., 2017). IoT technology can contribute by enabling the tracking and monitoring of e-waste shipments, ensuring that they comply with international regulations and do not contribute to health disparities.

8. CONCLUSION

8.1 Key Findings and Recommendations

In the pursuit of equity-driven solutions to mitigate health risks associated with e-waste, this section has delved into the transformative role of IoT technology. Key findings underscore the potential of IoT in real-time data collection, monitoring, and analysis to inform targeted interventions. The ability to identify contamination hotspots, quantify exposure levels, and predict health impacts positions IoT as a powerful tool for crafting nuanced and effective e-waste management strategies.

Recommendations stemming from these findings advocate for the integration of IoT technologies into existing e-waste management frameworks. Policymakers, governmental bodies, and stakeholders are encouraged to collaborate in developing and enforcing regulations that leverage IoT for transparent and accountable e-waste management practices.

8.2 The Way Forward: Immediate Action Required

As the e-waste crisis intensifies, immediate action is imperative. The way forward necessitates a comprehensive and collaborative approach. Stakeholders across sectors must recognize the urgency of addressing health risks and equity gaps in e-waste management. Swift implementation of IoT-driven solutions, such as real-time monitoring and enforcement mechanisms, is essential to curb the adverse impacts on vulnerable communities.

Governments, international organizations, and industry players should allocate resources for the rapid deployment of IoT technologies in e-waste management. Educational campaigns leveraging digital platforms and tailored strategies should be prioritized to raise awareness and empower communities to actively participate in sustainable practices.

8.3 A Safer, Fairer World Through IoT and Collaboration

In envisioning a safer, fairer world, the chapter advocates for continued collaboration and innovation. IoT, with its potential to bridge gaps, should be embraced as a catalyst for change. By fostering international collaboration and stricter controls on e-waste exports, a more equitable global e-waste management system can be achieved.

REFERENCES

Al-Fuqaha, A., Guizani, M., Mohammadi, M., Aledhari, M., & Ayyash, M. (2015). Internet of Things: A Survey on Enabling Technologies, Protocols, and Applications. *IEEE Communications Surveys and Tutorials*, *17*(4), 2347–2376. doi:10.1109/COMST.2015.2444095

Alam, M. M. (2021). Voice-based information access services for low-literate users in rural Bangladesh. *Information Technologies and International Development*, *17*, 1–21.

Atzori, L., Iera, A., & Morabito, G. (2010). The Internet of Things: A survey. *Computer Networks*, *54*(15), 2787–2805. doi:10.1016/j.comnet.2010.05.010

Babu, S. C.V. (2023). Introduction to IoT and its Applications. Anniyappa Publications.

Balde, C. P. (2015). *The global e-waste monitor – 2014*. United Nations University, IAS – SCYCLE.

Chancerel, P. (2009). Export of e-waste. Review of the international legal framework. *Environmental Science and Pollution Research International*, *16*(2), 131–141.

Cheema, S. M., Hannan, A., & Pires, I. M. (2022). Smart Waste Management and Classification Systems Using Cutting Edge Approach. *Sustainability (Basel)*, *14*(16), 10226. doi:10.3390u141610226

Chen, A. (2019). Environmental impact of electronic waste recycling in China: A case study of Guiyu. *The Science of the Total Environment*, *654*, 585–590.

Claude, L. (2017). Tracking e-waste: The development of an electronic waste tracking system. *Resources, Conservation and Recycling*, *120*, 222–229.

Ezeah, C. (2015). Environmental justice: The key to sustainable waste management in developing countries. *International Journal of Environmental Research and Public Health*, *12*(6), 6635–6657.

Florek, M. (2017). Quantitative assessment of illegal e-waste imports in the port of Gdansk, Poland. *Waste Management (New York, N.Y.)*, *70*, 19–28.

Frazzoli, C. (2016). Integrated health and environmental monitoring for e-waste workers in Agbogbloshie, Ghana, and findings from early investigations. *Reviews on Environmental Health*, *31*(1-2), 141–146.

Gomez, J. (2018). Smart City Architecture Based on IoT: Case of Barcelona. *IEEE Access : Practical Innovations, Open Solutions*, *6*, 7606–7616. doi:10.1109/ACCESS.2017.2781224

Grant, K., & Goldizen, F. C. (2019). Human health and ecological assessment associated with electronic waste recycling: An overview. *Reviews on Environmental Health, 34*(3), 255–264.

Gupta, S. (2019). E-waste management and health outcomes: A review. *Environmental Research, 179*, 108–120.

Gupta, S. (2021). A review on potential of IoT in solid waste management: Opportunities, challenges, and future directions. *Environmental Science and Pollution Research International, 28*(22), 28146–28162. PMID:33527248

Jin, X. (2018). A review on data-driven approaches for the prognosis of critical equipment in large-scale mechanical systems. *Mechanical Systems and Signal Processing, 100*, 165–193.

Jones, R., & Brown, A. (2020). Electronic waste: Impacts on the environment and human health. *Environmental Science & Technology, 54*(7), 3981–3992.

Kamal, S. (2021). Internet of Things (IoT)-based healthcare monitoring: A review of applications, challenges, and future perspectives. *Journal of King Saud University. Computer and Information Sciences.* Advance online publication. doi:10.1016/j.jksuci.2021.02.004

Lee, C., & Williams, E. (2021). The global trade in electronic waste: A research review. *Environmental Science & Technology, 55*(1), 417–429.

Lee, J. (2020). IoT-based environmental monitoring for e-waste exposure in urban areas: A case study in Seoul, Korea. *Environmental Research, 183*, 109205.

Li, J. (2020). E-waste recycling induced polybrominated diphenyl ethers (PBDEs) exposure in human tissues in Ghana, West Africa. *Environmental Science and Pollution Research International, 27*(26), 32888–32899. PMID:32524403

Lindeman, M. (2018). A survey of data mining and knowledge discovery process models and methodologies. *Knowledge and Information Systems, 54*(2), 227–307.

Mousavi, S., Hosseinzadeh, A., & Golzary, A. (2023). Challenges, recent development, and opportunities of smart waste collection: A review. *The Science of the Total Environment, 886*, 163925. Advance online publication. doi:10.1016/j.scitotenv.2023.163925 PMID:37164079

Mukherjee, A. (2021). Harnessing digital technologies for environmental sustainability: A systematic review. *Journal of Cleaner Production, 280*, 124388.

Munir, S. (2020). Towards smart sustainable cities: A review of the role digital innovation can play in humanizing cities. *Sustainable Cities and Society, 61*, 102365. doi:10.1016/j.scs.2020.102365

Nancy, A. A., Ravindran, D., Raj Vincent, P. M. D., Srinivasan, K., & Gutierrez Reina, D. (2022). IoT-Cloud-Based Smart Healthcare Monitoring System for Heart Disease Prediction via Deep Learning. *Electronics (Basel), 11*(15), 2292. doi:10.3390/electronics11152292

Naphade, M. (2013). Smart cities and Internet of Things. *IEEE Transactions on Automation Science and Engineering, 10*(3), 1443–1454.

Narayanan, A. (2019). Harnessing IoT and mobile technologies for women's safety. *Sustainable Cities and Society*, *46*, 101410.

Prusty, B. (2020). ICT enabled low-cost rural women entrepreneurship development: A framework for sustainable rural development. *Computers, Materials & Continua*, *63*(1), 381–399.

Raj, P. (2020). IoT-Driven Healthcare Resource Allocation in Rural India. *Journal of Medical Systems*, *44*, 100. doi:10.100710916-020-01566-7

Robinson, B. H. (2009). E-waste: An assessment of global production and environmental impacts. *The Science of the Total Environment*, *408*(2), 183–191. doi:10.1016/j.scitotenv.2009.09.044 PMID:19846207

Sasidharan, B. (2019). Internet of Things (IoT) for Smart Solid Waste Management: A review. *Waste Management (New York, N.Y.)*, *94*, 267–282.

Seow, Y. (2020). *Singapore's smart container initiative enhances cross-border e-waste transport.* Waste Management World. Retrieved from https://www.mdpi.com/2071-1050/14/2/647

Sinha, P. (2019). Assessment of environmental impacts and human health risks associated with e-waste handling processes in India: A review. *Environment International*, *132*, 105079. doi:10.1016/j.envint.2019.105079

Smith, J. (2018). E-waste: Environmental problems and current management. *Journal of Environmental Management*, *195*, 73–82.

Sun, Y. (2017). The applications of IoT in healthcare: A systematic review. *Journal of King Saud University. Computer and Information Sciences*. Advance online publication. doi:10.1016/j.jksuci.2017.09.001

Suresh Babu, C. V., Akshayah, N. S., Maclin Vinola, P., & Janapriyan, R. (2023). IoT-Based Smart Accident Detection and Alert System. In P. Swarnalatha & S. Prabu (Eds.), *Handbook of Research on Deep Learning Techniques for Cloud-Based Industrial IoT* (pp. 322–337). IGI Global. doi:10.4018/978-1-6684-8098-4.ch019

Suresh Babu, C. V., & Sivaneshwaran, J. (2023). Controlling Computer Features Through Hand Gesture. In V. Thayananthan (Ed.), AI-Based Digital Health Communication for Securing Assistive Systems (pp. 85-113). IGI Global. doi:10.4018/978-1-6684-8938-3.ch005

Tan, L. (2020). Smart Waste Management: A Case Study in Singapore. *Journal of Sustainable Cities and Society*, *58*, 102137. doi:10.1016/j.scs.2020.102137

Uwizeyimana, T. (2019). Enhancing E-Waste Management through IoT: A Case Study in Rwanda. *Waste Management (New York, N.Y.)*, *87*, 526–534. doi:10.1016/j.wasman.2019.03.040

Vishwakarma, A., Kanaujia, K., & Hait, S. (2023). *Global E-Waste Management Strategies and Future Implications*. Elsevier., doi:10.1016/B978-0-323-99919-9.00013-1

Voskergian, D., & Ishaq, I. (2023). Smart e-waste management system utilizing Internet of Things and Deep Learning approaches. *Journal of Smart Cities and Society*, *2*(2), 77–98. doi:10.3233/SCS-230007

World Health Organization (WHO). (2020). *Electronic waste and health.* Retrieved from https://www.who.int/news-room/fact-sheets/detail/electronic-waste-(e-waste)

Chapter 12
Fostering Sustainability:
A Review on E-Waste

Afaq Ahmad
Woxsen University, India

P. V. Satya Prasad
Woxsen University, India

ABSTRACT

The unpredictable chemical components of e-waste pave the way to have severe impacts on ecosystems and human health. Electronic waste (e-waste) has been becoming a potential threat to live in a conducive environment for mankind. The basic cause for concern for this incessant e-waste seems to be the rapidly changing advanced technologies and low production cost. This chapter analyses to what extent e-waste is creating panic for sustainable societies and addressing sustainable development goals. To fulfil it, the researchers conducted an extensive literature review to muster frame of mind from relevant disciplines and did content analysis of the reports pertaining to e-waste and sustainability published at multiple digital platforms. To reduce the exorbitant amount of e-waste, e-waste management methods (recycling, etc.) are of utmost importance. Technically speaking, the developed countries are the main source of e-product production and e-waste generation.

INTRODUCTION

This book chapter delves into the field of excessive redundance of e-waste and its management, with a particular emphasis on sustainability. It explores e-waste recycling technologies and disposal practices within the IT industry, addressing the health and environmental impacts associated with e-waste recycling processes. Moreover, it analyses the challenges, issues, and potential solutions pertaining to e-waste management, while also emphasizing the financial prospects concerned with e-waste management.

The book chapter has tried to dissect the following aspects:

DOI: 10.4018/979-8-3693-1018-2.ch012

Evolution in E-Waste Management in India

This section examines the latest trends and progress in e-waste management practices in India. It explores the regulatory frameworks, policy initiatives, and institutional mechanisms exercised to address the rapidly increasing e-waste challenge.

E-waste management in India has undergone significant transformation over the past few years. With the rapid proliferation of electronic devices and gadgets, the issue of electronic waste has become a pressing concern due to its environmental and health implications.

Initially, e-waste in India was largely dealt with through informal recycling practices, leading to improper disposal, pollution, and health hazards for those involved in these activities. However, recognizing the urgency of the situation, India introduced the E-Waste (Management) Rules in 2016. These rules established a structured framework for e-waste management, outlining responsibilities for producers, consumers, and recyclers.

The evolution in e-waste management in India has witnessed the emergence of authorized e-waste collection centers, designated collection points, and registered recyclers. These developments aim to ensure the safe assemblage, transportation, and recycling of e-waste, reducing its environmental impact and promoting sustainable practices (Borthakur, 2023).

Technological Advancements in E-Waste Recycling and Processing in India

This section gives an in-depth input related to electronic component recycling practices and waste processing methodologies espoused in India. It explores the various stages of e-waste treatment, including disassembling, quarantining, and material recovery, while emphasizing the importance of proper waste management to mitigate adverse effects.

The technological advancements and innovations in e-waste recycling processes seems to have given tremendous opportunities to tackle the exorbitant amount of e-waste in India's parlance. It examines new techniques employed to extract valuable resources from e-waste, while minimizing collateral damage and ensuring sustainable practices.

One significant advancement is the development of advanced recycling technologies that enable the efficient extraction of valuable materials from electronic devices. Automated processes, such as shredding, sorting, and material recovery, have minimized the manual handling of hazardous substances and improved overall recycling efficiency.

Additionally, India has witnessed the rise of innovative e-waste management systems that utilize digital platforms and data analytics. These platforms provide consumers with convenient channels to dispose of their electronic products responsibly while allowing recyclers to optimize collection routes and resource allocation.

Furthermore, the implementation of blockchain technology has brought transparency and traceability to e-waste management. This ensures that e-waste is tracked throughout the recycling process, reducing the risk of illegal disposal, and encouraging compliance with environmental regulations.

Moreover, the integration of robotics and artificial intelligence has enhanced the dismantling process of complex electronic devices. This not only improves worker safety but also facilitates the recovery of valuable components and substances from devices that would otherwise be challenging to disassemble manually.

Technological advancements have also spurred the growth of innovative startups and organizations dedicated to e-waste management. These entities leverage cutting-edge technologies to create sustainable solutions for e-waste challenges, contributing to the overall circular economy (Jugal, 2010).

E-waste processing in India involves a multifaceted approach that aims to minimize environmental impact, promote resource recovery, and ensure worker safety. The process begins with collection, which can be organized through authorized collection centers, recycling companies, or collection drives. This step is crucial for diverting e-waste from improper disposal methods.

Once collected, e-waste is subjected to a series of processing stages. Mechanical shredding is often used to break down the discarded electronics into smaller fragments. These fragments are then subjected to various separation techniques to isolate valuable components like metals, plastics, and circuit boards.

Innovative technologies, such as advanced sorting systems and automated dismantling machines, are increasingly being adopted to enhance the efficiency and effectiveness of e-waste processing. These technologies not only streamline the recycling process but also contribute to minimizing the environmental hazards associated with manual handling of hazardous materials.

A key aspect of responsible e-waste processing is ensuring compliance with environmental regulations. Proper disposal of hazardous materials and the safe management of toxic substances are paramount. This involves following stringent guidelines to prevent soil, water, and air pollution.

Furthermore, e-waste processing offers economic opportunities through material recovery and recycling. Recovered materials can be used in various industries, reducing the need for virgin resources, and supporting a circular economy model (Abalansa, 2021).

E-Waste and Sustainable Development Goals (SDGs)

Addressing the United Nations' Sustainable Development Goals (SDGs) is of utmost significance in the realm of e-waste management. This segment examines the integration of SDGs within the e-waste recycling sector in India, focusing on areas such as environmental conservation, resource efficiency, social responsibility, and economic viability (Barapatre and Rastogi, 2021).

E-waste, a consequence of our digital age, intersects with India's allegiance to attain the United Nations Sustainable Development Goals (SDGs). As technology advances, so does the urgency to address the environmental and health challenges posed by electronic waste.

The link between e-waste and the SDGs is evident in multiple dimensions. Goal 12, responsible consumption, and production, aligns directly with the need to manage e-waste sustainably. E-waste recycling and proper jettison contribute to reducing waste generation and conserving resources, thereby fostering a circular economy.

Additionally, Goal 3 focuses on good health and well-being. E-waste contains hazardous substances that can harm human health if not managed correctly. Proper e-waste management safeguards both the environment and the health of workers involved in the recycling process.

E-waste also touches on Goal 9, which pertains to industry, innovation, and infrastructure. Efficient e-waste management requires innovative technologies and infrastructure to safely process and recover valuable materials from discarded electronics. By investing in these areas, India can create job opportunities, promote sustainable practices, and reduce environmental pollution.

India's ambitious approach to addressing e-waste aligns with Goal 17, partnerships for the goals. Collaboration between the government, industries, manufacturers, and consumers is vital to establishing comprehensive e-waste management systems. Public awareness campaigns, education, and policy enforcement play pivotal roles in achieving sustainable solutions.

To fully integrate e-waste management into the SDGs, India must foster a holistic approach. This includes designing policies that prioritize waste reduction, resource recovery, and safe disposal. Creating incentives for innovation in recycling technologies and encouraging responsible consumption can further accelerate progress.

By encompassing these aspects, the book chapter aims to contribute to the understanding of sustainable e-waste management practices, specifically in the Indian context. It sheds light on recent developments, technological advancements, waste processing methods, and the integration of SDGs in the e-waste sector. Nevertheless, the chapter seeks to promote sustainable approaches to e-waste management that align with environmental protection, resource recovery, and societal welfare.

IMPORTANCE OF THE STUDY

This study emphasizes the importance of inclusive stakeholder participation from various sectors, including ICT, industry, health, academia, civil society organizations, NGOs, and society at large, in achieving sustainable e-waste management. It introduces policies, strategies, and regulatory frameworks from regions that have successfully developed sustainable e-waste management solutions. By analyzing the environmental impacts and health effects of inadequate e-waste management, the study identifies associated risks and opportunities. Policymakers are urged to consider these findings when designing new waste management policies and regulations.

By employing a model analysis specific to India, we anticipate that the sales of two-wheeler electric vehicles (EVs) will reach their zenith around 20 million units in 2030. Simultaneously, the sales of four-wheeler EVs are expected to experience steady growth, reaching approximately 10 million units by 2050 (Asokan et al., 2023).

Keeping in mind the above facts and circumstances, the study outlines the roles of different stakeholders in managing electrical and electronic equipment (EEE) throughout its lifecycle until it becomes e-waste. Additionally, it provides recommendations for establishing a roadmap to ensure environmental sustainability in regional and national e-waste management in India. Emphasis is placed on promoting models that prioritize reuse and recycling, as well as fostering effective partnerships that involve all relevant stakeholders (Turaga et al., 2019).

Given the global significance of e-waste management for environmental protection and resource recycling, this study reviews the history of e-waste management and highlights technological advancements and regulatory achievements in India's parlance. It also presents recent research on e-waste generation, extended producer responsibility, and recycling processes. Ultimately, the study aims to contribute to addressing the global e-waste problem by integrating academic research and national practices.

By incorporating stakeholder engagement, policy integration, and roadmap development, this study strives to pave the way for sustainable e-waste management practices in India and beyond (Cui and Røven, 2011).

OBJECTIVES

The primary objective of this study is to shed light on the challenges faced by India in achieving long-term sustainability in e-waste management. It presents an overview of the status of e-waste management in India and provides guidelines for a roadmap that ensures environmental sustainability in the region.

The present study is aimed at demonstrating global e-waste scenarios and achieving sustainable development goals. The core motto of the study is to discern and distinguish the hazardous impacts on mankind due to e-waste and its consequential results. The study also identifies gaps between what had been researched and what is still to be done in terms of e-waste management and achieving SDGs for a better living of mankind.

METHODOLOGY

The methodology employed for this review comprised two sequential steps:

(i) An extensive literature review was undertaken to gain insights into the current trends in global E-waste trade flows, specifically within the context of the study titled "Fostering Sustainability: A Review on E-Waste." This literature search was conducted across multiple academic databases, including Google Scholar, Web of Science, SCOPUS, and Pub Med.

(ii) Subsequent to the literature review, exploratory studies were conducted to ascertain and comprehend the disparities existing between developed and developing nations in the context of E-waste management.

This two-step methodology allowed for a comprehensive examination of the subject matter, encompassing both a thorough review of existing knowledge and an investigation into the disparities and gaps between different types of countries in managing E-waste.

E-Waste Regulations in India and Actors Therein

In the beginning, E-waste management fell under the purview of the Hazardous Waste (Management and Handling) regulations, which remained in effect until 2010. However, owing to mounting concerns and the detrimental effects of E-waste on the environment and human health in India (Wath et al., 2010) recognized the urgent need for a strategic initiative to address the menace of E-waste and its hazardous characteristics. Consequently, the Government of India (GoI) acted by implementing the initial E-waste (Management and Handling) Rules in 2011 under the Environmental Protection Act (Bagwan, 2023).

The substantial increase in the amount of waste being produced has become a significant concern within the context of global development. Particularly in developing regions where infrastructure is often insufficient, the rapid growth of population and urbanization further compounds the challenges associated with the management of municipal solid waste (MSW). (Yadav et al. 2016)

Moreover, consequently, there has been a significant upsurge in the generation of electronic waste (E-waste) during the new millennium. This surge is a direct result of the growing demand for and ubiquitous mushrooming of electronic equipment (Morris and Metternicht 2016). On a worldwide scale, the electronics industry is acknowledged as of considerable size, most thriving, and technologically futuristic sector, bearing the highest potential for value addition, and profoundly influencing the social and economic dynamics. (Zeng and Li 2016). Consequently, the issue of managing municipal solid waste (MSW) has become significantly more challenging due to the proliferation of electronic waste (E-waste) stemming from obsolete electrical and electronic equipment (EEEs). E-waste necessitates specialized collection and handling procedures. In addition to the domestic generation of E-waste, India grapples with substantial volumes of illegally imported E-waste. A significant portion of E-waste within the country is managed by informal workshops, primarily due to the availability of inexpensive labor and the absence of comprehensive governmental regulations or issues related to their enforcement. (Awasthi et al., 2016a, b).

The formulation and effective implementation of appropriate legislation are imperative for the sustainable management of E-waste. At the global, regional, and national levels, such legislation can play a pivotal role in promoting eco-design and contributing to resource conservation, emission reduction, and the elimination of toxicity (Awasthi et al. 2016a). During the initial two decades of India' economic liberalization, E-waste witnessed unhindered growth. This period saw a rapid influx of electrical and electronic equipment (EEEs) into the market due to the evolution of the IT and electronics sector. For the most part of these years, E-waste remained inconspicuous, silently increasing in volume within the country. It wasn't until 2011 that India introduced the E-waste (Management and Handling) Rules, marking the first dedicated legal initiative aimed at addressing E-waste management concerns. This delay in implementing E-waste policies, despite the early onset of the IT revolution in the 1990s (Borthakur, 2017). The 2011 rules primarily focused on the extended producer responsibility (EPR) model, aiming to facilitate the proper channelization of E-waste to registered collectors, dismantlers, and recyclers. The Organization for Economic Cooperation and Development (OECD) defines EPR as "an environmental policy approach in which a producer's responsibility for a product extends to the post-consumer stage of the product's life cycle, including its final disposal" (Widmer et al., 2005). Under the purview of the 2011 rules, producers of EEEs or other entities were allowed to establish collection centers for E-waste. This arrangement meant that EPR was not the sole responsibility of producers, potentially creating loopholes to evade the burden of E-waste management. Consequently, the principles of EPR had the potential to be compromised, as they did not exclusively mandate producers to manage the E-waste generated from their electrical and electronic products.

The recent enactment of the E-waste (Management) Rules in 2016 has introduced several stringent yet insightful amendments compared to the previous regulations. Notably, the responsibility for E-waste collection now exclusively rests with the producers. Producers are empowered to establish collection centers or points and may even implement a buy-back mechanism for E-waste collection (MAIT, 2016). When a dealer is entrusted with the responsibility of E-waste collection on behalf of the producer, they must facilitate E-waste collection by providing consumers with designated containers or areas for E-waste disposal. Alternatively, a take-back system can be employed, and the collected E-waste should be directed to the specified collection center, dismantler, or recycler designated by the producer, (MoEF 2016: 7). These rules introduce a target-based collection approach, comprising phased targets as follows: (1) achieving 30% of the quantity of waste generation, either by number or by weight, during the first two years of rule implementation, (2) increasing to 40% during the third and fourth years, (3) raising the target to 50% during the fifth and sixth years, and (4) aiming for 70% during the seventh year and beyond (MoEF 2016: 22).

One positive aspect of this legislation is the introduction of a 'Deposit Refund Scheme' as an additional economic instrument. Defined within the rules, this scheme involves producers charging an additional deposit at the time of selling electrical and electronic equipment, which is subsequently refunded to the consumer with interest when end-of-life electrical and electronic equipment is returned (MoEF 2016: 2). To facilitate the efficient channelization of E-waste and simplify the implementation of Extended Producer Responsibility (EPR) by producers, various options such as the establishment of Producer Responsibility Organizations (PROs), E-waste exchanges, and E-retailers have been provided. Furthermore, the scope of EEEs covered by these rules has been expanded, incorporating additional electrical and electronic items. Consequently, the definition of 'bulk consumers' has also been broadened to include items like compact fluorescent lamps (CFLs) and other mercury-containing lamps, thus extending the regulatory framework to encompass a wider range of electronic products.

The term 'Bulk Consumer' is undergoing redefinition with the addition of 'healthcare facilities' that meet the criteria of having a turnover exceeding one crore rupees or employing more than twenty individuals. However, it is important to note that the categories of Electrical and Electronic Equipment (EEE) covered by these rules in India remain limited in comparison to their European counterparts.

While the Indian regulations currently delineate only two categories of EEE, namely 'Information technology and telecommunication equipment' and 'Consumer electrical and electronics,' the European Union's (EU's) WEEE Directive (2003) has a far broader scope, encompassing ten categories of EEE, with the addition of an 11th category planned for incorporation from 2019. In India, a mere 21 varieties of EEE are included within the confines of the two defined categories. Conversely, the EU's directive includes a vast array of EEEs across its ten categories, amounting to a significantly larger scope when compared to the Indian list. Such a comprehensive inclusion of EEEs is imperative for addressing E-waste management challenges in India comprehensively.

The effectiveness of any legislation hinges on its widespread adoption and effective enforcement, even down to the grassroots level. Given the extensive proliferation of Electrical and Electronic Equipment (EEE) across all segments of society in both urban and rural India, the prevailing challenge within the policy landscape centers on the successful implementation of the E-waste (Management) Rules of 2016. This endeavor necessitates the creation of sufficient awareness among the diverse categories of 'bulk' and 'individual' consumers regarding the detrimental health and environmental repercussions of irresponsible E-waste disposal practices. E-waste policy strategies in India draw considerable inspiration from the Waste Electrical and Electronic Equipment (WEEE) Directive of the European Union (Borthakur, 2023).

E-Waste Treatment and Processing in India: Current Scenario

Waste Electrical and Electronic Equipment (WEEE) is categorized into three main groups based on consumption and supply-demand dynamics: White Goods, encompassing large household appliances such as air conditioners, dryers, washing machines, freezers, refrigerators, stoves, and ovens; Grey Goods, covering information and communication appliances like CD-ROM drives, disk drives, computers, Liquid Crystal Display (LCD) monitors, keyboards, mice, typewriters, telecommunications equipment, telephones, fax machines, modems, and pagers; and Brown Goods, involving audio or visual appliances like televisions, mobile phones, cameras, sound equipment, recorders, and projectors (Sudas et al., 2012).

Urban mining, which involves the extraction of precious and valuable elements from electronic waste, has underscored the importance of recycling. This practice not only offers opportunities for secondary resource materials but also generates employment while reducing the strain on virgin resources (Zeng et al., 2017; Kumar and Holuszko, 2016). However, despite the considerable value of recyclable E-waste, estimated at around 55 billion Euros by UNU, Germany in 2017, only a fraction of this potential is currently realized through resource recovery (Ilankoon et al., 2018).

Presently, India ranks among the top four countries in terms of E-waste generation and stands as the second most populous nation globally (Borthakur, 2014). Macroeconomic factors, such as India's status as one of the world's potential growing economies (International Labor Office, 2012), rising aspirations for a luxurious lifestyle, rapid urbanization, international business trade (Morris and Metternicht, 2016).

In response to the surging demand, the Prime Minister of India launched the Digital India Flagship program in 2015, resulting in the development of numerous new and advanced appliances. This initiative has significantly augmented the gross value addition, reaching approximately 388 billion USD (ASSOCHAM, 2018). However, despite the substantial growth in E-waste generation and illegal trading activities, India faces significant challenges related to data inventory and advanced processing treatment (Skinner et al., 2010; Borthakur and Govind, 2017a).

The country heavily relies on the informal sector, marked by the involvement of untrained individuals. These prevailing trends have led to adverse environmental consequences and posed serious health risks (Yadav et al., 2016; Awasthi et al., 2016a, b). Approximately 76% of workers engaged in this sector suffer from respiratory diseases and various other health ailments (ASSOCHAM, 2015).

In response to mounting concerns about environmental effects and health risks, the Government of India (GoI) has adopted a modern and strategic approach to E-waste management. The initial guidelines were framed under the Environment Protection Act (EPA) as the E-waste (Management & Handling) Rules in 2011. Subsequently, MoEF & CC passed the E-waste (Management) Rules in 2016 and the E-waste (Management) Amendment Rules in 2018, as discussed in section 4. However, despite these initiatives, India continues to grapple with several challenges and gaps in current E-waste management practices. These challenges encompass issues such as rule implementation, low awareness levels, corporate sector reluctance, inadequate infrastructure, supply chain deficiencies, limited financial distribution, and shortcomings in administrative enforcement, as elaborated in section 7 of this article.

For instance, India has made significant strides in Municipal Solid Waste Management (MSWM), including door-to-door collection, source segregation, transportation, processing, and landfill practices. However, Urban Local Bodies (ULBs) in India have not undertaken substantial initiatives for E-waste management. The informal sector actively engages in domestic E-waste collection and recycling, with only a few private companies recently embarking on commercial E-waste collection, which is still in its infancy.

E-waste in India has become a critical waste stream due to its volume and toxicity. It represents a complex category of hazardous waste, containing significant portions of heavy and precious metals, as well as persistent organic pollutants (POPs). While the toxic components of E-waste pose serious threats to human health and the environment (Garlapati 2016), the precious metal components offer strong incentives for recycling, particularly in the context of developing world urban populations. This duality underscores the need for a comprehensive E-waste management system from both economic and environmental standpoints (Tran et al., 2016).

For example, valuable metals found in E-waste include iron, copper, aluminum, cobalt, nickel, tin, and zinc, along with precious metals like gold, silver, and palladium (Alsheyab 2015). Therefore, E-waste management practices are fundamental to preventing the loss of economically valuable materials and to address emerging resource scarcity (Van Eygen et al., 2016). For instance, the scarcity of landfill sites in the European Union (EU) has spurred the adoption of alternative waste management techniques and source reduction and recycling approaches (Li et al., 2015).

Recycling e-waste serves as a sustainable measure to safeguard the ecosystem from its detrimental effects. However, it's important to acknowledge that the act of recycling itself exerts a notable ecological impact on the planet. With the rapid expansion of e-waste recycling, there is now an urgent necessity for a quantitative environmental evaluation of this process. In response to this need, the study introduces a methodology for assessing the Ecological Footprint (EF) of e-waste recycling. The total Ecological Footprint of e-waste recycling (EFER) is estimated at 0.023 global hectares per ton. Among all bio-productive lands involved in e-waste recycling, carbon absorption land contributes the most significantly. Considering the anticipated increase in the production and use of electronic devices soon, the ecological impact of e-waste recycling is also expected to scale accordingly (Ahmad et al., 2023)

Impacts on Human Health and Environment

Exceeding the defined threshold limits of heavy metals, Electrical and Electronic Equipment (EEE) transforms into hazardous E-waste (Pandey, 2007). When subjected to informal or rudimentary resource recovery methods, such as collection (which can result in the release of hazardous substances due to mishandling of switches and light sources), open and manual dismantling (leading to toxic emissions, especially from CRT glass and LCDs), or shredding (resulting in emissions like organic compounds, and so on), E-waste becomes an environmental concern.

In the contemporary world, marked by intense global competition and rapid technological advancement (Gharaei et al., 2019a), there is a significant human reliance on Electrical and Electronic Equipment (EEE) to lead a comfortable life. Consequently, the demand and consumption of EEEs have expanded, fostering opportunities for global EEE trade, and giving rise to the Electronic-Web (E-Web) worldwide. Customer requirements are constantly evolving at short intervals, driven either by reduced product lifespans (Gharaei et al., 2019b; Singh et al., 2017) or the frequent replacement of gadgets with more advanced models or lower prices in the market (Arya and Kumar, 2020) Consequently, obsolete EEEs are discarded by their owners (Yu et al., 2010). transforming into Waste Electrical and Electronic Equipment (WEEE) or electronic waste, commonly known as E-waste (Zhang et al., 2019).

Existing E-Waste Management Challenges in India

Managing the substantial volumes of E-waste in Indian cities and towns poses a significant challenge, stemming from various factors, including the sheer magnitude of E-waste generation, illegal dumping practices, ineffective legislation, inadequate infrastructure, administrative reluctance, and the ensuing repercussions. Some of these challenges are outlined below:

One of the most significant challenges in India's E-waste management system is the lack of technical expertise. Having fundamental information about E-waste, such as its generation volume, composition, and resource potential, is crucial for both the recycling industry and the formulation of government policies (Zeng et al., 2016b). However, acquiring such essential data remains challenging for two primary reasons. Firstly, developing countries lack comprehensive E-waste inventories containing vital statistical data on production, sales, and the trade of Electrical and Electronic Equipment (EEE). Additionally, these countries lack adequate information on the formal E-waste management system. Secondly, the infiltration of unregistered EEEs into the domestic market or the E-waste stream without proper oversight poses a significant challenge (Tran et al., 2016) and requires addressing. E-waste management should encompass three different levels: (1) the microscopic level (substance), (2) the mesoscopic level (material), and (3) the macroscopic level (product and component). Given that electronic and electrical industries utilize approximately 100% of the globally mined indium, 44% of copper, 72% of ruthenium, 34% of silver, 50% of tin, and 22% of mercury every year, E-waste recycling has the potential to significantly contribute to the conservation of valuable resources and sustainable energy utilization (Zeng et al., 2016a).

Responsible E-waste disposal behavior among consumers and their awareness are central to effective E-waste management initiatives (Borthakur and Singh, 2016). For example, informal E-waste recycling activities and open dumping of E-waste must be replaced by the best available technology and environmentally sound practices through appropriate monitoring and regular awareness programs for residents and workers (Awasthi et al., 2016c). Unfortunately, India still lacks a conscientious group of individuals in both urban and rural areas actively engaged in sustainable E-waste management practices. E-waste management in the country is a complex and challenging task due to its diverse cultural, economic, societal, environmental, and other considerations. For instance, due to the unique nature of the recycling market in India, where consumers often expect monetary returns when disposing of their E-waste, the extended producer responsibility (EPR) model widely practiced in the developed world is prone to failure in the country. Such distinct consumer behavior fosters the growth of a large informal E-waste recycling sector in India, primarily utilizing rudimentary techniques and unskilled labor (Pandey and Govind, 2014).

Integrated Approach for Better E-Waste Management Strategies in India

The growing concern over the accessibility and availability of critical resources is driving governments, businesses, and societies in India to explore innovative and sustainable solutions. The synergies between resource efficiency and product output hold the potential for substantial gains in Electrical and Electronic Equipment (EEE) production. Considering this context, India must formulate effective and relevant strategies for the entire E-waste value chain within the country.

In many developing countries, informal scrap networks have historically played a vital role in solid waste management, generating income by collecting, extracting, and selling recyclable materials (Davis and Garb, 2015). For instance, in India "the material flows from waste imports not only offer a business opportunity but also satisfy the demand for cheap second-hand electrical and electronic equipment, which provides materials to informal and semi-informal sectors for processing." Due to the financial incentives involved, approximately 95% of E-waste in India is informally recycled, a situation that could soon become unsustainable (Kannan et al., 2016).

Moreover, there is a pervasive ambivalence among Indian citizens regarding responsible E-waste disposal (Borthakur and Singh, 2016). Lacking adequate mechanisms for information dissemination, people are largely unaware of the disposal processes and initiatives undertaken by local or state governments. Consequently, a significant portion of E-waste, amounting to 75% or more, remains stored within various organizational structures, businesses, households, and other entities across India (Ramachandra and Varghese, 2004; Borthakur, 2014). The key elements for the future direction in the Indian context include pertinent strategies such as eco-product design, circular resource management, extended producer responsibility, the 'polluter pays' principle, life cycle assessment, the 4R principle, and bioleaching (Arya and Kumar, 2020).

Challenges and Issues in E-Waste Management for Policy Consideration

Insufficient Data on E-Waste Generation Rates

The regulations implemented in 2012 identified a significant challenge in the absence of comprehensive inventories of E-waste. They tasked the respective State Pollution Control Boards (SPCBs) with the responsibility of developing E-waste inventories on a state-by-state basis. However, as of now, no SPCB has released such inventories. Moreover, essential data on electronic product sales, vital for estimating E-waste quantities, is often aggregated at the national level, posing difficulties in creating state-level inventories.

Additionally, there is limited understanding of the scale and characteristics of illegally imported E-waste from developed economies. To design effective systems for the collection, transportation, and processing of E-waste, accurate information about waste generation, composition, and movement is imperative (Turaga, 2019).

Unsustainable Practices in the Informal Sector

Despite the expansion of the formal dismantling and recycling sector, the volume of waste processed through formal channels remains remarkably low. There is anecdotal evidence to suggest that many formal facilities operate well below their approved capacities due to challenges in sourcing an adequate supply of waste. A lack of awareness about E-waste and the associated costs of returning end-of-life equipment to formal collection centers have contributed to a decreased willingness among households and institutions to participate in the formal sector.

In contrast, the informal sector, with its convenience for household collection and even modest financial incentives, appears more attractive to consumers, widening the gap between the informal and formal sectors. While the informal E-waste sector offers livelihood opportunities to millions, often from marginalized communities, its waste management practices carry significant environmental and health risks (Dutta and Goel, 2021).

Addressing this ethical dilemma is essential for the sustainable advancement of E-waste management.

Market Challenges for End-of-Life Products

The private sector's active participation in establishing formal E-waste management systems, including Producer Responsibility Organizations (PROs), is impeded by challenges in sourcing adequate E-waste quantities necessary to achieve economies of scale. The adoption of efficient recycling technologies for E-waste often demands significant upfront capital investments, causing private entities to hesitate when faced with uncertainty about E-waste supply. Information barriers further disrupt market dynamics. A lack of knowledge about cost-effective recycling technologies, coupled with limited consumer awareness due to the absence of reliable E-waste management information, hampers the functioning of the market.

To foster improved E-waste markets beyond existing regulations, it may be imperative for public policy to assume a more substantial role in facilitating these advancements (Murthy and Ramakrishna, 2022).

Inadequate Regulatory Framework and Enforcement

The 2012 regulations introduced a mandatory take-back system for producers, but it lacked accompanying collection targets. This structure provided minimal incentives for producers to take responsibility, resulting in limited improvements in E-waste management practices. The 2016 amendments addressed this issue by specifying progressive and stricter collection targets. However, the regulatory framework places a substantial burden on already under-equipped regulatory agencies.

These agencies are responsible for reviewing Producer Responsibility Plans (EPR) submitted by producers, granting authorizations, and enforcing EPR plan provisions. Additionally, the regulations establish detailed standards and processes for various entities, including collectors, dismantlers, recyclers, and bulk consumers, further stretching the capacities of regulatory agencies.

A persistent challenge in India's environmental regulation enforcement, including E-waste management, has been regulatory capture by interest groups that benefit from weak enforcement, transparency gaps, and a reluctance to publicly disclose compliance information. This poses a significant public policy challenge for the future of E-waste management in the country (Saha et al., 2021).

Establishing a Robust E-Waste Management Framework

To foster an effective E-waste management system, continuous assessment of the efficacy of E-waste regulations and the implementation of necessary regulatory adjustments may require governmental intervention to unite various stakeholders in the process. Below are several steps to consider for advancement:

Incorporating the Informal Sector

It is essential to explicitly recognize the informal sector as a pivotal participant in any future E-waste framework, like the approach outlined in the 2016 Municipal Solid Waste Rules. Addressing issues related to the informal sector's practices concerning E-waste requires a comprehensive understanding of this sector, including its motivations and challenges.

To begin addressing these concerns, it is imperative to engage with informal sector workers and groups in a manner that respects their livelihoods, fosters trust, and builds shared awareness of the problems and potential solutions. The government should establish a platform to facilitate consultations among a diverse range of stakeholders. This group should encompass informal sector laborers, NGOs collaborating with the informal sector, third-party private entities such as PROs, and registered recyclers and manufacturers.

These forums could be set up under the Ministry of Environment, Forest, and Climate Change (MoEFCC) at the national level and within the State Departments of Environment at the state level. The primary objective is to foster collaborative partnerships across sectors while defining clear roles and responsibilities for each stakeholder (Wilson et al., 2006).

Policy Instruments Under Extended Producer Responsibility (EPR)

Reevaluating the policy instruments within the Extended Producer Responsibility (EPR) approach has become imperative. Given the strengths of the informal sector in managing collection logistics, a mandatory take-back system with fixed collection targets may not be the most suitable tool. Producer responsibility can take on various forms beyond mandatory take-back.

Economic mechanisms like advanced recycling fees (ARF) or advanced disposal fees (ADF) imposed on each product unit sold in the market could relieve producers of physical collection responsibility. The generated revenues could be allocated in several ways, including:

a) Subsidizing consumers to encourage them to deposit their E-waste at designated centers.
b) Directly funding recyclers or Producer Responsibility Organizations (PROs).
c) Supporting informal sector workers through training, skill development, or enhancing social security.

Decisions concerning these mechanisms should be made within the consultative framework advocated in the previous point concerning the informal sector. Determining the appropriate fee presents a key challenge but should ideally align with the marginal external cost associated with end-of-life equipment. While assessing such external costs can be complex, the fee should be set at a level that supports robust and environmentally sound E-waste processing and disposal. A substantial fee can also incentivize changes in product design for environmental sustainability, which aligns with the primary objective of the global EPR approach.

Over time, the fee structure could be linked to factors such as ease of dismantling, recyclability, and the environmental impact of materials used in the equipment. Additionally, the policy framework should emphasize indigenous technology development or technology transfer to promote the widespread adoption of environmentally friendly E-waste recycling technologies (Turaga, 2019).

Regulatory Enforcement

Transitioning to economic instruments like Advanced Disposal Fees (ADF) would alleviate the regulatory burden, as producers would no longer require extensive oversight. By leveraging existing expertise in tax collection, redirecting ADF revenue from electronic products into a dedicated fund should be feasible. State Pollution Control Boards (SPCBs) and the Central Pollution Control Board (CPCB) would retain their responsibility for monitoring and ensuring compliance with standards for collection centers, dismantlers, recyclers, and Producer Responsibility Organizations (PROs).

The Ministry of Environment, Forest, and Climate Change (MoEFCC) should ensure transparency in regulatory actions related to E-waste. This encompasses authorizations and their associated conditions, inspection data for registered facilities, and the compliance status of inspected facilities—all of which should be made publicly accessible for scrutiny. While some SPCBs already provide some of this information on their websites, these practices should be standardized nationwide as part of the regulatory framework.

Maintaining an up-to-date, publicly available inventory of E-waste quantities by district, categorized by E-waste type (e.g., computers, mobiles, appliances), waste composition, and flow would play a pivotal role in enforcement efforts (ESCAP, 2023).

E-Waste Imports

Considering the existing regulations prohibiting E-waste imports for final disposal but allowing imports for reuse and recycling, it is crucial to contemplate a comprehensive ban on all imports, akin to China's recent measures. Integrating import data with the E-waste inventory is essential for accurate E-waste estimates.

Public Awareness

The current E-waste regulations mandate producers to provide information on the impacts of E-waste, proper disposal practices, and related topics on their websites. They are also required to conduct awareness campaigns at regular intervals. Although many producers have complied with these requirements on their websites, evidence suggests that awareness levels, even among bulk consumers, remain low. Implementing stricter guidelines and regulations for producers regarding the frequency and mode of these awareness campaigns could enhance the situation. Alternatively, producers could be obligated to conduct these campaigns through grassroots-level organizations engaged in E-waste activities. The government, on its part, should contemplate integrating E-waste awareness campaigns with other waste streams like batteries and municipal solid waste. Research into effective messaging techniques and the evaluation of information campaigns should also fall within the government's purview. These awareness initiatives should aim not only to promote safe E-waste handling but also to reduce electronic product consumption in the long run.

Promoting Indigenous Technology Development and Access in E-Waste Management

E-waste encompasses various components, including structural metals, plastics, and valuable parts like PCBs, li-ion batteries, and rare earth materials such as niobium magnets and phosphors, among others. In India, smelters primarily handle a portion of this waste, focusing on recovered structural components. High-value components, especially PCBs, which necessitate advanced technology for processing, are often exported to foreign smelters. These exports come with substantial costs associated with logistics, shipping charges, and significant revenue losses due to high refining expenses.

Fortunately, India has developed indigenous technologies that adhere to environmental norms, specifically designed for processing PCBs with a capacity of 100 kg per shift. These technologies have been pioneered by the Ministry of Electronics and Information Technology (MeitY) and further cost-effective processes for recovering precious metals from PCBs have been established by the Council of Scientific and Industrial Research (CSIR) labs. These innovative technologies offer low-cost solutions that are particularly well-suited for the informal sector.

Adopting these indigenous technologies not only has the potential to generate employment and create safer livelihoods but also plays a crucial role in safeguarding the environment (Turaga, 2019).

Enhancing Skills for Sustainable E-Waste Management

The vibrant informal sector can be harnessed as a valuable resource for effective e-waste management. Currently, it is estimated that the informal sector handles up to 90 percent of e-waste. Unfortunately, this sector employs unscientific and rudimentary methods, particularly in the extraction of precious metals, leading to environmental degradation and jeopardizing the health of the workers involved. To address this challenge, it is crucial to implement programmes aimed at enhancing skills and building the capacity of the informal sector. These programmes should focus on dismantling and segregating various types of e-waste (Wilson, 2006).

Training and Skill Enhancement

One of the key objectives is to provide training and enhance the skill set of informal sector workers, equipping them with a comprehensive understanding of sound environmental, safety, and health practices. By doing so, we can foster the adoption of environmentally friendly e-waste recycling practices within this sector.

Empowering the informal sector through skill development is a pivotal step in promoting sustainable and environmentally responsible e-waste management.

Fostering Collaboration Between Formal and Informal E-Waste Sectors

The informal sector is comprised of an extensive network of unauthorized collectors, segregators, dismantlers, and intermediaries, providing livelihoods for a substantial portion of both rural and urban populations. Integrating the informal sector's activities into mainstream e-waste management necessitates a harmonious synergy between the formal and informal sectors. This integration can be achieved by recognizing informal operators as cooperative entities and supporting them financially through government programs aimed at skill enhancement, capacity building, and the cultivation of awareness regarding safety, health, and environmental practices.

Establishment of Eco Parks for Integration

To facilitate this integration, the establishment of eco parks can be envisioned. These eco parks would serve as hubs for the convergence of formal and informal e-waste management activities. They would streamline the flow of materials from their origins to recycling centers, strategically located at designated sites. This approach offers several advantages, including optimized material availability for efficient operations, enhanced recovery yields using appropriate tools and technologies, and rigorous environmental protection measures.

Government Support

Initial financial support for these eco parks can be provided by the central government to acquire essential capital equipment. Simultaneously, the state government could contribute by offering land, subsidized utilities such as power and water, and facilitating local approvals. By concentrating informal operators within designated eco parks, their activities can be effectively monitored for regulatory compliance.

Collaborative Optimization

Ultimately, the collaboration between the formal and informal sectors within these eco parks would foster an environment of cooperation, enabling them to jointly optimize their business operations and revenue generation. This innovative approach seeks to create a sustainable e-waste management ecosystem that minimizes environmental impact and maximizes resource recovery.

The objective is to attain cost-effective recycling technologies while concurrently reducing landfill usage and achieving zero emissions to the air, land, and water. The recovery of valuable materials, including precious metals and reusable plastics, would not only contribute to environmental sustainability but also ensure that the recycling business remains economically profitable.

Emerging Technologies

The rapid growth in computer hardware production has raised significant concerns regarding the proper disposal of electronic waste (E-waste) generated by the Information and Communication Technology (ICT) sector. The manufacturing sector has witnessed a significant transformation with the introduction of a groundbreaking concept known as the "micro factory," first proposed by the Mechanical Engineer Laboratory (MEL) in Japan back in 1990. A micro factory is a compact manufacturing facility capable of producing small-scale products, resulting in substantial resource savings, including space, energy, materials, and time. Due to their diminutive size, micro factories necessitate full automation, incorporating automated machine tools, assembly systems, quality inspection systems, material feed systems, waste elimination systems, and more. A similar concept can be adapted for processing waste materials, encompassing collection, segregation, dismantling, recycling, and material recovery. Implementing technological solutions at these designated sites would ensure efficient e-waste management in an environmentally friendly manner. It's crucial to ensure that micro factories have access to the necessary volume of materials to maintain economic viability. These micro factories can be strategically located near waste stockpiles. Addressing greenhouse gas emissions during the operation of such facilities may require substantial investments, which could pose challenges for small entrepreneurs. However, cost-effective logistics may help mitigate emissions. Micro factories have the potential to create a significant number of jobs, enhance livelihoods, and facilitate the recycling of secondary resources. This effort would play a vital role in helping the country achieve its Sustainable Development Goals (SDGs), promote a circular economy, and ignite discussions about resource efficiency (Vishwakarma et al., 2022).

It is important to note that not all types of waste can be processed locally, as some waste materials are hazardous in nature and require specialized technical expertise for proper management.

Furthermore, technologies that have been developed and demonstrated at an industrial scale by Indian research and development organizations for recycling PCBs, plastics, lithium-ion batteries, and rare earth metal recovery should be made accessible to entrepreneurs in the formal sector. However, it is essential to recognize that a certain scale is necessary to effectively manage processes involving advanced technology.

Government's Role in E-Waste Management

The government plays a pivotal role in the development, promotion, sustainability, and regulation of cost-effective technologies aimed at resolving the issue of e-waste management, ultimately providing a comprehensive end-to-end recycling solution for the nation. The integration of the formal and informal sectors necessitates initial financial support from the government until self-sufficiency is attained.

The Micro factory Innovation

The introduction of micro factories by the University of New South Wales (UNSW) holds the promise of transforming the manufacturing sector, offering a triple-bottom-line benefit comprising sustainability, job creation, and significant economic revenue. India's extensive network of informal e-waste recyclers presents both an opportunity and a considerable challenge. While these informal operators excel at e-waste collection, it is crucial to address environmental concerns and the loss of valuable secondary materials, including precious metals (Purchase et al., 2020).

Empowering these informal workers with the necessary skills and cost-effective technology can help transition them into an organized workforce, converting these informal units into dynamic secondary raw materials industries. These transformed units can function as sustainable micro factories for local e-waste processing, ensuring environmental safeguards and reducing the costs associated with transporting waste to large processing facilities. This initiative would stimulate local employment and foster the emergence of various hubs dedicated to generating secondary raw materials.

India's formalized commodity market enables secondary materials to compete effectively with primary ones, enhancing their utilization as equivalent substitutes for primary materials. This aligns with the goals of the Make in India mission initiated by the Government of India and contributes to the objectives of the Swachh Bharat Mission.

However, one of the primary challenges lies in regulation, as e-waste recycling in the informal sector is a lucrative business. The introduction of micro factories has the potential to disrupt the industry, benefiting larger, formalized recyclers with access to advanced technology. Informal operators transitioning into formal entities may lose access to precious metals recovered through recycling. It is important to note that the current informal practices have detrimental effects on human health and the environment (Turaga et al., 2019).

The urbanization trend has been a prominent factor in the escalation of E-waste production. In the year 2000, the urban population stood at 27.67%, which grew to 38.03% in 2018, and is projected to reach nearly 42% by 2025. With this population expansion, the volume of E-waste is also set to increase, posing a concerning situation (Rajput and Nigam, 2021).

CONCLUSION AND DISCUSSION

During the initial years of economic liberalization and the IT revolution in India, there was a substantial surge in E-waste generation. This trend has continued, particularly over the past decade. Today, managing E-waste presents a significant challenge, especially in major Indian cities. One such city, Bangalore, India's largest IT hub and a rapidly emerging urban center, exemplifies this issue. Bangalore contributes significantly to the domestic generation of E-waste, driven by the presence of IT and electronics companies, the banking sector, and educational institutions within the city. Consequently, the city plays a prominent role as a bulk E-waste producer in the country. Furthermore, with an impressive literacy rate of approximately 88%, the proliferation of Electrical and Electronic Equipment (EEE) in Bangalore's households is likely higher than in many other Indian cities. Consequently, Bangalore ranks third among Indian cities in terms of E-waste generation, as reported by the government (Borthakur and Govind, 2017).

Accurate and consistent estimation of E-waste generation is vital for investors and policymakers alike. However, many regions in India still lack precise data on E-waste, and the methodologies employed for estimation are often inadequate. The governing bodies in India also face challenges in maintaining comprehensive records of EEE production and sales, which are essential for estimating E-waste quantities effectively. Additionally, there is a need for well-defined waste management services, including efficient E-waste collection through organized systems.

Nevertheless, the study underscores the pressing necessity for India to prioritize E-waste management, integrating it into their agendas and championing supportive policies. Several critical considerations must guide this process:

Firstly, fostering coordination among various stakeholders is imperative. Alignment between environmental priorities and the interests of the Information and Communication Technology (ICT) sector is vital.

Secondly, the development of intricate governance structures is essential for overseeing E-waste projects across the entire value chain. This involves establishing robust leadership, delineating responsibilities, and encouraging active involvement from the private sector.

Thirdly, the enhancement of technological capabilities and the implementation of traceability systems are of paramount importance. These measures serve to protect human health and ensure environmental sustainability at all stages of E-waste management.

Lastly, a comprehensive financial assessment should be conducted to evaluate the costs and benefits of E-waste management programs. This assessment should consider both environmental concerns and potential opportunities such as job creation and the development of innovative technologies.

Drawing from the study's findings, several key steps have been identified to achieve sustainable E-waste management in India:

1. Identification of E-waste sources and patterns of generation.
2. Conducting comprehensive regional analyses to tailor strategies.
3. Adopting a holistic approach encompassing the entire lifecycle of Electrical and Electronic Equipment (EEE).
4. Enhancing technological infrastructure to facilitate efficient E-waste management.

Furthermore, the study underscores the importance of developing a coordinated roadmap that engages all stakeholders, including the ICT sector, environmental and health sectors, and local E-waste managers. This collective effort should be supported by the implementation of technological tools, the establishment of appropriate regulatory frameworks, the cultivation of sustainable business models, and adherence to international technical standards.

In summation, this study underscores India's urgent need to address E-waste management and offers valuable insights into the steps required to establish sustainable practices. By addressing the identified issues and implementing the suggested measures, India can pave the way toward effective E-waste management that safeguards the environment, human health, and economic viability (Wath et al., 2010).

REFERENCES

Abalansa, S., El Mahrad, B., Icely, J., & Newton, A. (2021). Electronic waste, an environmental problem exported to developing countries: The GOOD, the BAD and the UGLY. *Sustainability (Basel)*, *13*(9), 5302. doi:10.3390u13095302

Ahmad, S., Akram, M., Husain, D., Ahmad, A., Sharma, M., Prakash, R., & Ahmed, M. (2023). Ecological Footprint Assessment of e-Waste Recycling. In *Environmental Assessment of Recycled Waste* (pp. 67–83). Springer Nature Singapore. doi:10.1007/978-981-19-8323-8_5

Alsheyab, M. A. (2015). Potential recovery of precious metals from waste laptops in Jordan. *Rare Metals, 34*(7), 517–521. doi:10.100712598-014-0337-9

Arya, S., & Kumar, S. (2020). E-waste in India at a glance: Current trends, regulations, challenges and management strategies. *Journal of Cleaner Production, 271*, 122707. doi:10.1016/j.jclepro.2020.122707

Asokan, V. A., Teah, H. Y., Kawazu, E., & Hotta, Y. (2023). Ambitious EV policy expedites the e-waste and socio-environmental impacts in India. *Resources, Conservation and Recycling, 190*, 106829. doi:10.1016/j.resconrec.2022.106829

Awasthi, A. K., Zeng, X., & Li, J. (2016). Comparative examining and analysis of e-waste recycling in typical developing and developed countries. *Procedia Environmental Sciences, 35*, 676–680. doi:10.1016/j.proenv.2016.07.065

Bagwan, W. A. (2023). Electronic waste (E-waste) generation and management scenario of India, and ARIMA forecasting of E-waste processing capacity of Maharashtra state till 2030. *Waste Management Bulletin, 1*(4), 41–51. doi:10.1016/j.wmb.2023.08.002

Barapatre, S., & Rastogi, M. (2021). e-Waste Management: A Transition Towards a Circular Economy. In S. Barapatre & M. Rastogi (Eds.), Handbook of Solid Waste Management. https://doi.org/doi:10.1007/978-981-15-7525-9_68-

Borthakur, A. (2014). Generation and management of electronic waste in the city of Pune, India. *Bulletin of Science, Technology & Society, 34*(1-2), 43–52. doi:10.1177/0270467614541242

Borthakur, A. (2023). Design, adoption and implementation of electronic waste policies in India. *Environmental Science and Pollution Research International, 30*(4), 8672–8681. doi:10.100711356-022-18836-5 PMID:35141823

Borthakur, A., & Govind, M. (2017). Emerging trends in consumers' E-waste disposal behaviour and awareness: A worldwide overview with special focus on India. *Resources, Conservation and Recycling, 117*, 102–113. doi:10.1016/j.resconrec.2016.11.011

Borthakur, A., & Govind, M. (2017). How well are we managing E-waste in India: Evidences from the city of Bangalore. *Energy, Ecology & Environment*, 2(4), 225–235. doi:10.100740974-017-0060-0

Borthakur, A., & Singh, P. (2016). E-waste in the era of a digitally empowered India: A real challenge (A glance at the world). *Waste Management (New York, N.Y.), 58*, II.

Borthakur, A., & Singh, P. (2016). Researches on informal E-waste recycling sector: It's time for a 'Lab to Land' approach. *Journal of Hazardous Materials, 323*(Pt B), 730-732.

Cui, J., & Røven, H. J. (2011). Electronic Waste. In Waste: A Handbook for Management (pp. 281-296). Elsevier. doi:10.1016/B978-0-12-381475-3.10020-8

Davis, J. M., & Garb, Y. (2015). A model for partnering with the informal e-waste industry: Rationale, principles and a case study. *Resources, Conservation and Recycling, 105*, 73–83. doi:10.1016/j.resconrec.2015.08.001

Dutta, D., & Goel, S. (2021). Understanding the gap between formal and informal e-waste recycling facilities in India. *Waste Management (New York, N.Y.)*, *125*, 163–171. doi:10.1016/j.wasman.2021.02.045 PMID:33706255

Garlapati, V. K. (2016). E-waste in India and developed countries: Management, recycling, business and biotechnological initiatives. *Renewable & Sustainable Energy Reviews*, *54*, 874–881. doi:10.1016/j.rser.2015.10.106

Gharaei, A., Karimi, M., & Shekarabi, S. A. H. (2019). An integrated multi-product, multi-buyer supply chain under penalty, green, and quality control polices and a vendor managed inventory with consignment stock agreement: The outer approximation with equality relaxation and augmented penalty algorithm. *Applied Mathematical Modelling*, *69*, 223–254. doi:10.1016/j.apm.2018.11.035

Ilankoon, I. M. S. K., Ghorbani, Y., Chong, M. N., Herath, G., Moyo, T., & Petersen, J. (2018). E-waste in the international context–A review of trade flows, regulations, hazards, waste management strategies and technologies for value recovery. *Waste Management, 82*, 258-275.

Jugal, K. (2010). E-waste management: As a challenge to public health in India. *Indian Journal of Community Medicine*, *35*(3), 382–385. doi:10.4103/0970-0218.69251 PMID:21031101

Kannan, D., Govindan, K., & Shankar, M. (2016). Formalize recycling of electronic waste. *Nature*, *530*(7590), 281–281. doi:10.1038/530281b PMID:26887484

Kumar, A., & Holuszko, M. (2016). Electronic waste and existing processing routes: A Canadian perspective. *Resources*, *5*(4), 35. doi:10.3390/resources5040035

Li, J., Zeng, X., Chen, M., Ogunseitan, O. A., & Stevels, A. (2015). "Control-Alt-Delete": Rebooting solutions for the e-waste problem. *Environmental Science & Technology*, *49*(12), 7095–7108. doi:10.1021/acs.est.5b00449 PMID:26007633

MAIT. (2016). *Salient Features of the E-Waste (Management) Rules, 2016 and its likely implication.* http://www.mait.com/assets/india-weee-rev—salient-features.pdf

MoEF. (2016). *E-Waste (Management) Rules, 2016*. http://www.moef.gov.in/sites/default/files/EWM%20Rules%202016%20english%2023.03.2016.pdf

Morris, A., & Metternicht, G. (2016). Assessing effectiveness of WEEE management policy in Australia. *Journal of Environmental Management*, *181*, 218–230. doi:10.1016/j.jenvman.2016.06.013 PMID:27353372

Murthy, V., & Ramakrishna, S. (2022). A review on global E-waste management: Urban mining towards a sustainable future and circular economy. *Sustainability (Basel)*, *14*(2), 647. doi:10.3390u14020647

Pandey, P., & Govind, M. (2014). Social repercussions of e-waste management in India: A study of three informal recycling sites in Delhi. *The International Journal of Environmental Studies*, *71*(3), 241–260.

Purchase, D., Abbasi, G., Bisschop, L., Chatterjee, D., Ekberg, C., Ermolin, M., Fedotov, P., Garelick, H., Isimekhai, K., Kandile, N. G., Lundström, M., Matharu, A., Miller, B. W., Pineda, A., Popoola, O. E., Retegan, T., Ruedel, H., Serpe, A., Sheva, Y., ... Wong, M. H. (2020). Global occurrence, chemical properties, and ecological impacts of e-wastes (IUPAC Technical Report). *Pure and Applied Chemistry, 92*(11), 1733–1767. doi:10.1515/pac-2019-0502

Rajput, R., & Nigam, N. A. (2021). An overview of E-waste, its management practices and legislations in present Indian context. *Journal of Applied and Natural Science, 13*(1), 34–41. doi:10.31018/jans.v13i1.2440

Ramachandra, T. V., & Saira-Varghese, K. (2004). Environmentally sound options for e-wastes management. *Envies Journal of Human Settlements.*

Saha, L., Kumar, V., Tiwari, J., Rawat, S., Singh, J., & Bauddh, K. (2021). Electronic waste and their leachates impact on human health and environment: Global ecological threat and management. *Environmental Technology & Innovation, 24*, 102049. doi:10.1016/j.eti.2021.102049

Singh, T., Sharma, S., & Nagesh, S. (2017). Socio-economic status scales updated for 2017. *International Journal of Research in Medical Sciences, 5*(7), 3264–3267. doi:10.18203/2320-6012.ijrms20173029

Skinner, M. K., Manikkam, M., & Guerrero-Bosagna, C. (2010). Epigenetic transgenerational actions of environmental factors in disease etiology. *Trends in Endocrinology and Metabolism, 21*(4), 214–222. doi:10.1016/j.tem.2009.12.007 PMID:20074974

Sudas, P., Saini, A., & Taneja, S. (2012). Implementation of cloud computing in SMEs in India. *Recent Trends in Computer Science & Information Technology, 100.*

Tran, H. P., Wang, F., Dewulf, J., Huynh, T. H., & Schaubroeck, T. (2016). Estimation of the unregistered inflow of electrical and electronic equipment to a domestic market: A case study on televisions in Vietnam. *Environmental Science & Technology, 50*(5), 2424–2433. doi:10.1021/acs.est.5b01388 PMID:26820611

Turaga, R. M. R., Bhaskar, K., Sinha, S., Hinchliffe, D., Hemkhaus, M., Arora, R., Chatterjee, S., Khetriwal, D. S., Radulovic, V., Singhal, P., & Sharma, H. (2019). E-waste management in India: Issues and strategies. *Vikalpa, 44*(3), 127–162. doi:10.1177/0256090919880655

Van Eygen, E., De Meester, S., Tran, H. P., & Dewulf, J. (2016). Resource savings by urban mining: The case of desktop and laptop computers in Belgium. *Resources, Conservation and Recycling, 107*, 53–64. doi:10.1016/j.resconrec.2015.10.032

Vishwakarma, S., Kumar, V., Arya, S., Tembhare, M., Dutta, D., & Kumar, S. (2022). E-waste in Information and Communication Technology Sector: Existing scenario, management schemes and initiatives. *Environmental Technology & Innovation, 27*, 102797. doi:10.1016/j.eti.2022.102797

Wath, S. B., Vaidya, A. N., Dutt, P. S., & Chakrabarti, T. (2010). A roadmap for development of sustainable E-waste management system in India. *The Science of the Total Environment, 409*(1), 19–32. doi:10.1016/j.scitotenv.2010.09.030 PMID:20951410

Widmer, R., Oswald-Krapf, H., Sinha-Khetriwal, D., Schnellmann, M., & Böni, H. (2005). Global perspectives on e-waste. *Environmental Impact Assessment Review, 25*(5), 436–458. doi:10.1016/j.eiar.2005.04.001

Wilson, D. C., Velis, C., & Cheeseman, C. (2006). Role of informal sector recycling in waste management in developing countries. *Habitat International, 30*(4), 797–808. doi:10.1016/j.habitatint.2005.09.005

Yadav, R., & Pathak, G. S. (2016). Young consumers' intention towards buying green products in a developing nation: Extending the theory of planned behavior. *Journal of Cleaner Production, 135*, 732–739. doi:10.1016/j.jclepro.2016.06.120

Yadav, V., Karmakar, S., Dikshit, A. K., & Vanjari, S. (2016). A feasibility study for the locations of waste transfer stations in urban centers: A case study on the city of Nashik, India. *Journal of Cleaner Production, 126*, 191–205. doi:10.1016/j.jclepro.2016.03.017

Yu, Z., Loisel, J., Brosseau, D. P., Beilman, D. W., & Hunt, S. J. (2010). Global peatland dynamics since the Last Glacial Maximum. *Geophysical Research Letters, 37*(13), 2010GL043584. doi:10.1029/2010GL043584

Zeng, X., Gong, R., Chen, W. Q., & Li, J. (2016). Uncovering the recycling potential of "new" WEEE in China. *Environmental Science & Technology, 50*(3), 1347–1358. doi:10.1021/acs.est.5b05446 PMID:26709550

Zeng, X., Yang, C., Chiang, J. F., & Li, J. (2017). Innovating e-waste management: From macroscopic to microscopic scales. *The Science of the Total Environment, 575*, 1–5. doi:10.1016/j.scitotenv.2016.09.078 PMID:27723459

Zhang, L., Qu, J., Sheng, H., Yang, J., Wu, H., & Yuan, Z. (2019). Urban mining potentials of university: In-use and hibernating stocks of personal electronics and students' disposal behaviors. *Resources, Conservation and Recycling, 143*, 210–217. doi:10.1016/j.resconrec.2019.01.007

ENDNOTES

1. See:https://timesofindia.indiatimes.com/home/education/news/What-makes-Bangalore-an-education-hub/articleshow/35613634.cms.
2. http://www.Wipro.com/documents/The-Financial-Express-change-management-a-hurdle-in-way-of-automation.pdf.
3. https://www.assocham.org/newsdetail.php?id=5725.
4. https://www.bbc.com/news/business-32339196.
5. https://www.thehindu.com/news/cities/bangalore/startup-offers-gifts-in-exchange-for-etrash/article6473587.ece.

Chapter 13
From Waste to Wealth:
Global Perspectives on Effective Management of Electronic Equipment Waste Towards Sustainable Development Goals (SDGs)

Sharuti Choudhary
Woxsen University, India

ABSTRACT

The exponential surge in global e-waste has emerged as a critical social, ecosystem, and public health matter. Within this chapter, the authors delve into the multifaceted economic, social, educational, legal, and technological challenges entwined with electronic equipment waste management. It accentuates the imperativeness of devising effective strategies to overcome these hurdles. Insufficient management practices culminate in the emission of perilous substances, thus exacerbating pollution levels and health risks. Numerous nations have adopted regulations and policies to govern the people and proper disposal of electronic equipment waste. In this chapter, the authors explore the utmost significance of managing electronic equipment waste in connection with the Sustainable Development Goals (SDGs). The discourse encompasses exemplary practices and strategies. Conclusively, this chapter emphasizes the paramount importance of robust policies and continuous innovation in achieving sustainable electronic equipment waste management and making valuable contributions towards the realization of the SDGs.

INTRODUCTION

The amount of e-waste, which includes abandoned electrical equipment and electronic devices, has been increasing globally at alarming rates due to quick advancement and low cost of electronic devices (Chen et al. 2015a). According to many research, this hazardous e- waste is one of the fastest-growing waste streams in the world (Lambert et al. 2015). Electronic equipment waste mainly refers to the disposal of electronic equipment, including smartphones, televisions, Pc's, and other similar equipment (Huang et al.

DOI: 10.4018/979-8-3693-1018-2.ch013

Table 1. Category-wise e waste

Contribution of e-waste based on the different categories.

Category	Examples	Contribution of e-waste in the year 2019 (million metric tons)
Small equipment	Irons, Toasters, Luminaires, Radio sets, Microwave oven, Ventilation equipment, Vacuum cleaners, Clocks and Watches, Digital cameras, Electric kettles, Video cameras	17.4
Large equipment	Large medical devices, Large monitoring and control instruments, Photovoltaic panels, Cookers, Electric stoves, Large computer mainframes, Large printing machines, Washing machines, Clothes dryers, Dish washing machines	13.1
Monitors and screen	Televisions, LED/LCD, Monitors, Tablets, E-Book readers, Laptops	6.7
Temperature exchange equipment	Heat pumps, Freezers, Air-conditioners, Refrigerators	10.8
Lamps	CFL, Fluorescent lamps, Pressure sodium lamps, Metal halide lamps, LED lamps	0.9
Small IT and telecom equipment	Mobile phones, Printers, Routers, Personal computers, Telephones, GPS and navigation equipment	4.7

2014). Table 1 depicts the item-wise e-waste (Forti et al., 2018). These devices are considered electronic waste when they reach their expiry or are no longer desired or needed. The global increase in electronic equipment waste can be attributed to the rapid proliferation of electronic devices and their shorter lifespan, which represents an important 21 percent increase over the last five years (Chen et al. 2015b).

In 2016, an astonishing 44.8 million metric tons of electronic waste were produced across the world. Further, Data also depicts the toxic items growth by 3 to 5 percent annually (European Parliament Briefing, 2015). In the year 2019, the Global electronic waste generation was approximately 54 million tonnes in which only 17% has been managed and rest 83% not accounted (Forti et al., 2020). As per the Global Electronic Equipment Waste Monitor 2020 statistics report "the prediction is, by 2030 this will touch approximately 75 million tonnes (Statista, 2023). However, It is shocking to see that only approximately 18% of the world's electronic waste has been properly disposed of in an environmentally friendly way.

There are multiple reasons behind this growth. Among them, one is an average lifetime of Electronic products. Previous research has revealed a worrying trend: the average life expectancy of the latest computer has slowed considerably. In 1992, these electronic miracles boasted an average lifetime of 4.5 years. However, by 2005, this number had fallen to about two years, and this trend seems to persist (Widmer et al., 2005). This reduced lifespan leads to a significant increase in the number of computers that need to be disposed of or, in some cases, exported to developing countries.

Adding to this concern is the fact that the longevity of central processing units (CPUs) has also witnessed a decline. Back in 1997, CPUs could be expected to function optimally for a span of 4 to 6 years (Shamim et al., 2015). Fast forward to 2015, and this duration had further shrunk to a range of 2 to 3 years (Yazici and Deveci, 2013). According to a comprehensive report, the annual influx of obsolete computers, monitors, and televisions in the United States alone surpasses a staggering 130 million, and this number is on an upward trajectory (Bushehri, 2010). To put this into perspective, the period between 1997 and 2007 saw an eye-opening 500 million computers rendered obsolete within the United States. Notably, Japan was not far behind, with a tally of 610 million discarded computers by the close of December 2010.

Table 2. Country-wise e-waste production

Global and selected country e-waste production in 2014 and 2016. Source: The global e-waste monitor 2017 (Baldé et al., 2017); StEP, e-waste world map (StEP, 2018).

Country/Region	Per capita production (kg)	Total e-waste production (million tonnes)
World (2014 data)	5.9	41.8
World (2016 data)	6.1	44.7
World (2021 estimated)	6.8	52.2
United Kingdom	23.5	1.50
USA	22.1	7
Germany	21.7	1.77
Hong Kong	21.5	0.16
Canada	20.4	0.72
Australia	20.1	0.46
Singapore	19.6	0.11
European Union	18.7	9.50
Taiwan	18.6	0.44
Japan	17.3	2.20
South Korea	15.9	0.80
Malaysia	7.6	0.23
Brazil	7.0	1.40
Argentina	7.0	0.29
South Africa	6.6	0.35
China	4.4	6.00
Sri Lanka	4.2	0.09
India	1.3	1.60
Nigeria	1.3	0.22
Zambia	0.9	0.01

Similar alarming circumstances exist in China, As per the study of Hicks (2005), In 2003 annually an average of five million new PCs and approximately ten million new TVs have been purchased by the public. Studies show approx 1.2 million tonnes of e-waste are produced annually as a result of this enormous consumption of electronics; the main sources of this trash are imports from other countries, the production of electronic equipment, and the assembly of home appliances and information technology products (Rucevska et al.,2015). It is also plausible to assume that many other nations are producing a similar amount of e-waste.

In particular, the global problem of electronic equipment waste is caused by the high-tech adoption and consumption of developed and developing countries such as the US, Japan, Norway, China, India, and the United Kingdom (Song et al., 2014). The e-waste country-wise is displayed in Table 2.

Economic, Social, Educational, Legal, and Technological Challenges in E-Waste Management

Inadequate management of electronic equipment waste presents substantial risks to both humanity and the environment, with particularly profound implications for human health (Bhaskar et al., 2018). The problem of e-waste disposal has become a pressing concern. This concern is supported by an array of well-documented case studies, primarily originating from China (Zhao et al., 2008). These studies emphasize the immediate need for enhanced protocols in handling and disposing of electronic waste to protect both mankind and the environment from the detrimental consequences of e-waste mismanagement (Huo et al., 2007).

The inappropriate disposal and management of electronic waste result in the release of hazardous substances into the environment. These substances, including lead, mercury, cadmium, and brominated flame retardants, exert adverse effects on the environment (Wang et al., 2009b). These toxic elements have the capability to contaminate soil, water sources, and the atmosphere, consequently exacerbating pollution levels and contributing to ecosystem degradation (Chan et al., 2007; Chen et al., 2015b).

Exposure to pollutants stemming from electronic equipment waste can lead to severe health problems in both human and animal populations (Bald et al.,2015). Such health issues encompass respiratory problems, reproductive disorders, neurological damage, and an elevated risk of cancer (Qu et al., 2007). Reports substantiate the peril posed by the toxicity of e-waste to human health, encompassing both enduring and acute conditions, evolving into a significant societal concern (Zheng et al., 2008).

This concern finds support in a multitude of case studies from various regions, including China (Wang et al., 2009), Japan (Zhao et al., 2008), India (Ha et al., 2009), Korea (Eguchi et al., 2012), and Ghana (Asante et al., 2012). These comprehensive studies vividly illustrate the gravity of the issue, emphasizing the critical necessity for implementing measures to mitigate the health risks associated with e-waste. Economic Challenges: A foremost economic hurdle in Electronic equipment waste management stems from the high costs associated with collection, recycling, and proper disposal. Electronic equipment waste comprises complex materials that necessitate specialized technologies and processes for safe handling (Xing et al., 2009; Chan et al., 2007). The economic practicability of Electronic equipment waste management initiatives can prove challenging, particularly in developing nations with limited financial resources and infrastructure (Shittu et al., 2021; Qu et al., 2007; Xing et al., 2009).

- Social Challenges: Social challenges arise because of a lack of awareness and misconceptions surrounding Electronic equipment waste management. Many individuals remain unaware of the social, ecological and health hazards linked to inappropriate Electronic equipment waste disposal. Additionally, societal behaviors that prioritize acquiring new electronic devices over repair or refurbishment contribute to the burgeoning Electronic equipment waste crisis. Addressing prevailing social attitudes and fostering behavioral changes are pivotal to the effectiveness of Electronic equipment waste management efforts (Borthakur et al., 2018; Mudali et al.,2021).
- Technological Challenges: Rapid technological advancements result in shorter product lifecycles and a growing diversity of electronic devices. Effectively managing various types of Electronic equipment waste necessitates advanced recycling technologies capable of handling diverse materials and components. Technological challenges include the development of efficient and cost-effective recycling processes that can keep pace with evolving electronic devices and their intricate compositions (Wong et al.,2007a; Lu et al., 2015)
- Lack of Awareness and Education on Electronic equipment waste Management: One major hindrance to effective Electronic equipment waste management is the limited awareness and education surrounding the subject matter. Both consumers and producers often possess limited knowledge concerning the environmental impact and appropriate disposal practices for Electronic equipment waste. It is crucial to promote public awareness regarding the repercussions of incorrect Electronic equipment waste disposal, the significance of recycling, and the accessibility of collection systems. By enhancing awareness, we can encourage responsible consumer practices and cultivate a culture that prioritizes appropriate Electronic equipment waste management (Puckett and Smith, 2002; Rucevska et al., 2015).

- Policy and Regulatory Gaps: IT has been seen over a period of time that challenges in Electronic equipment waste management emerge due to lack of comprehensive legislation and enforcement mechanisms. In certain regions, inadequate policies or inconsistent implementation create loopholes in Electronic equipment waste management systems. Effective policies should encompass the entire Electronic equipment waste lifecycle, spanning from production to disposal, and ensure diligent monitoring and enforcement to curb illegal disposal practices and encourage responsible recycling. (Robinson, 2009; Basel Convention, 2018b)
- Illegal Electronic equipment waste Trade and Dumping: In global level, Illegal trade and dumping of Electronic equipment waste present significant challenges to sustainable Electronic equipment waste management. In many instances, it has been experienced that Electronic equipment waste is unlawfully exported to developing nations with lenient regulations, resulting in hazardous and environmentally detrimental recycling practices. This illicit trade circumvents proper disposal processes and contributes to pollution and health risks in the recipient countries (Abalansa et al., 2021).
- Strengthening international cooperation, implementing stricter regulations, and enhancing monitoring systems are vital in combatting illegal Electronic equipment waste trade and dumping. The economic and social challenges related to Electronic equipment waste management, including issues of awareness and policy gaps, highlight the urgency for collective action to overcome these obstacles and effectively pursue the Sustainable Development Goals (SDGs) (Patil, R. A., & Ramakrishna, S.,2020).

Recognizing the pressing urgency for effective management of electronic waste, many countries have implemented rules and regulations and framed policies aimed at controlling Electronic equipment waste generation, promoting recycling practices, and ensuring environmentally sound disposal. For instance, the European Union initiative in waste management. The enactment of the Waste Electrical and Electronic Equipment (WEEE) Directive (Pan, X. et al., 2022) has established guidelines for the collection, recycling, and recovery of electronic waste. In the United States, the Resource Conservation and Recovery Act (RCRA) incorporates Electronic equipment waste within its framework for managing hazardous waste (Ali et. al., 2023). Many research reports depict the concern of several developing nations relating to both societal and human health issues that have garnered recognition. Notably, Various Asian nations countries like India, and China, have recently introduced amendments to their legislation aimed at addressing e-waste management and its imports (Widmer et al., 2005). Additionally, the developed and developing countries together proactively ventured into employing advanced technologies for the safe disposal of e-waste (US Government Accountability Office, 2008; Widmer et al., 2005). Furthermore, on the international front, collaborative initiatives such as the Basel Convention and the Stockholm Convention are working diligently to regulate the cross-border transport and appropriate disposal of hazardous waste, with a specific focus on electronic equipment waste (Wingfield, S., & Lim, M., 2022). These collective efforts signify a growing awareness of the e-waste issue and the commitment to finding sustainable solutions on a global scale.

Sustainable Development Goals (SDGs) and Their Relevance to E-Waste Management

The SDGs comprising 17 global objectives endorsed by the United Nations, serve as a comprehensive framework to address diverse social, economic, and environmental challenges on a global scale, with the ultimate aim of fostering sustainable development worldwide. These goals direct a roadmap for countries and companies to work together towards a more equitable, prosperous, and environmentally conscious future. Among these goals, "SDG 12- Responsible Consumption and Production" is closely focused on management of electronic waste. By effectively handling Electronic equipment waste, we can actively contribute toward the control of environmental pollution, the resources preservation, the promotion of sustainable consumption patterns, and the cultivation of a Resource-efficient economy. In the same line, SDG 12.5, in particular, emphasizes the other aspects such as reduction in waste generation through reuse, recycling, reduction, and prevention methods. The management of Electronic equipment waste is intricately linked to this target, as it involves the advocacy of sustainable practices and the mitigation of the environmental impact caused by discarded electronic devices (AlQattan, N et al., 2018). Furthermore, on global platform Electronic equipment waste management is crucial role to facilitate the Resource-efficient economy transition, which strives to prolong the lifespan of products and materials through mechanisms such as recycling, repair, and remanufacturing (Bhaskar, K., & Kumar, B., 2019). By adopting principles of the Resource-efficient economy, nations can effectively address Electronic equipment waste challenges while simultaneously advancing their SDG's.

Studies proved that electronic equipment waste management plays a pivotal role in advancing sustainable development across multiple dimensions:

- Environmental Sustainability: Proper Electronic equipment waste management significantly reduces environmental pollution and the reduction of natural assets. Through the implementation of recycling and recovery processes, valuable materials can be extracted, reducing the demand for raw material extraction and promoting resource conservation. Moreover, this strategy minimizes the secretion of harmful substances into the environment, mitigating ecosystem contamination and preserving biodiversity.
- Social Well-being: Effective Electronic equipment waste management serves to protect human health and foster a safe living environment. By preventing toxic substances excessive exposure, communities can experience improved well-being and reduced health risks. Additionally, Electronic equipment waste management initiatives have the potential to raise awareness and educate individuals, empowering them to make responsible choices regarding the electronic waste proper disposal and consumption.
- Economic Development: Better management of Electronic equipment waste lead to economic opportunities through resource recovery. The extraction of valuable materials from Electronic equipment waste substances not only reduces the environmental burden but also generates economic value. The recycling and refurbishment of electronic devices can create employment opportunities, support local industries, and contribute to sustainable economic growth.

The implementation of effective Electronic equipment waste management strategies is of utmost importance for several reasons. Firstly, Electronic equipment waste confines valuable materials that can be extracted from the products and recycled for further purposes, thereby reducing the demand of raw materials and minimizing resource depletion. Secondly, proper management practices help prevent environmental pollution and the associated health risks. Lastly, the mentioned practices is also aligned with other UN SDG goals which cover smart cities and responsible consumption. The implementation of efficient Electronic equipment waste management strategies may guide to decrease the negative impacts of Electronic equipment waste and increase the sustainable bright future.

GLOBAL ELECTRONIC EQUIPMENT WASTE MANAGEMENT STRATEGIES

Electronic equipment waste legislations and policies vary across different regions worldwide. In Asia, countries including Singapore, China, India, Korea, and Japan have acknowledged the urgency of efficient Electronic equipment waste management and executed legislation, taking inspiration from the European Union's directives. The Americas, with notable contributors such as the United States, Japan Canada, Germany Brazil, and Mexico, generate substantial volumes of Electronic equipment waste; however, legislative inconsistencies persist, and illicit practices continue to undermine progress. Europe, on the other hand, has established comprehensive Electronic equipment waste management regulations through directives like the Waste Shipment Regulation, RoHS Directive, and WEEE Directive, championing the principles of Resource-efficient economy and extended producer responsibility. Africa faces distinct challenges in Electronic equipment waste management due to limited regulation and the predominance of informal collectors and recyclers, while Oceania, spearheaded by Australia, has made significant headway in Electronic equipment waste management through legislative frameworks and product stewardship initiatives.

Numerous countries and regions have championed commendable electronic equipment waste management initiatives, providing successful models for sustainable electronic equipment waste handling(Kim et al., 2001). Apart from this improved energy efficient Air Conditioning units (Prek, 2004) play a critical role in reducing the e-waste. Other studies also support new advance products like cloth washer machines (Park et al., 2006), and electronic gadgets (Muñoz et al., 2009) have created a countable reduction in e-wate. Research proves Life Cycle Assessment tool has shown to be a powerful means of identifying potential environmental impacts.

Numerous studies have shown how widely used Life Cycle Assessment is in e-waste management. This adaptable tool offers a number of advantages in supporting e-waste management. Numerous research has demonstrated that life cycle assessment calculates the environmental effects of material consumption during the eco-design and development of products.

In order to help with more informed decisions regarding the disposal of electronic waste, further life cycle assessment also evaluates the economic and environmental aspects of electronic device end-of-life disposal. Table 3 demonstrate the application of various tool across different countries.

In addition to life cycle assessments, multi-criterion analyses are used for environmental challenges, including electronic waste management, to create alternative strategies for electronic waste management (Queiruga et al., 2008). Although it is limited to the treatment of electronic waste, MCA is widely used for the treatment of solid waste and hazardous waste. In addition, some countries use material movement analysis (MFA) techniques to manage and monitor electronic waste movements. The study of e-waste creation is another application of MFA technology. Another important tool is the extension of producer responsibility (EPR). The leading e-waste management EPR programs are concentrated in rich countries

Table 3. Tools for e-waste management approached

Tools for e-waste management approaches being used or proposed.

Tools	Application	Aspects	Country/ region	References
LCA	Recycling potential	Environment and economic	Korea	Kim et al. (2004)
LCA	Six EoL treatment scenarios: case study of mobile phones	Environment	Switzerland	Scharnhorst et al. (2005)
LCA	Recycling of end-of-life of personal computers	Environment	Korea	Choi et al. (2006)
LCA	Recycling systems: case study of notebook computers	Environment and economic	Taiwan	Lu et al. (2006)
LCA	Recycling of end-of-life	Economic	Japan	Nakamura and Kondo (2006)
LCA	Decision makers for managing computer waste	Environment and economic	India	Ahluwalia and Nema (2007)
LCA	Compare different disposal methods (recycle and non-recycle): case study of fluorescent lamps	Environment	Thailand	Apisitpuvakul et al. (2008)
LCA	Compare different waste scenarios	Environment	Germany	Barba-Gutiérrez et al. (2008)
LCA and EPR	Four EoL treatment scenarios: case study of printers	Environment and economic	United Kingdom	Mayers et al. (2005)
LCA and MFA	The supply of computers to schools	Environment, economic and social	Colombia	Streicher-Porte et al. (2009)
LCA and MFA	Recycling systems	Environment	Switzerland	Hischier et al. (2005)
LCA and MFA	Collection and recovery systems	Environmental impacts	Switzerland	Wäger et al. (2011)
MFA	The flow of e-waste	Generation	China	Liu et al. (2006a)
MFA	E-waste trade value chain	Life span and market supply	India	Jain and Sareen (2006)
MFA	The flow of personal computers and the pathways of recycling	Economic value	India	Streicher-Porte et al. (2007)
MFA	The flow of mobile phones	Environmentally sound and eco-efficient management	Nigeria	Osibanjo and Nnorom (2008)
MFA	The flow of e-waste and e-waste trade	Law and environmental pollution	Asia	Shinkuma and Nguyen Thi Minh (2009)
MFA	The flow of used personal computers	Compare the results from before and after the introduction of personal computers recycling system	Japan	Yoshida et al. (2009)
MFA	E-waste quantities	Generation	Chile	Steubing et al. (2010)
MCA	Decision making for optimisation of product disassembly	Environment and economic	United State	Hula et al. (2003)
MCA	Decision making for e-waste management	Environment, economic and social	Cyprus	Rousis et al. (2008)
MCA	Decision making for the location of e-waste recycling plants	Economic, infrastructural and legal	Spain	Queiruga et al. (2008)

Table 4. E-waste management approaches towards the EPR

E-waste management approaches to EPR.

Country	Policy	Target	References
The Netherlands	- Take back (large household appliances and IT equipment)	Recycling rate 45–75% by weight	Tojo (2001)
United Kingdom	- Take back (electronic appliances)	Recycle and recovery 50–80%	Gottberg et al. (2006) and Mayers et al. (2005)
Germany	- Take back (electronic appliances)	-	Roller and Furhr (2008)
Switzerland	- Take back (electronic appliances) - Disposal ban in landfill - Advance recycling fees	-	Khetriwal et al. (2009), Nnorom and Osibanjo (2008), and Widmer et al. (2005)
Japan	- Take back (four large household appliances: TV sets, refrigerators, air conditioners and washing machines) - Product re-design (lead free solders and bromine free printed circuit boards)	Recycling rate 50–60% by weight	Nnorom and Osibanjo (2008) and Tojo (2001)
United States	- Take back household appliances in some states, such as Maine (take back only televisions and computer monitors)	-	Sachs (2006) and Wagner (2009)
Canada	- Take back household appliance in some provinces, including Alberta and Ontario - Develop advanced EPR program	-	McKerlie et al. (2006)
India	- Feasibility study	-	Manomaivibool (2009)
Thailand	- Developing legal framework	Collection and recycling	Manomaivibool and Vassanadumrongdee (2011)

like the United States (Balde and al., 2015). In the same league countries such as Canada, US, Switzerland, India, Japan Korea, and the European Union show trust in LCA and MCA tools.

In addition to life cycle assessment, multi-crime analysis (MCA) was used to solve environmental problems, including the management of electronic waste, and to create alternative strategies for the management of electronic waste (Queiruga, 2008). Table 4 shows the different EPRs in each country.

In addition to that World Organisation for Economic Cooperation and Development supports e waste management initiatives and helps nations to improve the guidance manuals to promote acceptable electronic waste management practices globally (OECD, 2001).

Table 5. Different features of LCA, MFA, MCA, and EPR

The distinctive features of LCA, MFA, MCA and EPR for e-waste management.

Tools	Benefits	Country/ region	References
LCA	- Estimates the effects of materials consumption. Conducts assessment of eco-design and product development - Allocates the impacts of the examined product or process of environmental interest - Evaluate the environmental and economic aspects related to the end of life disposal of electronic devices - Takes better decisions regarding e-waste disposal	Columbia	Streicher-Porte et al. (2009)
		Germany	Barba-Gutiérrez et al. (2008)
		Japan	Nakamura and Kondo (2006)
		Korea	Choi et al. (2006) and Kim et al. (2004)
		India	Ahluwalia and Nema (2007)
		Switzerland	Hischier et al. (2005), Scharnhorst et al. (2005), and Wäger et al. (2011)
		Taiwan	Lu et al. (2006)
		Thailand	Apisitpuvakul et al. (2008)
		United Kingdom	Mayers et al. (2005)
MFA	- Investigates the flow of e-waste - Estimates e-waste generation - Used for environmental decision making	Asia	Shinkuma and Nguyen Thi Minh (2009)
		Chile	Steubing et al. (2010)
		China	Liu et al. (2006a)
		Columbia	Streicher-Porte et al. (2009)
		India	Jain and Sareen (2006) and Streicher-Porte et al. (2007)
		Japan	Yoshida et al. (2009)
		Nigeria	Osibanjo and Nnorom (2008)
		Switzerland	Hischier et al. (2005) and Wäger et al. (2011)
MCA	- Used for environmental decision making	Cyprus	Rousis et al. (2008)
		Spain	Queiruga et al. (2008)
		United State	Hula et al. (2003)
EPR	- Solves e-waste problems in national scale - Enforce producers based on polluter-pays principal	Canada	McKerlie et al. (2006)
		Germany	Roller and Furhr (2008)
		Japan	Nnorom and Osibanjo (2008) and Tojo (2001)
		India	Manomaivibool (2009)
		Switzerland	Khetriwal et al. (2009), Nnorom and Osibanjo (2008) and Widmer et al. (2005)
		Thailand	Manomaivibool and Vassanadumrongdee (2011)
		The Netherlands	Tojo (2001)
		United Kingdom	Gottberg et al. (2006) and Mayers et al. (2005)
		United States	Sachs (2006) and Wagner (2009)

Though there are many tools available, in our study we focus on the mainly four major tools such as EPR, LCA, MFA, MCA and suggest the best practices followed across the nations to manage the e-waste.

As shown in Table 5 each tool has special qualities and benefits when used for e-waste management.

In addition to above tools, Norway, for instance, has emerged as a frontrunner in Electronic equipment waste management, attaining high collection rates and emphasizing resource recovery. The country's well-established deposit-refund system has contributed to an impressive collection rate of approximately 80% for electronic waste. Norway's approach aligns seamlessly with SDG goals, promoting responsible consumption and production while curbing waste generation. Similarly, Japan has made remarkable progress in Electronic equipment waste management, harnessing advanced recycling technologies and conducting extensive public awareness campaigns. Leveraging a sophisticated recycling infrastructure, Japan has achieved a recycling rate of over 90%, facilitating the extraction of valuable materials from Electronic equipment waste products. Japan's initiatives strongly resonate with SDG 12, particularly in terms of promoting responsible consumption and reducing environmental footprints. In a similar line, Switzerland has launched a highly successful Technological waste management system that is based on latest EPR policy. The country boasts impressive collection rates and has established streamlined recycling processes. Switzerland's EPR model holds manufacturers accountable for managing the entire lifecycle of their products, harmonizing with SDG 12 and fostering sustainable production and consumption patterns. Table 6 represent the various international initiative for e-waste management.

Table 6. Various policies and initiative for e-waste management

International legislation and initiatives on e-waste management.

Legislation and initiatives	Key features
The Basel Convention[1]	Reduction of hazardous waste generation and the promotion of sustainable handling techniques at the place of disposal The restriction of trans-boundary movements of hazardous wastes Form legislative frameworks where trans-boundary movements are permissible
The European Union's WEEE Directive[2]	Free take-back schemes for consumers Reduce WEEE to landfills through collection and recovery targets WEEE to be collected separately from other wastes Stimulate 'eco-design' – product design that reduces WEEE and increases its ease of recovery Producer responsibility for end of life treatment of their products
StEP (Solving the E-waste Problem)[3]	Research and piloting on e-waste treatment Strategy and goal-setting to eliminate the e-waste problem Training and development on e-waste issues Establish communication and awareness among members and throughout the industry
G-8 3Rs Initiative[4]	Promoting 3Rs: Reduce, Reuse, Recycle Building a "sound-material-cycle society"
National Strategy for Electronics Stewardship (NSES), USA[5]	Improve design of electronic products The Federal government leads by example Enhance the handling and management of used or discarded electronic equipment in the USA Reduce harm caused by used or discarded electronic equipment in developing countries
GeSI (Global e-Sustainability Initiative)[6]	Bring together ICT companies, industry associations and non-governmental organisations One of the key focus area is e-waste Achieving sustainable objectives and manage risks associated with ICT through innovative technology
IETC (International Environmental Technology Centre)[7] - UNEP	Application of environmentally sound technologies (ESTS) in developing countries and countries in transition on waste management

[1] Basel Convention (2017).
[2] European Commission (2017b).
[3] StEP (2018).
[4] UNCRD – 3R initiative (2017).
[5] National Strategy for Electronics Stewardship (NSES) (2011).
[6] GeSI (2017).
[7] IETC, UN Environment 2017.

Other Best Practices and Strategies for Digital Devices Waste Management

Recycling and recovery technologies assume a crucial role in effective Electronic equipment waste management. These technologies strive to extract valuable materials from discarded electronic devices, curtail waste generation, and minimize environmental impacts. Key facets of recycling and recovery technologies include:

- Segregation and Sorting: Electronic equipment waste is initially segregated and sorted into distinct categories based on material types, facilitating the identification of components amenable to recycling or recovery.
- Dismantling and Disassembly: Electronic devices undergo dismantling processes to access individual components and separate them for subsequent processing. The disassembly techniques employed vary depending on the complexity of the devices.
- Material Recovery: Recycling processes involve the various valuable materials, such as gold etc. and plastics from Electronic equipment waste. Techniques such as shredding, smelting, and hydrometallurgical methods are employed to retrieve these materials.
- Proper Disposal of Non-Recyclable Components: Certain components of Electronic equipment waste may be deemed non-recyclable owing to their composition or technological constraints. Environmentally responsible disposal methods, such as secure landfilling or controlled incineration, are implemented to manage these non-recyclable components effectively.
- Achieving effective Electronic equipment waste management necessitates collaboration among various stakeholders, including governments, manufacturers, recyclers, consumers, and civil so-

ciety. The following key aspects of stakeholder engagement and collaboration contribute to the success of Electronic equipment waste management endeavors.

- Government Policy and Implementation: Governments play a pivotal role in formulating and implementing policies and regulations related to Electronic equipment waste management. Collaboration between policymakers, industry representatives, and environmental experts is essential to ensure the development of effective and enforceable regulations that address the complexities of Electronic equipment waste challenges.
- Public-Private Partnerships: Collaboration between public and private entities fosters the exchange of knowledge, resources, and expertise, leading to mutually beneficial outcomes. Public-private partnerships can significantly enhance Electronic equipment waste collection systems, recycling infrastructure, and awareness campaigns, thus promoting sustainable Electronic equipment waste management practices.
- Consumer Engagement: Educating and engaging consumers are crucial components of promoting responsible Electronic equipment waste management practices. Awareness campaigns, consumer incentives, and the establishment of easily accessible collection points are effective means of encouraging consumer participation in recycling and proper disposal, ensuring their active involvement in sustainable Electronic equipment waste management efforts.

To address the key challenges we can work collaboratively and achieve sustainable Electronic equipment waste management, protect the planet, and preserve our finite natural resources. Better e waste management practices across the globe navigate the complex landscape of Digital waste management, mitigate the adverse effects, and forge a path toward a better future backed up by a Resource-efficient economy.

INDIA ACTION TOWARDS E-WASTE MANAGEMENT

By enacting the Electronic Equipment Waste (Management) Rules, 2016, the Indian government has made tremendous progress in addressing the problem of electronic equipment waste, or "e-waste." These laws provide a thorough framework for the appropriate handling, disposal, and collection of e-waste throughout the nation. They impose particular obligations on recyclers, producers, and large consumers to guarantee the ethical handling of e-waste(Huo et al., 2007).

In addition to regulatory measures, India has also embraced technology-driven solutions for e-waste management. Robust online platforms have been established to facilitate the exchange and recycling of electronic equipment waste. These platforms serve as a bridge between consumers, producers, and recyclers, effectively streamlining the management and directed flow of e-waste to authorized recycling facilities.

Furthermore, The various awareness campaigns and initiatives aimed at educating the public about e-waste management launched by Indian government and companies. These initiatives emphasize the significance of recycling, responsible disposal, and the various health, social, and environmental impacts associated with improper handling of electronic equipment waste. The collaborative efforts between the public and private sectors have led to the establishment of collection centers, recycling facilities, and capacity-building programs that effectively address the e-waste challenge.

An additional strategy involves harnessing technology to extend the lifespan of electronic products. This approach shifts the focus towards maximizing usage and minimizing premature disposal. Collaboration with e-commerce platforms enables the collection and direct routing of discarded electrical and electronic equipment (EEE) to authorized recyclers, ensuring proper handling and reducing the prevalence of informal or unauthorized recycling practices.

Moreover, functioning electronic equipment discarded by consumers who seek the latest versions can be collected, refurbished, and redistributed to underprivileged communities, promoting a more sustainable approach to product lifecycle management.

It's crucial to recognize that effective e-waste management is not only an environmental concern but also aligns with Sustainable Development Goals (SDGs). Achieving these goals necessitates ongoing efforts, collaboration, and innovation from all stakeholders, including governments, manufacturers, consumers, and recycling industries. These initiatives can also be seamlessly integrated into corporate social responsibility (CSR) efforts within various industries. By implementing these strategies and aligning them with the SDGs, India and other nations can effectively manage e-waste, minimize environmental impact, safeguard public health, and contribute to a more sustainable and responsible future.

Future Recommendations

To advance electronic equipment waste management and align it with the Sustainable Development Goals (SDGs), the following policy recommendations are of paramount importance:

- Comprehensive Legislation: Governments should proactively devise and enforce comprehensive legislation that spans the entire life cycle of electronic equipment waste. These regulations should encompass diverse facets, including robust collection systems, stringent recycling standards, restrictions on hazardous substances, and the implementation of extended producer responsibility (EPR) programs. Such measures are essential in fostering responsible electronic equipment waste management practices.
- International Collaboration: Given the global nature of the electronic equipment waste problem, enhanced international cooperation is indispensable. Nations must collaborate to harmonize their electronic equipment waste management policies. By sharing best practices and strategies to combat illegal electronic equipment waste trade and dumping, the international community can more effectively address this issue. Moreover, facilitating technology transfer and capacity-building initiatives in developing nations is crucial.
- Public Awareness Campaigns: Governments, NGO's, and industry stakeholders, should launch extensive public awareness campaigns. These campaigns should aim to educate the public about the multifaceted impacts of electronic equipment waste, including environmental, social, and health consequences. Furthermore, they should underscore the significance of responsible disposal practices and the importance of recycling. Raising awareness is instrumental in fostering a culture of responsible electronic equipment waste management and prompting behavioral change.
- Educational Initiatives:Integrating electronic equipment waste management education into school curricula and vocational training programs is pivotal. By imparting knowledge about the adverse effects of improper electronic equipment waste disposal, recycling methods, and the promotion of sustainable consumption patterns, educational initiatives empower future generations to confront the electronic equipment waste challenge.

- Producer Responsibility: Governments should establish clear guidelines and regulations regarding extended producer responsibility (EPR) for electronic device manufacturers. This approach serves as an incentive for eco-design, encourages product durability, and promotes the development of more sustainable electronic products.
- Voluntary Initiatives and Partnerships: Collaboration among industry associations, businesses, and civil society organizations can lead to the creation of voluntary initiatives that advocate responsible electronic equipment waste management. These initiatives can emphasize the sharing of best practices, the implementation of sustainable supply chains, support for recycling and recovery efforts, and contributions to funding research and innovation in electronic equipment waste management.
- Circular Design and Material Innovation: Encouraging research and innovation in circular design and the development of eco-friendly materials can result in more sustainable electronic devices. Products designed with considerations for repair, upgradability, and recyclability can extend their lifespan and minimize waste generation.
- Advanced Recycling Technologies: Continuous research into advanced recycling technologies is imperative for enhancing the efficiency and effectiveness of electronic equipment waste recycling processes. Innovative methods for recovering valuable materials, such as rare metals and critical minerals, can contribute to resource conservation and reduce the environmental impact associated with electronic equipment waste recycling.

CONCLUSION

Electronic equipment waste management aligns with various SDGs that focus on responsible consumption and production patterns. By promoting extended producer responsibility (EPR), eco-design, and sustainable supply chains, electronic equipment waste management encourages the adoption of Resource-efficient economy principles, resource efficiency, and waste reduction. Strong policies and regulations are essential for creating an enabling environment for effective electronic equipment waste management. Governments should collaborate with industry stakeholders, civil society organizations, and the public to develop comprehensive legislation, enforce regulations, and raise awareness. Stakeholder engagement facilitates knowledge-sharing, innovation, and the development of replicable models. Continuous research and innovation are necessary to address emerging challenges and improve electronic equipment waste management practices. This includes developing advanced recycling technologies, exploring sustainable materials and design strategies, and enhancing monitoring and tracking systems for proper disposal and traceability of electronic equipment waste. By implementing these strategies, raising awareness, engaging stakeholders, and fostering innovation, societies can transition towards sustainable electronic equipment waste management and make significant contributions to the SDGs. It is imperative that governments, businesses, communities, and individuals work together to create a future where electronic equipment waste is effectively accomplished, resources are conserved, and the environment and human well-being are safeguarded.

REFERENCES

Abalansa, S., El Mahrad, B., Icely, J., & Newton, A. (2021). Electronic waste, an environmental problem exported to developing countries: The GOOD, the BAD and the UGLY. *Sustainability (Basel)*, *13*(9), 5302. doi:10.3390u13095302

Ahluwalia, P. K., & Nema, A. K. (2007). A life cycle based multi-objective optimization model for the management of computer waste. *Resources, Conservation and Recycling*, *51*(4), 792–826. doi:10.1016/j.resconrec.2007.01.001

Ali, A., Malik, S. A., Shafiullah, M., Malik, M. Z., & Zahir, M. H. (2023). Policies and regulations for solar photovoltaic end-of-life waste management: Insights from China and the USA. *Chemosphere*, *340*, 139840. doi:10.1016/j.chemosphere.2023.139840 PMID:37597624

AlQattan, N., Acheampong, M., Jaward, F. M., Ertem, F. C., Vijayakumar, N., & Bello, T. (2018). Reviewing the potential of Waste-to-Energy (WTE) technologies for Sustainable Development Goal (SDG) numbers seven and eleven. *Renewable Energy Focus*, *27*, 97–110. doi:10.1016/j.ref.2018.09.005

Arya, S., & Kumar, S. (2020). E-waste in India at a glance: Current trends, regulations, challenges and management strategies. *Journal of Cleaner Production*, *271*, 122707. doi:10.1016/j.jclepro.2020.122707

Asante, K. A., Agusa, T., Biney, C. A., Agyekum, W. A., Bello, M., Otsuka, M., Itai, T., Takahashi, S., & Tanabe, S. (2012). Multi-trace element levels and arsenic speciation in urine of e-waste recycling workers from Agbogbloshie, Accra in Ghana. *The Science of the Total Environment*, *424*, 63–73. doi:10.1016/j.scitotenv.2012.02.072 PMID:22446112

Baldé, C. P., Forti, V., Gray, V., Kuehr, R., & Stegmann, P. (2017). The global e-waste monitor – 2017. United Nations University (UNU), International Telecommunication Union (ITU) and International Solid Waste Association, Bonn/Geneva/Vienna.

Barba-Gutiérrez, Y., Adenso-Díaz, B., & Hopp, M. (2008). An analysis of some environmental consequences of European electrical and electronic waste regulation. *Resources, Conservation and Recycling*, *52*(3), 481–495. doi:10.1016/j.resconrec.2007.06.002

Basel Convention – Controlling transboundary movements of hazardous wastes and their disposal. (2017). https://www.basel.int/TheConvention/Overview/tabid/1271/Default.aspx

Belboom, S., Renzoni, R., Deleu, X., Digneffe, J.-M., & Leonard, A. (2011). Electrical waste management effects on environment using life cycle assessment methodology: the fridge case study. *SETAC EUROPE 17th LCA Case Study Symposium Sustainable Lifestyles*, 2.

Bhaskar, K., & Kumar, B. (2019). Electronic waste management and sustainable development goals: Is there a business case for linking the two? *Journal of Indian Business Research*, *11*(2), 120–137. doi:10.1108/JIBR-01-2018-0051

Bhat, V. (2023, April). E-waste management and Achieving SDG-Challenges in Indian Context. *IOP Conference Series. Earth and Environmental Science*, *1161*(1), 012008. doi:10.1088/1755-1315/1161/1/012008

Borthakur, A., & Govind, M. (2018, February). Management of the challenges of electronic waste in India: An analysis. *Proceedings of the Institution of Civil Engineers-Waste and Resource Management, 171*(1), 14–20. doi:10.1680/jwarm.17.00035

Bushehri, F. I. (2010). *UNEP's role in promoting environmentally sound management of e-waste.* In 5th ITU Symposium on ''ICTs, the Environment and Climate Change, Cairo, Egypt.

Chan, J. K. Y., Xing, G. H., Xu, Y., Liang, Y., Chen, L. X., Wu, S. C., Wong, C. K. C., Leung, C. K. M., & Wong, M. H. (2007). Body loadings and health risk assessment of polychlorinated dibenzo-p-dioxins and dibenzofurans at an intensive electronic waste recycling site in China. *Environmental Science & Technology, 41*(22), 7668–7674. doi:10.1021/es071492j PMID:18075072

Chen, M., Zhang, F.-S., & Zhu, J. (2011). Leaching toxicity of Pb and Ba containing in cathode ray tube glasses by SEP-TCLP. In *Recycling of Electronic Waste II, Proceedings of the Second Symposium, TMS.* The Minerals, Metals & Materials Society. 10.1002/9781118086391.ch15

Choi, B.-C., Shin, H.-S., Lee, S.-Y., & Hur, T. (2006). Life cycle assessment of a personal computer and its effective recycling rate. *The International Journal of Life Cycle Assessment, 11*(2), 122–128. doi:10.1065/lca2004.12.196

Duan, H., Eugster, M., Hischier, R., Streicher-Porte, M., & Li, J. (2009). Life cycle assessment study of a Environment Canada, 2000. *Understanding the Environmental Aspects of Electronic Products: A Life Cycle assessment Case Study of a Business Telephone.*

Eguchi, A., Nomiyama, K., Devanathan, G., Subramanian, A., Bulbule, K. A., Parthasarathy, P., Takahashi, S., & Tanabe, S. (2012). Different profiles of anthropogenic and naturally produced organohalogen compounds in serum from residents living near a coastal area and e-waste recycling workers in India. *Environment International, 47*, 8–16. doi:10.1016/j.envint.2012.05.003 PMID:22717641

Environment Canada. (n.d.). Chinese desktop personal computer. *The Science of the Total Environment, 407*, 1755–1764.

Faist Emmenegger, M., Frischknecht, R., Stutz, M., Guggisberg, M., Witschi, R., & Otto, T. (2006). Life cycle assessment of the mobile communication system UMTS: Towards eco-efficient systems. *The International Journal of Life Cycle Assessment, 11*(4), 265–276. doi:10.1065/lca2004.12.193

Forti, V., Baldé, C. P., Kuehr, R., & Bel, G. (2020). *The global e-waste monitor 2020. United Nations University (UNU), International Telecommunication Union (ITU) & International Solid Waste Association.* ISWA.

Ghosh, B. K., Mekhilef, S., Ahmad, S., & Ghosh, S. K. (2022). A review on global emissions by E-products based waste: Technical management for reduced effects and achieving sustainable development goals. *Sustainability (Basel), 14*(7), 4036. doi:10.3390u14074036

Ha, N. N., Agusa, T., Ramu, K., Tu, N. P. C., Murata, S., Bulbule, K. A., Parthasaraty, P., Takahashi, S., Subramanian, A., & Tanabe, S. (2009). Contamination by trace elements at e-waste recycling sites in Bangalore, India. *Chemosphere, 76*(1), 9–15. doi:10.1016/j.chemosphere.2009.02.056 PMID:19345395

Hicks, C., Dietmar, R., & Eugster, M. (2005). The recycling and disposal of electrical and electronic waste in China-legislative and market responses. *Environmental Impact Assessment Review*, *25*(5), 459–471. doi:10.1016/j.eiar.2005.04.007

Hischier, R., Wäger, P., & Gauglhofer, J. (2005). Does WEEE recycling make sense from an environmental perspective? The environmental impacts of the Swiss takeback and recycling systems for waste electrical and electronic equipment (WEEE). *Environmental Impact Assessment Review*, *25*(5), 525–539. doi:10.1016/j.eiar.2005.04.003

Hossain, M. S., Al-Hamadani, S. M., & Rahman, M. T. (2015). E-waste: A challenge for sustainable development. *Journal of Health & Pollution*, *5*(9), 3–11. doi:10.5696/2156-9614-5-9.3 PMID:30524771

Huo, X., Peng, L., Xu, X., Zheng, L., Qui, B., Qi, Z., Zhang, B., Han, D., & Piao, Z. (2007). Elevated blood lead levels of children in Guiyu, an electronic waste recycling town in China. *Environmental Health Perspectives*, *115*(7), 1113–1117. doi:10.1289/ehp.9697 PMID:17637931

Ilankoon, I. M. S. K., Ghorbani, Y., Chong, M. N., Herath, G., Moyo, T., & Petersen, J. (2018). E-waste in the international context–A review of trade flows, regulations, hazards, waste management strategies and technologies for value recovery. *Waste Management (New York, N.Y.)*, *82*, 258–275. doi:10.1016/j.wasman.2018.10.018 PMID:30509588

Kiddee, P., Naidu, R., & Wong, M. H. (2013). E-waste management in India: A strategy for the attainment of SDGs 2030. *Materials Today: Proceedings*, *60*, 811–814.

Kuehr, R. (2019). E-waste seen from a global perspective. In Waste Electrical and Electronic Equipment (WEEE) Handbook (pp. 1-16). Woodhead Publishing. doi:10.1016/B978-0-08-102158-3.00001-X

Lambert, F., Gaydardzhiev, S., Leonard, G., Lewis, G., Bareel, P.-F., & Bastin, D. (2015). Copper leaching from waste electric cables by biohydrometallurgy. *Minerals Engineering*, *76*, 38–46. doi:10.1016/j.mineng.2014.12.029

Lu, C., Lin, Z., Zhong, Y., Ren, W., Tobias, M., Mu, Z., Ma, Z., Geng, Y., & Xue, B. (2015). An overview of e-waste management in China. *Journal of Material Cycles and Waste Management*, *17*(1), 1–12. doi:10.100710163-014-0256-8

Lu, Y., & Xu, Z. (2016). Precious metals recovery from waste printed circuit boards: A review for current status and perspective. *Resources, Conservation and Recycling*, *113*, 28–39. doi:10.1016/j.resconrec.2016.05.007

Minashkina, D., & Happonen, A. (2022). Analysis of the Past Seven Years of Waste-Related Doctoral Dissertations: A Digitalization and Consumer e-Waste Studies Mystery. *Energies*, *15*(18), 6526. doi:10.3390/en15186526

Mudali, U. K., Patil, M., Saravanabhavan, R., & Saraswat, V. K. (2021). Review on e-waste recycling: Part I—a prospective urban mining opportunity and challenges. *Transactions of the Indian National Academy of Engineering : an International Journal of Engineering and Technology*, *6*(3), 547–568. doi:10.100741403-021-00216-z

Muñoz, I., Gazulla, C., Bala, A., Puig, R., & Fullana, P. (2009). LCA and ecodesign in the toy industry: Case study of a teddy bear incorporating electric and electronic components. *The International Journal of Life Cycle Assessment, 14*(1), 64–72. doi:10.100711367-008-0044-6

Murthy, V., & Ramakrishna, S. (2022). A review on global E-waste management: Urban mining towards a sustainable future and circular economy. *Sustainability (Basel), 14*(2), 647. doi:10.3390u14020647

Nakamura, S., & Kondo, Y. (2006). A waste input-output life-cycle cost analysis of the recycling of end-of-life electrical home appliances. *Ecological Economics, 57*(3), 494–506. doi:10.1016/j.ecolecon.2005.05.002

Pan, X., Wong, C. W., & Li, C. (2022). Circular economy practices in the waste electrical and electronic equipment (WEEE) industry: A systematic review and future research agendas. *Journal of Cleaner Production, 365*, 132671. doi:10.1016/j.jclepro.2022.132671

Parajuly, K., Kuehr, R., Awasthi, A. K., Fitzpatrick, C., Lepawsky, J., Smith, E., ... & Zeng, X. (2019). *Future e-waste scenarios*. Academic Press.

Park, P.-J., Tahara, K., Jeong, I.-T., & Lee, K.-M. (2006). Comparison of four methods for integrating environmental and economic aspects in the end-of-life stage of a washing machine. *Resources, Conservation and Recycling, 48*(1), 71–85. doi:10.1016/j.resconrec.2006.01.001

Pollock, D., & Coulon, R. (1996). Life cycle assessment of an inkjet print cartridge. *Proceedings of the 1996 IEEE International Symposium on Electronics and the Environment, 1996, ISEE-1996*, 154–160. 10.1109/ISEE.1996.501870

Prek, M. (2004). Environmental impact and life cycle assessment of heating and air conditioning systems, a simplified case study. *Energy and Building, 36*(10), 1021–1027. doi:10.1016/j.enbuild.2004.06.008

Qu, W., Bi, X., Sheng, G., Lu, S., Fu, J., Yuan, J., & Li, L. (2007). Exposure to polybrominated diphenyl ethers among workers at an electronic waste dismantling region in Guangdong, China. *Environment International, 33*(8), 1029–1034. doi:10.1016/j.envint.2007.05.009 PMID:17618686

Rucevska, I., Nellemann, C., Isarin, N., Yang, W., Liu, N., Yu, K., Sandnæs, S., Olley, K., McCann, H., Devia, L., Bisschop, L., Soesilo, D., Schoolmeester, T., Henriksen, R., & Nilsen, R. (2015). *Waste crime – waste risks: Gaps in meeting the global waste challenge. A UNEP Rapid Response Assessment.* United Nations Environment Programme and GRID-Arendal, Nairobi and Arendal. http://www.grida.no

Schischke, K., & Spielmann, M. (2001). Environmental assessment in production of electronic components – possibilities and obstacles of LCA methodology. *13th Discussion Forum on Life Cycle Assessment*, 9.

Shittu, O. S., Williams, I. D., & Shaw, P. J. (2021). Global E-waste management: Can WEEE make a difference? A review of e-waste trends, legislation, contemporary issues and future challenges. *Waste Management (New York, N.Y.), 120*, 549–563. doi:10.1016/j.wasman.2020.10.016 PMID:33308953

Socolof, M. L., Overly, J. G., & Geibig, J. R. (2005). Environmental life-cycle impacts of CRT and LCD desktop computer displays. *Journal of Cleaner Production, 13*(13-14), 1281–1294. doi:10.1016/j.jclepro.2005.05.014

StEP Green Paper (Solving the e-Waste Problem). (2015). *E-waste prevention, take-back system design and policy approaches*. Academic Press.

StEP (Solving the e-Waste Problem). (2018). http://www.step-initiative.org

StEP (Solving the e-Waste Problem) Annual Report, 2015/2016.

Syafa Bakri, S. N., Surif, S., & Ramasamy, R. K. 2008. A case study of life cycle assessment (LCA) on ballast for fluorescent lamp in Malaysia. *IEEE International Symposium on Electronics and the Environment*, 1–4.

US Government Accountability Office. (2008). *Electronic Waste EPA Needs to Better Control Harmful U.S. Exports through Stronger Enforcement and More Comprehensive Regulation*. United States Government Accountability Office.

Wäger, P. A., Hischier, R., & Eugster, M. (2011). Environmental impacts of the Swiss collection and recovery systems for waste electrical and electronic equipment (WEEE): A follow-up. *The Science of the Total Environment*, *409*(10), 1746–1756. doi:10.1016/j.scitotenv.2011.01.050 PMID:21342702

Wang, D., Cai, Z., Jiang, G., Leung, A., Wong, M. H., & Wong, W. K. (2005). Determination of polybrominated diphenyl ethers in soil and sediment from an electronic waste recycling facility. *Chemosphere*, *60*(6), 810–816. doi:10.1016/j.chemosphere.2005.04.025 PMID:15946727

Wang, F., Leung, A. O. W., Wu, S. C., Yang, M. S., & Wong, M. H. (2009a). Chemical and ecotoxicological analyses of sediments and elutriates of contaminated rivers due to e-waste recycling activities using a diverse battery of bioassays. *Environmental Pollution*, *157*(7), 2082–2090. doi:10.1016/j.envpol.2009.02.015 PMID:19299057

Wang, J.-P., & Guo, X.-K. (2006). Impact of electronic wastes recycling on environmental quality. *Biomedical and Environmental Sciences*, *19*, 137–142. PMID:16827186

Wang, T., Fu, J., Wang, Y., Liao, C., Tao, Y., & Jiang, G. (2009b). Use of scalp hair as indicator of human exposure to heavy metals in an electronic waste recycling area. *Environmental Pollution*, *157*(8-9), 2445–2451. doi:10.1016/j.envpol.2009.03.010 PMID:19346038

Wang, Y., Luo, C., Li, J., Yin, H., Li, X., & Zhang, G. (2011). Characterization of PBDEs in soils and vegetations near an e-waste recycling site in South China. *Environmental Pollution*, *159*(10), 2443–2448. doi:10.1016/j.envpol.2011.06.030 PMID:21763040

Widmer, R., & Lombard, R. (2005). *E-waste Assessment in South Africa a Case Study of the Gauteng Province*. EMPA.

Widmer, R., Oswald-Krapf, H., Sinha-Khetriwal, D., Schnellmann, M., & Böni, H. (2005). Global perspectives on e-waste. *Environmental Impact Assessment Review*, *25*(5), 436–458. doi:10.1016/j.eiar.2005.04.001

Wingfield, S., & Lim, M. (2022). *The United Nations Basel convention's global plastic waste partnership: history, evolution and progress. In Microplastic Environment Pattern Process*. Springer.

Xing, G. H., Chan, J. K. Y., Leung, A. O. W., Wu, S. C., & Wong, M. H. (2009). Environmental impact and human exposure to PCBs in Guiyu, an electronic waste recycling site in China. *Environment International*, *35*(1), 76–82. doi:10.1016/j.envint.2008.07.025 PMID:18804865

Yanagitani, K., & Kawahara, K. (2000). LCA study of air conditioners with an alternative refrigerant. *The International Journal of Life Cycle Assessment*, *5*(5), 287–290. doi:10.1007/BF02977581

Zhao, G., Wang, Z., Dong, M. H., Rao, K., Luo, J., Wang, D., Zha, J., Huang, S., Xu, Y., & Ma, M. (2008). PBBs, PBDEs, and PCBs levels in hair of residents around e-waste disassembly sites in Zhejiang Province, China, and their potential sources. *The Science of the Total Environment*, *397*(1-3), 46–57. doi:10.1016/j.scitotenv.2008.03.010 PMID:18439655

Zheng, L., Wu, K., Li, Y., Qi, Z., Han, D., Zhang, B., Gu, C., Chen, G., Liu, J., Chen, S., Xu, X., & Huo, X. (2008). Blood lead and cadmium levels and relevant factors among children from an e-waste recycling town in China. *Environmental Research*, *108*(1), 15–20. doi:10.1016/j.envres.2008.04.002 PMID:18514186

Chapter 14
Nanotechnology for Sustainability

Yashaswini Reddy Challa
Woxsen University, India

Nandaram Priyanka
Woxsen University, India

S. Preethi Lahari
Woxsen University, India

Elesetty Surya
Woxsen University, India

ABSTRACT

Nanotechnology is one of the most relevant technical factors that have applications in different types of industrial sectors, and it seems to have a consistent scope of technical innovation that focuses on technical derivation with sustainable outcomes. In this study, the perception of nanotechnology and its prominent usage with potential outcomes in different industrial activities has been derived. In addition to this, the study has also reflected on the advantages along with some of the negative factors that come with nanotechnology and its integration. The study helps to evaluate the perception of understanding nanotechnology along with the purpose of evaluating the correlation between the social specifications and commercial attributes.

INTRODUCTION

The perception of nanotechnology is eventually regarded as the branch of science that helps to focus on the purpose of producing design along with the derivation of structural factors that helps to evaluate systems by manipulating the atoms and molecules. The process of manipulation incorporated in nanotechnology is more derived at a nanoscale level. Nanotechnology is often regarded as a small solution to different and big problems that help to derive the perception of producing nanoparticles that can be of great use

DOI: 10.4018/979-8-3693-1018-2.ch014

in different industrial sectors (Kirtane *et al.* 2021). Nanotechnology is heavily used in different types of industrial activities and it is extremely beneficial in terms of improving the manufacturing process and it supports the development and infrastructure revaluation of the production process. This helps to make the use of nanotechnology highly prominent in different industrial areas such as the engineering sector, medical sector, food industry, Information technology and communication sector, electricity industry and so on. The specification of nanotechnology is the branch that helps to manipulate the structure of the molecules related to different materials that have a revolutionary application with positive outcomes that makes the purpose of various industrial

activities and infrastructural factors even more sustainable (Sahoo *et al.* 2021). It is important to specify that the economic group and consistent competition easily achieved with nanotechnology in different

ways seem to have a positive impact. It is important to realise that nanotechnology comes at a cost of certain negative attributes that are also important to evaluate in the fields where it is typically used.

LITERATURE REVIEW

The prominent use of quantum nanoscience is one of the theoretical factors that has typically been incorporated in deriving man of science development and its technical specifications that help to focus on the quantum mechanism and the use of the coherent quantum effects that helps to focus nanostructure development. In addition to this, the nanotechnology theory is also specific with the conceptualisation of bottom-up and top-down approaches that helps to focus on nano-objects and their construction which is eventually helpful for deriving the purpose of nanotechnology (Heinrich *et al.* 2021). Therefore, it can be easily derived that the quantum science theory is one of the prominent theories that are incorporated in the field of nanotechnology that helps to understand its mechanism and approach of deriving the quantum development that helps to understand the purpose of using nanotechnology as a whole (Shen and Dierking, 2019). Nanotechnology also focuses on the approach of engineering specifications that helps to derive structurization, production process and designing of attributes that help to focus on atom manipulation at the nanoscale level.

BENEFITS

The worldwide market of nanotechnology is one of the major markets that seem to have prominent aspects and consistent commercial usage. It is important to specify that there are some consistent advantages that come with nanotechnology. It has been found that the use of nanotechnology has consistently changed the electronic goods and their development which is creating more opportunities for commercialisation and human access with sustainable outcomes (Weiss *et al.* 2020). Nanotechnology is one of those technical frameworks that is constantly being used in the medical field work and the healthcare sector is heavily facilitated by the potential uses of nanotechnology. Nanotechnology is making benefits for the medical industry as it is affecting the process of diagnosis and medication evaluation following a treatment process that remains extremely beneficial for focusing on patient care specifications. Nanotechnology in the medical field is also helping to eliminate the trial and error method in terms of drug prescription to the patients which is making healthcare services even more sustainable and accurate (Zhou *et al.* 2020). Nanotechnology has even more advantages as it helps to focus on the feasibility to cure different types of illnesses to a great extent as it incorporates various equipment and tools that are specific for diagnostic procedures.

The engineering and the manufacturing industry is considering the framework of nanotechnology which is eventually extremely beneficial in terms of producing nano-products that are more resilient and powerful and along with that their equivalent to other non-nanotechnology products. Nanotechnology has made a favourable transformation and manufacturing process in a great way that is developing sustainable products (Larsson *et al.* 2019). Coming to the field associated with energy production it has been observed that nanotechnology is one of the most prominent techniques that has been extremely beneficial in terms of developing energy technologies that are more sufficient and sustainable with long term usage. Therefore the energy production along with the conceptualization of nanotechnology has typically increased the scope of energy saving technology and their development which can be easily availed by humans along with the approach of increasing an ability which is more sufficient in terms of deriving economic sustainability as well (Ghiasi *et al.* 2021). Nanotechnology makes the approach of energy storage in the devices even more sustainable which helps to specify effective operation. In the field of computing and electronics, the potential use of nanotechnology has helped to transform the electronic industry in such a way that it is creating more changes and possibilities that makes computing and electronic service highly sustainable (Babatunde *et al.* 2019). These are some of the prominent advantages that come with nanotechnology in different fields and industries.

DISADVANTAGES

The conceptualization of nanotechnology seems to have considerable disadvantages that are important to understand in order to specify its application. During the process of evaluating nanotechnology the production of nanoparticles is usually taken into consideration and this is eventually affecting the pollution level in barren rent which has increased majorly because of the nanoparticle. These nanoparticles are coming from various drugs and pharmaceutical activities following engineering works and the atomic bomb development process which is evaluated as a major environmental concern (Porcari *et al.* 2019). The implementation of nanotechnology is evaluated in such a manner that its technical factors are eventually reducing the use of human labour which is eventually creating complications in terms of deriving employment opportunities for individuals leading to a major economic imbalance that remains impactful. In most of the health care attributes and treatment related services nanoparticles or nanotechnology are consistently being used which is eventually creating the scope of inhaling nanoparticles by humans. This is eventually creating the white suitability of nanoparticles which is eventually reflecting a slow impact on the human host which is one of the most complicated factors that comes with the incorporation of nanotechnology that has serious health implications. It has also been observed that nanotechnology is one of the most risky and highly dangerous technologies that lead to major consequences if it is not handled accurately. In addition to this it is more costly which is making various medical processes, engineering activities and environmental factors more expensive and the average man is unable to access such technology for their own benefit (Sportelli *et al.* 2020). The perception of nanotechnology is also creating weapons that are highly dangerous and can be easily accessed by the common man which is typically going to have a social complication and can lead to susceptible human error that seems to be extremely negative for the environment.

METHODOLOGY

Statement of Problem

There have been several resources that helped to understand the positive factors that come with nanotechnology and its incorporation in different sectors and industrial frameworks. There have been notable prominent changes that have helped to increase the sustainability of various industries that have optimised for nanotechnology and its comprehensive usage. However, it is important to specify that the research additionally needs to focus on the practical and comprehensive negative factors that come with nanotechnology and the way it is seriously having an impact on social and commercial specification (Shang *et al.* 2019). More essentially it is important to evaluate that the study needs to understand the ethical concerns as well that come with the implementation of nanotechnology which is eventually reflecting various complications that come with the incorporation of nanotechnology in various industrial activities. The main agenda of this study is to synthesise the factual impact of nanotechnology and its integration in different fields that helps to focus on both the positive and negative factors along with the ethical attributes that are associated with it.

RESEARCH QUESTIONS

In association with the selected study, there are some prominent research questions that must be answered and are mentioned below.

- What are the main specifications that help to perceive the conceptualization of nanotechnology and the way it can be utilised for future growth and development in different industrial frameworks?
- What are the factors that potentially influence the application of nanotechnology evaluated services and make them more sustainable?
- How effectively is it important to utilise the perception of nanotechnology and its technical factors that will help to reduce the complications of environmental aspects and focus on the ethical value and its maintenance as well?
- Why is it important to use nanotechnology in the field of healthcare and derived by medical approaches along with a diagnosis of diseases and providing relevant treatment?
- What are the positive factors that come with the evaluation of nanotechnology in different industrial activities and make it more optimal for deriving sustainable outcomes?
- What are the different factors that seem to be extremely impactful in terms of deriving nanotechnology in various industrial activities and reducing its capability of acceptance due to ethical concerns?
- What are the social and ethical concerns that are associated with the functionality and integration of nanotechnology in different sectors and industrial factors?

RESEARCH OBJECTIVES

The following are the objectives that are associated with this study.

- To specify the prominent factors those are associated with the perception of nanotechnology and its performance manifestation.
- To evaluate the considerable advantages those come with the incorporation of nanotechnology in the different industrial activities specially.
- To recognise the concerning areas that are influencing the negative factors those are associated with the perception of nanotechnology in different industrial frameworks.
- To understand the ethical issues that come with the integration of nanotechnology and the way it is affecting the environment is sustainability in several aspects facilitating its after effects as well.
- The aim of the study is to investigate the conventional factors that facilitate nanotechnology and focus on the prominent concerns that will help to reduce the concerns of nanotechnology and use it significantly without any restrictions in different industrial activities.
- The focus of the study is to evaluate the purpose of understanding nanotechnology in the field of biomedical aspects.

RESEARCH METHODOLOGY

The perception of research methodology helps to deal with the prominent techniques and tools that have been typically integrated with the selected study. In association with this study the incorporation of both the qualitative research methodology along with the quantitative research methodology can be easily used that would help to focus on the derivation of nanotechnology (Strijker *et al.* 2020). The incorporation of qualitative research methodology will help to focus on the secondary resources that have already been established and have authentic information that is associated with them that would help to understand the perception of nanotechnology. The quantitative research methodology will help to focus on the evaluation of statistical analysis along with data evaluation and this will help to focus on the primary data collection process related to nanotechnology and its specification in different (Taherdoost, 2022). The combination of both the methodology has been selected reflecting on a mixed methodological approach that will help to specify the paper.

HYPOTHESIS

The following is the hypothesis that is relevant to the study:

- Nanotechnology has prominent advantages in different production processes and manufacturing aspects in various engineering and electricity industries which are going to enhance performances. This is the first hypothesis of the study.
- The second hypothesis of the study is that nanotechnology can immediately change the perception of industrial activities and make it more relevant to the sustainable outcomes that it comes with them.
- The final hypothesis associated with this study helps to specify the understanding of nanotechnology and its possibilities that will make the biomedical industry even more sustainable and effective with their approaches by considering the environment to consequences at the same time.

SCOPE AND LIMITATIONS

In association with the chosen study, it is important to specify that the subject matter of this paper helps to focus on the scope of the varied uses of nanotechnology in different fields and industrial activities. The effectiveness of understanding nanotechnology along with the negative factors is also important and remains an important part of the selected research and its scope that helps to focus on the evaluation of relevant information that helps to understand the conceptualization of nanotechnology and its application in specifications (Chauhan *et al.* 2020).

Through the study, it has been observed that the different aspects related to nanotechnology have been used for understanding the conceptualization of this technology. However, the study does not account for the perception of the negative consequences in a given manner that would have helped to focus on both the positive and negative factors in a sustainable manner which is eventually becoming a limitation related to this time (Martinelli *et al.* 2020). This has also been reflected in terms of accessing nanotechnology in specific fields and its application process which is also a major gap.

DISCUSSION

Nanotechnology in the Field of Engineering

In the field of engineering nanotechnology is one of the most prominent technical factors that is important and it is used for the purpose of manipulation related to the nanomaterials. The technical conceptualisation of nanotechnology is useful in terms of converting vital materials and structures into manipulated nanomaterials and this branch of engineering was eventually specified as NanoEngineering. Especially when it comes to the industrial commercialisation process nanotechnology is more specified with the approach of deriving the procedural activity that helps to manifest with the composition and derivation that supports modification and diversification of molecular factors (Singh *et al.* 2019). The most primary reason that helps to derive nanotechnology in the field of engineering is that it helps to enhance the overall quality of materials and it eventually helps to reduce various issues in terms of material diversification as it makes it more efficient and sustainable. There are different attributes that are associated with the field of engineering and have various subsections. The purpose of deriving nanotechnology has helped to focus on the development of materials that are more sustainable and self-repairing and have self-cleaning properties in them.

Nanotechnology in the field of engineering is typically used for Information technology engineering and mechanical activities along with other engineering facilities that help to derive the conceptualization of product consumption and diversification process in the most sustainable manner. Moreover in the field of electricity derivation nanotechnology is also used as it helps to focus on the internal engineering required for electronic specifications and helps to put outputs that are more conductive and strong for electrical usage. In addition to this, it is predicted that in the field of engineering and its different sections, nanotechnology is eventually regarded as the prominent feature that helps to enhance supportive developments in the field of engineering. The department of civil engineering heavily uses nanotechnology and it focuses on the scope of sustaining the structural and production and development in civil developments (Abdussalam-Mohammed, 2019). More essentially it can be evaluated that the importance of nanotechnology has increased massively as there are consistent developments that are taking in its approach which is eventually making the integration of this technology in the field of engineering even more compulsory. These are some of the prominent factors that are associated with nanotechnology and engineering.

Biomedicine Industry Using Nanotechnology

The prominent specifications that are associated with nanotechnology are a useful branch of technology that helps to manipulate the structural evaluation of molecular materials which eventually helps to change the properties of a substance that helps to integrate revolutionary applications in different arenas. In the field of biomedicine, nanotechnology helps to focus on the improvement of various diagnosis processes and insurance for the treatment of different types of neurodegenerative diseases and specifically for cancer treatment it is used. Moreover, the nanotechnologies that are typically used for deriving nanoparticles seem to have a consistent impact in terms of improving the pharmaceutical products that are used for medicinal purposes (Wu *et al.* 2021). More essentially the combination of biomedical technology is more specific with biology and nanotechnology incorporation that helps to focus on innovative technology that is beneficial for the improvement of biological processes by considering the nanoscale that helps to enhance the medical practices and eventually, it specifies the different types of biomaterials with accurate measurement evaluation. Moreover, the perception of increasing therapeutic development and focusing on a stable approach to deriving sustainable medical Services is heavily facilitated by the application of nanotechnology that helps to focus on biomedical development.

The perception of deriving micro-scalability is made possible by integrating biotechnology and this helps to focus on different types of diseases such as neurological diseases, cancer, different types of chronic diseases and various other viral and common diseases that can be easily reduced. Therefore it can be easily derived that the application of nanotechnology is more constructive and it helps to focus on the consistent scope of development using technical specifications that possibly focuses on the evaluation of accurate treatment. Especially when it comes to the field of diagnostics, nanotechnology is helpful in terms of focusing on the nano scaling purpose that eventually helps to derive nanomedicine specification which is also a part of nanotechnology (Javed *et al.* 2020). This is eventually specific that nanotechnology is also helpful in deriving drug delivery and this plays an important role in terms of the delivery of drugs for the improvement of pharmaceutical development which is eventually one of the core parts associated with the biomedicine industry. These are some of the prominent aspects that are associated with the industry of biomedical factors to that of nanotechnology.

Industrial Use of Nanotechnology

The industrial specifications of nanotechnology are regarded as one of the most influencing techniques that have helped to specify in different types of technology such as the electricity industry, engineering industry, medical sector and many more. It is important to understand that the perception of utilising nanotechnology is more specific and it is important to focus on certain factors that help to ensure that nanotechnology is used in a sustainable manner (Mandeep and Shukla, 2020). The perception of nanotechnology is typically used in different types of industries and it is majorly evaluated in different types of industrial work functions and this makes the purpose of deriving various opportunities that supports the consistent scope of evaluation and production process that is more composite in nature. When it comes to the industry of food and cosmetics nanotechnology is one of the acute technical factors that has helped to specify on the approach of food processing and packaging (Garcia-Segura *et al.* 2020). Nanotechnology has helped to enhance the bioavailability of different types of nutrients and helps to preserve the food in a sustainable way that can be easily consumed by individuals once the food has been processed accordingly. Therefore it can be easily evaluated that biotechnology is eventually creating more sustainable opportunities for the food sector that helps to focus on the development of food and derive the mechanical consistency which is eventually creating the scope of improving the food production and its quality maintenance as well.

In the industry of cosmetics and skin care, biotechnology is heavily used as it helps to focus on the drug related derivation and also helps to derive some medicinal attributes which help to develop the cosmetics and evaluate its production process accordingly. This is eventually one of the most core factors that help to focus on the advantages of integrated nanotechnology in different types of industrial works that helps to focus on the conceptualization of biodegradability along with compatibility of various types of nanomaterials that has a prominent impact on commercial usage. This technology has a prominent aspect and is constantly being used for biotechnology purposes and eventually helps to focus on the development of various medication opportunities as well (Sim and Wong, 2021). More specifically in most of the industry activities nanotechnology acts as a modification process that helps to the right manner of structuralisation and focuses on the derivation of functional compounds that is highly required for establishing industrial activities in different types of industries specifically. The textile industry is also consistently using nanotechnology that helps to derive their activities and operational areas.

Nanotechnology and Environmental Science

Environmental science is one of the four industrial activities where the perception of nanotechnology is consistently being used and it is one of the most influential factors that helps to stabilise environmental activities. More essential the impact of nanotechnology is typically used in the field of environmental science and development to focus on the purpose that helps to reduce the generation of pollution and eventually derive the manufacturing process in such a manner that will help to generate sustainable opportunities that support environmental development (Zhang *et al.* 2019). It is important to realise that there are various factors that facilitate the scope of driving nanotechnology. However, in the field of environmental science nanotechnology is typically used for deriving the scope of sustainability and eventually, it helps to understand the devastated nature of the ecosystem based on which prominent factors can be derived that will help to support environmental sustainability at its best. Nanotechnology also helps to focus on electricity generation in a sustainable way which is also a prominent environmental issue that can be easily reduced if used consistently (Gupta, 2020). The positive factors that help to focus on the development of recycling and maintaining the ecological balance can be easily optimised by considering nanotechnology evaluation that helps to derive nanoparticles that are more sustainable for environmental treatment and development.

It is important to specify that there are positive factors that come with nanotechnology on environmental science and ecological factors but there are some prominent negative aspects that come with nanotechnology. It has been observed that nanotechnology can have numerous impacts in terms of producing toxicity because of the nanoparticles that are produced during the process of nanotechnology. Nanoparticles and their exposure to the environment eventually reduce the sustainable derivation of the ecological factors which is eventually producing more toxic materials (Zhang *et al.* 2021). It has also been observed that the particles that are evaluated from nanotechnology have certain complications in terms of deriving their recycling ability and the perception of sustainability which is eventually making the society have prominent concerns that come with the incorporation of nanotechnology in various industrial activities.

CONFLICT OF INTEREST

While discussing the different aspects that are associated with the perception of nanotechnology and its impact on different industrial activities, a prominent conflict of interest can occur if the research is eventually creating certain economical aspects or considering personal attributes that are eventually addressing nanotechnology. The scope of neutralisation in association with the chosen topic of the study can eventually come at a point of biasness which is eventually one of the critical conflicts of interest that eventually needs complications in terms of establishing the outcome of the study (Gottardo *et al.* 2021). In addition to this another significant conflict of interest is associated with the lack of transparency maintenance with the financial factors and other social activities that are associated with nanotechnology and its specification in different types of industries. However, it is important to focus on the neutral approach of the research that will help to focus on the derivation of nanotechnology without addressing biases.

CONCLUSION AND RESULTS

On the conclusion of the selected study it can be easily derived that there are some prominent advantages and the consistent use of nanotechnology in different fields seems to have both benefits and negative impact that needs to be evaluated in the commercial and social context. Nanotechnology is one of the most valuable factors that has predominantly been used in different types of industrial activities that have helped to improve the sustainability of those industries and their operation and activities that has made it more strategic and well oriented with the outcomes specifications (Kumhar, 2020). However, there are some after effects that are associated with the nanotechnology evaluation which is eventually regarded as highly complex and it can lead to various consequences which must be taken into consideration before the implementation of nanotechnology.

The research helps to indicate that there are some prominent advantages that come with nanotechnology and it is heavily used in different types of fields that make it more sustainable for the scope of social development that supports commercialisation. Nanotechnology has used technical attributes that make things more effective and focuses on the durability and strong evaluation that has benefits for various industrial activities. These are the main factors that have been evaluated in this study and it helps to understand the responsibilities along with the social, commercial and ethical factors that help to coordinate the understanding of nanotechnology in different types of fields (Sridhar *et al.* 2021). More essentially the study has helped to support that nanotechnology is becoming one of the most powerful technical factors that is eventually having various changes in the environment and industrial specifications focusing on the scope of sustainability.

REFERENCE

Abdussalam-Mohammed, W. (2019). Review of therapeutic applications of nanotechnology in medicine field and its side effects. *Journal of Chemical Reviews*, *1*(3), 243–251. doi:10.33945/SAMI/JCR.2019.3.5

Babatunde, D. E., Denwigwe, I. H., Babatunde, O. M., Gbadamosi, S. L., Babalola, I. P., & Agboola, O. (2019). Environmental and societal impact of nanotechnology. *IEEE Access : Practical Innovations, Open Solutions*, *8*, 4640–4667. doi:10.1109/ACCESS.2019.2961513

Chauhan, G., Madou, M. J., Kalra, S., Chopra, V., Ghosh, D., & Martinez-Chapa, S. O. (2020). Nanotechnology for COVID-19: Therapeutics and vaccine research. *ACS Nano*, *14*(7), 7760–7782. doi:10.1021/acsnano.0c04006 PMID:32571007

Garcia-Segura, S., Qu, X., Alvarez, P. J., Chaplin, B. P., Chen, W., Crittenden, J. C., Feng, Y., Gao, G., He, Z., Hou, C. H., Hu, X., Jiang, G., Kim, J.-H., Li, J., Li, Q., Ma, J., Ma, J., Nienhauser, A. B., Niu, J., ... Westerhoff, P. (2020). Opportunities for nanotechnology to enhance electrochemical treatment of pollutants in potable water and industrial wastewater–a perspective. *Environmental Science. Nano*, *7*(8), 2178–2194. doi:10.1039/D0EN00194E

Ghiasi, T. S., Kaverzin, A. A., Dismukes, A. H., de Wal, D. K., Roy, X., & van Wees, B. J. (2021). Electrical and thermal generation of spin currents by magnetic bilayer graphene. *Nature Nanotechnology*, *16*(7), 788–794. doi:10.103841565-021-00887-3 PMID:33958763

Gottardo, S., Mech, A., Drbohlavová, J., Małyska, A., Bøwadt, S., Sintes, J. R., & Rauscher, H. (2021). Towards safe and sustainable innovation in nanotechnology: State-of-play for smart nanomaterials. *NanoImpact*, *21*, 100297. doi:10.1016/j.impact.2021.100297 PMID:33738354

Gupta, S. K. (2020). Study of Nanotechnology and Its Application. *J Phys Opt Sci*, *2*(1), 1–7.

Heinrich, A. J., Oliver, W. D., Vandersypen, L. M., Ardavan, A., Sessoli, R., Loss, D., Jayich, A. B., Fernandez-Rossier, J., Laucht, A., & Morello, A. (2021). Quantum-coherent nanoscience. *Nature Nanotechnology*, *16*(12), 1318–1329. doi:10.103841565-021-00994-1 PMID:34845333

Javed, R., Zia, M., Naz, S., Aisida, S. O., Ain, N. U., & Ao, Q. (2020). Role of capping agents in the application of nanoparticles in biomedicine and environmental remediation: Recent trends and future prospects. *Journal of Nanobiotechnology*, *18*(1), 1–15. doi:10.118612951-020-00704-4 PMID:33225973

Kirtane, A. R., Verma, M., Karandikar, P., Furin, J., Langer, R., & Traverso, G. (2021). Nanotechnology approaches for global infectious diseases. *Nature Nanotechnology*, *16*(4), 369–384. doi:10.103841565-021-00866-8 PMID:33753915

Kumhar, B. L. (2020). Concept and Approaches of Nano Technology. *Agri Mirror: Future India*, *1*(2), 19–23.

Larsson, S., Jansson, M., & Boholm, Å. (2019). Expert stakeholders' perception of nanotechnology: Risk, benefit, knowledge, and regulation. *Journal of Nanoparticle Research*, *21*(3), 1–17. doi:10.100711051-019-4498-1

Mandeep, & Shukla, P. (2020). Microbial nanotechnology for bioremediation of industrial wastewater. *Frontiers in Microbiology*, *11*, 590631. doi:10.3389/fmicb.2020.590631 PMID:33224126

Martinelli, C., Pucci, C., Battaglini, M., Marino, A., & Ciofani, G. (2020). Antioxidants and nanotechnology: Promises and limits of potentially disruptive approaches in the treatment of central nervous system diseases. *Advanced Healthcare Materials*, *9*(3), 1901589. doi:10.1002/adhm.201901589 PMID:31854132

Porcari, A., Borsella, E., Benighaus, C., Grieger, K., Isigonis, P., Chakravarty, S., Kines, P., & Jensen, K. A. (2019). From risk perception to risk governance in nanotechnology: A multi-stakeholder study. *Journal of Nanoparticle Research*, *21*(11), 1–19. doi:10.100711051-019-4689-9

Sahoo, M., Vishwakarma, S., Panigrahi, C., & Kumar, J. (2021). Nanotechnology: Current applications and future scope in food. *Food Frontiers*, *2*(1), 3–22. doi:10.1002/fft2.58

Shang, Y., Hasan, M. K., Ahammed, G. J., Li, M., Yin, H., & Zhou, J. (2019). Applications of nanotechnology in plant growth and crop protection: A review. *Molecules (Basel, Switzerland)*, *24*(14), 2558. doi:10.3390/molecules24142558 PMID:31337070

Shen, Y., & Dierking, I. (2019). Perspectives in liquid-crystal-aided nanotechnology and nanoscience. *Applied Sciences (Basel, Switzerland)*, *9*(12), 2512. doi:10.3390/app9122512

Sim, S., & Wong, N. K. (2021). Nanotechnology and its use in imaging and drug delivery. *Biomedical Reports*, *14*(5), 1–9. doi:10.3892/br.2021.1418 PMID:33728048

Singh, A., Dubey, S., & Dubey, H. K. (2019). Nanotechnology: The future engineering. *Nanotechnology*, *6*(2), 230–233.

Sportelli, M. C., Izzi, M., Kukushkina, E. A., Hossain, S. I., Picca, R. A., Ditaranto, N., & Cioffi, N. (2020). Can nanotechnology and materials science help the fight against SARS-CoV-2? *Nanomaterials (Basel, Switzerland)*, *10*(4), 802. doi:10.3390/nano10040802 PMID:32326343

Sridhar, A., Ponnuchamy, M., Kumar, P. S., & Kapoor, A. (2021). Food preservation techniques and nanotechnology for increased shelf life of fruits, vegetables, beverages and spices: A review. *Environmental Chemistry Letters*, *19*(2), 1715–1735. doi:10.100710311-020-01126-2 PMID:33192209

Strijker, D., Bosworth, G., & Bouter, G. (2020). Research methods in rural studies: Qualitative, quantitative and mixed methods. *Journal of Rural Studies*, *78*, 262–270. doi:10.1016/j.jrurstud.2020.06.007

Taherdoost, H. (2022). What are different research approaches? Comprehensive review of qualitative, quantitative, and mixed method research, their applications, types, and limitations. *Journal of Management Science & Engineering Research*, *5*(1), 53–63. doi:10.30564/jmser.v5i1.4538

Weiss, C., Carriere, M., Fusco, L., Capua, I., Regla-Nava, J. A., Pasquali, M., Scott, J. A., Vitale, F., Unal, M. A., Mattevi, C., Bedognetti, D., Merkoçi, A., Tasciotti, E., Yilmazer, A., Gogotsi, Y., Stellacci, F., & Delogu, L. G. (2020). Toward nanotechnology-enabled approaches against the COVID-19 pandemic. *ACS Nano*, *14*(6), 6383–6406. doi:10.1021/acsnano.0c03697 PMID:32519842

Wu, D., Zhou, J., Creyer, M. N., Yim, W., Chen, Z., Messersmith, P. B., & Jokerst, J. V. (2021). Phenolic-enabled nanotechnology: Versatile particle engineering for biomedicine. *Chemical Society Reviews*, *50*(7), 4432–4483. doi:10.1039/D0CS00908C PMID:33595004

Zhang, P., Guo, Z., Ullah, S., Melagraki, G., Afantitis, A., & Lynch, I. (2021). Nanotechnology and artificial intelligence to enable sustainable and precision agriculture. *Nature Plants*, *7*(7), 864–876. doi:10.103841477-021-00946-6 PMID:34168318

Zhang, T., Lowry, G. V., Capiro, N. L., Chen, J., Chen, W., Chen, Y., Dionysiou, D. D., Elliott, D. W., Ghoshal, S., Hofmann, T., Hsu-Kim, H., Hughes, J., Jiang, C., Jiang, G., Jing, C., Kavanaugh, M., Li, Q., Liu, S., Ma, J., ... Alvarez, P. J. J. (2019). In situ remediation of subsurface contamination: Opportunities and challenges for nanotechnology and advanced materials. *Environmental Science. Nano*, *6*(5), 1283–1302. doi:10.1039/C9EN00143C

Zhou, J., Kroll, A. V., Holay, M., Fang, R. H., & Zhang, L. (2020). Biomimetic nanotechnology toward personalized vaccines. *Advanced Materials*, *32*(13), 1901255. doi:10.1002/adma.201901255 PMID:31206841

Chapter 15
Opportunities and Challenges in Solar Photovoltaic Waste Management

Bhanu Prakash Saripalli
Woxsen University, India

Adip Krishna Guduru
Woxsen University, India

Jayaditya Reddy Yeruva
Woxsen University, India

ABSTRACT

With the current global overexploitation and depletion of coal resources, it has become necessary to shift toward renewable energy sources to fulfil the world's energy requirements. Among these alternatives, solar energy stands out as the most promising solution. The sun, known for its life-enhancing properties, showers the Earth with an abundant amount of energy. In the case of India, for instance, the country receives an impressive 5000 trillion kWh of energy annually, with a substantial portion of the nation enjoying daily solar radiation ranging from 4 to 7 kWh per square meter. India has demonstrated worthy expertise in harnessing solar energy. Following coal extraction, solar energy developed as the most dependable and widespread energy source capable of meeting the world's energy demands. The study addresses the essential facts of solar waste management as it currently exists in India in addition to the challenges related to managerial policy. Waste management will improve as solar panel installations expand.

1. INTRODUCTION

Solar energy has emerged as a prominent solution to provide for the rising energy needs of expanding populations. Solar photovoltaic technology efficiently converts solar energy into electricity, contributing

DOI: 10.4018/979-8-3693-1018-2.ch015

to climate change mitigation. While the progress and adoption of solar photovoltaics have positively impacted energy system decarbonization, it is crucial to address the proper disposal of end-of-life solar panels to prevent potential toxic waste concerns. As solar installations continue to expand globally, there is a growing emphasis on effectively managing the end-of-life phase of solar PV panels. These panels house valuable metals like silicon, silver, and copper, which can be salvaged and reused, thus lessening the need for new raw materials. This trend aligns with the principles of a circular economy, where materials are recycled and repurposed, minimizing waste. Moreover, the domain of solar PV waste management carries the potential to generate employment opportunities, spanning roles in collection, recycling, and component reclamation. The ultimate concern is environmental protection. Currently, India is projected to generate around 200,000 tonnes of solar photovoltaic waste by 2030, and this figure is anticipated to escalate to 1.8 million tonnes by 2050. Globally, solar waste might surge to an estimated 60 million tonnes by that time. To curb the adverse impact of ongoing expansion, solar waste has been newly classified under waste electrical and electronic equipment, aiming to mitigate its potential repercussions (Rathore, N., & Panwar, N. L., 2022). India has only recently started using solar panels; therefore, the nation now has little waste building up. However, predictions show that the amount of this waste is expected to increase dramatically, ranging from 4.4 to 7.5 million tonnes by the year 2050. Despite India's well-developed infrastructure for recycling electronic waste, there are currently no explicit standards for the end-of-life handling of solar PV panels. Instead, the General Waste Law's guidelines for waste management apply to how their waste is handled (Majewski, P., et al., 2021). Sufficient disposal and recycling processes are imperative to prevent the release of hazardous substances, such as cadmium and lead, inherent in solar panels, into the ecosystem. Establishing a robust recycling infrastructure remains a significant hurdle in various regions, delaying the efficient handling of accelerating waste volumes. Due to the hazardous elements present, careful management and disposal techniques are required to avert environmental contamination. A lack of uniform regulatory frameworks and policies for PV waste management across differing jurisdictions complicates the recycling endeavor. The intricate composition of solar panels, incorporating diverse materials, necessitates innovative technologies to effectively extract and segregate components. The economic feasibility of recycling hinges on factors like recycling costs versus the value of the reclaimed materials. There exists a notable gap in awareness and education among consumers and industries regarding the significance of proper PV waste management, which has led to improper disposal practices. Despite their impressive 25–30-year lifespan of producing clean energy, the improper handling of end-of-life solar photovoltaic modules can pose environmental risks. This concern becomes more critical as the global utilization of photovoltaics expands rapidly, contributing substantial clean energy.

India will face significant challenges in the future when it comes to handling the management and disposal of solar PV waste. This area has been neglected for a long time and is known for its informal, unscientific approach, necessitating immediate attention (Sheoran, Sharma, and Kumar, 2020). The implementation of an online PV waste monitoring system, whether in the early phases or during the end-of-life phase, could be helpful in easing these worries. A system like this would make it easier to track discarded modules extensively, giving regulatory agencies and other stakeholders the information, they need to create useful policies and procedures. Solar PV waste is still substantially excluded from these rules, despite India having adopted the E-waste Management Rules of 2016 (CPCB 2018), which significantly differ from the 2011 version by including the idea of extended producer responsibility (EPR). This raises questions regarding the likelihood of improper disposal, including the use of open dumping grounds or unregulated landfills (EU 2021). The appropriate handling of PV waste is hampered by the lack of incentives, legal frameworks, and awareness of the value of extractable components and their corresponding environmental effects.

Only 10% of all solar PV waste is now recycled globally, which is concerning and indicates a serious lack of operational infrastructure availability (Lunardi et al. 2018). The management of PV waste is a critical matter as the solar industry expands exponentially, which is made worse by the lack of effective legal and regulatory frameworks. Solar technology clearly has a positive impact on the environment, but if adequate disposal procedures are not in place when it reaches the end of its useful life, this impact could become negative. Open dumping of solar panels could result in serious environmental pollution if done improperly. Heavy metals including lead, tin, and cadmium are included in these panels, which is a cause for concern. The waste is harmful because of the presence of these heavy metals. For instance, a typical polycrystalline PV cell has a hazardous material content of 0.15 g/W. This would amount to about 62.26 tonnes of hazardous materials for an expected 33,205.36 tonnes of PV cell waste (Tasnim, S. S., et al. 2022). The environment and public health are at risk since this hazardous waste has the potential to contaminate not only soil but also groundwater and surface water.

1.1 Solar PV Technologies

Solar PV technology can be categorized into three generations: first, second, and third. Crystalline silicon (c-Si) wafer-based cells, which are readily accessible on the consumer market, make up the first generation. The second generation comprises a few thin-film technologies, such as single-junction Gallium Arsenide (GaAs) cells, amorphous Silicon (a-Si) cells, Copper indium gallium di-selenide (CIGS) cells, and Cadmium telluride (CdTe) cells. Because they are produced in large quantities and have several applications, these two generations currently dominate the solar PV market. With a remarkable 93% production capacity, c-Si cells outperform other second-generation technologies, with 24% of that capacity attributable to mono-crystalline and 69% to multi-crystalline technologies. Amorphous silicon and cadmium telluride make up 3% and 2.5%, respectively, of the entire manufacturing of thin films, while copper indium gallium di-selenide makes up less than 2% (Shubbak, M. H. 2019). The third generation of PV technologies, which is geared towards industrialization, is still in the research stage. Multijunction cells, Dye-Sensitized Solar Cells (DSSC), Perovskite Solar Cells (PSC), and Organic Solar Cells (OPV) are among the promising technologies in this area. Future solar PV applications may benefit from these developments, although they have not yet been commercialized due to ongoing research and development activities (Oteng, D., Zuo, J., & Sharifi, E., 2022). Solar PV panels have an average lifespan, therefore if left untreated, those put between 1980 and 1990 are currently in their end-of-life phase. This issue is anticipated to get worse over time. This emphasises the requirement for researching workable strategies for controlling solar panels towards the end of their useful lives. Understanding the construction of solar panels is a vital first step when thinking about recycling them. Although silicon-based and thin-film-based PV panels are the two main varieties based on material composition, the dominance of silicon-based panels—which account for over 95% of the PV market—shifts the primary focus towards the recycling of crystalline silicon-based PV panels (Majewski, P., et al., 2021). Among photovoltaic technologies, silicon modules are the most prevalent. Unfortunately, these modules have demonstrated limited economic viability for recycling due to challenges like material mixture, separation difficulties, and low material value. As a result, the recycling cost is often not offset (Deng, R., et al., 2019). Mono- or poly-crystalline silicon makes up the majority (> 80%) of the installed panels' predominant composition. In addition to silicon, these panels also contain glass, copper, aluminium, tin, and silver.

1.2 End-of-Life (EoL) of PV

Naturally, the production of electrical energy decreases as the efficiency of PV panels ages. The efficiency gradually drops until it reaches a minimum of around 14%, at which point the photovoltaic unit is no longer able to pay its running expenses and eventually reaches the end of its useful life. According to reports from the National Renewable Energy Laboratory (NREL), a cell typically loses 0.5% of its energy annually. After around 25 years of use, the efficiency of the unit approaches a limit where it can no longer be used, which causes the PV units to stop working and become waste. The reduction in efficiency is greatest during the first year, where it typically averages around 2.5%, and then it gradually gets smaller as time goes on. This annual efficiency decline is a result of several variables, including exposure to ultraviolet rays, different weather patterns, dust buildup, probable failures, and poor ability (Jordan, D. C., & Kurtz, S. R., 2013).

Figure 1 illustrates the Impact of India's solar production process on the environment. Waste from solar modules can be categorized based on its economic value, recyclability, and environmental impact. Due to the incorporation of potentially hazardous components during manufacturing and the lack of suitable recycling facilities, managing module waste is a considerable difficulty. The solar industry needs to adopt a thorough closed-loop life-cycle management plan if it wants to back up its promises to be a source of green energy. This strategy is necessary to address important issues including the effects of module production on the environment and the recycling of end-of-life modules.

Figure 1. Impact of India's solar production process on the environment

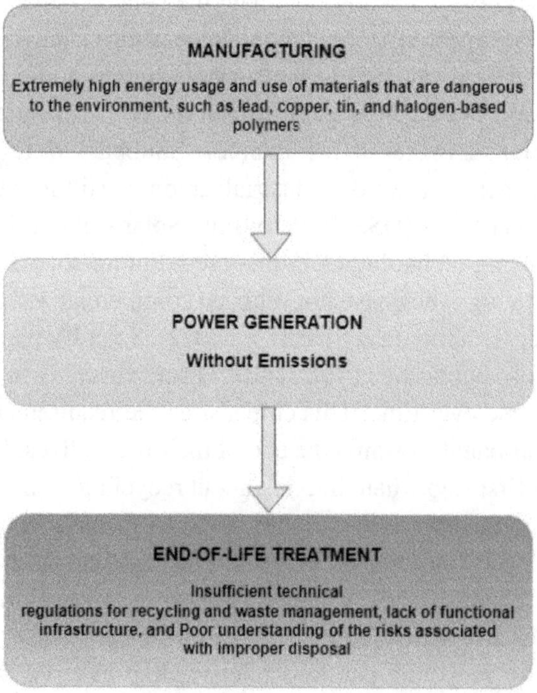

1.3 PV Waste

Solar PV modules invariably reach their end-of-life stage and need to be replaced because all technologies have a finite lifespan. The average usable life of these PV modules is 30 years. PV installations are growing quickly, and as a result, more modules will eventually reach the end of their useful lives. This will result in a corresponding increase in waste. A report from the International Renewable Energy Agency (IRENA) and the International Energy Agency Photovoltaic Power Systems Programme (IEA PVPS) Task12 in 2016 predicted that the cumulative global waste from PV modules would be between 1.7 and 8.0 million tonnes by 2030 and 60 to 78 million tonnes by 2050. Electrical and electronic garbage, or "e-waste," produced a record-breaking 41.8 million metric tonnes annually in 2014. Comparatively, during the same year, the annual global buildup of PV panel waste was 1,000 times smaller. According to predictions, by 2050, the annual growth in PV panel waste might approach 10% of the remarkable worldwide e-waste surge witnessed in 2014 (Weckend, S., Wade, A., & Heath, G. A. 2016).

China, a world leader in the production and use of PV modules, is predicted to gather approximately 3.5 million tonnes of garbage from malfunctioning PV modules by 2030. According to research, carbon steel will make up the majority of PV waste, amounting to about 1.45 million tonnes, followed by glass, amounting to 1.1 million tonnes, and smaller amounts of plastic, aluminium, copper, silicon, and silver. These figures provide a serious environmental concern, too, as there aren't many recycling facilities that can adequately handle PV waste. About 70% of PV modules are made of recyclable glass, and 30% are made of various recyclable metals like silver, copper, and aluminium. However, to properly extract these invaluable metals, suitable dismantling and recycling techniques are required. In addition to having value economically, recycling these metal elements helps manage PV waste in a way that is environmentally sustainable overall (Ghaleb Al-Aqqad, A. O., & Menyhart, J., 2023).

Figure 2. End-of-life solar PV waste for the world's top five nations by 2050

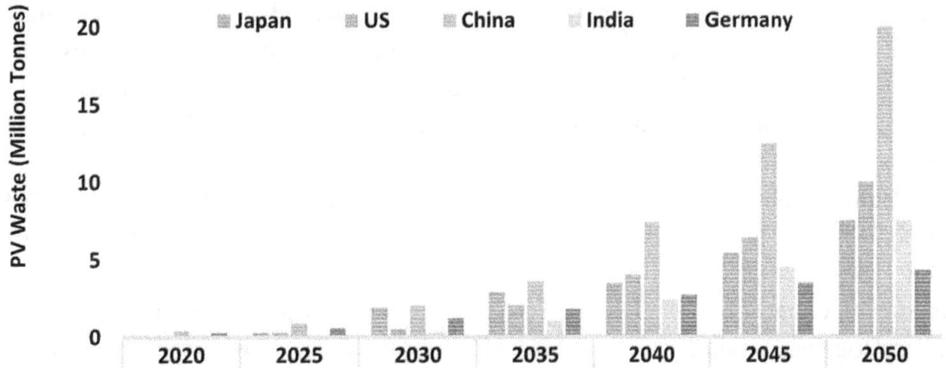

	2020	2025	2030	2035	2040	2045	2050
Japan	0.07	0.3	1.9	2.9	3.5	5.4	7.5
US	0.03	0.3	0.5	2	4	6.4	10
China	0.4	0.9	2	3.6	7.4	12.5	20
India	0	0.09	0.3	1	2.4	4.5	7.5
Germany	0.3	0.6	1.2	1.8	2.7	3.5	4.3

Year

Figure 2 illustrates the End-of-life Solar PV waste for the world's top five nations by 2050. Cumulative PV waste volume data taken from source IRENA 2016 (Weckend, S., Wade, A., & Heath, G. A. 2016). According to this analysis, 0.4 tonnes of China PV panel waste will have accumulated by the year 2020. The forecasts show a sharp increase to between 0.9 and 2.0 million tonnes by 2030, then a sharp increase to 12.5-20 million tonnes by 2050. The government and PV producers are paying more attention to PV panel recycling because of the PV industry's fast growth in China. As a result, China established its own national framework for managing PV panel waste. China, the US, and Japan are likely to be the top three waste originators, followed by India and Germany.

2. OPPORTUNITIES

2.1 Resource Recovery

Solar PV panels contain valuable components including silicon, metals, and glass that can be removed and used to make new circuit boards or for other purposes. Reducing the need for new resources can save money and benefit the environment by properly adopting recycling and material recovery techniques. The field of recycling is currently dominated by mechanical, chemical, and thermal processes. Delamination of the sheets is followed by separation of the integrated parts and refinement to extract the metals. The experimental stage of these approaches is still ongoing, nevertheless. Solar waste recycling is still in its early stages on a worldwide scale. It is necessary to develop a variety of business models on an economic level to support the commercialization of recycling. Only a small handful of economically viable technologies have been able to make a significant impact on recycling business models. Mechanical, thermal, chemical, or a combination of these techniques are the main methods for controlling solar waste. Due to its cost-effectiveness, the mechanical approach is preferred in the modern environment. However, because silicon is extracted in powder form with significant contamination from mechanical grinding, this approach does not allow for the valorization of silicon (Cerchier, P. et al., 2022). The integrity of silicon is also seriously threatened by pure chemical processes, primarily because of ethylene/vinyl acetate copolymer (EVA) swelling. The most promising method for recovering and reusing silicon is the use of thermo-chemical treatment. The polymeric portion is burned during this novel process, which was tested on a small scale prior to industrialization in 2019. To achieve effective recovery and reintegration of the silicon component, this is supplemented by careful material separation and chemical treatment of the silicon cells.

The recyclable components used in PV panels and their composition are outlined in Table 1 (Source Xu, Y. et al., 2018). Front covers, electrical circuit connections for solar cells, a protective envelope made up of two encapsulant layers, and a rear cover (which can be a back sheet or tempered glass) make up the five main components of the widely used c-Si technology. In the context of c-Si technology, glass, and aluminium together account for about 80% of the total weight. Lead, copper, and tin are included in the remaining portion (Rathore, N., & Panwar, N. L., 2022). The present-day handling of c-Si and CdTe PV modules results in a minor portion (<5%) of the overall environmental impacts associated with residential rooftop PV systems. Among the impact categories, PV module recycling has the most significant contribution to climate change due to factors like transport, electricity supply, and waste disposal. Recycling processes for extracting glass, metals, and semiconductor materials from PV modules

Table 1. Materials used in crystalline silicon solar panels

Recyclable Products	Content (kg/kWp
Frame	Al 12.27
Silver bar line	Ag 0.03
Polycrystalline silicon chips	Si 3.10
Adhesive sealant	10%
Glass	54.72
Tinned wire	Cu 0.45
Background	Plastic 17.09

have lower environmental impacts compared to the extraction, refinement, and supply of these materials from primary resources (Garvin Heath, 2019).

Various recycling methods for crystalline silicon PV waste were analysed in a study by Ardente and colleagues (Ardente, F., et al. 2019), from a life cycle perspective. The analysis compared the environmental advantages of recovering secondary raw materials with the life cycle effects of recycling in several scenarios. According to the analysis, typical or "base-case" recycling techniques were inefficient and occasionally fell short of statutory goals. On the other hand, recycling procedures with a high level of efficiency showed the ability to not only satisfy legal requirements but also recover priceless high-grade resources like silicon, glass, and silver. These priceless components, which are frequently lost in base-case recycling, could be successfully recovered using more efficient methods.

2.1.1 Approaches to Recycling Solar Panels

Within a solar energy system, solar panels are the basic power producing devices that can operate on their own. An aluminium alloy frame, tempered glass, an energy storage device, EVA, and a rear panel are all included in a standard panel. The recovery of silicon and the recycling of precious metals have been the main areas of inquiry for most solar-panel recycling studies. The article by Deng, Chang, Ouyang, and Chong (2019) provides a comprehensive review of the technological and economic aspects of recycling silicon PV modules. The study explores the recycling methods, techniques, and processes used for silicon PV module waste. It analyzes the environmental impact, resource recovery, and cost-effectiveness of different recycling approaches (Deng, R., et al., 2019). Author Fernández L. (Fernández, L. et al. 2011) describes the possibility of using leftover solar cell residue (SCR) from PV panels to make building materials. This novel strategy attempts to cut costs while also providing carbon dioxide (CO_2) cement, helping to reduce the waste PV systems produce in the process. Furthermore, the conversion of calcium aluminate cement (CAC) is avoided by including silicon cell waste as an active ingredient. Solar PV waste management could undergo a revolution thanks to artificial intelligence (AI), which will increase productivity, accuracy, and sustainability. Robots and machines with AI capabilities can automatically sort the various materials found in scrap panels. Metals, glass, polymers, and other components can be identified and separated, expediting the recycling process, and increasing recovery rates. With discarded panels, AI can find valuable minerals like rare metals. Techniques based on AI can be applied to improve the collection and recycling of waste (Fang, B. et al. 2023). AI can increase the effectiveness of exploiting these resources by optimizing the recovery process, obviating the necessity for mining, and minimizing the environmental impact. Throughout their lifespan, solar panels may be monitored for health and performance using AI-powered sensors. This data can be used to anticipate when panels are getting close to the end of their useful lives, enabling prompt replacement and recycling.

There are now three different methods for dealing with used solar panels: component restoration, module disassembly, and silicon and other precious materials extraction from the components.

2.1.1.1 Component Repair

The component replacement does not include component separation or material recycling but strives to renew systems and prevent electrical breakdowns within board clusters. To examine junction box flaws and increase the power output of old solar panels, (Lin et al. 2011) used two distinct ways to assemble and disassemble. However, these methods are only effective when the outer layer encapsulation film and exterior junction box show signs of deterioration. Greater complexity in the methods is necessary when components age.

2.1.1.2 Module Separation

Material segregation is not included in physical separation, which also includes the mechanical segmentation of cracked panels. Author Dong (Dong 2009) investigated a method that involved disassembling solar panels and separating parts according to their physical and chemical characteristics, structural makeup, and materials. Thermal treatment was determined to be the most efficient option after considering the advantages and disadvantages of the three disposal methods for solar panels (manual disassembly, use of organic solvents, and thermal treatment). The process of processing crystalline silicon involves three distinct steps. To start with, a sodium hydroxide solution is used to remove the aluminum electrodes that are situated at the back of the panels. Saltpeter solution is used to retrieve the silver positive grid line. "Anti-reflective coatings on high purity polysilicon are removed from its surface using a hydrogen fluoride (HF) solution to recover and recycle the material. ' Due to the presence of antireflection film coatings and panel electrodes, converting crystalline silicon panels into pure silicon wafers necessitates several chemical etchings. Upon removal of the metal layer, reflector stripping, and doped junction preparation, chemical etching is followed by refinement, washing-up ("cleaning"), dehydration, and refinery.

2.1.1.3 Removal of Silicon and Rare Metals From Components

Metals and silicon that are of high value can be recycled from solar panels. There have been several studies conducted on the recycling of thin film solar panels (Berger et al. 2010). A study was carried out by Fernandez (Fernandez et al. 2011) to investigate the recycling of silicon panels in combination with cement in a thermal insulation system. A chemical process that involved acids, alkali solvents, and organic solvents was employed to dissolve or remove the EVA film. In their study on recycling thin film solar panels, Berger (Berger et al. 2010) examined wet mechanical methods such as grinding and flotation, as well as dry Mechanical techniques like vacuum blasting, including CdTe and CIGS recycling.

2.2 Job Creation

A wide range of employment opportunities that result from various stages of the waste management process are fostered by the field of solar PV waste management. These options cover a wide range of activities, including collecting and categorization, recycling, research, creative problem-solving, and policy development. The growth of solar industry jobs and the management of PV waste involves a broad range of duties, including collection, recycling, research, the creation of policies, instruction, and innovation (Weckend, S., Wade, A., & Heath, G. A. 2016). In addition to addressing the environmental issues raised by PV waste, the growth of this industry also promotes economic development by creating job opportunities at various points throughout the value chain. The first step in managing PV waste entails gathering wasted panels and organizing them according to the material type, condition, and possibility of recovery. A team experienced in logistics, transportation, and skilled sorting techniques is needed for this stage. Roles include transportation specialists, sorting operatives, and employees at collection centers. The need for skilled technicians who can handle PV panels and carry out disassembly processes is increasing as panels are disassembled to recover important components. The relevance of professional personnel following environmental and safety guidelines is emphasized by the need to ensure the safe and contamination-free disposal of hazardous substances. Recycling plants require a qualified team with knowledge of various material extraction procedures. Experts in mechanical processing, chemical separation, and metallurgy are essential to successfully recovering priceless components from scrap panels, including metals, silicon, and glass. Researchers and innovators are needed to advance recycling technology, develop novel material extraction techniques, and investigate sustainable utilisation pathways. For more efficient and sustainable management of PV waste, this field needs scientists, engineers, and researchers who can create and test cutting-edge approaches.

2.3 Circular Economy

The circular economy (CE) is based on the idea that by taking practical steps, creating clever products, and using novel business models, it is possible to increase the lifespan of electrical and electronic equipment while reducing their constant creation (Naik, S., & Eswari, J. S. 2022). The effectiveness of these operations is hampered by a few issues, including inadequate collection systems, inferior technology, a lack of training in the unorganised sector, and financial limitations. The circular economy's fundamental belief promotes the development of clean, renewable, and sustainable technologies. Encouraging recycling and optimising product materials, entails developing strategies, policies, and developmental frameworks with the goal of minimising waste and maximising the usability and value of products. Implementing effective waste management strategies for solar PV panels aligns with the principles of a circular economy, where materials are utilized for as long as possible. By recycling and reusing components of PV panels, the materials cycle can be closed, minimizing waste generation, and promoting sustainability. The notion of a circular economy, particularly when applied to the context of solar PV waste, embodies the concept of maximizing the utility of resources and materials while minimizing waste, environmental pollution, and the necessity for new resources. Rather than adhering to the conventional linear model of "take-make-dispose," the circular economy strives to establish a self-sustaining system where materials and products undergo continuous cycles of reuse, refurbishment, remanufacturing, and recycling. This comprehensive strategy ultimately alleviates the pressure on natural resources and mitigates the adverse ecological consequences associated with waste generation.

In many developing nations, the domain of recycling, refurbishing, and remanufacturing (RRR) within the solar panel industry is categorized as small and medium-sized enterprises (SMEs). The pivot from a conventional 'take-make waste' business approach to the progressive 'reduce-reuse-recycle' model serves as the driving force behind circular economy (CE) initiatives, particularly in the wake of substantial environmental challenges. In this transformative journey, SMEs emerge as pivotal players (Gautam A. et al., 2022).

The focal point of CE efforts is currently centered on end-of-life waste concerns, encompassing enhancements in waste collection, improved recycling methodologies, and the repurposing of materials and items that would otherwise be discarded. This study underscores the viability and profitability of SMEs adopting a circular economy-rooted model for the reuse and recycling of EoL solar PV waste within the context of a developing nation. SMEs are now progressively occupying crucial roles in domains like product restoration, disassembly, and remanufacturing. These smaller enterprises are proving the practicality of business models based on reuse, remanufacturing, and recycling. The overarching goal is the elimination of waste and the elongation of product lifespans. For the efficient management of solar PV resources, strategic placement of RRR SMEs becomes imperative. This localization is designed to ensure the optimal recovery of critical and rare earth metals contained within electronic waste. Raw resources are processed to make products, which are subsequently transported to consumers and eventually disposed to waste at the end of the cycle. This is a common linear economic model in India. A circular economy model, on the other hand, aims to convert waste streams through diverse processes into essential raw resources. This creates an ongoing closed-loop cycle that successfully reintegrates waste into the supply chain for materials.

From a larger perspective, the idea of a circular economy can be considered a crucial part of sustainable development, aligning with several of the Sustainable Development Goals (SDGs) of the United Nations. Among these targets, SDG 12 stands out as being especially relevant because it emphasises responsible consumption and production. About 10 of the seventeen SDGs are relevant in the context of improving the effective recycling of PV modules. This suggests that attaining sustainability within a framework for the recycling of PV modules has positive effects that extend far beyond the limitations of this industry. Such accomplishments might encourage other sectors to follow accordingly, making the PV industry an example in advancing more extensive sustainable practises (Farrell, C. C et al. 2020).

An organized approach to handling solar waste has a great deal of promise to benefit the environment and advance the idea of a circular economy. As a result, recycled materials will be used instead of new ones in manufacturing. The following section explores possible waste mitigation techniques in the context of India (Sharma, A. et al., 2023). The "reduce, reuse, and recycle" concept in solar PV waste management follows the principles of sustainable waste management and resource conservation. Here's how it applies to solar PV waste:

- **Reduce:** This idea emphasizes reducing waste formation in the first place. Reducing waste in solar PV entails designing and manufacturing panels and components to have the least possible negative environmental impact while having the longest possible lifespan. This can entail utilizing stronger and more long-lasting materials, optimizing production procedures to cut waste, and creating panels that are simple to disassemble and repair.
- **Reuse:** Finding new uses for parts or materials that are still useful is the process of recycling. This can entail refurbishing and fixing solar PV panels with minor problems so they can be used once again. For instance, even if a solar panel has lost a little amount of efficiency, it may still be used in off-grid systems or other situations where 100% efficiency is not required.
- **Recycle:** Recycling is the process of disassembling components and panels that have been discarded and utilizing those parts again to make new items. Recycling in the context of solar PV would entail removing valuable components from the panels, including silicon, metals (such as

Figure 3. Indian waste management strategies

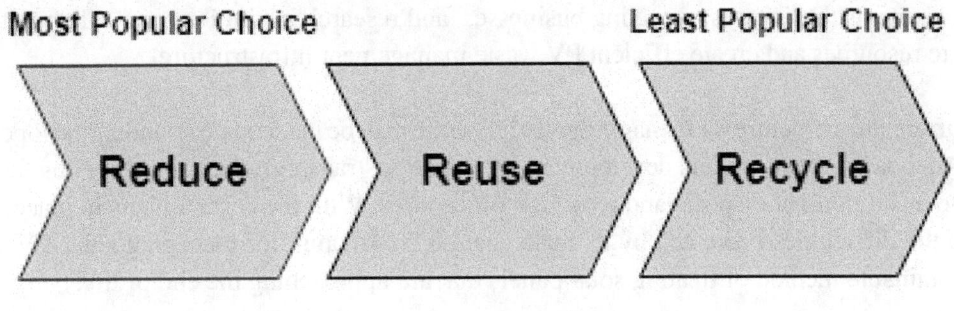

silver, copper, and aluminum), and even rare elements like indium and gallium. New solar panels or other items can then be produced using these materials.

Figure 3 demonstrates the Indian waste management strategies. Solar PV waste management can lessen its negative environmental effects while also conserving precious resources and lowering the need for new raw materials. This idea is consistent with the overall goal of developing a circular economy, in which goods and resources are used and reused repeatedly while reducing waste and enhancing long-term environmental health. Recycling plays a significant role in efficient e-waste management. Recycling has the potential to significantly reduce the number of dangerous materials in the environment and conserve energy from priceless natural resources when done properly. Local governments must play a crucial part in promoting and supporting recycling activities. The effectiveness of these initiatives depends on raising awareness and educating the community at the same time (Naik, S., & Eswari, J. S. 2022).

3 CHALLENGES

3.1 Limited Infrastructure

One major issue that requires immediate action is India's inadequate infrastructure for managing solar PV waste (Sharma, A. et al., 2023). The challenge of managing end-of-life solar PV waste becomes more pressing as the nation quickly adopts solar energy to satisfy its renewable energy targets. However, the infrastructure needed to treat this waste is still in its infancy, creating difficulties like Insufficient Recycling Facilities, Lack of Awareness, Technological Challenges, Regulatory Frameworks, and Economic Viability. The following are the solutions to address these difficulties:

i. Consumers, producers, and the public can all benefit from public awareness efforts that highlight the value of good PV waste management and its effects on the environment.
ii. Investments by the public and private sectors in the development of recycling facilities, collection locations, and research into cutting-edge recycling technology are crucial.
iii. Creating and enforcing legislation expressly for managing solar PV waste can provide the sector a clear direction and promote ethical behaviour.

iv. Collaborations with nations that have sophisticated recycling techniques can aid in the transfer of knowledge and technology to effectively handle PV waste.
v. Governments, businesses, recycling businesses, and research institutions can collaborate to coordinate resources and create efficient PV waste management infrastructure.

The current infrastructure for managing solar PV waste may be inadequate or underdeveloped in many regions. Adequate investment and development in collection, transportation, and recycling facilities are required to ensure proper disposal and recycling procedures. With the correct plans in place, India can overcome the difficulties presented by its inadequate infrastructure for managing solar PV waste and build a sustainable method of treating solar panels that are approaching the end of life.

3.2 Hazardous Materials

Materials having potential threats to the environment and human health due to their poisonous, corrosive, combustible, or other harmful characteristics fall under the category of hazardous substances in the context of PV waste. Solar PV panels are essential for producing clean energy; however, they can include some dangerous elements, mostly during production and in particular component parts. As solar panels get closer to the end of their useful lives, careful handling of these dangerous materials becomes necessary to prevent negative effects on both human health and the ecosystem. All different kinds of PV panels primarily consist of aluminium and glass. Indium and germanium are rare metals, with indium being more common than in the world of amorphous silicon. However, there is inadequate regulation over the disposal of materials like cadmium, lead, and polymer. As a result, these compounds have the potential to have a negative impact on both human health and the environment. For instance, authors (Ramos-Ruiz et al. 2017) stated that Cadmium telluride (CdTe) can cause significant lung inflammation and fibrosis when improperly controlled. Both the causes and effects of lead leaching are concerning. Loss of biodiversity can come from this due to reduced development and decreased reproduction rates in both plants and animals. Furthermore, it can result in a variety of health problems, compromising important processes like renal function as well as the immunological and brain systems (Leccisi, E. et al. 2016). CdTe panels are one type of thin-film solar panel that contains this significant heavy metal, which has the potential to be hazardous to both the environment and human health. Cadmium that is not properly handled may contaminate the soil and water, endangering both human health and ecosystems.

When these panels are retired or damaged, proper handling and disposal procedures are essential to prevent cadmium contamination. Lead is inherently hazardous and has the potential to severely impact people's health. When lead is exposed to the environment, it can seep into the soil and water, causing pollution (Xu, Y. et al. 2108). Hazardous waste may enter the environment if end-of-life solar panels are not properly disposed of. By the end of 2017, there were over 400 GW of installed PV capacity worldwide. By the year 2050, this number is expected to increase to almost 4500 GW. Given that solar panels typically last 25 years, it is predicted that by 2030, EoL solar PV waste will account for anywhere between 4% and 14% of total generation capacity, rising to an overwhelming 80% (or nearly 78 million tonnes) by 2050. As a result, in the decades to come, disposing of PV panels will likely become a substantial environmental issue (Chowdhury, M. S. et al. 2020). It's important to note that during the conversion from silicon, about 80% of the raw silicon is lost, which is relevant to the manufacture of crystalline silicon. In addition, silicon tetrachloride, a highly poisonous contaminant, is produced during this process. In addition, solar panels caused the release of nitrogen trifluoride (NF_3), a substance that has a greenhouse

Figure 4. Photovoltaic panel waste categories and the impact they have on the environment

Photovoltaic panel waste categories and the impact they have on the environment			
Components	**Material Used**	**Recyclability**	**Environmental Impact**
Frame	**Aluminium**		
Modular cover	**Glass**		
Solar cell	**Silicon**		
Solar cell	**Cadmium, tellurium, indium, gallium, selenium compounds**		
Solar cell coating	**Silver, copper, lead, gallium /boron/phosphorus, aluminium**		
Cell and module interconnections	**Lead, copper, tin**		
Backsheet, encapsulants	**Polymer**		

effect that is more than 17,000 times stronger than CO_2 for more than a century. Due to their extremely poisonous characteristics, lead (Pb) and cadmium (Cd) in solar panels raise environmental issues. Nickel (Ni) exposure, among other toxic metals, can have serious health effects, including cancer and damage to organs like the kidneys, liver, lungs, and brain. Furthermore, connected to harmful effects on the respiratory, prostate, and blood systems are these hazardous substances. Additionally, their presence has been linked to death in both humans and wildlife as well as problems with reproduction, behavioural problems, and mortality. Extended Producer Responsibility (EPR), also known as Product Stewardship, is a policy strategy that addresses e-waste issues during these worries. According to this policy, product manufacturers must continue to be accountable for their goods during the post-consumer stage of the product lifecycle, which is in line with the rules established by the Organisation for Economic Cooperation and Development (Bajagain, R. et al. 2020).

Figure 4 illustrates the Photovoltaic panel waste categories and the impact they have on the environment (Data from Source Suresh S et al. 2019). PV panels' back sheets and encapsulants are made of polymer parts made of fluorinated and cross-linked polymers. Unfortunately, these materials can't be recycled effectively. When these polymers are burned, they release toxic fumes and other harmful compounds that are harmful to your health. The ecology may suffer serious harm if incorrect disposal takes place.

3.3 End-of-Life Collection

End-of-life collection for solar PV waste entails the systematic collecting, transport, and organization of retired or discarded PV panels as they approach the end of their functional lifespan. The amount of PV waste produced is increasing along with the utilization of solar energy. To avoid causing environmental harm and to maximize the recovery of priceless resources, this calls for precise management. This procedure is crucial for ensuring responsible disposal, effective recycling, and the reduction of negative effects related to improper treatment of PV waste. The process of end-of-life collection for solar PV waste is a pivotal stage in the lifecycle of PV panels. By constructing robust collection networks, heightening awareness, incentivizing responsible disposal, and fostering cooperation, the industry can ensure that retired PV panels are accomplished in a manner that is both ecologically sustainable and resource efficient. Proper end-of-life collection stands as a cornerstone on the path toward realizing a circular economy within the solar energy sector. After collection, the precise sorting and segregation of PV panels based on their material type, condition, and potential for recovery are crucial. This step streamlines subsequent recycling procedures, ensuring that materials are effectively extracted and reprocessed. The following are the two types of PV waste recycling technologies (Suresh S, Singhvi S, and Rustagi V. 2019):

3.3.1 Conventional Recycling Technologies

To separate laminated structures inside standard c-Si modules, traditional glass and metal recyclers frequently use mechanical techniques all over the world. This procedure includes a series of steps like the human or automated extraction of copper from junction boxes and aluminium from module frames. The next processes involve separating the glass and polymer fractions using crushers, sieves, magnets, eddy-current devices, and inductive and optical sorters. Following extraction, the metal fraction is sent to metal recyclers for smelting. Glass particles, however, are still mixed in with metal, silicon, and polymer shards. The glass becomes of inferior quality because of this mixing, making it unfit for use in the production of new modules. Instead, the construction industry uses this degraded glass to produce low-value glass fibre. Approximately 70–80% of the raw materials in a c-Si module can be recovered by weight using this traditional recycling method. Glass is the main component here, followed by aluminium and copper. High-value metals like silver and solar-grade silicon, as well as environmentally hazardous substances like lead and polymers, are noticeably absent from this procedure.

3.3.2 Advanced Recycling Technologies

Many materials, including glass, aluminium, copper, silicon, silver, lead, and tin, are recovered using a combination of mechanical, thermal, and chemical processes in advanced or specialised PV recycling. The 8–10% weight of the polymer fraction that cannot be recovered must be disposed of in a landfill or burned.

Governments and industry stakeholders can stimulate end-of-life collection through mechanisms like extended producer responsibility programs. EPR programs hold manufacturers accountable for the appropriate disposal and recycling of their products, motivating them to establish efficient collection and recycling systems (Sheoran M. et al., 2022). Figure 5 describes the associations to tackle the Solar PV End-of-Life Issues in the Indian framework.

Figure 5. Associations to address solar PV end-of-life issues

For the effective technical and regulatory management of solar PV end-life, a collaborative framework involving both the public and private sectors is proposed.

This collaboration intends to unite solar manufacturers with select governmental bodies, fostering a collective approach. This initiative aims to incorporate representatives from various public and private institutions, fostering a diverse and inclusive perspective. The operational body responsible for overseeing this endeavour operates under the purview of the organization dedicated to technical and regulatory management of solar PV end-life. This entity shoulders the responsibility of implementing the Extended Producer Responsibility (EPR) framework, designed to address the challenges of managing waste from expired solar modules. This encompasses various activities, from transportation to recovery, storage, and extraction of valuable resources from decommissioned solar panels or their components. Collaborative partnerships are anticipated between solar module manufacturers and recycling facilities, concluding in joint ventures. These partnerships aim to synergize efforts and expertise, ensuring a holistic approach to waste management.

The governing body overseeing the management of end-of-life solar panel waste is comprised of representatives from key central ministry departments.

These include MNRE (Ministry of New and Renewable Energy), MoEF & CC (Ministry of Environment, Forest & Climate Change), and CEA (Central Electricity Authority). This merging of expertise ensures a multifaceted perspective in policy formulation for establishing effective end-of-life management for solar module waste. The core responsibility of this organization lies in guiding the functional and practical implementation of the solar waste management policy. It takes on the pivotal role of not only creating public awareness about responsible waste management but also enhancing the capabilities of state-level partners to effectively execute waste management strategies. It actively initiates programs within the education system that align with the ethos of sustainable development goals, ensuring that the principles of responsible waste management are deeply rooted in the educational framework. To ensure the smooth governance of end-of-life proceedings for solar panels, such institutions also extend support by sponsoring the infrastructural and financial requirements

3.4 Regulatory Frameworks

When it comes to the recovery, transportation, recycling, and proper disposal of PV panels at the end of their useful lives, regulatory procedures controlling the management of waste from solar PV systems in India include a complicated framework. These frameworks are essential for promising the responsible and sustainable disposal of PV waste in India due to the country's rapid expansion of solar energy usage. It is apparent that the Indian context and the WEEE directives are important. A strong policy and regulatory framework for the nation can be created using the European Union's directives. The WEEE guidelines and the 2016 Indian E-waste handling standards both share the Extended Producer Responsibility (EPR) principle, which is the framework's central principle. This EPR-driven strategy focuses on consumer-centered electronic waste management and mandates that producers oversee the full product lifecycle. The integration of Indian and European rules demonstrates a shared commitment to environmentally responsible disposal of waste. It's interesting to note that both the EU and India view solar panels from the perspective of "one product, one equipment". The perspective contributed to certain waste management policies to exclude photovoltaic (PV) component systems. This strategic choice not only demonstrates compliance with EU rules (EU 2021) but also captures the core of circularity in solar panel waste management. Setting up a take-back mechanism, which is the duty of manufacturers, is essential to this circular approach. This method demands a planned sequence of tasks involving solar module recycling, collecting, storage, recovery, and disintegration (Sheoran, M. et al., 2022). The following is the analysis of India's regulatory procedures regarding solar PV waste:

i. India's environmental laws serve as the foundation for managing PV waste. Under the guidelines of the Environment Protection Act of 1986, the 2016 Hazardous and Other Wastes (Management and Transboundary Movement) Rules were formed, and they include requirements for the environmentally responsible management of hazardous waste, including the hazardous components included in PV panels.
ii. Due to their electronic components, PV panels fall within the 2016 E-Waste (handling) Rules' comprehensive legal framework for the handling of electronic waste. These regulations define standards for the collection, transportation, recycling, and disposal of electronic waste and include Extended Producer Responsibility clauses that hold producers accountable for the collection and appropriate management of end-of-life items. EPR plays a significant role in India's regulatory strategy for handling PV waste.
iii. Principles for the advancement of solar energy in India are outlined in the Jawaharlal Nehru National Solar Mission (JNNSM), also referred to as the National Solar Mission. The importance of proper PV waste management, including the development of infrastructure for collection and recycling, is emphasized by these recommendations.
iv. PV panel manufacturers are responsible for adhering to all legal guidelines for end-of-life management. This entails creating procedures for recycling, constructing systems for collecting, and making sure that hazardous waste is disposed of safely.
v. The Central Pollution Control Board (CPCB 2018), which operates under the auspices of the Ministry of Environment, Forestry, and Climate Change, issues directives for the management of PV waste. Technical standards for the secure handling, gathering, transferring, and recycling of PV panels are included in these suggestions.

vi. The Indian regulatory environment supports research and development related to PV waste management. To efficiently manage PV waste, this entails developing cutting-edge recycling processes and sustainable practises.
vii. Regulations emphasize public education and knowledge of the importance of sustainable PV waste management. These frameworks also stimulate the improvement of stakeholder capabilities in the collection, transportation, and recycling of PV waste.

Globally, a significant percentage of solar manufacturing companies are now in the position of handling end-of-life solar waste, mostly through the adoption of frameworks for extended producer responsibility. As a result, industrialized countries are adopting more sophisticated solar panel technology less frequently. A robust regulatory framework focused on the management of end-of-life solar waste must be established concurrently. Except for the European Union, solar PV waste disposal lacks comprehensive regulation across other significant countries as highlighted in the International Journal of Environmental Science and Technology. Instead, for the management of solar PV waste, these countries primarily use non-regulatory solutions. A concise overview of developments in key global regions is presented in Table 2 (Komoto, K. et al. 2018, Sharma, A. et al., 2023). As seen in Table 2, different countries implement varied regulatory strategies.

It's interesting to note that the European Union primarily uses non-regulatory techniques to control end-of-life solar waste. Particularly, through mandates like WEEE, nations like Italy, Germany, and the UK have accepted the principles of Extended Producer Responsibility. The 2012/19/UE directive served as the first WEEE directive adopted by the European Union. This important initiative has created a solid foundation for encouraging resource utilization through recycling techniques. As a result, the casual disposal of waste in areas with ecological risks has significantly decreased. The recovery, collection, and recycling of solar panels are the main areas of concentration, serving as the cornerstone of a more environmentally friendly approach to the management of solar waste at its end-of-life stage.

Table 2. National and worldwide scenarios for waste management

S. No.	Different Nations	Policy on Recycling Solar PV
1	India	Since there is no centrally administered regulation in India, only standard waste regulations that apply to household electronics are used to handle PV waste.
2	United States of America	Since there are no specific federal laws or regulations, only general waste laws are now in effect; however, several states, like California and Washington, are drafting their own regulations or legislation for the use and recycling of PV modules.
3	European Union	In 2003 the WEEE (Waste Electric and Electronic Equipment) directive of the European Union did not mention PV waste. The modification to WEEE involves the management of PV modules by EoL, and as per the latest WEEE directive, PV manufacturers will bear the main responsibility for collecting, handling, and treating PV waste regardless of where the production facility is situated. Nevertheless, there are legal provisions in place that limit this liability.
4	China	At the moment, neither a federal law nor a method of disposal exists, but the National High-tech R&D program's research on PV recycling and safe disposal has made recommendations for formulating guidelines for handling PV waste issues.
5	Japan	The nation is still formulating policies for recycling PV. However, there has been significant advancement in the research and testing of technology for the recycling of PV panels.

A thorough structure is needed to give stakeholders in solar PV waste management explicit instructions to increase resource efficiency. Figure 6 illustrates the theoretical structure of waste management systems for PV panels. The framework should address possibilities for recycling, export, and landfill disposal, as well as how to handle solar panels that are already in place (Oteng, D. et al. 2022). The current lack of tracking of PV waste both within and outside of the State borders emphasizes the necessity for better monitoring and management. Understanding the infrastructure and recycling needs is essential for sustainable waste management. Although certain States provide incentives for investigators and recyclers to create cutting-edge recycling technology, the majority of PV waste still goes to landfills in various States.

Figure 6. Theoretical structure: Waste management scheme for solar PV system

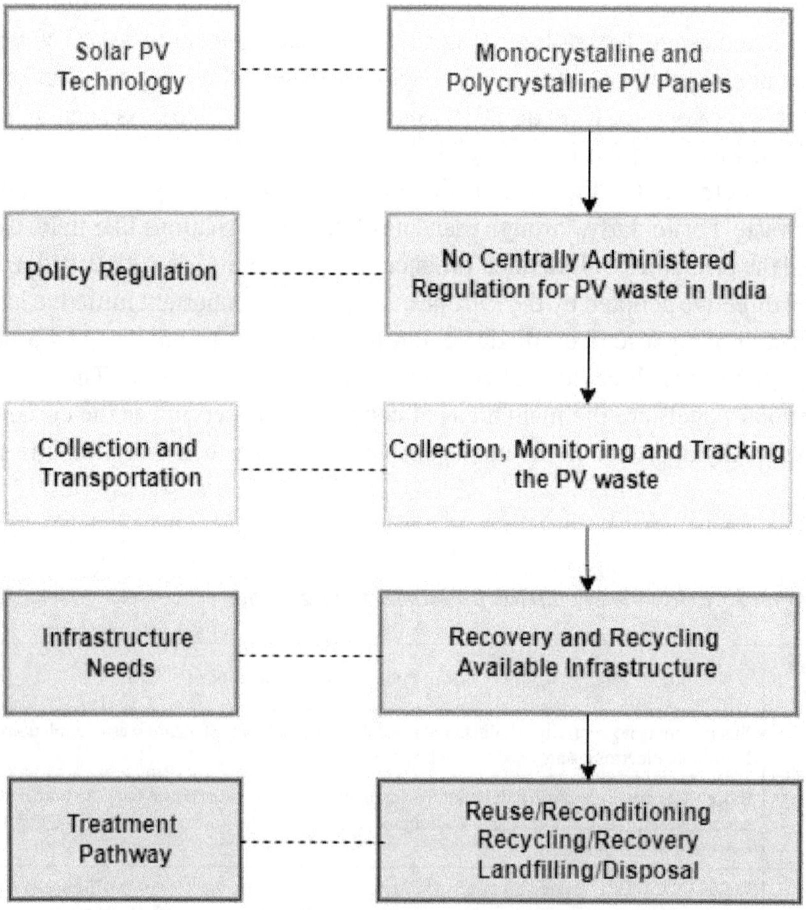

4 SWOT Analysis

SWOT analysis is a well-known technique utilized to assess and manage strategic objectives. The acronym SWOT refers to Strengths, Weaknesses, Opportunities, and Threats. While advantages and disadvantages are internal, opportunities and risks are external. SWOT analysis is a strategic planning method that is used to evaluate both internal and external elements that may have an impact on an initiative, organiza-

Figure 7. Management of end-of-life PV waste in India: SWOT evaluation

STRENGTHS	WEAKNESS
I. There is a growing awareness among the Indian population about the significance of managing end-of-life PV waste. II. The Indian government is actively promoting renewable energy and sustainable practices. III. The rapid growth of the solar industry in India presents a substantial opportunity for PV waste generation, creating possibilities for recycling and resource recovery.	I. The current recycling infrastructure for PV waste management in India is underdeveloped. II. There is a lack of specialized knowledge and expertise in PV waste. III. Involvement of the informal sector in PV waste management can lead to improper disposal and potential environmental risks.
OPPORTUNITIES	**THREATS**
I. It can generate employment opportunities in collection, recycling, and component recovery, contributing to economic growth. II. Emphasizing circular economy principles can encourage the reuse and recycling of materials from PV waste. III. Advancements in recycling technologies and innovative techniques can enhance the efficiency of extracting and separating valuable materials from PV waste.	I. Inadequate disposal of PV waste can lead to environmental pollution and health hazards due to hazardous materials in solar panels. II. The lack of standardized regulations and policies for PV waste management may create hurdles in ensuring proper recycling practices.

tion, or effort. SWOT analysis provides a thorough grasp of the internal and external landscape, acting as a compass to direct solar PV waste management solutions. The creation of a strategic plan demands the precise identification and utilization of all available resources, as well as an effective approach to dealing with potential weaknesses (Ghaleb Al-Aqqad, A. O., & Menyhart, J. 2023, Tasnim, S. S., et al. 2022). The SWOT analysis is utilized in PV waste management, as demonstrated in Figure 7. All available opportunities can be fully capitalized with the use of existing strengths.

CONCLUSION

PV waste is expected to increase significantly in the next years due to the explosive growth of solar panel installations. The immense volume of this upcoming waste surge presents an enormous challenge for proper management. If the problem of solar waste disposal is not properly resolved, it could have disastrous effects on the ecosystem. A problem in hazardous waste management might develop during the disposal of used PV materials. PV panel components could potentially contaminate the environment and present safety risks by dissolving into the soil and groundwater. There are still some barriers in the way of recycling solar PV modules, despite advancements. Comprehensive material recovery, process effectiveness, and financial viability are challenges. As a result, both during recycling and during the initial manufacturing of PVs, the recycling process continues to be complex, energy-intensive, and frequently requires the use of harmful materials. It is necessary to significantly increase the amount of money allocated to projects related to solar waste, such as PV recycling, strategic planning, and monitoring. To prevent renewable energy initiatives from unintentionally burdening future generations with waste accumulation, end-of-life (EoL) solutions must be developed. A circular economy, a framework that minimises potential environmental effects and maximises resource use, has the potential to be sparked by the effective recovery of EoL PV materials. Artificial intelligence driven marketing efforts can educate the public about the value of recycling and properly disposing of solar PV waste, which promotes ethical consumer behaviour. The use of Artificial intelligence in solar PV waste management has the potential to improve productivity dramatically, lessen the negative effects on the environment, and encourage sustainable practices throughout the whole lifecycle of solar panels. The current study of the literature highlights important problems with solar PV waste and its handling, providing policymakers and recycling stakeholders with useful information, SWOT analysis, and India's regulatory procedures. Stakeholders may successfully go through problems, exploit opportunities, and create strategies that are in line with goals for sustainable waste management by recognizing their strengths, weaknesses, opportunities, and threats. In summary, resource recovery, employment creation, and the advancement of a circular economy are all possible outcomes of solar PV waste management. However, there are difficulties in developing infrastructure, managing hazardous chemicals, setting up efficient collecting systems, and putting regulatory frameworks in place. Maximizing the environmental and financial advantages of solar PV waste management requires tackling these issues through investment, innovation, and stakeholder cooperation.

REFERENCES

Ardente, F., Latunussa, C. E., & Blengini, G. A. (2019). Resource efficient recovery of critical and precious metals from waste silicon PV panel recycling. *Waste Management (New York, N.Y.)*, *91*, 156–167. doi:10.1016/j.wasman.2019.04.059 PMID:31203937

Bajagain, R., Panthi, G., An, Y. J., & Jeong, S. W. (2020). Current practices on solar photovoltaic waste management: an overview of the potential risk and regulatory approaches of the photovoltaic waste. 대한환경공학회지, *42*(12), 690-708.

Berger, W., Simon, F. G., Weimann, K., & Alsema, E. A. (2010). A novel approach for the recycling of thin film photovoltaic modules. *Resources, Conservation and Recycling*, *54*(10), 711–718. doi:10.1016/j.resconrec.2009.12.001

Central Pollution Control Board (CPCB). (2018). *E-waste*. Available at: https://cpcb.nic.in/e-waste/

Cerchier, P., Dabalà, M., Pezzato, L., Tammaro, M., Zucaro, A., Fiorentino, G., ... Brunelli, K. (2022). Silicon-PV panels recycling: Technologies and perspectives. *La Metallurgia Italiana*, 16.

Chowdhury, M. S., Rahman, K. S., Chowdhury, T., Nuthammachot, N., Techato, K., Akhtaruzzaman, M., Tiong, S. K., Sopian, K., & Amin, N. (2020). An overview of solar photovoltaic panels' end-of-life material recycling. *Energy Strategy Reviews*, *27*, 100431. doi:10.1016/j.esr.2019.100431

Deng, R., Chang, N. L., Ouyang, Z., & Chong, C. M. (2019). A techno-economic review of silicon photovoltaic module recycling. *Renewable & Sustainable Energy Reviews*, *109*, 532–550. doi:10.1016/j.rser.2019.04.020

Dong, L. (2009). *Research on waste crystalline silicon solar panels resource recovery*. Southwest Jiaotong University.

European Union (EU)-India Technical Cooperation Project: Energy. (2021). *PV waste management in India: Comparative analysis of state of play and recommendations*. Available at https://www.cecp-eu.in/uploads/documents/events/pv-waste-management-report-25-01-2021.pdf

Fang, B., Yu, J., Chen, Z., Osman, A. I., Farghali, M., Ihara, I., ... Yap, P. S. (2023). Artificial intelligence for waste management in smart cities: A review. *Environmental Chemistry*. PMID:37362015

Farrell, C. C., Osman, A. I., Doherty, R., Saad, M., Zhang, X., Murphy, A., Harrison, J., Vennard, A. S. M., Kumaravel, V., Al-Muhtaseb, A. H., & Rooney, D. W. (2020). Technical challenges and opportunities in realising a circular economy for waste photovoltaic modules. *Renewable & Sustainable Energy Reviews*, *128*, 109911. doi:10.1016/j.rser.2020.109911

Fernández, L. J., Ferrer, R., Aponte, D. F., & Fernandez, P. (2011). Recycling silicon solar cell waste in cement-based systems. *Solar Energy Materials and Solar Cells*, *95*(7), 1701–1706. doi:10.1016/j.solmat.2011.01.033

Garvin Heath. (2019). *PV Modules End of Life Management Setting the Stage. NREL*. National Renewable Energy Laboratory.

Gautam, A., Shankar, R., & Vrat, P. (2022). Managing end-of-life solar photovoltaic e-waste in India: A circular economy approach. *Journal of Business Research*, *142*, 287–300. doi:10.1016/j.jbusres.2021.12.034

Ghaleb Al-Aqqad, A. O., & Menyhart, J. (2023). Solar panels problem after end-of-life and waste management (SWOT analysis for the global directives of PV's waste management). *International Review of Applied Sciences and Engineering*, *14*(2), 220–229. doi:10.1556/1848.2022.00511

Heath, G. A., Silverman, T. J., Kempe, M., Deceglie, M., Ravikumar, D., Remo, T., Cui, H., Sinha, P., Libby, C., Shaw, S., Komoto, K., Wambach, K., Butler, E., Barnes, T., & Wade, A. (2020). Research and development priorities for silicon photovoltaic module recycling to support a circular economy. *Nature Energy*, *5*(7), 502–510. doi:10.103841560-020-0645-2

International Energy Agency Photovoltaic Power Systems Programme (IEA PVPS). (2016). *Task12 and the International Renewable Energy Agency (IRENA), End of Life Management of Photovoltaic Panels Trends in PV Module Recycling Technologies*. Available at: https://iea-pvps.org/key-topics/end-of-life-management-of-photovoltaic-panels-trends-in-pv-module-recycling-technologies-by-task-12/

Jordan, D. C., & Kurtz, S. R. (2013). Photovoltaic degradation rates—An analytical review. *Progress in Photovoltaics: Research and Applications*, *21*(1), 12–29. doi:10.1002/pip.1182

Komoto, K., Lee, J. S., Zhang, J., Ravikumar, D., Sinha, P., Wade, A., & Heath, G. A. (2018). *End-of-life management of photovoltaic panels: Trends in PV module recycling technologies (No. NREL/TP-6A20-73847). National Renewable Energy Lab*. NREL. doi:10.2172/1561523

Leccisi, E., Raugei, M., & Fthenakis, V. (2016). The energy and environmental performance of ground-mounted photovoltaic systems—A timely update. *Energies*, *9*(8), 622. doi:10.3390/en9080622

Lunardi, M. M., Alvarez-Gaitan, J. P., Bilbao, J. I., & Corkish, R. (2018). A review of recycling processes for photovoltaic modules. In B. Zaidi (Ed.), *Solar panels and photovoltaic materials* (pp. 9–27). Intechopen. doi:10.5772/intechopen.74390

Majewski, P., Al-shammari, W., Dudley, M., Jit, J., Lee, S. H., Myoung-Kug, K., & Sung-Jim, K. (2021). Recycling of solar PV panels-product stewardship and regulatory approaches. *Energy Policy, 149*, 112062. doi:10.1016/j.enpol.2020.112062

Naik, S., & Eswari, J. S. (2022). Electrical waste management: Recent advances challenges and future outlook. *Total Environment Research Themes*, 100002.

Oteng, D., Zuo, J., & Sharifi, E. (2022). An expert-based evaluation on end-of-life solar photovoltaic management: An application of Fuzzy Delphi Technique. *Sustainable Horizons, 4*, 100036. doi:10.1016/j.horiz.2022.100036

Ramos-Ruiz, A., Wilkening, J. V., Field, J. A., & Sierra-Alvarez, R. (2017). Leaching of cadmium and tellurium from cadmium telluride (CdTe) thin-film solar panels under simulated landfill conditions. *Journal of Hazardous Materials, 336*, 57–64. doi:10.1016/j.jhazmat.2017.04.052 PMID:28472709

Rathore, N., & Panwar, N. L. (2022). Strategic overview of management of future solar photovoltaic panel waste generation in the Indian context. *Waste Management & Research, 40*(5), 504–518. doi:10.1177/0734242X211003977 PMID:33885343

Sharma, A., Mahajan, P., & Garg, R. (2023). End-of-life solar photovoltaic panel waste management in India: Forecasting and environmental impact assessment. *International Journal of Environmental Science and Technology*, 1–20. doi:10.100713762-023-04953-2 PMID:37360560

Sheoran, M., Kumar, P., Sharma, S., & Bukya, M. (2022). Current situation analysis of solar PV waste management in India. *Materials Today: Proceedings, 58*, 773–782. doi:10.1016/j.matpr.2022.03.118

Sheoran, M., Sharma, S., & Kumar, P. (2020). Compatible standard policy measure to tackle solar photovoltaic waste in Indian scenario. *Journal of Physics: Conference Series, 1504*(1), 012012. doi:10.1088/1742-6596/1504/1/012012

Shubbak, M. H. (2019). Advances in solar photovoltaics: Technology review and patent trends. *Renewable & Sustainable Energy Reviews, 115*, 109383. doi:10.1016/j.rser.2019.109383

Suresh, S., Singhvi, S., & Rustagi, V. (2019). *Managing India's PV module waste*. Available at: https://bridgetoindia.com/backend/wp-content/uploads/2019/04/BRIDGE-TO-INDIA-Managing-Indias-Solar-PV-Waste-1.pdf

Tasnim, S. S., Rahman, M. M., Hasan, M. M., Shammi, M., & Tareq, S. M. (2022). Current challenges and future perspectives of solar-PV cell waste in Bangladesh. *Heliyon, 8*(2), e08970. doi:10.1016/j.heliyon.2022.e08970 PMID:35243090

Weckend, S., Wade, A., & Heath, G. A. (2016). *End-of-life management: solar photovoltaic panels (No. NREL/TP-6A20-73852). National Renewable Energy Lab*. NREL.

Xu, Y., Li, J., Tan, Q., Peters, A. L., & Yang, C. (2018). Global status of recycling waste solar panels: A review. *Waste Management, 75*, 450-458.

Chapter 16
Sustainable E-Waste Reduction Methods and Models

B. Swathi
G. Pulla Reddy Engineering College, Kurnool, India

G. Yamini
National College (Autonomous), Tiruchirappalli, India

R. Sasikala
National College (Autonomous), Tiruchirappalli, India

Mary Teresa
Gurunanak Institution Technical Campus, India

D. Jagadeesan
The Apollo University, Chittoor, India

Venkata Reddy Yasa
https://orcid.org/0009-0005-5775-2806
Woxsen University, India

ABSTRACT

This study aims to examine strategies and approaches that are designed to reduce the production of electronic waste. The primary methodologies encompass the development of electronic goods with enhanced durability and repairability, the facilitation of consumer education and awareness, the implementation of circular economy initiatives such as refurbishing and remanufacturing, the establishment of take-back and recycling programs, and the enactment of legislative and policy measures. These methods collectively contribute to the reduction of electronic waste by prolonging the lifespan of products, fostering responsible consumption, advocating for reuse and recycling, and establishing a regulatory framework that supports these efforts. Through the implementation of these strategies, it is possible to mitigate the environmental consequences associated with electronic waste, thereby fostering a more sustainable and resource-efficient electronic industry.

DOI: 10.4018/979-8-3693-1018-2.ch016

INTRODUCTION

Electronic garbage, or e-waste, is a global problem caused by the disposal of obsolete electronics. This category includes cellphones, laptops, desktops, tablets, refrigerators, washing machines, TVs, and more. E-waste generation has increased due to the rapid spread of electronic gadgets in modern society, their short lifespan, and frequent replacement, making it a major environmental and health issue. E-waste contains a huge variety of materials. Material types include valuable and harmful. Circuit boards, connections, and cables contain valuable materials like gold, silver, copper, and palladium. E-waste also contains glass from screens, monitors, cathode ray tubes and polymers from casings, wires, and insulation (Bakhiyi et al., 2018; Bhat & Patil, 2014). The dangers of e-waste are also serious. Electronics often include lead, mercury, cadmium, and flame retardants. These materials represent serious environmental and health dangers if not appropriately managed and disposed of. Incineration or landfilling e-waste can release dangerous compounds into the soil and water, damaging ecosystems and perhaps entering the food chain. E-waste management is difficult. The massive amount of electronics discarded each year strains waste management systems. Due to rapid technological improvement, electronic gadgets are frequently upgraded and replaced, increasing e-waste. Consumers typically choose new models, leaving behind functioning but outdated equipment. Health concerns are also associated with e-waste handling and disposal, especially in informal recycling industries in underdeveloped nations. When dismantling and processing e-waste, workers may be exposed to harmful chemicals, causing respiratory, skin, and neurological concerns. Managing e-waste requires recycling and responsible disposal. Recycling e-waste recovers valuable materials while protecting the environment. Recycling is complicated and requires specialized equipment and knowledge. E-waste must be recycled safely and ethically to protect workers and the environment (Ahmad et al., 2021; Davis & Wolski, 2009).

E-waste laws differ by country. Certain regions have extended producer responsibility (EPR) programs that require manufacturers and importers to properly dispose and recycle. Others have e-waste collection and recycling infrastructure to help people safely dispose of electronics. Electronic waste (e-waste) has become a global issue due to its numerous and detrimental environmental effects. E-waste includes smartphones, laptops, TVs, and refrigerators that have reached the end of their lifespan. These devices are crucial in modern civilization, but improper disposal and mismanagement have caused a variety of environmental issues that require immediate attention. E-waste contains hazardous compounds and heavy metals as lead, mercury, cadmium, brominated flame retardants, and polyvinyl chloride(Gupta, 2011). When improperly handled and disposed of, these toxic substances can permeate soil and water systems, causing environmental damage. Lead, which is found in CRT monitors and lead-acid batteries, pollutes groundwater, altering aquatic ecosystems and harming aquatic life. Informal e-waste recycling sometimes involves cremation, which releases dangerous chemicals. Dioxins and furans, which are highly poisonous to humans, are produced during electronic component combustion. These pollutants cause air pollution and harm human health. High levels of these pollutants can cause respiratory issues, cardiovascular illness, and cancer, highlighting their health risks. Poor landfill disposal of electronic waste can pollute soil owing to device degradation. The toxic components in electronic waste (e-waste) leak into the soil, affecting soil quality and plant health. The introduction of contaminants can disrupt ecosystems, harming insects, and microorganisms, which are essential for nutrient cycling. Cadmium and mercury can enter groundwater and surface water through inappropriate electronic device disposal. Pollutants then enter rivers, lakes, and oceans, endangering aquatic ecosystems and animals. Water pollution alters food webs, endangering aquatic organisms. This contamination may also threaten human communities that depend on these water bodies for food, agriculture, and healthy drinking water. Rare earth elements, precious metals (gold, silver), and base metals (copper) are essential to the electronics manufacturing sector, which drives the global economy. Discarding electronics improperly wastes valuable resources. Due to the growing need for new electronics, mining and extraction have degraded habitats, eroded soil, and polluted water (V. Kumar et al., 2011; Roy et al., 2022; Wath et al., 2010).

Electronics manufacturing and disposal need a lot of energy. Electronic waste recycling and incineration take a lot of energy. Early disposal of electronics wastes energy used in their production. Recycling procedures that use a lot of energy may unintentionally harm the environment. For several reasons, efficient electronic waste management is crucial given the above concerns. Ecosystems are its primary responsibility, protecting biodiversity and decreasing air, land, and water pollution. Responsible electronic waste management reduces hazardous chemicals and heavy metals that affect the environment and humans. By making precious materials easier to retrieve, this method reduces the need for resource-intensive mining and extraction, preserving scarce resources. It reduces greenhouse gas emissions from incineration and energy-intensive gadget production. It also safeguards the physical and emotional health of those who manage and dispose of electronic garbage (e-waste), particularly those in informal recycling sectors, who are more at risk. Finally, e-waste laws encourage proper disposal to reduce environmental damage. Additionally, these policies aim to hold producers accountable for their products' whole lifecycle. For environmental and public health reasons, these criteria must be followed. E-waste is a global issue that countries are addressing. The idea that inappropriate electronic device disposal can harm the environment and health drives these initiatives. Some significant regional approaches:

Opinion Mining on E-Wastage From Various Countries

European Union: The EU has a comprehensive e-waste reduction and management policy. The Waste Electrical and Electronic Equipment (WEEE) Directive, which requires manufacturers, importers, and retailers to dispose of electronics responsibly, is its foundation. This directive promotes the recyclability and lifetime of electronic gadgets through eco-design. The EU promotes recycling and resource recovery. It has collection and recycling stations in member states to properly dispose of e-waste. Additionally, the Restriction of Hazardous compounds (RoHS) Directive bans hazardous compounds in electronic equipment, protecting the environment. The EU uses law, eco-design, recycling infrastructure, and public awareness to reduce e-waste's environmental impact and maximize resource efficiency (Lee & Na, 2010).

United States: E-waste reduction in the US is mostly addressed at the state level, resulting in fragmented legislation and efforts. Several jurisdictions require manufacturers to develop e-waste recycling programs or participate in Extended Producer Responsibility (EPR) efforts to recycle and properly dispose of their electronics. The Federal Electronics Challenge and other federal initiatives promote appropriate electronic asset management throughout federal departments. Many charities and recycling organizations gather and recycle electronic waste. The above efforts show electronic waste reduction progress. To ensure nationwide e-waste treatment, a comprehensive federal policy is needed (Kang & Schoenung, 2005).

Japan: Japan has successfully reduced e-waste by prioritizing collection, recycling, and resource recovery. E-waste management in the country includes mandatory recycling of some electronics. An Extended Producer Responsibility (EPR) framework requires producers to collect and recycle their products, creating an effective product take-back and recycling system. This policy has increased electronic garbage recycling in Japan, where advanced recycling methods extract valuable materials. Public awareness efforts and eco-labeling encourage responsible garbage disposal and sustainable consumer choices. Japan's e-waste reduction efforts demonstrate its environmental and resource efficiency (Lee & Na, 2010).

India: India's comprehensive e-waste policy includes government actions, public awareness campaigns, and sustainable practices . The E-Waste (Management) Rules require producers, importers, and dealers to sustainably manage e-waste. Through the Extended Producer obligation (EPR) framework, environmentally friendly product design is promoted. The nation offers many e-waste collection facilities to help consumers and businesses safely dispose of electronic debris. Public awareness campaigns have also been created to warn about the environmental and health risks of unlawful disposal and encourage appropriate recycling. India is integrating the informal e-waste sector into the formal management system to promote safer practices and fund recycling technology research. By exchanging best practices and sustainable solutions, international collaboration helps India's efforts. Finally, India's holistic approach to electronic waste's environmental and health impacts is commendable (V. Kumar et al., 2011; Wath et al., 2010).

China: China has taken significant steps to reduce and manage electronic waste. Electronics manufacturers must take responsibility for their goods' whole life cycle, including disposal, under the Extended Producer Responsibility (EPR) framework. This encourages eco-friendly product design and recycling. China also provides subsidies for recycling and disposing of electronic waste. Manufacturers fund these funds, encouraging responsible product design and management. To ensure environmental and safety compliance, government regulation and supervision of e-waste recycling companies has increased. The above projects demonstrate China's commitment to reducing electronic waste's environmental and health impacts and promoting sustainable electronics practices (Ignatuschtschenko, 2018; Lee & Na, 2010).

Australia: Australia has implemented several e-waste prevention schemes. The National Television and Computer Recycling Scheme (NTCRS) made electronic device trash collection facilities more accessible. Producers and importers must collect and recycle electronic waste under this system. The shipment of electronic waste (e-waste) to poorer nations has been limited by government regulations to reduce improper disposal. Electronic waste disposal is stressed in public awareness and education activities. Australia is improving its recycling infrastructure to recover valuable elements from electronic devices. These efforts demonstrate the nation's commitment to reducing electronic waste and promoting sustainable electronic practices (Cucchiella et al., 2015; Janani et al., 2023; Xavier et al., 2021).

These examples show that governments are using extended producer responsibility, tough legislation, and recycling programs to limit e-waste's environmental and health implications. Combating the growing global problem of electronic waste requires global action. Electronic waste reduction is essential to reduce environmental and health implications. Different methods are used to reduce e-waste and promote sustainable management. Manufacturers should build electronics with longevity and repairability in mind. Easy-to-repair and upgrade products last longer and require fewer replacements. Extended Producer Responsibility (EPR) programs, which provide producers control over electronic product lifecycle management, are another key technique. This implies producers must collect, recycle, and safely dispose of their products after use. EPR promotes green product design and disposal. Consumer education and awareness initiatives also help people understand the environmental and health risks of inappropriate e-waste disposal. These activities encourage consumers to return outdated electronics through authorized channels for recycling and responsible disposal.

To make recycling and disposal easy, governments and electronics manufacturers often offer take-back and recycling programs. These programs recover valuable resources, decreasing resource-intensive mining and extraction. Worldwide, e-waste management laws set requirements for safe disposal, recycling, and environmental responsibility. These requirements must be followed to reduce e-waste's environmental and health implications. E-waste recycling efficiency is being improved by researching and developing new recycling technologies like robotic sorting and sustainable materials recovery. These e-waste reduction approaches are reducing the environmental and health implications of e-waste while preserving resources.

LITERATURE REVIEW

Electronic garbage, also referred to as e-waste, is experiencing rapid growth and has emerged as a significant waste stream on a global scale. This escalating issue presents a range of substantial concerns pertaining to both the environment and human health. Electronic garbage, commonly referred to as e-waste, encompasses a diverse array of perilous substances, including lead, mercury, cadmium, and arsenic. These hazardous compounds have the potential to permeate the soil and water, polluting the environment and presenting a significant threat to human well-being. The incineration of e-waste and the manufacturing of new electronic equipment can also have an impact on climate change. There exist various sustainable approaches and frameworks for mitigating the environmental consequences associated with electronic trash (e-waste) that can be effectively implemented. The methodologies and concepts discussed in this study can be classified into three primary domains, namely reduction, reuse, and recycling (Dasgupta et al., 2017; V. Kumar et al., 2011; Olla & Toth, 2010; Roy et al., 2022).

Reduction: Reduced waste is key to e-waste management. These initiatives promote proper use, repair, and disposal of discarded electronics to minimize environmental effects. Buying fewer electronics is one of the best strategies to reduce e-waste. This technique promotes buying electronics when needed. Before buying something new, people should see if an old item will do. Choosing robust, repairable gadgets is key. Long-lasting, easy-to-maintain electronics reduce e-waste by decreasing replacements. Repairing electronics is key to minimizing e-waste. Repairing electronics is greener than tossing them away. Electronics repair Online and library resources are available to DIYers. Repair shops for electronics are accessible. Electrical waste can be reduced and money saved by enabling electronic repair. Electronic waste management needs proper disposal of broken or outdated devices. Give electronics to charities or sell them to consumers and businesses to reuse. Recycling useless electronics is moral. Many local and online electronic waste recycling programs recycle components safely. Users can join e-waste recycling schemes from several electronics manufacturers. Using and disposing of electronics responsibly reduces e-waste's environmental impact (Hossain et al., 2015; Misra et al., 2021).

Recycling: Recycling is a good way to reduce electronic waste. Recycling electronic garbage saves precious components from landfills and incinerators. E-waste contains gold, silver, copper, and aluminum. Recycling electrical waste lets these components be used to make new products. This conserves natural resources and reduces manufacturing's environmental impact. Recycling e-waste reduces pollution. The combustion of electronic waste releases harmful pollutants. E-waste recycling prevents toxic pollution. Electronic waste recycling has many methods. A nearby e-waste recycling center or drop-off location can properly dispose of e-waste. Some electronics manufacturers and retailers offer mail-back initiatives to recycle e-waste. Recycling electronic garbage (e-waste) is crucial. Some e-waste contains lead, mercury, and cadmium. To protect people and the environment, these materials must be properly handled and disposed of (Chen et al., 2021; Kang & Schoenung, 2005; Nnorom & Osibanjo, 2008).

Reuse: Reusing electronics reduces electronic waste. Electronic devices can be reused by using them yourself or giving them to others. Electronics can be reused in many ways. Examples are given:

1. Optimal electronics maintenance is essential. This will increase their durability. Repair and maintain electronic gadgets that break. Internet and library resources provide complete electronic repair education. Repair shops can be found locally.
2. Donate electronics to charities. Electronic gadgets are accepted by many philanthropic groups, including schools, hospitals, and homeless shelters. Sell electronics to consumers and businesses. If your electrical gadgets are in good condition, you can sell them to consumers or businesses.
3. Electronics have many uses. An old laptop can be used as a media center or digital photo frame.
4. Reusing electronics reduces electronic waste's environmental impact and saves money. It extends the lifespan of electronic devices and reduces their manufacturing.

Recycling electronics offers several benefits beyond reducing trash. The environmental impact of electrical gadget production has decreased significantly. Recycling helps us meet our resource needs sustainably by saving important natural resources. Electronic recycling also reduces harmful substances in the environment, increasing environmental quality. This technological invention preserves energy and reduces greenhouse gas emissions, helping to mitigate climate change. Electronic device recycling is vital to environmental preservation and global sustainability (Pariatamby & Victor, 2013; Xavier et al., 2021).

METHODS

Electronic waste reduction is essential to reduce environmental and health implications. Different methods are used to reduce e-waste and promote sustainable management. Manufacturers should build electronics with longevity and repairability in mind. Easy-to-repair and upgrade products last longer and require fewer replacements. Extended Producer Responsibility (EPR) programs, which provide producers control over electronic product lifecycle management, are another key technique. This implies producers must collect, recycle, and safely dispose of their products after use. EPR promotes green product design and disposal. Consumer education and awareness initiatives also help people understand the environmental and health risks of inappropriate e-waste disposal. These activities encourage consumers to return outdated electronics through authorized channels for recycling and responsible disposal. To make recycling and disposal easy, governments and electronics manufacturers often offer take-back and recycling programs. These programs recover valuable resources, decreasing resource-intensive mining and extraction. Worldwide, e-waste management laws set requirements for safe disposal, recycling, and environmental responsibility. These requirements must be followed to reduce e-waste's environmental and health implications. E-waste recycling efficiency is being improved by researching and developing new recycling technologies like robotic sorting and sustainable materials recovery. These e-waste reduction approaches are reducing the environmental and health implications of e-waste while preserving resources (Chatterjee, 2012; Fytili & Zabaniotou, 2008).

Design for Durability and Repairability

Durable and repairable electronics reduce e-waste by prolonging their lives and reducing premature disposal. These concepts are gaining hold in India as consumers and authorities recognize the environmental and economic benefits of durable, repairable products. Modular design is important for durability. Electronic products are becoming more modular, allowing users to replace or update individual parts. Instead of replacing the complete phone, smartphones with modular battery packs or camera modules let consumers change individual sections. E-waste is reduced, consumers save money, and resources are conserved. Standardization is crucial to repairability design. Electronic equipment producers in India are being encouraged to use standardized components and connectors. Standardization makes suitable replacement parts easily available, simplifying repair. Repair specialists may easily obtain standardized components, making gadget repairs cheaper and easier for consumers, reducing the incentive to abandon them prematurely. Repair documentation access is also crucial. India wants manufacturers to give consumers and repair technicians repair manuals and schematics. This encourages DIY electronics repair by empowering people to mend them. Consumers can avoid discarding repairable gadgets by having repair information readily available, reducing e-waste. For optimal repairability, spare parts must be available. Indian initiatives are lobbying for legislation requiring manufacturers to keep spare parts on hand after a product is retired. This dedication to spare parts availability ensures that users and repair professionals can get genuine replacement parts, lowering the probability of devices being rejected due to part shortages (Das et al., 2021; A. Kumar & Dixit, 2018).

India is advancing "right to repair" laws to support these initiatives. Manufacturers would have to provide diagnostic and repair information, tools, and parts to consumers and independent repair shops under such laws. E-waste reduction is supported by this legislative framework, which allows consumers to make educated decisions about fixing rather than replacing their equipment. Consumer awareness is crucial to implementing design for durability and repairability in India. Education and advocacy efforts are educating people about their rights to repair their electronics and the environmental benefits of doing so. By raising awareness, consumers can understand the consequences of their choices and the benefits of repair over replacement. India is making progress in e-waste reduction through design for durability and repairability. Through modular design, standardization, repair documentation, spare parts availability, "right to repair" legislation, and consumer awareness programs, the government may reduce e-waste and encourage sustainable electronic device usage. These efforts improve the environment, save consumers money, and conserve resources.

Consumer Education and Awareness

Increasing public awareness to reduce technological waste and promote sustainable products requires a multifaceted approach. Education and information campaigns are vital. These efforts should use television, radio, social media, and community workshops to educate people about e-waste's environmental impact and the benefits of sustainable alternatives. Public awareness efforts should teach correct electronic garbage disposal, recycling, and ecologically friendly buying habits. Integrating e-waste and sustainability into schooling is crucial. By teaching youngsters about proper gadget use, recycling, and the environmental impact of electronic trash, we can instill sustainable values early on. Educational institutions can promote environmental stewardship by recycling e-waste and promoting sustainability. Electronic equipment eco-labeling and certifications influence consumers to choose sustainable options. These labels indicate a product's energy efficiency, recyclability, and environmental impact. These labels can be mandated by the government to give consumers easy access to information that affects their purchases. Successful agreements with electronics stores allow for extensive information transmission to a diverse audience. Retailers can educate customers on recycling, take-back, and energy-efficient products. Discounts and trade-in programs can help recycle obsolete equipment safely. Community workshops, seminars and e-waste recycling initiatives raise awareness. We empower individuals to take grassroots activities by engaging community members in sustainability efforts and providing easily available information. Digital tools like websites and smartphone apps make it easy to find e-waste recycling facilities, green product reviews, and sustainable lifestyle advice. Access to nearby recycling centers and valuable resources are key to encouraging public responsibility. Government regulations are crucial for e-waste management and sustainable product marketing. Promoting strict laws that hold manufacturers accountable for recycling and eco-design can alter the business. Penalties for non-compliance may encourage businesses to adopt sustainable practices. Governments, NGOs, corporations, and educational institutions collaborate to promote awareness campaigns and sustainable products. Multi-stakeholder alliances can combine resources and talents to create impactful awareness campaigns.

Tax benefits or incentivized systems may encourage people to buy eco-friendly products or recycle e-waste. Positive reinforcement can greatly boost sustainable decisions. Ultimately, public advocacy and active engagement in electronic waste mitigation efforts are essential. Environmental activists, NGOs, and concerned citizens who promote responsible electronics use and sustainability should be supported and recognized. Their goal is to promote environmental awareness (Bhat & Patil, 2014; Garlapati, 2016; Li & Achal, 2020; Whitman & Mattord, 2021).

Circular Economy Initiatives

The implementation of circular economy initiatives holds significant significance in the mitigation of electronic waste and the advancement of sustainable practices in the management of electronic waste. The primary objective of these endeavors is to cultivate a heightened sense of responsibility and promote an ecologically viable approach across the whole lifecycle of electronic products. One notable initiative concerns the use of product reuse and refurbishment practices. Rather than engaging in the disposal of obsolete electronic devices, organizations and organizations possess the alternative to collect such goods, rectify any operational issues, and refurbish them in order to prolong their lifespan. This activity not only leads to a decrease in electronic waste but also provides economically viable options for consumers. Remanufacturing is a prominent technique that holds substantial importance within the context of the circular economy. This process goes beyond simple refurbishment and involves the systematic dismantling of electronic devices, followed by their reassembly to attain a state that is equivalent to that of newly manufactured items. Remanufactured products often come with warranty protection and can serve as an environmentally sustainable alternative to newly manufactured devices. Extending the longevity of electronic products is a critical approach within the context of the circular economy. There exists a widely accepted suggestion for manufacturers to give precedence to the integration of durability, modularity, and repairability in their product designs. This concept facilitates the active involvement of consumers in the upkeep and improvement of their equipment, hence reducing the need for frequent replacements. Take-back and recycling activities are of utmost importance in the successful execution of circular economy endeavors. Manufacturers and merchants employ strategies that facilitate consumer participation in the process of returning their obsolete electronic equipment for the goal of environmentally responsible recycling and disposal. This method ensures the acquisition and subsequent exploitation of significant resources, hence diminishing the requirement for new raw materials. Moreover, the enactment and enforcement of legislation and regulations play a crucial role in strengthening and maintaining the fundamental concepts and practices of the circular economy. Governments enact legislation to establish regulations pertaining to the management of electronic waste, establish specific criteria for recycling, and promote the adoption of suitable practices for disposal. It is imperative to adhere to these guidelines to mitigate the environmental and health consequences linked to electronic waste.

The integration of circular economy initiatives assumes a pivotal role in promoting a more sustainable and conscientious approach to the management of electronic waste. The primary objective of these schemes is to effectively preserve resources and mitigate the environmental consequences linked to electronic devices (Ignatuschtschenko, 2018).

Take-Back and Recycling Programs

Take-back activities bolster global efforts to reduce electronic waste and promote sustainable electronics. Manufacturers and retailers often lead these initiatives to encourage consumers to return their old devices for recycling and disposal. This discussion examines how take-back programs promote a more sustainable electronics sector. The safe and correct disposal of old electronics is the main goal of take-back programs. These initiatives divert electronic waste from landfills and informal recycling sectors, which pose serious environmental and health risks, by providing consumers with easy ways to return their old devices. Conversely, returned gadgets are recycled at responsible facilities to extract useful components and dispose of dangerous chemicals. Take-back programs also support circular economy principles of resource efficiency and waste reduction. Electronics manufacturers and dealers are realizing the value of salvaging parts from old devices. These programs collect and reuse metals, polymers, and rare earth materials for new electronics. This conserves finite resources and reduces mining and extraction environmental impacts. Discounts on future purchases and trade-in programs are used to encourage consumer participation in take-back activities. Encouragement to dispose of electronic waste responsibly promotes a culture of sustainability and responsible consumption. Regulatory laws in several countries require electronics makers to create take-back schemes and meet recycling and recovery goals. These guidelines hold electronic device makers accountable for their products' entire life cycle, encouraging them to incorporate recycling and sustainability into their design processes.

Take-back programs steer electronic waste away from ecologically hazardous disposal techniques, promote resource efficiency, and encourage ethical consumer behavior. Electronic waste management systems must be developed and improved as the electronics industry grows. Electronic waste management programs must be expanded and improved to be more sustainable and ecologically friendly.

Legislative and Policy Measures

Laws and policies help manage electronic trash (e-waste) by providing a framework for addressing its environmental and social impacts. From manufacturing to disposal, the above actions have been designed to promote responsible electronic device use. Extended Producer Responsibility (EPR) projects are crucial. EPR states that electronics manufacturers must efficiently manage and fund the retrieval, recycling, and correct disposal of their gadgets when they have reached the end of their functional lifespan. The transfer of duty from customers and local governments to producers encourages manufacturers to incorporate sustainability into product designs. Extension Producer Responsibility (EPR) projects encourage environmentally friendly product design and provide financial incentives for manufacturers to decrease e-waste. Several countries have also banned harmful chemicals in electronics. The European Union's Restriction of Hazardous compounds (RoHS) directive restricts the use of some hazardous compounds in electronic products. This reduces harmful materials in consumer electronics. These rules emphasize environmental protection and promote the use of less harmful products.

The WEEE guidelines are essential to sustainable electronics management. Common across the EU, the above directives set high electronic waste collection and recycling targets. Companies must provide collection and recycling infrastructure for their products to control electronic waste and recover valuable materials. Governments encourage manufacturers to create energy-efficient devices that last a long time to promote sustainability. Implementing these standards reduces the environmental impact of electronic items and consumer energy costs, fostering a sustainable and energy-efficient society. Product labelling and certification procedures also help shoppers choose sustainable products. Energy Star designation helps consumers choose energy-efficient electronics. These labels empower consumers to make educated purchases, promote green technologies, and reward manufacturers for sustainable product development. Many regions have adopted eco-design guidelines that define electrical products' environmental impact. Standards consider energy efficiency, material use, and recyclability. Government rules encourage producers to make products with higher environmental and social sustainability.

Thus, legislative and policy actions are crucial to electronics sustainability. These institutions' regulatory structure promotes environmentally sensitive sector practices through a variety of ways. Responsible disposal and recycling rules and sustainable electronic equipment design requirements are included. To reduce the environmental and social impacts of electronic waste and promote a sustainable electronics sector, several initiatives are needed.

RESEARCH AND DEVELOPMENT MODELS

Durable Design Model

The durable design model revolutionizes electronic device design, development, and production. This approach prioritizes the design of durable, long-lasting, and easy-to-maintain electrical equipment. Durable design challenges consumer electronics' disposability culture. In this culture, purposeful design reduces product lifespan, encouraging frequent replacements. The durable design concept involves developing electrical equipment to withstand physical deterioration, environmental factors, and time. Material selection, building, and product testing are all part of this process. Manufacturers prioritize high-quality components, materials, and finishes that withstand degradation, corrosion, and physical damage. In various conditions, products are tested for durability and endurance. This architecture emphasizes modularity. Electronic devices are designed for easy disassembly and replacement. If a battery or screen fails, you can replace them independently, extending the device's lifespan. This method improves repairability and reduces electronic waste by allowing customers and professionals to fix flaws without discarding the device.

The robust design approach also encourages interoperability with prior generations and software support for older devices. Manufacturers provide software upgrades and security patches for long periods, ensuring ageing equipment's functioning and security. This method reduces electronic waste by prolonging device lifespans and promotes sustainable and conscientious electronic product management. The durable design paradigm could transform consumer electronics. Sustainability and resource preservation are promoted by this strategy, which reduces the need for new electrical devices and raw materials. It also lets consumers make eco-friendly purchases by buying products designed for long-term use and repair. The durable design concept improves brand reputation, consumer loyalty, and environmental sustainability, although producers may face initial challenges. The durable design model is a compelling way to alter the electronic sector and reduce the environmental impact of e-waste due to society's rising emphasis on sustainability and responsible consumption.

Circular Economy Model

The circular economy model represents a transformative approach to the production, consumption, and disposal of electronic goods, aiming to reduce electronic waste (e-waste) and foster sustainability throughout the product lifecycle. Unlike the traditional linear economy, where products are manufactured, used, and discarded, the circular economy model promotes a closed-loop system where resources are conserved, and waste is minimized. At the heart of the circular economy model is the concept of product longevity and reusability. Electronics are designed with durability in mind, allowing them to have longer lifespans and remain functional over time. Manufacturers prioritize robust construction and modular designs that facilitate easy repair and component replacement. This approach discourages the disposability of electronic devices and encourages consumers to keep their products in use for extended periods. Refurbishing and remanufacturing are key components of the circular economy model. Instead of discarding used electronic devices, they are collected, inspected, and, if necessary, repaired or upgraded to meet current performance standards. Refurbished electronics are then reintroduced to the market, extending their useful life and reducing the demand for new products. This not only reduces e-waste but also conserves valuable resources and energy that would have been used to manufacture new devices. Take-back and recycling programs play a crucial role in the circular economy model. Manufacturers and retailers often establish collection systems where consumers can return their old electronics for proper recycling and disposal. This ensures that valuable materials, such as precious metals and plastics, are recovered and reused in the manufacturing of new products. By closing the materials loop, the circular economy model reduces the need for resource-intensive mining and extraction. Furthermore, the circular economy model fosters innovation in sustainable materials and eco-friendly manufacturing processes. It encourages the use of recycled and recyclable materials, reducing the environmental footprint of electronic products. Manufacturers are incentivized to adopt eco-design principles, emphasizing resource efficiency and waste reduction throughout the product's life.

In essence, the circular economy model represents a holistic and sustainable approach to electronic product management. It not only reduces e-waste and conserves resources but also aligns with the principles of responsible consumption and environmental stewardship. As consumers and industries increasingly recognize the importance of sustainability, the circular economy model emerges as a powerful strategy to address the challenges posed by e-waste and create a more resilient and eco-conscious electronic industry.

Eco-Legislation Model

Sustainable electronics management requires legislative and regulatory frameworks to ensure ecologically acceptable practices throughout the lifecycle of electronic devices. By requiring producers, users, and other parties to comply with legislation, criteria, and suggestions, this project aims to reduce the environmental impact of electronic rubbish (e-waste). Product design that prioritizes durability, repairability, and recyclability underpins the Eco-Legislation Model. Legal requirements often require electrical equipment manufacturers to incorporate eco-design concepts. Use of sustainable materials, minimization of harmful substances, and guarantee of repairable and upgradeable equipment are included. This regulation promotes long-lasting electronics, reducing the need for replacement and electrical waste. The approach also promotes Extended Producer duty (EPR) frameworks, which shift environmental duty for electronic product management from customers to manufacturers. Under the Extended Producer Responsibility (EPR) principle, manufacturers must efficiently manage their products' end-of-life disposal. This involves product collection, recycling, and safe disposal. This technique emphasizes eco-friendly product design and recycling. The Eco-Legislation Model emphasizes consumer knowledge. Legal requirements often require producers and sellers to disclose the environmental impacts of their electronic purchases. Transparency lets people choose eco-friendly products. Educating and labelling consumers helps reduce e-waste. Recycling and waste laws support the model. Many nations have e-waste recycling laws that require manufacturers to collect and dispose of old electronics in a sustainable manner. These schemes help recover valuable materials and safely dispose of hazardous substances, lowering the environmental effect of improper electronic waste disposal.

DISCUSSIONS

The examination of strategies and methodologies aimed at mitigating electronic waste in a sustainable fashion assumes a pivotal role in addressing the urgent environmental concerns linked to e-waste. As we engage in contemplation of the future, it becomes apparent that the topic being discussed holds considerable importance and presents a promising opportunity for innovative concepts and beneficial transformations. Within the realm of scientific advancements, the forthcoming progress in materials science, electronics, and production techniques presents a compelling prospect. Researchers will conduct an inquiry into novel materials that are ecologically friendly, with the aim of enhancing the durability and maintainability of electronic gadgets. The potential to revolutionize the repairability of products is present by using emerging technologies, such as 3D printing. This technology empowers users to fabricate replacement components within the comfort of their own residences. The expected progression of consumer-centric methodologies is foreseen, placing significant focus on enhancing education and awareness initiatives. The advent of artificial intelligence and data analytics can provide customized recommendations and important insights on sustainable consumption practices, hence improving their accessibility for individuals. The incorporation of interactive platforms and virtual reality experiences holds promise in making a substantial contribution to consumer education on the tangible environmental impacts linked to their decision-making.

The subject of circular economy innovations, encompassing the practices of refurbishing and remanufacturing, holds substantial potential for future breakthroughs. The primary focus of research endeavors is anticipated to revolve around the improvement of these procedures through the implementation of automation and advanced diagnostic techniques. Circular business models, namely those that revolve on the notion of product-as-a-service, whereby individuals compensate for the usage of electronics rather than possessing them outright, possess the capacity to attain substantial appeal and widespread adoption. Given the global magnitude of the issue, it is imperative to underscore the importance of international collaboration in addressing the formidable problem posed by electronic waste (e-waste). Possible future attempts may involve the adoption of comprehensive global standards that regulate the sustainable design of electronic items and the management of electronic waste. The potential for fostering collaborative research projects among nations and organizations to yield creative solutions with positive global societal benefits is substantial. The domain of policy and legislation is anticipated to experience simultaneous modifications in tandem with the dynamic landscape of electronic waste. There is an expectation that governments would enact increasingly stringent legislation mandating businesses to adhere to sustainable product design principles and promote responsible recycling practices. Economic incentives and taxation policies can be employed as instruments to foster environmentally conscious decision-making. The anticipated effect of incorporating data analytics and Internet of Things (IoT) technologies is poised to be substantial in mitigating the issue of electronic waste, as it facilitates data-driven decision-making processes. The dissemination of current information pertaining to the life cycles, consumption patterns, and rates of recycling for electronic items will have a significant impact on decision-making processes in various fields, including consumer decision-making and policy interventions. The application of predictive analytics is anticipated to enable the detection of emerging patterns and opportunities for strategic intervention.

CONCLUSION

Electronic garbage (e-waste) is a global issue in the digital era. Rapid technological advancement and consumer electronics' pervasiveness have led to an increasing flood of discarded devices, posing environmental and health risks. To address this growing challenge, sustainable e-waste reduction models are needed. E-waste output and impact are reduced by a variety of methods in this comprehensive study. Designing durable and repairable electronics is a key method. We can greatly minimize obsolescence and e-waste by encouraging manufacturers to provide gadgets that last longer and are easy to fix or upgrade. Consumer education and awareness are also crucial because knowledgeable consumers recycle responsibly and embrace eco-friendly practices. Circular economy activities like refurbishing and remanufacturing extend product lifecycles and reduce waste. Sustainable e-waste management requires strong take-back and recycling programs and legislative and policy changes. These technologies and models could revolutionize e-waste reduction, creating a more sustainable and resource-efficient electronic industry that protects the environment and public health.

REFERENCES

Ahmad, S., Jha, S., Eljialy, A. E. M., & Khan, S. (2021). A Systematic Review on e-Wastage Frameworks. *International Journal of Advanced Computer Science and Applications*, *12*(12), 701–709. doi:10.14569/IJACSA.2021.0121287

Bakhiyi, B., Gravel, S., Ceballos, D., Flynn, M. A., & Zayed, J. (2018). Has the question of e-waste opened a Pandora's box? An overview of unpredictable issues and challenges. *Environment International*, *110*, 173–192. doi:10.1016/j.envint.2017.10.021 PMID:29122313

Bhat, V., & Patil, Y. (2014). E-waste consciousness and disposal practices among residents of Pune city. *Procedia: Social and Behavioral Sciences*, *133*, 491–498. doi:10.1016/j.sbspro.2014.04.216

Chatterjee, S. (2012). Sustainable electronic waste management and recycling process. *American Journal of Environmental Engineering*, *2*(1), 23–33. doi:10.5923/j.ajee.20120201.05

Chen, J., Huang, S., BalaMurugan, S., & Tamizharasi, G. S. (2021). Artificial intelligence based e-waste management for environmental planning. *Environmental Impact Assessment Review*, *87*, 106498. doi:10.1016/j.eiar.2020.106498

Cucchiella, F., D'Adamo, I., Koh, S. C. L., & Rosa, P. (2015). Recycling of WEEEs: An economic assessment of present and future e-waste streams. *Renewable & Sustainable Energy Reviews*, *51*, 263–272. doi:10.1016/j.rser.2015.06.010

Das, P., Gabriel, J.-C. P., Tay, C. Y., & Lee, J.-M. (2021). Value-added products from thermochemical treatments of contaminated e-waste plastics. *Chemosphere*, *269*, 129409. doi:10.1016/j.chemosphere.2020.129409 PMID:33388566

Dasgupta, D., Debsarkar, A., Hazra, T., Bala, B. K., Gangopadhyay, A., & Chatterjee, D. (2017). Scenario of future e-waste generation and recycle-reuse-landfill-based disposal pattern in India: A system dynamics approach. *Environment, Development and Sustainability*, *19*(4), 1473–1487. doi:10.100710668-016-9815-6

Davis, G., & Wolski, M. (2009). E-waste and the sustainable organisation: Griffith University's approach to e-waste. *International Journal of Sustainability in Higher Education*, *10*(1), 21–32. doi:10.1108/14676370910925226

Fytili, D., & Zabaniotou, A. (2008). Utilization of sewage sludge in EU application of old and new methods—A review. *Renewable & Sustainable Energy Reviews*, *12*(1), 116–140. doi:10.1016/j.rser.2006.05.014

Garlapati, V. K. (2016). E-waste in India and developed countries: Management, recycling, business and biotechnological initiatives. *Renewable & Sustainable Energy Reviews*, *54*, 874–881. doi:10.1016/j.rser.2015.10.106

Gupta, S. (2011). E-waste management: Teaching how to reduce, reuse and recycle for sustainable development-need of some educational strategies. *Journal of Education and Practice*, *2*(3), 1735–2222.

Hossain, M. S., Al-Hamadani, S. M. Z. F., & Rahman, M. T. (2015). E-waste: A challenge for sustainable development. *Journal of Health & Pollution*, *5*(9), 3–11. doi:10.5696/2156-9614-5-9.3 PMID:30524771

Ignatuschtschenko, E. (2018). Electronic waste in China, Japan, and vietnam: A comparative analysis of waste management strategies. *Vienna Journal of East Asian Studies*, *9*(1), 29–58. doi:10.2478/vjeas-2017-0002

Janani, S., Preethi, K., Gowtham, S., & Kulanthaivel, P. (2023). A sustainable reuse of e waste as a partial replacement material for aggregate. *Materials Today: Proceedings*. Advance online publication. doi:10.1016/j.matpr.2023.05.740

Kang, H.-Y., & Schoenung, J. M. (2005). Electronic waste recycling: A review of US infrastructure and technology options. *Resources, Conservation and Recycling*, *45*(4), 368–400. doi:10.1016/j.resconrec.2005.06.001

Kumar, A., & Dixit, G. (2018). An analysis of barriers affecting the implementation of e-waste management practices in India: A novel ISM-DEMATEL approach. *Sustainable Production and Consumption*, *14*, 36–52. doi:10.1016/j.spc.2018.01.002

Kumar, V., Garg, R., Rahman, Z., & Kazmi, A. A. (2011). Sustainability and E-waste management scenario in India. *The First International Conference on Interdisciplinary Research and Development*, *31*, 31.

Lee, S., & Na, S. (2010). E-waste recycling systems and sound circulative economies in East Asia: A comparative analysis of systems in Japan, South Korea, China and Taiwan. *Sustainability (Basel)*, *2*(6), 1632–1644. doi:10.3390u2061632

Li, W., & Achal, V. (2020). Environmental and health impacts due to e-waste disposal in China—A review. *The Science of the Total Environment*, *737*, 139745. doi:10.1016/j.scitotenv.2020.139745 PMID:32516663

Misra, N. R., Kumar, S., & Jain, A. (2021). A review on E-waste: Fostering the need for green electronics. *2021 International Conference on Computing, Communication, and Intelligent Systems (ICCCIS)*, 1032–1036. 10.1109/ICCCIS51004.2021.9397191

Nnorom, I. C., & Osibanjo, O. (2008). Overview of electronic waste (e-waste) management practices and legislations, and their poor applications in the developing countries. *Resources, Conservation and Recycling*, *52*(6), 843–858. doi:10.1016/j.resconrec.2008.01.004

Olla, P., & Toth, J. (2010). E-waste education strategies: Teaching how to reduce, reuse and recycle for sustainable development. *International Journal of Environment and Sustainable Development*, *9*(1–3), 294–309. doi:10.1504/IJESD.2010.029977

Pariatamby, A., & Victor, D. (2013). Policy trends of e-waste management in Asia. *Journal of Material Cycles and Waste Management*, *15*(4), 411–419. doi:10.100710163-013-0136-7

Roy, J. J., Rarotra, S., Krikstolaityte, V., Zhuoran, K. W., Cindy, Y. D.-I., Tan, X. Y., Carboni, M., Meyer, D., Yan, Q., & Srinivasan, M. (2022). Green recycling methods to treat lithium-ion batteries E-waste: A circular approach to sustainability. *Advanced Materials*, *34*(25), 2103346. doi:10.1002/adma.202103346 PMID:34632652

Wath, S. B., Vaidya, A. N., Dutt, P. S., & Chakrabarti, T. (2010). A roadmap for development of sustainable E-waste management system in India. *The Science of the Total Environment*, *409*(1), 19–32. doi:10.1016/j.scitotenv.2010.09.030 PMID:20951410

Whitman, M. E., & Mattord, H. J. (2021). *Principles of information security*. Cengage Learning.

Xavier, L. H., Giese, E. C., Ribeiro-Duthie, A. C., & Lins, F. A. F. (2021). Sustainability and the circular economy: A theoretical approach focused on e-waste urban mining. *Resources Policy*, *74*, 101467. doi:10.1016/j.resourpol.2019.101467

Chapter 17
Sustainable E-Waste Management:
Present Situation, Emerging Solutions, and Future Trends

Ishaan Dawar
https://orcid.org/0000-0003-4217-0182
DIT University, India

Sakshi Negi
https://orcid.org/0000-0002-7798-0110
DIT University, India

ABSTRACT

E-waste, which is generated rapidly owing to technological advancements in information technology, has led to a significant rise in the utilization of electronic devices such as laptops, mobile phones, headphones, and tablets. As these devices become outdated and less useful, they contribute to the growing volume of electronic waste and pose significant threats to the environment and human health. To mitigate the negative impacts of increasing global electronic waste (e-waste), it is crucial to implement proper e-waste management. However, these approaches require substantial labour and resources and are not considered perfect solutions. Fortunately, technological progress has provided new opportunities for more efficient e-waste management. Many countries are now exploring advanced methods, such as the internet of things (IoT) and artificial intelligence (AI), to manage e-waste. However, they face various challenges when implementing these technologies. This study aimed to portray the current situation, emerging trends, and future of smart e-waste management.

DOI: 10.4018/979-8-3693-1018-2.ch017

INTRODUCTION

Electronic devices that are no longer in use, at the end of their lifespans, or damaged are referred to as e-waste. These include everyday items, such as computers, TVs, printers, and battery cells. Despite technological advancements leading to frequent upgrades, the longevity of electronic products is compromised by the presence of toxic substances, such as Polychlorinated Nickel Biphenyls (PCBs), lithium, cobalt, cadmium, lead, and copper. As a result, all Electronic and Electrical Parts (EEPs) eventually become unusable, resulting in e-waste (StEP Solving the E-Waste Problem - StEP Initiative, n.d.). E-waste, also known as technical waste or EEP waste, represents a specific category of discarded electronic and electrical components (Ganguly, 2014). The global accumulation of E-waste or Waste Electrical and Electronic Equipment (WEEE) is estimated to reach 30-50 million tons annually (Cucchiella et al., 2015). Although WEEE has significant recycling value, many of these items also contain hazardous substances that must be separated from the regular waste stream (Oguchi et al., 2011).

The responsibility for collecting and transporting waste materials or airborne particles from the streets falls on either municipal workers or private organizations. This responsibility includes the handling of waste from commercial, industrial, and residential sources. Waste is subsequently moved to the next phase of waste management for processing. During this process, waste is divided into many categories according to its qualities, such as recyclability, biodegradability, and non-degradability. Waste management is crucial for pollution reduction and ecosystem preservation; hence, its importance in environmental protection cannot be emphasized. Furthermore, it is an essential step in preventing threats to public health, as poor waste management may contaminate water and encourage the spread of pests that spread the illness. Waste management also promotes employment opportunities and local economic growth from the perspective of the economy.

The E-waste domain also serves as a platform for business ventures focused on extracting precious metals and harnessing energy through biochemical methods. Therefore, the identification, retrieval, and appropriate recycling of e-waste are important steps for a sustainable future. The growing problem of illegal e-waste exportation to developing nations to reduce recycling expenses is a significant concern in terms of self-sufficiency.

The disposal of e-waste in landfills is not advised because of its hazardous nature. Toxic substances and metal-containing components such as halogenated flame retardants in the nonmetal section are commonly present in e-waste. If it is dumped in a landfill, there's a chance that long-term metal leaching will occur. Leachates from this type of disposal could contain dangerous metals and other compounds that exceed permissible thresholds in the soil and water, endangering both human health and the ecosystem. This emphasizes the significance of appropriately handling this type of solid waste, which has been the focus of investigation and analysis in the study of solid waste management.

In the past few years, dedicated treatment and recovery facilities have emerged in certain countries established by politicians, policymakers, recyclers, producers, and other stakeholders. These facilities are specifically designed to gather and process e-waste from owners, ensuring that the waste is handled appropriately according to the specific needs of each case.

AI and other advanced technologies are widely used in various domains such as education and healthcare, and are relatively new in the environmental domain. Addressing environmental pollution effectively involves implementing an advanced waste management system that utilizes Deep Learning (DL), the Internet of Things (IoT), and Machine Learning (ML). This integrated approach can offer real-time waste information and optimize waste collection routes, ultimately reducing costs and time. Current waste management systems face challenges such as inefficient scheduling, where waste collectors lack awareness of collection needs and drop-off locations. By harnessing these technologies, it is possible to classify waste, predict its volume, and identify the most suitable disposal methods, thereby assisting policymakers in making informed decisions. E-waste management in developing nations requires more comprehensive strategies than that in developed countries to align with Sustainable Development Goals (SDGs).

MOTIVATION

The motivation for this book chapter on sustainable e-waste management can be drawn from the increasing attention given to the sustainable management of electronic waste. As the world continues to grapple with the challenges of e-waste generation and disposal, there is a growing need to understand the present situation, emerging solutions, and future trends in e-waste management. This is evident from the global efforts to develop sustainable production, consumption, and disposal patterns for electrical and electronic equipment, as well as the significant gap between the generation of e-waste and its proper collection and recycling. The hazards caused by improper handling of e-wastes and the limited implementation and enforcement of e-waste policies and regulations further underscore the importance of addressing this issue. Therefore, the book chapter can serve as a valuable resource for understanding the complexities of e-waste management and exploring potential solutions to ensure a sustainable and circular economy.

UNITED NATIONS SUSTAINABLE GOALS AND E-WASTE

The sudden increase in E-waste levels and the use of illegal and hazardous methods for handling it, such as dumping it openly in landfills or extracting precious metals through incineration, present a huge risk to both human health as well as the environment. This hinders progress toward achieving Sustainable Development Goals (SDGs) (*Greenpeace - Where Does E-Waste End Up? - English Version*, n.d.).

E-waste management is closely linked to several SDGs including:

SDG 3: Ensuring well-being and good health.
SDG 6: Promoting clean sanitation and water.
SDG 8: Fostering economic growth and decent work.
SDG 11: Building communities and sustainable cities.
SDG 12: Encouraging responsible production and consumption.
SDG 14: Protecting life below water.

In essence, addressing E-waste issues is essential for the environment as well as human health and for advancing the broader goals of sustainable development.

The rising demand for raw materials in the Electronics and Electrical Equipment (EEE) manufacturing industry is closely related to the Sustainable Development Goal (SDG) 12, particularly its indicator of the origin and domestic utilization of raw materials (Ahirwar & Tripathi, 2021). To track the growth of E-waste more accurately, a specific SDG indicator has been established, which is of significant concern because of its potential hazards and residual value. This indicator has been incorporated into the design plan for the SDG indicator 12.5.1, which measures the national recycling rate and amount of material recycled. The importance of addressing E-waste is further emphasized in SDG indicator 12.4.2, which focuses on waste treatment, generation of hazardous waste, and waste management, categorized by treatment type. These indicators and targets were clearly defined as well as measured according to guidelines developed for monitoring purposes, with the ultimate goal of achieving these objectives over 15 years.

DIFFERENT TYPES OF E-WASTE

Multiple sources contribute to the generation of E-waste in the environment. These may include automobile industries, our households, the medical sector, IT industries, and other electrical and electronic materials, as depicted in Figure 1.

Figure 1. Different categories of e-waste

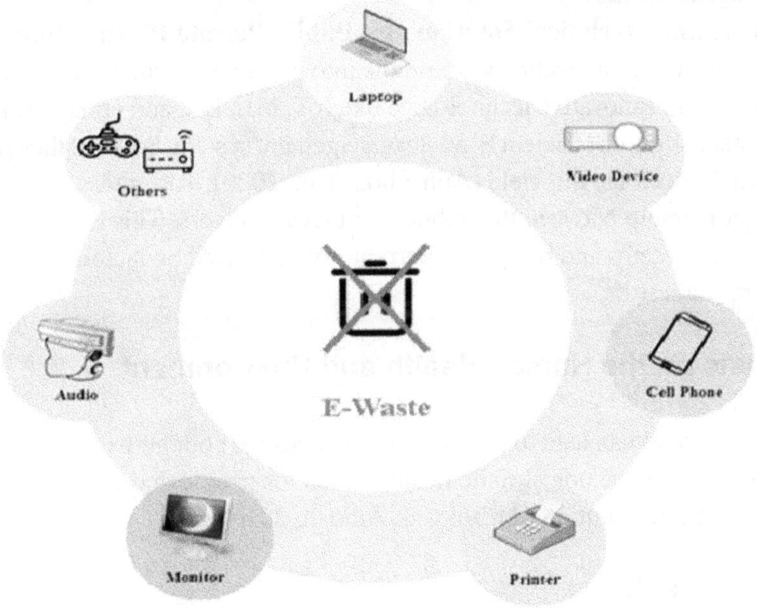

PROBLEMS ASSOCIATED WITH E-WASTE MANAGEMENT

In urban areas, the management of waste produced presents a significant challenge, especially when it comes to E-waste in developing countries. In India, there is a substantial workforce dedicated to handling waste, including waste collectors, scrap dealers, and recyclers, certain groups within the waste disposal system, such as unrecognized garbage collectors, are particularly impacted (María Fernández-González et al., 2020). These unacknowledged individuals make up the largest informal sector in the country. However, there are some problems when it comes to E-waste management.

1. **Less Awareness of E-waste Management:** Many individuals lack awareness of the environmental pollution caused by inefficient waste management and collection. With the rapid digitization experienced in recent decades, the challenge of managing the increasing volume of E-waste has become even more daunting. In developing countries, there is often a lack of knowledge on the various categories of waste, leading to improper mixing of E-waste with other types of waste or inadequate disposal methods.
   To address this issue and draw attention to the importance of sanitation and garbage segregation in every city, it is crucial to raise awareness among the public. This can be achieved through the organization of workshops, the use of slogans, and the implementation of social media campaigns. By spreading awareness among the community, there is a potential for a positive change in E-waste management.
2. **Improper Classification of E-waste:** Improper disposal and categorization of E-waste can pose significant dangers, impacting not only the well-being of humans and animals but also hindering plant growth due to the presence of various chemical elements and metals. The only effective remedy lies in appropriately categorizing, disposing of, and treating E-waste ("Handbook of Environmental Materials Management," 2019).
3. **Gap in Implementing Technical Solution and Public-Private Partnership:** There's a notable gap when it comes to putting technical solutions into action, particularly in the realm of effective E-waste management. Innovative technologies like IoT, ML, DL, and other cutting-edge solutions are crucial in establishing an efficient E-waste management system. Some of these new technologies have the potential to excel in this field (Anh Khoa et al., 2020). To address this gap, it's imperative to establish a partnership between the public and private sectors. This collaboration is necessary to implement such systems and bridge the current divide caused by factors like a lack of technical expertise and resources.

Effects of E-Waste on the Human Health and Environment

Improperly managed E-waste can lead to severe consequences for both human and environmental well-being. These negative effects include chronic health issues for people and environmental degradation, including air, water, and soil pollution (Amankwah-Amoah, 2016).

Effects on Human Health

Improper management, inadequate recycling, and processing methods for E-waste can have life-threatening consequences for human health (Orlins & Guan, 2016). Hazardous materials released from e-waste might enter the body through ingestion, dermal contact, or inhalation, posing significant risks (Awasthi et al., 2018). Workers involved in E-waste handling or recycling are particularly vulnerable to direct occupational exposure owing to a lack of proper safety measures. Identifying E-waste exposure is complex because of the diverse sources of exposure, various routes of entry, individual characteristics (such as age, sex, immune status, and body weight), and various exposure durations (Alabi & Bakare, 2017).

Metals can enter the human body via inhalation, dermal contact, or ingestion. Ingestion involves the consumption of water and food contaminated with E-waste (Zheng et al., 2013). It is the main route of exposure to heavy metals in the human body, which accounts for approximately 90% of cases (Weerasundara et al., 2020). Additionally, heavy metals can accumulate in plant tissues from contaminated air, soil, and water sources (Singh et al., 2010).

The consumption of food items grown in heavy-metal-contaminated soil is a significant factor responsible for the increased concentrations of heavy metals in animal-based products. Various studies conducted worldwide have reported that this metal exposure leads to several human disorders, such as the nervous system, cardiovascular system, urinary system, blood, kidneys, liver, learning ability, urine-related, and cancer (Yan et al., 2013). Considering possible health hazards, local authorities and government agencies need to carry out comprehensive studies to assess exposure and evaluate risks in areas where E-waste processing and recycling operations are conducted.

Effects on Environment

Improper disposal of E-waste results in significant environmental contamination, affecting water, soil, air, and various organisms. The disposal sites and recycling facilities have been pinpointed as the origins of heavy metal and particulate matter emissions into the environment. Materials released from E-waste can be broadly divided into two types: hazardous and non-hazardous materials. These waste releases pose serious risks to the environment and can lead to long-term ecological damage if not properly managed.

Both categories of pollutants emanating from E-waste can have detrimental effects on human health and the environment if their concentrations exceed the allowable thresholds(Saini et al., 2023) . These hazardous substances, when introduced into the environment, get entangled in different processes like food contamination, bioaccumulation, and extensive exposure within ecological systems. This, in turn, leads to substantial and far-reaching risks. Hence, it is imperative to tackle and regulate the discharge of these harmful substances to mitigate their influence on both human and environmental welfare (Pradhan et al., 2014).

Improper disposal and processing of E-waste have multifaceted consequences, extending beyond local environmental damage and human health risks. Due to the large volumes of carbon dioxide that these practices release into the atmosphere, they also contribute to climate change and global warming.

Effects on Air Quality

E-waste contaminants are released in the air in the form of dust or fumes, posing risks to human health through ingestion, inhalation, and skin absorption. In the processing and recycling of E-waste, several hazardous primary contaminants like Hg, Pb, Cd, and PCBs, as well as secondary toxicants such as PAHs, difurans, and dioxins are produced and released into the environment. The presence of these contaminants depends on the specific type and method of recycling employed.

These metals remain suspended as particulate matter in the air and get deposited on open water resources and the soil surface. As a result, they can accumulate in aquatic life, soil, and plants, leading to serious health consequences when humans consume contaminated water or food products (Sthiannopkao & Wong, 2013). Therefore, it is important to adopt safe and environmentally responsible E-waste recycling practices to reduce the production of these harmful substances and protect both humans and the environment.

Effects on Water Quality

E-waste contaminants find their way into the aquatic environment through leaching and dispersal processes, originating from the dumping of EEE in industries and other activities related to E-waste. Water resources located near various E-waste disposal facilities and recycling fail to meet the established water quality standards concerning parameters like total dissolved solids, hardness, pH, chlorides, and conductivity (Dharini et al., 2017).

Effects on Soil Quality

The rapid escalation of human activities results in the discharge of contaminants that are hazardous into the soil, which further leads to pollution of farmland soil (Zhang et al., 2015). Waste recycling plants are identified as significant contributors to soil contamination, primarily due to the release of elevated concentrations of various metals and pollutants. These harmful toxins, including heavy metals, can be readily absorbed by plant roots, and taken to different parts of the plant, such as the leaves, and fruits. Consequently, soil contaminated with E-waste is recognized as a major reason for crop and vegetable contamination.

Figure 2. Impacts of e-waste

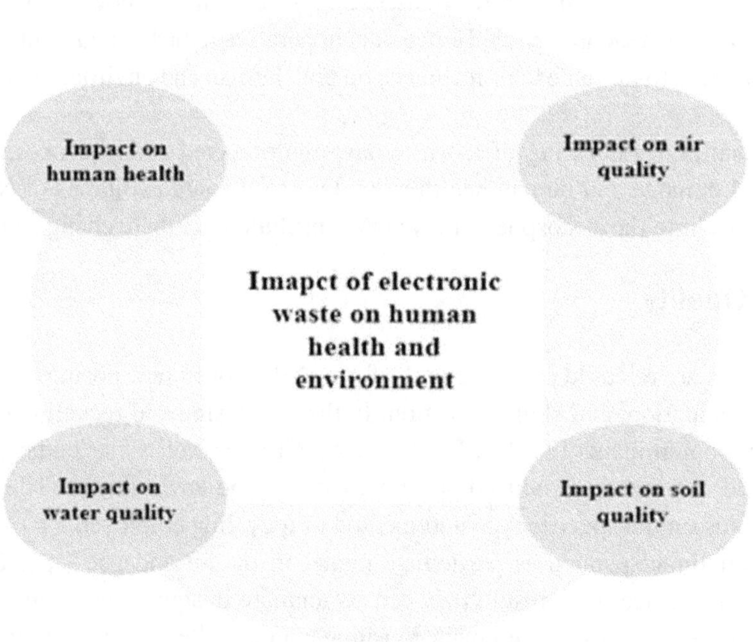

E-WASTE MANAGEMENT USING ADVANCED TECHNOLOGIES

Artificial intelligence (AI)-based approaches enable machines to replicate human capabilities such as learning from examples, recognizing patterns in data, and making decisions based on those patterns. AI is widely used across the globe to assist individuals in making informed decisions because it can process vast amounts of data much faster than humans. AI plays a crucial role in helping humans tackle various challenges such as waste detection, space debris management, and sustainability efforts. Its rapid rise to prominence is owing to its previously unimaginable applications (Dawar et al., 2023). For instance, AI has been employed by companies such as Xcel Energy to accurately forecast energy consumption trends and optimize their systems for enhanced efficiency, even in industries traditionally associated with coal and emissions. Moreover, AI is used to address a wide array of environmental engineering issues, including water and wastewater treatment, air pollution reduction, soil contamination, municipal waste management planning, and remediation.

The implementation of AI plays a significant role in optimizing the routes for garbage trucks for waste collection. This was achieved by identifying the locations of waste management facilities, predicting trends in waste generation, and simulating the processes involved in converting waste. Precise predictions regarding the quality of E-waste enable the effective gathering, treatment, and disposal of waste (Dhir et al., 2021). Numerous studies have explored the potential of AI for managing E-waste. In this context, the primary applications of AI and ML techniques are segregation and forecasting in E-waste management.

IoT is being adopted across various industries to reduce energy consumption and carbon emissions. In this century, the proliferation of Internet-connected devices has rapidly increased. According to predictions from the statistics service Gardiner and Statista, the number of devices connected is expected to reach 30 billion by 2020 (Pardini et al., 2019). IoT involves the integration of physical or virtual devices with the Internet, enabling functions such as detection, data gathering, storage, and processing. In this context, IoT offers highly interconnected and efficient network applications. Combining IoT with cloud computing allows E-waste management to be automated and streamlined, providing a mechanized approach to handling electronic waste (Sharma et al., 2021).

Advanced ML and DL technologies have been utilized to gather and analyze E-waste data, contributing significantly to smart-city waste management. By applying these techniques, waste analysis, determination of waste bin sizes, waste collection truck selection, and waste material analysis become more accessible, providing valuable information for effective decision-making. These intelligent methods are highly effective in waste identification and categorization and facilitate more informed and efficient steps for waste handling.

RECENT STUDIES RELATED TO E-WASTE MANAGEMENT USING ADVANCED TECHNOLOGIES

A specific article or research paper is regarded as a document or a review unit. The databases of Google Scholar, Wiley, Scopus, Elsevier, IEEE, Springer, and Web of Science are used to obtain all the documents. These are the best libraries of peer-reviewed works from throughout the world in several fields that authors frequently use for SLR. Numerous studies related to EWM, ML, and DL were found in the initial search. Numerous articles that used the traditional model (such as the statistical model) in EWM were also among the top searches. This retrieved over 550 documents between 2019 and 2023 from the online search databases. Subsequently, a structured query of the results of searches using inclusion and exclusion criteria collects pertinent literature and limits the amount of literature suited.

To ensure the accuracy, dependability, and relevance of the literature, a manual inspection was performed. Exclusion and inclusion criteria were established to make the scrutiny procedure simple, controllable, and impartial. Publication accessibility, document language, type, and subject/title were among these classifications. Subsequently, the relevance of each document that had been chosen during the search and data collection procedures was examined. The first screening included an English-language, fully accessible journal, or conference research. Additionally, their titles and subjects were examined to assess and select the best research.

E-waste management challenges stem from skill gaps, infrastructure issues, and financial constraints. The toxic effects on health and the environment underscore the need for a holistic approach, including regulatory frameworks, education, safe conditions, and sustainable practices. Bridging formal and informal sectors can foster sustainability, job creation, and economic growth, aligning with global development goals. The approach presented in (Rautela et al., 2021) benefits both developed and developing countries.

In (Ahirwar et al., n.d.; Ahirwar & Tripathi, 2021; Akram et al., 2019; Jangre et al., 2022; Srivastav et al., 2023; Tunali et al., 2021; Xu et al., 2023), studies have underscored the urgency of sustainable E-waste management, advocating for strategic interventions across various aspects of its life cycle. The surge in E-waste demands effective recycling while considering its environmental impact. To address toxic releases during recycling, a holistic approach involving product design, recycling rates, and reduced hazardous emissions is vital. E-waste trends, processes, and health implications are covered, suggesting strategies like regulations, informal sector support, and sustainable techniques (bioleaching, biosorption). Integrating recycling in device design can ensure safe management.

Despite E-waste's toxic content, management policies are insufficient. Categories and pollutants such as metals and plastics were highlighted, underscoring their air, water, and soil impact. Global management demands regulations, education, and 3R practices. Collaboration among governments, industries, and citizens is vital to curbing E-waste through responsible buying, innovation, and reuse.

Aiming to prioritize barriers in E-waste management, (Jangre et al., 2022) employed Fuzzy Decision-Making and Interpretive Structural Modeling, identifying awareness, policy gaps, and public insensitivity as main obstacles. Fuzzy logic aids policymakers by revealing interconnections for informed decisions. Technological advancement leads to increasing E-waste, raising environmental risks. (Srivastav et al., 2023) advocated for sustainable strategies, including recycling, circular economy, and policies.

Aligning with sustainable development goals minimizes harm. (Xu et al., 2023) delved into E-waste's rise due to gadget demand, emphasizing its harmful disposal effects. Approaches like secondary resource use, 3Rs, urban mining, and green electronics were suggested. Regulatory policies, producer responsibility, and international cooperation are crucial. Green tech and monitoring enhance formal E-waste management. (Tunali et al., 2021) examines heavy, precious, and rare earth metals in E-waste components. Digestion methods' comparison revealed major components like Cu, Fe, Zn, and precious metals. Recovery techniques are encouraged, addressing metal-specific methods like U.S.EPA.3050B for certain metals and MMDM for others. The study advances E-waste composition understanding and suggests focusing on metal recovery.

Studies (S. C. Chakraborty et al., 2022; Dixit et al., 2022; Dutta et al., 2023; Liu et al., 2023; Marimuthu et al., 2022; Moheb-Alizadeh et al., 2023; Nande et al., 2023; Seif et al., 2023) showcased the critical proposed international frameworks, regulations, and innovative techniques to manage E-waste effectively, reduce environmental impacts, and promote sustainable practices across industries and sectors.

Addressing the global issue of E-waste requires collective efforts. (Liu et al., 2023) suggested adding E-waste to the Basel Convention, creating global management associations, and enforcing strict product standards. Extended Producer Responsibility (EPR) and safe recycling practices are crucial, along with monitoring pollutants. The goal was green E-waste recycling via advanced separation techniques, emphasizing international collaboration, strong regulations, and responsible actions across all sectors.

E-waste's environmental impact has led to an increased focus on end-of-life electronic product recycling. (Moheb-Alizadeh et al., 2023) introduced two models, user-optimum and system-optimum, to analyze the ecological and economic effects of take-back legislation. These models assess impacts during different stages of E-waste's life cycle, particularly transportation from residences to collection sites. Results highlighted transportation as a major cost and emissions contributor, suggesting the need for efficient algorithms to address this challenge.

The surge in WEEE demand due to renewable energy tech has led to shorter lifespans and disposal challenges. (Seif et al., 2023) discussed how governments, scientists, and international bodies strive to recycle WEEE for valuable materials. E-waste management policies promote material recovery to meet sustainability goals. Hydrometallurgy techniques for materials extraction from circuit boards and batteries are discussed. The transition to renewable energy and energy efficiency relies on technology, education, and material innovation. Recovered materials could catalyze eco-friendly energy processes, and future research should target material retrieval for broader renewable applications.

Improper disposal of E-waste harms the health and environment. (Dutta et al., 2023) explored metal recovery methods—pyrometallurgy, hydrometallurgy, metallurgy—each with its limitations. It offers insights into these processes, their advantages, and their limits. It also emphasizes the importance of the circular economy and socio-economic aspects of sustainable E-waste management, benefiting students, researchers, industry, and policymakers.

Industries that produce harmful by-products and waste contribute to pollution. (Nande et al., 2023) discussed how technological progress, seen in evolving gadgets and electronics, brings comfort but also generates E-waste. Disposing of such waste is a global challenge. This chapter explored various waste types, emphasizing industrial and E-waste's detrimental effects and disposal methods. It also addressed the balance between development and environmental impact.

(Marimuthu et al., 2022) discussed the utilization of PCB fibers from E-waste in concrete to improve durability, reduce waste, and enhance properties. Adding 1-3% fibers to M40-grade concrete increases strength and rebound hammer tests align with compressive strength. Recycling E-waste in construction manages waste and prevents environmental issues.

Hydrometallurgy is effective but challenging due to E-waste complexity. Innovative methods like phytochemical adsorption and bio-adsorbents offer sustainable gold retrieval from E-waste. Rising E-waste generation threatens the environment with toxic metals like lead (Pb) and mercury (Hg). (S. C. Chakraborty et al., 2022) Highlighted the impact, sources, and the need for proper management of E-waste. Combining biological, physical, and chemical methods offers effective remediation.

(Dixit et al., 2022) explored using certified recycled E-waste as a substitute for coarse aggregates in concrete. Replacing 7-10% of coarse aggregates with E-waste maintains compressive strength. This eco-friendly approach could mitigate the negative environmental impacts of E-waste disposal. (Singh et al., n.d.) emphasized the need for a shift in the electronics industry toward sustainable practices. It covers waste PCB management evolution, eco-friendly materials, and resource-efficient manufacturing.

The goal was to reduce waste through improved materials, processing, and circular-economy strategies. Green electronics, alternative materials, and rigorous testing can transform the industry, despite economic and regulatory challenges.

(Almulhim, 2022; Asibey et al., 2022; M. Chakraborty et al., 2022; Ismail & Hanafiah, 2021; Moyen Massa & Archodoulaki, 2023; Nuwematsiko et al., 2021a; Xu et al., 2021a)collectively underscore the global urgency to address E-waste challenges through sustainable practices, policy changes, and enhanced awareness. The issues raised span across various nations and regions, revealing the intricate interplay between environmental, social, and economic factors in effective E-waste management. Addressing these concerns requires collaborative efforts, education, technological innovation, and stringent regulations to mitigate the adverse impacts on both human health and the environment. Inadequate waste management practices in developing nations lead to environmental pollution and health concerns.

(M. Chakraborty et al., 2022)focused on E-waste management in Africa, addressing generation, treatment, recycling, and metal recovery. Stakeholder roles, sustainable design, and awareness campaigns are discussed. Given low-income lifestyles in African populations, awareness campaigns through education, media, and workshops are vital. Improved governance and eco-design could create green jobs and combat pollution. Strict global E-waste regulation led by an entity like UNEP is recommended for enhanced transparency, awareness, and proper management worldwide. Despite being illegal, the export of used European Electric and Electronic Equipment (UEEE) to Nigeria raised sustainability, circularity, and fairness concerns.

(Moyen Massa & Archodoulaki, 2023) involved 523 respondents in Saudi Arabia and revealed a lack of awareness among 70.1%, yet 88.35% expressed willingness to be a part of proper E-waste management after education. Limited recycling options and awareness have led to the treatment of E-waste as regular waste. Practical management approaches were suggested to raise public knowledge and align with sustainability goals, including Saudi Vision 2030, emphasizing community interventions and education for effective E-waste management.

(Almulhim, 2022) in Kampala assessed consumer knowledge and practices, revealing that 67.7% had poor E-waste management knowledge, but 79.1% had positive perceptions. Consumer suggestions ranged from burning to general waste disposal. Urgent awareness campaigns led by relevant authorities were deemed necessary to bridge the knowledge gap, emphasizing proper handling, legislation, and disposal practices. The environmental effectiveness of E-waste management strategies in Southeast Asian countries like Malaysia was evaluated, focusing on infrastructure and practicality.

In (Nuwematsiko et al., 2021b), four options were assessed: direct incineration with energy recovery, landfill disposal, materials recovery with energy recovery, and materials recovery without energy recovery. Surprisingly, direct incineration with energy recovery emerged as the optimal choice, challenging initial expectations and previous studies.

(Ismail & Hanafiah, 2021) investigated informal E-waste activities in the state of Kumasi in Ghana, identifying political, spatial, economic, and socio-cultural, factors as influencers. Poverty, financial gain, landlessness, and peer influence were found to drive engagement. Recommendations included policy changes to foster stable socio-economic conditions for sustainable and inclusive development.

E-waste, a rapidly growing waste stream globally, has posed challenges for China, a major processor, as highlighted in (Asibey et al., 2022). The study examined barriers to sustainable E-waste management in a major Chinese province, offering an integrated planning approach. This structured approach was seen as contributing to managing E-waste barriers effectively and achieving sustainable practices, providing a valuable methodological contribution to multi-criteria evaluation research. E-waste, often sent to developing nations like China for disposal, has posed severe environmental and health hazards.

(Xu et al., 2021b) has highlighted the urgent need for controlled contamination sources, eco-friendly remedies, and efficient E-waste management practices to safeguard public health as well as the environment while addressing the challenges informal E-waste sectors face. Ho Chi Minh City, a vital urban area in Vietnam, faced a growing E-waste challenge.

Studies (Arya & Kumar, 2020; Asokan et al., 2023; Awasthi et al., 2022; Borthakur, 2023; Li & Achal, 2020; Panchal et al., 2021; Shreyas Madhav et al., 2022; Vijayan et al., 2023; Wath et al., 2011) depicted the challenges India had faced in E-waste and battery waste management. The rapid pace of technological advancement and EV adoption has necessitated coordinated efforts, innovative solutions, and policy adjustments to achieve sustainable waste management practices. Collaborative approaches between formal and informal sectors, coupled with enhanced awareness and efficient collection methods, have been essential to mitigate the unfavorable environmental as well as health impacts of E-waste.

In India, the rapid growth of Electric Vehicle (EV) adoption has been accompanied by a concerning lag in battery waste management, as predicted by (Li & Achal, 2020). Significant EV sales by 2030 and 2050, with lead-acid and lithium-ion battery waste posing challenges, had been projected. An EPR system and waste crime tracking had been deemed crucial for worker and community safety. The global surge in E-waste, exacerbated by the COVID-19 pandemic, has posed governance complexities, particularly in countries like India.

The E-waste policies influenced by The EU's WEEE directive implemented by India have been analyzed in (Asokan et al., 2023). The study advocated for a collaborative approach between informal and formal sectors to achieve sustainable E-waste governance, tailoring policy transfer to local contexts. Amid the technological revolution, India produced 3.2 million tons of E-waste in the year 2020.

(Borthakur, 2023) had introduced a novel solution: a mobile robot using transfer learning to identify and segregate E-waste. The system was attached to garbage trucks, enhancing collection efficiency while reducing costs. (Shreyas Madhav et al., 2022) had identified hazardous substances in E-waste, emphasizing the need for eco-design in electronic manufacturing and responsible recycling practices to mitigate environmental issues. Within the circular economy framework, E-waste had been seen as holding potential as an urban mine for valuable materials.

(Awasthi et al., 2022) had estimated India's E-waste generation, assessed embedded materials, and evaluated their economic recovery potential. Strategies like eco-design, EPR, circular resource management, and bioleaching have been proposed for a sustainable E-waste value chain. The challenge regarding E-waste has lain in its dual nature as a source of valuable metals and a potential toxic threat. (Panchal et al., 2021) proposed strategies for India, incorporating informal recyclers transparently, improving waste collection systems, and promoting eco-friendly technologies. (Arya & Kumar, 2020) had highlighted the need for advanced treatment systems, public-private partnerships, and investment in modern facilities to combat outdated recycling methods in India. Motives and outcomes of E-waste recycling intentions have been explored using methods like the Theory of Planned Behavior (TPB). (Wath et al., 2011) had established a model investigating attitude, norms, and control's impact on recycling intention and behavior. The insights had the potential to drive awareness campaigns by local governments to encourage 'green behavior'. The electronics industry's rapid growth has led to a sudden increase in E-waste production(Vijayan et al., 2023).

Figure 3. Countries considered in the literature

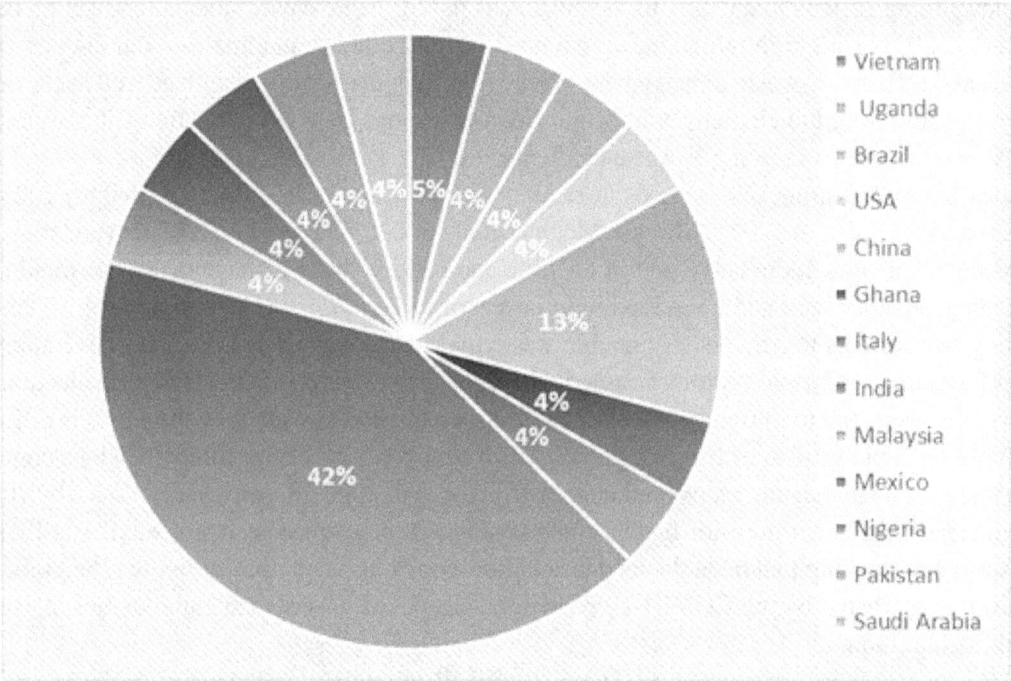

OPEN CHALLENGES IN E-WASTE MANAGEMENT

The growing production of E-waste and the need for sustainable disposal methods are likely to be the main trends in E-waste management in the future. The key difficulties and trends are as follows:

1. Growing volume: Even greater amounts of E-waste will be produced because of the widespread use of electronic gadgets and their accelerating rate of obsolescence. This amount of waste will be difficult to manage without proper management techniques.
2. Shift towards Circular Economy: More emphasis will be placed on the transition to a circular economy, where goods and resources are recycled, upgraded, and reused. E-waste management will increasingly emphasize extending the life of electronic components and equipment.
3. Awareness and innovative recycling techniques: The demand for products that are simpler to repair, upgrade, and recycle may increase as consumers become more environmentally concerned. This may be due to changes in consumer behavior and product design. Technological developments in recycling, such as better ways to extract valuable elements from E-waste, will continue. These developments may lessen the consequences of E-waste disposal on the environment lessened by these developments.
4. Use of Advanced Techniques and Eco-design Principles: While lot of scope in the usage of AI-based techniques, other techniques such as IOT and Cloud may also be used in management and disposal. In addition, manufacturers can use eco-design concepts to make items that have less impact on the environment and are simpler to recycle and dispose of properly.

5. Global Collaborations: International cooperation will become more crucial as E-waste becomes a widespread problem, especially in developing and large nations. Effectively tackling this issue will be made possible through the cross-border sharing of best practices, technologies, and resources.

FUTURISTIC STRATEGIES FOR E-WASTE

The rapid advancement of technology is driven by growing consumer demands, leading to a significant increase in E-waste worldwide. It often contains dangerous substances like cadmium, lead, polybrominated diphenyl esters, mercury, and more. The improper handling and escalating production of E-waste pose significant environmental as well as health risks. Consequently, it is essential to prioritize the development of safe E-waste management practices. Additionally, these efforts can be linked to the principles of a circular economy, aiming to maximize resource utilization by recycling and reusing waste materials.

Extraction of Precious Metals from E-Waste

It can potentially serve as a medium of revenue for individuals and industries. To illustrate, in Bangalore, India, an annual E-waste generation of approximately 18,000 metric tons is observed. This E-waste contains valuable materials like platinum, gold, silver, aluminum, copper, and rare earth metals, which can later be recycled and reused. Moreover, there is room to explore new economic opportunities stemming from these elements.

Recycling of E-Waste

Similar to other types of waste, certain components of E-waste, like capacitors, plastics, and circuit boards, can undergo recycling processes. Incineration of E-waste can also be a medium to produce energy, particularly through the pyrolysis of the plastic components. Thermo-chemical methods for generating energy from waste materials have garnered global interest, not only for energy production but also for waste recycling. Furthermore, the recovered materials from E-waste can be used to create new materials, aligning with the initial or the very first goal of recycling, which is to obtain valuable resources while mitigating environmental and health risks. Additionally, recycling efforts can create various opportunities for society, including employment and business prospects.

Microrecycling Concept for Recycling E-Waste

Extracting metals from E-waste presents a lucrative business opportunity for private companies. However, the transportation of large quantities of such waste can be costly. A solution to this challenge is the establishment of micro-recycling units, which can be set up in smaller areas, approximately 50 square meters, at a relatively low cost. The idea behind micro-recycling is to employ small-scale facilities to collect and sort E-waste items from nearby areas. This methodology contributes to cost savings in transportation, minimizes raw material use, preserves fossil fuels, and improves environmental outcomes. It also results in the production of value-added recycled materials for future utilization.

Creation of Efficient E-Waste Management Policies

Policies that have been introduced and successfully worked have consistently proven to be a powerful instrument for achieving success in various fields. Using India as an example, despite its position as a major producer of electronic devices, e-waste management had difficulties in the early phases due to the lack of adequate policies. Nevertheless, the Indian Ministry of Environment, Forest and Climate Change (MoEFCC) has introduced guidelines to improve E-waste management.

Indian policymakers have introduced several waste management regulations and guidelines spanning from 1989 to 2016. The first set of rules for handling E-waste was formulated in 2009, followed by the issuance of E-waste rules in 2011, which also included regulations for plastic waste. In India, the policies related to E-waste management are currently underdeveloped and need significant revision to ensure the safety of both the environment as well as human health. Updated policies are urgently needed, though, because of the increasing amount of E-waste being produced.

Usage of Advanced Technologies

The effectiveness, sustainability, and safety of the E-waste management process can be considerably increased by utilizing AI, robotics, and IoT technologies. Due to the rising quantity of electronic trash generated globally, which poses threats to the environment as well as human health, E-waste management is a crucial issue. Here are a few applications for this technology:

1. **E-Waste Identification and Sorting:** AI-based techniques can be used to automatically identify and categorize electronic waste items from images or video feeds. This can help in sorting and classifying E-waste items more accurately and quickly. Also, Robots equipped with computer vision systems can efficiently sort E-waste items based on their category and material composition. They can be programmed to pick and place items into different bins or conveyor belts for further processing.
2. **Inventory Management:** To keep track of the fill levels in electronic trash collection bins and containers, IoT sensors can be installed. This data can be used to optimize collection routes, reduce unnecessary pickups, and ensure timely E-waste collection.
3. **Recycling Process:** Robots can be used to disassemble electronic devices efficiently and safely, separating reusable components from hazardous materials. They can also help in removing valuable materials like precious metals from old devices. Further AI-based ML algorithms can optimize the recycling process by predicting the most efficient way to recover valuable materials while minimizing waste. This can lead to higher recycling rates and reduced environmental impact.
4. **E-Waste Awareness and Education:** Virtual assistants and AI-driven chatbots can be used to provide information and guidance to consumers and businesses on proper E-waste disposal practices. They can answer queries and raise awareness of the importance of recycling electronic trash, especially in public places like railway stations, malls, and colleges.

Integrating these advanced technologies into E-waste management can improve recycling rates, reduce environmental harm, increase worker safety, and create a more sustainable approach to handling electronic waste. These technologies enable smarter, data-driven decision-making throughout the E-waste lifecycle, from collection and sorting to recycling and disposal.

CONCLUSION

One of the world's most difficult issues is the management of e-waste. E-wastes such as PCs, mobile phones, laptops, and TVs are disposed of unscientifically in landfills and open dumps, creating problems for the environment and public health. With E-waste generation rapidly becoming a significant aspect of solid waste management, coupled with the scarcity of mineral resources and their growing consumption rates, efficient e-waste collection has become crucial, particularly in urban areas. While certain countries, such as Japan, have adopted sustainable household e-waste management practices such as EPR schemes, the full implementation of such approaches is still lacking in many developing countries.

The primary objective of EPR is to inspire companies to seamlessly integrate waste management into every phase of their product manufacturing processes, while simultaneously increasing public awareness. EPR strategies have been proven to significantly reduce waste associated with consumer products and enhance their recyclability. Furthermore, EPR plays a crucial role in improving e-waste management by minimizing the generation of excessive waste, reducing the consumption of raw materials during manufacturing, and including sustainability costs in product pricing.

E-waste can contain radioactive substances, which pose a threat to plant growth and tissues. Soils containing radioactive elements can also disrupt the ability of plants to absorb nutrients. Effective E-waste management faces various challenges including irresponsible consumption, inadequate infrastructure, and financial constraints. However, implementing good practices can mitigate these issues, such as inspiring and involving society, promoting behavioral changes, providing sufficient funds and incentives, establishing proper e-waste management facilities, and embracing eco-friendly approaches.

To address these challenges, advanced computational tools, such as the IoT, and AI, play a crucial role. By leveraging these techniques, improved E-waste management practices can be developed. In the future, multi-disciplinary research can be conducted using knowledge and techniques from different domains to develop integrated solutions for making the process economically feasible and contributing to achieving various sustainable development goals.

REFERENCES

Ahirwar, R., & Tripathi, A. K. (2021). E-waste management: A review of recycling process, environmental and occupational health hazards, and potential solutions. *Environmental Nanotechnology, Monitoring & Management, 15*, 100409. doi:10.1016/j.enmm.2020.100409

Ahirwar, R. (n.d.). *E-waste management: A review of recycling process, environmental and occupational health hazards, and potential solutions.* Elsevier. Retrieved November 20, 2023, from https://www.sciencedirect.com/science/article/pii/S2215153220303925

Akram, R. (2019). *Trends of electronic waste pollution and its impact on the global environment and ecosystem.* Springer. doi:10.1007/s11356-019-04998-2

Alabi, A. O., & Bakare, A. A. (2017). Genetic damage induced by electronic waste leachates and contaminated underground water in two prokaryotic systems. *Toxicology Mechanisms and Methods, 27*(9), 657–665. doi:10.1080/15376516.2017.1349228 PMID:28670935

Almulhim, A. I. (2022). Household's awareness and participation in sustainable electronic waste management practices in Saudi Arabia. *Ain Shams Engineering Journal, 13*(4), 101729. doi:10.1016/j.asej.2022.101729

Amankwah-Amoah, J. (2016). Global business and emerging economies: Towards a new perspective on the effects of e-waste. *Technological Forecasting and Social Change, 105*, 20–26. doi:10.1016/j.techfore.2016.01.026

Anh Khoa, T., Phuc, C. H., Lam, P. D., Nhu, L. M. B., Trong, N. M., Phuong, N. T. H., Van Dung, N., Tan-Y, N., Nguyen, H. N., & Duc, D. N. M. (2020). Waste management system using IoT-based machine learning in university. *Wireless Communications and Mobile Computing, 2020*, 1–13. doi:10.1155/2020/6138637

Arya, S., & Kumar, S. (2020). E-waste in India at a glance: Current trends, regulations, challenges and management strategies. *Journal of Cleaner Production, 271*, 122707. doi:10.1016/j.jclepro.2020.122707

Asibey, M. O., Lykke, A. M., & King, R. S. (2022). Understanding the factors for increased informal electronic waste recycling in Kumasi, Ghana. *International Journal of Environmental Health Research, 32*(2), 305–320. doi:10.1080/09603123.2020.1755016 PMID:32393050

Asokan, V. A., Teah, H. Y., Kawazu, E., & Hotta, Y. (2023). Ambitious EV policy expedites the e-waste and socio-environmental impacts in India. *Resources, Conservation and Recycling, 190*, 106829. doi:10.1016/j.resconrec.2022.106829

Awasthi, A. K., Awasthi, M. K., Mishra, S., Sarsaiya, S., & Pandey, A. K. (2022). Evaluation of E-waste materials linked potential consequences to environment in India. *Environmental Technology & Innovation, 28*, 102477. doi:10.1016/j.eti.2022.102477

Awasthi, A. K., Wang, M., Awasthi, M. K., Wang, Z., & Li, J. (2018). Environmental pollution and human body burden from improper recycling of e-waste in China: A short-review. *Environmental Pollution, 243*, 1310–1316. doi:10.1016/j.envpol.2018.08.037 PMID:30268981

Borthakur, A. (2023). Design, adoption and implementation of electronic waste policies in India. *Environmental Science and Pollution Research International, 30*(4), 8672–8681. doi:10.100711356-022-18836-5 PMID:35141823

Chakraborty, M., Kettle, J., & Dahiya, R. (2022). Electronic Waste Reduction Through Devices and Printed Circuit Boards Designed for Circularity. *IEEE Journal on Flexible Electronics, 1*(1), 4–23. doi:10.1109/JFLEX.2022.3159258

Chakraborty, S. C., Qamruzzaman, M., Zaman, M. W. U., Alam, M. M., Hossain, M. D., Pramanik, B. K., Nguyen, L. N., Nghiem, L. D., Ahmed, M. F., Zhou, J. L., Mondal, M. I. H., Hossain, M. A., Johir, M. A. H., Ahmed, M. B., Sithi, J. A., Zargar, M., & Moni, M. A. (2022). Metals in e-waste: Occurrence, fate, impacts and remediation technologies. *Process Safety and Environmental Protection, 162*, 230–252. doi:10.1016/j.psep.2022.04.011

Cucchiella, F., D'Adamo, I., Koh, S. C. L., & Rosa, P. (2015). Recycling of WEEEs: An economic assessment of present and future e-waste streams. *Renewable & Sustainable Energy Reviews, 51*, 263–272. doi:10.1016/j.rser.2015.06.010

Dawar, I., Gupta, S. D., Singh, R., Kothari, Y., & Layek, S. (2023). Forest Fire Detection using Deep Learning Techniques. *2023 2nd International Conference on Vision Towards Emerging Trends in Communication and Networking Technologies (ViTECoN)*, 1–6.

Dharini, K., Cynthia, J. B., Kamalambikai, B., Arul, J. P., Celestina, S., & Muthu, D. (2017). Hazardous E-waste and its impact on soil structure. *Iopscience*. 10.1088/1755-1315/80/1/012057

Dhir, A., Koshta, N., Goyal, R. K., Sakashita, M., & Almotairi, M. (2021). Behavioral reasoning theory (BRT) perspectives on E-waste recycling and management. *Journal of Cleaner Production*, *280*, 124269. doi:10.1016/j.jclepro.2020.124269

Dixit, S., Arora, R., Kumar, K., Bansal, S., Vatin, N., Araszkiewicz, K., & Epifantsev, K. (2022). Replacing E-waste with coarse aggregate in architectural engineering and construction industry. *Materials Today: Proceedings*, *56*, 2353–2358. doi:10.1016/j.matpr.2021.12.154

Dutta, D., Rautela, R., Gujjala, L. K. S., Kundu, D., Sharma, P., Tembhare, M., & Kumar, S. (2023). A review on recovery processes of metals from E-waste: A green perspective. *The Science of the Total Environment*, *859*, 160391. doi:10.1016/j.scitotenv.2022.160391 PMID:36423849

Ganguly, S. (2014). E-Waste, generation, disposal and its associated environmental impact: A Review. *Indian Journal of Scientific Research and Technology*, *2*(5), 5–6.

Greenpeace - Where Does E-Waste End Up? - English Version. (n.d.). Retrieved November 21, 2023, from https://media.greenpeace.org/archive/Where-Does-E-Waste-End-Up----English-Version-27MZIF2D3F8X.html

Ismail, H., & Hanafiah, M. M. (2021). Evaluation of e-waste management systems in Malaysia using life cycle assessment and material flow analysis. *Journal of Cleaner Production*, *308*, 127358. doi:10.1016/j.jclepro.2021.127358

Jangre, J., Prasad, K., & Patel, D. (2022). Analysis of barriers in e-waste management in developing economy: An integrated multiple-criteria decision-making approach. *Environmental Science and Pollution Research International*, *29*(48), 72294–72308. doi:10.100711356-022-21363-y PMID:35696062

Li, W., & Achal, V. (2020). Environmental and health impacts due to e-waste disposal in China–A review. *The Science of the Total Environment*, *737*, 139745. doi:10.1016/j.scitotenv.2020.139745 PMID:32516663

Liu, K., Tan, Q., Yu, J., & Wang, M. (2023). A global perspective on e-waste recycling. *Circular Economy*, 100028.

María Fernández-González, J., Díaz-López, C., Martín-Pascual, J., & Zamorano, M. (2020). *Recycling organic fraction of municipal solid waste: Systematic literature review and bibliometric analysis of research trends*. doi:10.3390/su12114798

Marimuthu, V., & Hazardous, A. R.-J. (2022). Investigation of the mechanical properties of M40-grade concrete with PCB fiber from recycled electronic waste. *Journal of Hazardous, Toxic and Radioactive Waste*, *27*(1), 04022034. Advance online publication. doi:10.1061/(ASCE)HZ.2153-5515.0000725

Moheb-Alizadeh, H., Sadeghi, A. H., Sahebi Fakhrabad, C. A., Jaunich, M. K., Kemahlioglu-Ziya, E., & Handfield, R. B. (2023). *Reverse Logistics Network Design to Estimate the Economic and Environmental Impacts of Take-back Legislation: A Case Study for E-waste Management System in Washington State*. https://arxiv.org/abs/2301.09792v1

Moyen Massa, G., & Archodoulaki, V.-M. (2023). Electrical and Electronic Waste Management Problems in Africa: Deficits and Solution Approach. *Environments (Basel, Switzerland)*, *10*(3), 44. doi:10.3390/environments10030044

Nande, A., Raut, S., Swart, H. C., & Dhoble, S. J. (2023). Heavy metal waste management—Side products of industries and electronic waste. *360-Degree Waste Management, Volume 2. Biomedical, Pharmaceutical, Industrial Waste, and Remediation*, *2*, 203–219. doi:10.1016/B978-0-323-90909-9.00012-5

Nuwematsiko, R., Oporia, F., Nabirye, J., Halage, A. A., Musoke, D., & Buregyeya, E. (2021). Knowledge, perceptions, and practices of electronic waste management among consumers in Kampala, Uganda. *Journal of Environmental and Public Health*, *2021*. doi:10.1155/2021/3846428 PMID:33628280

Orlins, S., & Guan, D. (2016). China's toxic informal e-waste recycling: Local approaches to a global environmental problem. *Journal of Cleaner Production*, *114*, 71–80. doi:10.1016/j.jclepro.2015.05.090

Panchal, R., Singh, A., & Diwan, H. (2021). Economic potential of recycling e-waste in India and its impact on import of materials. *Resources Policy*, *74*, 102264. doi:10.1016/j.resourpol.2021.102264

Pardini, K., Rodrigues, J. J. P. C., Kozlov, S. A., Kumar, N., & Furtado, V. (2019). IoT-based solid waste management solutions: A survey. *Journal of Sensor and Actuator Networks*, *8*(1), 5. doi:10.3390/jsan8010005

Pradhan, J. (2014). *Informal e-waste recycling: environmental risk assessment of heavy metal contamination in Mandoli industrial area, Delhi, India*. doi:10.1007/s11356-014-2713-2

Rautela, R., Arya, S., Vishwakarma, S., Lee, J., Kim, K.-H., & Kumar, S. (2021). E-waste management and its effects on the environment and human health. *The Science of the Total Environment*, *773*, 145623. doi:10.1016/j.scitotenv.2021.145623 PMID:33592459

Saini, S., Chauhan, A., Thakur, G., & Sapra, L. (2023). Challenges and Opportunities in Secure Smart Cities for Enhancing the Security and Privacy. *Enabling Technologies for Effective Planning and Management in Sustainable Smart Cities*, 1–27.

Seif, R., Salem, F. Z., & Allam, N. K. (2023). E-waste recycled materials as efficient catalysts for renewable energy technologies and better environmental sustainability. *Environment, Development and Sustainability*. Advance online publication. doi:10.100710668-023-02925-7 PMID:36691418

Sharma, H. B., Vanapalli, K. R., Samal, B., Cheela, V. R. S., Dubey, B. K., & Bhattacharya, J. (2021). Circular economy approach in solid waste management system to achieve UN-SDGs: Solutions for post-COVID recovery. *The Science of the Total Environment*, *800*, 149605. doi:10.1016/j.scitotenv.2021.149605 PMID:34426367

Shreyas Madhav, A. V., Rajaraman, R., Harini, S., & Kiliroor, C. C. (2022). Application of artificial intelligence to enhance collection of E-waste: A potential solution for household WEEE collection and segregation in India. *Waste Management & Research, 40*(7), 1047–1053. doi:10.1177/0734242X211052846 PMID:34726090

Singh, A., & Sharma, R. (n.d.). *Health risk assessment of heavy metals via dietary intake of foodstuffs from the wastewater irrigated site of a dry tropical area of India.* Elsevier. Retrieved November 19, 2023, from https://www.sciencedirect.com/science/article/pii/S0278691509005572

Singh, A., Sharma, R. K., Agrawal, M., & Marshall, F. M. (2010). Health risk assessment of heavy metals via dietary intake of foodstuffs from the wastewater irrigated site of a dry tropical area of India. *Food and Chemical Toxicology, 48*(2), 611–619. doi:10.1016/j.fct.2009.11.041 PMID:19941927

Srivastav, A. L., Markandeya, Patel, N., Pandey, M., Pandey, A. K., Dubey, A. K., Kumar, A., Bhardwaj, A. K., & Chaudhary, V. K. (2023). Concepts of circular economy for sustainable management of electronic wastes: Challenges and management options. *Environmental Science and Pollution Research International, 30*(17), 48654–48675. doi:10.100711356-023-26052-y PMID:36849690

Sthiannopkao, S., & Wong, M. H. (2013). Handling e-waste in developed and developing countries: Initiatives, practices, and consequences. *The Science of the Total Environment, 463*, 1147–1153. doi:10.1016/j.scitotenv.2012.06.088 PMID:22858354

Tunali, M., Tunali, M. M., & Yenigun, O. (2021). Characterization of different types of electronic waste: Heavy metal, precious metal and rare earth element content by comparing different digestion methods. *Journal of Material Cycles and Waste Management, 23*(1), 149–157. doi:10.100710163-020-01108-0

Vijayan, R. V., Krishnan, M. M., Parayitam, S., Duraisami, S. P. A., & Saravanaselvan, N. R. (2023). Exploring e-waste recycling behaviour intention among the households: Evidence from India. *Cleaner Materials, 7*, 100174. doi:10.1016/j.clema.2023.100174

Wath, S. B., Dutt, P. S., Chakrabarti, T., Wath, S. B., & Dutt, P. S. (2011). *E-waste scenario in India, its management and implications*. doi:10.1007/s10661-010-1331-9

Weerasundara, L., & Mahatantila, K. (2020). *E-waste as a challenge for public and ecosystem health.* Elsevier. doi:10.1016/B978-0-12-817030-4.00003-6

Xu, Y., Yeh, C. H., Liu, C. G., Ramzan, S., & Zhang, L. (2021a). Evaluating and managing interactive barriers for sustainable e-waste management in China. *The Journal of the Operational Research Society, 72*(9), 2018–2031. doi:10.1080/01605682.2020.1759381

Xu, Y., Yeh, C. H., Liu, C. G., Ramzan, S., & Zhang, L. (2021b). Evaluating and managing interactive barriers for sustainable e-waste management in China. *The Journal of the Operational Research Society, 72*(9), 2018–2031. doi:10.1080/01605682.2020.1759381

Xu, Y., Zhang, L., Ghulam, S. T., & Abushammala, H. (2023). *Challenges and Opportunities in the Management of Electronic Waste and Its Impact on Human Health and Environment*. doi:10.3390/su15031837

Yan, C. H., Xu, J., & Shen, X. M. (2013). Childhood lead poisoning in China: Challenges and opportunities. *Environmental Health Perspectives*, *121*(10). Advance online publication. doi:10.1289/ehp.1307558 PMID:24218672

Zhang, X., Zhong, T., Liu, L., & Ouyang, X. (2015). Impact of soil heavy metal pollution on food safety in China. *PLoS One*, *10*(8), e0135182. Advance online publication. doi:10.1371/journal.pone.0135182 PMID:26252956

Zheng, J., Chen, K., Yan, X., Chen, S.-J., Hu, G.-C., Peng, X.-W., Yuan, J., Mai, B.-X., & Yang, Z.-Y. (2013). Heavy metals in food, house dust, and water from an e-waste recycling area in South China and the potential risk to human health. *Ecotoxicology and Environmental Safety*, *96*, 205–212. doi:10.1016/j.ecoenv.2013.06.017 PMID:23849468

KEY TERMS AND DEFINITIONS

Artificial Intelligence: The theory governing the development of computer systems that can perform tasks that normally require human intelligence, such as visual perception, speech recognition, decision-making, and translation between languages.

Deep Learning: A part of a broader family of machine learning methods based on learning data representations.

E-Waste: It refers to discarded electronic devices, including computers, smartphones, and appliances, that have reached the end of their usable life. Improper disposal of e-waste can lead to environmental pollution and health hazards due to the presence of hazardous materials.

IoT: It refers to the concept of designing and developing artificial intelligence systems in a way that allows humans to understand and interpret the decisions and reasoning behind the AI's outputs.

Machine Learning: A field of artificial intelligence that uses statistical techniques to give computer systems the ability to learn.

Sustainable Development: Sustainable development is a concept and an approach that seeks to balance economic, environmental, and social considerations to achieve long-term well-being for current and future generations.

Chapter 18
Technological Advancements in E-Waste Management and Resource Recovery

Jayaditya Reddy Yeruva
Woxsen University, India

Adip Krishna Guduru
Woxsen University, India

Bhanu Prakash Saripalli
Woxsen University, India

ABSTRACT

As the world digitizes rapidly, there has been an exponential surge in the generation of electronic waste (e-waste), and it poses a grave threat to our planet due to its hazardous components, improper disposal methods, and the sheer volume being generated. Adding to the problem, e-waste recycling is highly complex and tedious, making it difficult to keep up with the pace of generation as well as leading to wastage of various precious metals. The solution involves using advanced technology to develop and optimise a multi-step resource recovery system. This chapter emphasizes the importance of engineers in tackling e-waste and how cutting edge technology can have a significant positive impact on society without necessarily requiring on-site involvement. The research aims to explore innovative strategies and technologies which will play a vital role in revolutionizing e-waste management, with AI-driven diagnostics, repair guidance, component identification, and value assessment being the future of efficient, effective, and sustainable e-waste recycling.

INTRODUCTION

As the world digitizes rapidly, there has been an exponential surge in the generation of electronic waste (E-Waste), and it poses a grave threat to our planet due to its hazardous components, improper disposal

DOI: 10.4018/979-8-3693-1018-2.ch018

methods, and the sheer volume being generated. Adding to the problem, e-waste recycling is highly complex and tedious, making it difficult to keep up with the pace of generation as well as leading to wastage of various precious metals.

The solution involves using advanced technology to develop and optimise a multi-step resource recovery system. This chapter emphasizes the importance of engineers in tackling e-waste and how cutting edge technology can have a significant positive impact on society without necessarily requiring on-site involvement. Two AI-based software utilising computer vision and deep learning algorithms are involved in the recovery process:

1. Identification of the different components present in a printed circuit board along with their physical health status to calculate recyclability.
2. Accurate assessment of the value of a printed circuit board by analysis of its composition and precious metals present.

The research aims to explore innovative strategies for making genuine factory parts more affordable and accessible, thus encouraging smartphone repair and prolonging their useful lifespan. Artificial intelligence will play a vital role in revolutionizing e-waste management, with Al-driven diagnostics, repair guidance, component identification, and value assessment, being the future of efficient, effective, and sustainable e-waste recycling.

The Problem at Hand

The rise in popularity of electronic devices brings with it an escalating global concern: the surge of electronic waste (e-waste), especially pronounced in nations like India. The burden posed by e-waste is significant due to its harmful components, inadequate disposal strategies and overwhelming quantities being produced. Statistics further attest to the severity of this predicament. For instance, during 2019 alone, India stood as Earth's third highest producer of e-waste at roughly 3.2 million metric tons—with a mere tenth successfully recycled or properly disposed of. Regrettably, most end up strewn across landfills or processed through unofficial and perilous deconstruction methods (World Economic Forum, 2017).

The perils related to improper handling are highlighted via real-world occurrences; informal recycling endeavors within locations such as Ghana and Guiyu from China expose laborers to lethal elements without adequate protection—resulting in polluting soil & and water resources detrimentally affecting local ecological systems and communities alike (Chinese Academy of Sciences, 2018). Exactly so occurs also within Seelampur-Delhi, a region synonymous with unregulated e-waste recycling has impacted residents' health adversely with increased concentrations of heavy metals found throughout air, soil and even water bodies.

Concerns mount when coupled with lackluster awareness about correctly disposing of E-waste considered signs show large sectors oblivious regarding potential risks associated with the importance of maintaining responsible recycling habits. Therefore hordes continue to pile unwanted electronics onto landfill sites which invariably leads to loops dangerous pollution resource exhaustion.

Addressing this complex situation requires adopting all-encompassing sustainable action plans which include initiating educational outreach programs, spearheading policy change procedures, promoting infrastructure growth besides encouraging mindful consumer production behaviors. Roll-out responsibilities would mean setting up operative collection points and reengineering existing technologies used until now ensuring manufacturers develop easily recyclable designs designed essential. As these coordinated efforts persist we ensure better environmental control safeguard well-being meanwhile nurturing a circular economy around utilization.

What Is E-Waste Recycling?

In the current digital age, the rapid development of technology has caused a commensurate rise in the production of electronic trash (e-waste). E-waste recycling has become a vital part of sustainable waste management techniques in order to reduce the environmental and health risks connected with incorrect disposal. In order to promote a circular economy and lessen the environmental effect of electronic equipment, this part focuses on the procedure and importance of recycling e-waste.

Current Recycling Process

The safe and effective recovery of valuable elements from e-waste requires a number of clearly defined stages. Usually, the procedure entails:

1. **Collection and Transportation**: Appropriate methods for gathering abandoned electronic devices have been created. To make it easier to collect e-waste from homes, businesses, and authorized drop-off locations, cooperation between manufacturers, recycling enterprises, and governmental organizations is essential.
2. **Disassembly and Sorting**: After being gathered, the electronic trash is meticulously disassembled by professional specialists. In this stage, hazardous components, such batteries and circuit boards, are separated for specialist treatment. E-waste is categorized based on material composition and recyclable status with the use of cutting-edge technology and automated sorting systems.
3. **Processing and Material Recovery**: The e-waste is treated using a variety of methods in order to recover valuable materials after sorting. The e-waste is broken down into tiny bits by mechanical shredding, and precious metals may be extracted more easily via smelting and hydrometallurgical processes. Chemical processes ensure the purity and quality of the recovered materials by further refining them.
4. **Refining and Reintroduction**: In order to purge impurities and improve their recycling potential, the extracted materials, including priceless metals like copper, gold, and silver, go through refining operations. The industrial supply chain may then include these materials once again, eliminating the need for virgin resources and lowering the environmental effect.

Importance of E-Waste Recycling

Recycling e-waste is essential to sustainable waste management since it has several positive environmental and financial effects. Significant points include:

1. Resource Conservation: By recovering metals and materials that may be used again in industrial processes, e-waste recycling protects scarce resources. This lessens the requirement for raw material extraction, which often includes harmful environmental practices and energy-intensive methods.

2. Environmental Protection: Effective e-waste recycling prevents harmful compounds from entering landfills and damaging ecosystems, such lead, mercury, and brominated flame retardants. Recycling aids in the preservation of natural resources and the reduction of climate change by reducing environmental pollution.
3. E-waste recycling offers economic prospects by generating work in the collection, disassembly, sorting, and processing of e-waste. Additionally, it promotes the development of a circular economy in which resources are recycled in order to increase their usefulness and decrease waste production.
4. Compliance with Regulations: Recycling of e-waste adheres to legal frameworks put in place by national and international organizations to address the ethical management of e-waste. Recycling facilities assure correct handling, treatment, and disposal of hazardous components by adhering to certain rules.

E-waste recycling is essential for reducing the negative environmental effects of electronic gadgets and improving environmentally friendly trash disposal methods. We can build a more sustainable future where e-waste is responsibly recycled, resources are conserved, and the negative effects of improper disposal are reduced by putting in place efficient collection systems, investing in cutting-edge recycling technologies, and raising awareness among consumers, manufacturers, and policymakers (World Economic Forum, 2017).

LITERATURE REVIEW AND BACKGROUND INFORMATION

The section that follows offers a summary of the literature and background data on e-waste management, covering significant research results, international efforts, and contemporary practices:

E-Waste Generation Trends

E-waste is a global issue that is becoming worse. Around 53.6 million metric tons of e-waste were produced globally in 2019, with a 21% annual increase rate over the previous five years (World Economic Forum, 2017). As technology becomes more available and inexpensive, increasing consumption and reducing product lifecycles, this tendency is anticipated to continue.

Primary Causes of Generation

- People are purchasing more electronic devices as technology becomes more accessible and inexpensive, expanding the market for electronic goods. As a result, there is an increasing supply of outdated electronic items that are ultimately thrown away.
- Electronic items are becoming more complicated and containing more harmful components, which results in a shorter lifetime. This leads to their reduced lifetime and increases the difficulty and cost of recycling them.
- Lack of knowledge on e-waste's effects on the environment and human health: Many individuals are not aware of e-waste's effects on the environment and human health. This results in incorrect e-waste disposal, which may have detrimental effects on both human health and the environment.

Effects on the Environment and Health

Hazardous compounds including lead, mercury, cadmium, and flame retardants are found in e-waste. If not handled appropriately, these compounds may pose serious dangers to both human health and the environment. These chemicals may be released into the environment by improper disposal methods including open burning and landfill disposal. This may contaminate the air, water, and soil and have a negative impact on human health by increasing the risk of cancer, developmental disabilities, and respiratory issues. Additionally, there may be health dangers associated with the informal recycling of e-waste. Informal recyclers often disassemble e-waste by hand, which exposes them to potentially dangerous materials. Serious health issues including cancer, skin conditions, and respiratory disorders may result from this.

E-Waste Management Techniques in Use Today

Different nations and areas have different methods for managing e-waste. The infrastructural and legal frameworks are often more advanced in developed countries, which results in greater rates of formal recycling. Contrarily, underdeveloped nations often have difficulties handling e-waste as a result of few resources, insufficient laws, and the proliferation of illegal recycling methods. Disassembling and processing e-waste in an ecologically friendly and safe way is required for formal recycling of e-waste. This may aid in the recovery of priceless materials and assist stop the environmental discharge of dangerous compounds. On the other side, informal recycling of e-waste is often done in a risky and uncontrolled way. Hazardous compounds may be exposed to the environment and personnel as a result (Kuehr & Wang, 2015).

International Programs and Rules

Several international efforts and legislation have been created to combat the expanding e-waste issue. E-waste transboundary movement will be regulated under the Basel Convention on the Control of Transboundary Movements of Hazardous Wastes and Their Disposal, which also intends to encourage ecologically responsible management. In addition, a number of nations have put in place Extended Producer Responsibility (EPR) frameworks that oblige manufacturers to manage the end-of-life of their goods. EPR frameworks may aid in boosting formal recycling rates and lowering the quantity of e-waste incinerated or disposed of in landfills (Kuehr & Wang, 2015).

TECHNOLOGICAL ADVANCEMENTS

General Overview of Current Technologies Used

Figure 1 shows a simplified illustration of the stages required in gathering, preparing, analyzing, and visualizing data, as well as applying machine learning to learn from the data and generate predictions. The mentioned techniques will be used in the execution of different applications of e-waste management technologies (International Telecommunications Union, 2019; Kumar et al., 2021a).

Figure 1. Overview of current technologies

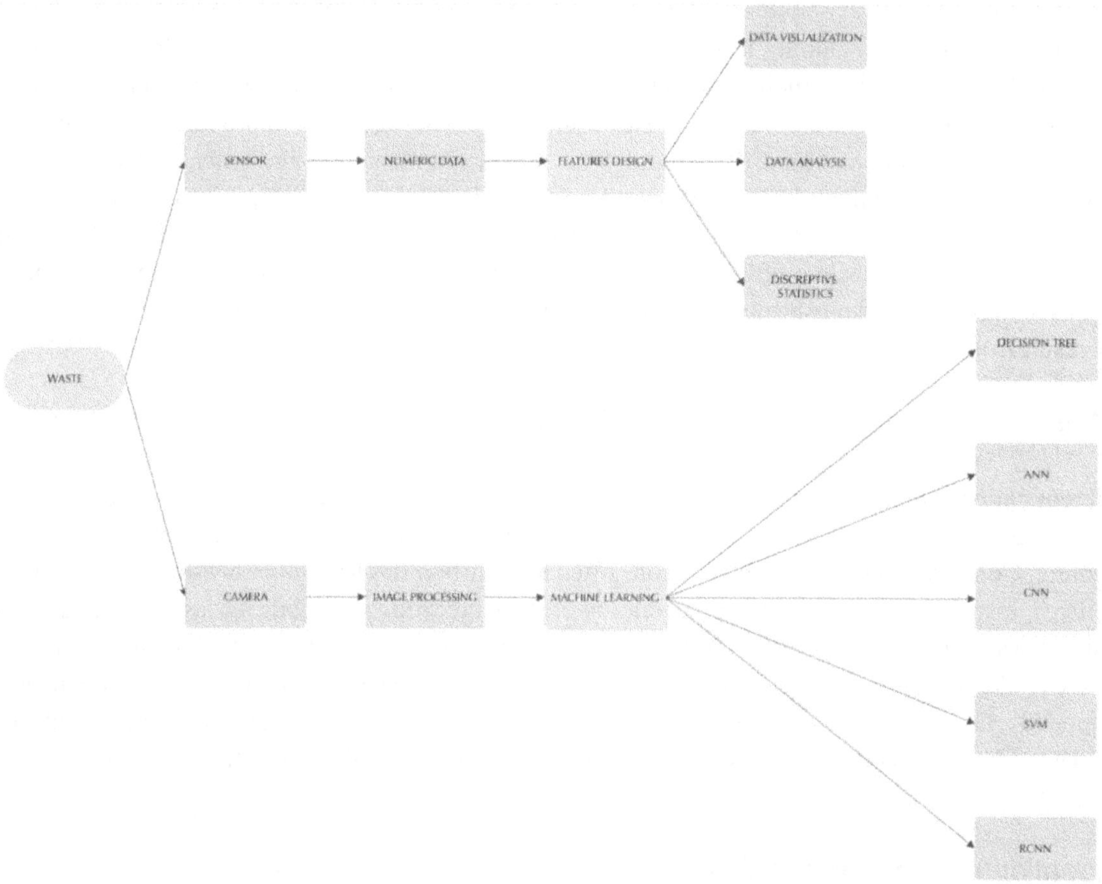

Image Classification to Distinguish E-Waste

Image classification is a method for recognizing and categorizing various items in photographs. AI may be used to construct photo classification algorithms that can be trained on vast amounts of pictures from electronic waste. Once trained, these models may be used to automatically recognize and categorize different e-waste components in photos. This makes it feasible for e-waste to be automatically separated into several sorts, including plastics, metals, and electronics. This could assist to boost the efficiency and speed of the process while lowering the amount of human work necessary to sort the rubbish.

The development of robotic sorting equipment powered by AI is one example of how picture classification is being utilized to enhance e-waste management. Robotic sorting systems employ cameras and image classification algorithms powered by AI to recognize and categorize various sorts of waste. This information may subsequently be used to classify the waste into numerous classes. Robotic sorting machines that apply artificial intelligence (AI) may be used, for instance, to separate e-waste into categories like plastics, metals, and electronics. As a consequence, the recovery of key components from e-waste may be boosted in terms of accuracy and efficiency.

Systems for classifying photos using AI frequently employ machine learning methods. Machine learning algorithms used for e-waste sorting are taught using a huge dataset of e-waste images. The computer system picks up on specific components in the photos, like plastics, metals, and electronics. After being taught, the machine learning system may be used to categorize current images of e-waste. The program first collects attributes from the picture to perform this. Features are elements of the picture that may be exploited to locate particular items within it. The algorithm utilizes the characteristics it has just obtained to categorise the picture and compares the picture's properties to the training dataset's feature set. Following that, the algorithm puts the picture in the class that most closely reflects the attributes of the photos in the training dataset (Liu et al., 2022).

For categorization of photos, numerous machine learning approaches may be utilized, including:

- CNNs (convolutional neural networks): CNNs are machine learning algorithms that are used for classification by pulling characteristics out of pictures and applying those features to categorize the images.
- Support Vector Machines (SVMs): SVMs are a form of machine learning approach that may be applied to a number of tasks, such as classification and regression.
- Random Forests: A form of machine learning approach, random forests may be used for a number of tasks, such as classification and regression. Building a lot of decision trees, then averaging their estimates, is how random forests function.

Identification of E-Waste Through Object Detection

Object detection is a technique for finding particular components in photographs. AI may be used to construct object identification models that can be trained on vast photobanks of e-waste. Once trained, these models may be used to automatically identify particular things in e-waste pictures. With the usage of this, harmful e-waste such as batteries, circuit boards, and capacitors may be detected. These objects may be handled or disposed of appropriately when they are isolated from other e-waste using this information.

One example of how object recognition is being used to enhance e-waste management is the development of AI-powered e-waste screening machines. By combining cameras and AI-powered object identification algorithms, e-waste screening devices may be applied to check for the presence of harmful compounds in e-waste. Using this information, it may later be chosen how to treat and dispose of the e-waste. For instance, if toxic compounds are identified in e-waste, they may be processed separately from other e-waste or appropriately disposed of.

Object detection systems powered by AI generally employ a two-stage technique. Making a list of probable sites for the shot is the first step. The next step is to categorize each potential area to evaluate if it includes a relevant item. The item identification system generally utilizes a deep learning model to offer a list of possible locations. A form of machine learning technology called deep learning models may uncover complicated patterns from data. The deep learning model is trained on a huge dataset of photos of e-waste in the instance of e-waste identification. The model acquires the capacity to discern between numerous things in the photos, including capacitors, circuit boards, and batteries.

The object identification system utilizes a classifier to categorize each candidate area after the deep learning model has created a collection of candidate regions. A machine learning approach known as a classifier may be used to separate data into multiple groups. The classifier is trained to identify candidate sites as either having an item of interest or not having an item of interest in the instance of e-waste identification (Kumar et al., 2021b).

Machine learning approaches that could be utilized for object detection include:

- RPNs (Region Proposal Networks) are a form of machine learning technology that may be used to discover prospective photo locations.
- Faster R-CNN (region-based convolutional neural networks): Faster R-CNN is a form of machine learning technology that may be applied to both the classification and identification of objects. To recognize and categorize objects in pictures, quicker R-CNN combines an RPN with a classifier.
- Single Shot MultiBox Detector (SSD) is a machine learning approach that may be applied to both the detection and categorization of items. SSD can create and identify candidate areas in a single step because it is a one-stage object detector.

Prediction of Hazardous Materials for Handling E-Waste

A technique called hazardous materials prediction may be used to forecast the existence of toxic compounds in e-waste. Using massive datasets of e-waste samples with known hazardous materials composition, AI may be applied to develop hazardous materials prediction models.

These models may be used to predict the presence of hazardous components in fresh e-waste samples once they have been trained. This may assist guarantee that e-waste is handled and disposed of responsibly, preserving both the environment and human health. The advent of AI-powered e-waste screening systems is one example of how hazardous materials prediction is being applied to assist e-waste management. E-waste screening devices may be used to check for the presence of hazardous materials by employing sensors and AI-powered hazardous materials prediction algorithms.The treatment and disposal of the e-waste may later be determined upon utilizing this information. For instance, if toxic compounds are identified in e-waste, it may be segregated from other e-waste and processed or disposed of securely.

Machine learning methods are commonly utilized in hazardous materials prediction systems powered by AI. A sort of artificial intelligence that can learn from data are machine learning algorithms. The machine learning algorithm is trained on a large dataset of e-waste samples with known hazardous materials composition in the case of hazardous materials prediction.The system develops the capacity to recognize data patterns linked with the presence of harmful compounds.

After being taught, the machine learning system may be used to anticipate the occurrence of dangerous compounds in fresh e-waste samples. The program first obtains characteristics from the sample of e-waste in order to perform this. The e-waste sample features characteristics that may be utilized to predict the presence of hazardous compounds.

The algorithm takes use of the attributes after they have been gathered to anticipate the occurrence of dangerous components in the e-waste sample. The qualities of the training dataset's e-waste samples are compared to the characteristics of the e-waste sample in the algorithm. The e-waste sample is then allocated to the class that best matches the features of the e-waste samples in the training dataset.

The technique known as Gradient Boosting Machines (GBMs) may be utilized to produce this prediction. A form of machine learning approach called GBMs may be applied to a range of tasks, such as classification and regression. GBMs function by developing a sequence of weak learners, merging them, and producing a strong learner (Tiwari et al., 2020).

Process Improvement for Recycling E-Waste

AI may be applied to design new, more effective, and efficient recycling procedures. AI may be applied, for instance, to create inventive solutions for recycling lithium-ion batteries. This may be accomplished by utilizing AI to simulate different recycling processes and determine the most feasible and effective techniques.

AI may be used to design techniques for enhancing the purity of recycled materials, which would increase the quality of e-waste recycling. AI may be used, for instance, to design ways for purging contaminants from recovered polymers. This may be done by employing AI to find and design unique separation approaches. Reduce the environmental effect of recycling: Artificial intelligence (AI) may be used to create solutions for minimizing the energy and water use involved with recycling. AI may be utilized, for instance, to boost the efficiency of recycling procedures. This may be done by employing AI to continually monitor and manage recycling efforts.

Reinforcement learning is one approach that AI may utilize to discover innovative ideas for recycling e-waste. Through trial and error, reinforcement learning, a subset of machine learning, enables agents to learn how to act in a given environment.

The agent in the instance of recycling e-waste may be a robot that is learning how to divide e-waste into numerous categories. The agent would begin by interacting with its surroundings at random. The agent would gain incentives as it interacted with the environment for activities that helped it accomplish its goals, such as classifying recyclable electronic waste. Additionally, the agent would be punished for behaviors like inappropriately classifying e-waste, which had the undesired impact.The agent would gradually learn to engage in behaviors that result in the desired outcome and abstain from acts that result in the unwanted one. This procedure would run on until the agent could properly classify e-waste into the relevant categories.

Using deep learning is another method AI may be utilized to enhance e-waste recycling. Artificial neural networks are used in deep learning, a sort of machine learning, to learn from data. Artificial neural networks are designed after how the human brain functions and is structured.

Building computer vision models that can detect and categorize items in photos may make use of deep learning. These models may be used to construct AI-powered e-waste sorting systems that can classify e-waste automatically into multiple types (Wang et al., 2021).

PCB Value Assessment

One of the most significant parts of managing e-waste is estimating the value of PCBs by calculating their metal content. The procedure's steps are as follows:

1. **Gathering of Samples:** A wide variety of PCB samples representing diverse electrical devices and manufacturing standards were obtained. The examination of the quantity and distribution of precious metals was based on these samples.
2. **Analysis of the metal content:** To test the concentration of precious metals contained in the PCBs, including copper, gold, silver, platinum, and others, X-ray fluorescence (XRF) spectroscopy and inductively coupled plasma mass spectrometry (ICP-MS) were applied.
3. **Calibration and Standardization:** Calibration standards were employed to establish a relationship between the measured signals and the actual metal concentrations. Standardization processes were adopted in order to assure measurement accuracy and repeatability.
4. **Data Evaluation:** The obtained metal concentration data were examined to assess the value of the PCBs based on the amount and kinds of precious metals present. It was crucial to correlate the measurements with the current metals market values in order to assess the board's worth.

5. **Reporting and Visualization:** Reports or visual representations that comprised the findings of the metal content study and PCB valuation were prepared to give a clear appraisal of the board's worth based on the precious metals it contains.
6. **Validation and Quality Control:** Quality control measures, such as sample duplicates, validated reference materials, and periodic equipment calibration checks, were employed in order to assure the reliability and accuracy of the metal content analysis.

In conclusion, the approach for constructing AI software to analyze printed circuit boards involved data collecting, AI model construction, training, validation, testing, and the establishment of a compatibility database for reuse.

Additionally, the approach applied to assess the value of PCBs comprised sample collection, metal content analysis, data processing, and reporting. These methodologies offer the framework for the design of AI-driven systems that can analyze and optimize printed circuit board reuse and value assessment in the context of e-waste management.

ECONOMIC AND SOCIAL IMPLICATIONS

Creating Jobs

E-waste management and resource recovery technological improvements have the potential to generate employment in a variety of ways. For instance, modern e-waste recycling and remanufacturing processes need the creation and upkeep of specialized machinery. The need for personnel in the engineering, manufacturing, and maintenance industries may rise as a result.

Additionally, the expansion of the e-waste recycling and remanufacturing businesses may result in the development of new positions in fields like customer service, logistics, and transportation.

Pollution Reduction and Enhanced Public Health

By limiting the discharge of harmful elements from e-waste into the environment, technological advances in e-waste management and resource recovery may contribute to a decrease in pollution and an improvement in public health.

To recover valuable metals from e-waste, for instance, innovative e-waste recycling methods may be employed without the use of dangerous chemicals. This may lessen water and air pollution.

Additionally, dangerous elements from e-waste may be securely disposed of using innovative e-waste treatment technology. This may aid in defending against the negative effects of exposure to these elements on human health.

Enhanced Access to Low-Cost Electronics

The availability of recycled and remanufactured goods may be increased because of technological developments in e-waste management and resource recovery. For instance, modern e-waste remanufacturing technologies may be utilized to create high-quality recycled electronic items that, in terms of performance and dependability, are on par with new products. Additionally, as the e-waste recycling and remanufacturing sectors expand, costs for recycled and remanufactured electronic items may fall as a result of increasing competition.

Promotion of Social Justice

By lessening the adverse consequences of e-waste on disadvantaged populations, technological advances in e-waste management and resource recovery may aid in the promotion of social justice. For instance, the pollution brought on by e-waste recycling plants often has a disproportionately negative impact on vulnerable areas. These populations may be shielded from the negative consequences of exposure to dangerous chemicals with the use of new e-waste recycling technologies that decrease pollution. Additionally, individuals in underprivileged areas may have access to new employment possibilities in the recycling and remanufacturing of e-waste.

Reduced Waste Disposal Costs

The cost of garbage disposal for companies and governments may be decreased with the aid of technological improvements in e-waste management and resource recovery.

For instance, new technology for recycling e-waste may be employed to turn it into useful resources like metals and polymers. This may lessen the quantity of electronic trash that has to be burned or dumped in landfills. Additionally, the amount of e-waste that has to be carried and disposed of may be decreased by using innovative e-waste management technology. This may lower garbage disposal costs even further.

Recycled and Remanufactured Products: New Markets

Resource recovery and e-waste management technological advances may open up new markets for recycled and remanufactured goods. For instance, modern e-waste remanufacturing technologies may be utilized to create high-quality recycled electronic items that, in terms of performance and dependability, are on par with new products. This may open up a new market for individuals and companies to purchase recycled electrical devices. The demand for recycled materials from other sectors may also rise as a result of the expansion of the e-waste recycling and remanufacturing businesses. For instance, recycled plastic from e-waste may be utilized to create brand-new items like furnishings, clothes, and construction supplies.

Conservation of Natural Resources

By lowering the need to mine new raw materials, technological developments in resource recovery and e-waste management may aid in the preservation of natural resources. For instance, valuable elements like metals and polymers may be extracted from e-waste using innovative e-waste recycling technology.

Without having to mine fresh raw materials from the ground, these resources may subsequently be utilized to make new goods. In addition, less e-waste may be dumped in landfills with the help of modern e-waste management technology. Groundwater contamination and greenhouse gas emissions are risks associated with landfills. The environment may be protected by innovative e-waste management systems by lowering the quantity of e-waste that is dumped in landfills.

Accelerated Economic Growth

E-waste management innovations and resource recovery technologies may contribute to tax income and economic development. For instance, the expansion of the recycling and remanufacturing sectors for e-waste might result in more jobs and more tax income for the government. As well as opening up new markets for enterprises, the development of novel e-waste management and resource recovery technologies may spur economic growth (World Bank, 2020).

WHAT DOES THAT MEAN FOR US?

We now face both possibilities and problems in the field of managing e-waste as a result of the data and future scopes that have been explored. Inadequate recycling rates, poor disposal methods, and worrying rates of e-waste creation are all highlighted by the statistics, on the one hand. Along with serious threats to the environment and public health, this also results in the waste of priceless resources.

The possibilities for the future, however, provide opportunities for improvement. Potential legislative and regulatory changes may result in stronger rules and better governance for the handling of e-waste. Campaigns for awareness and education may encourage people and organizations to act responsibly and dispose of and recycle e-waste in the right ways. Collaborations within the sector and the development of sustainable recycling infrastructure may result in more ethical and effective business practices. technological innovations, like AI software, may improve e-waste component identification, evaluation, and value, increasing the effectiveness of recycling and resource recovery. Additionally, the idea of a circular economy provides a framework for converting e-waste into a useful resource and lowering dependency on raw materials. This idea is reinforced by statistics on valuable components and metals.

It takes coordinated efforts by governments, companies, organizations, and people to take full use of these possibilities. To solve the issues raised by e-waste and create a more sustainable future, collaboration, innovation, and a change to sustainable habits are required. We may work toward efficient e-waste management, environmental preservation, and the development of a circular economy that benefits both society and the environment by taking action based on the facts and future scopes.

CONCLUSION

The paper emphasizes the serious problem of e-waste and its harmful impacts on the environment, human health, and resource depletion in its conclusion. The report emphasizes the critical need for efficient e-waste management by highlighting the worrisome pace of e-waste creation, incorrect disposal practices, and inadequate recycling efforts.

The results also provide chances for innovative fixes that take use of technological improvements and support sustainable behaviors. The use of AI software in e-waste management provides exciting characteristics that have the potential to completely transform the sector. AI software may make it easier to refurbish and repair cell phones, prolonging their useful lives and minimizing electronic waste, by recognizing various components from e-waste and giving a compatibility list for reuse.

Additionally, using AI software to evaluate printed circuit boards (PCBs) offers two beneficial characteristics. In order to promote the circular economy and reduce waste, the program can first identify different components from e-waste and present a list of appropriate devices where they may be reused. Second, by precisely estimating the value of the board by counting the quantity of precious metals in PCBs, the program promotes fair trade practices and ensures that intermediaries are not kept in the dark about the real value of PCBs.

It is essential to make research and development investments, work together with industry partners, and encourage the wider use of these technical characteristics if we are to fully realize their potential. Additionally, addressing unfair trade practices and improving the effectiveness of e-waste management may be accomplished by educating intermediaries about the importance of PCBs and encouraging the availability of affordable factory components for smartphone repairs.

In order to implement comprehensive e-waste management policies, it is crucial for governments, businesses, and people to collaborate. To do this, strong legislation and regulations must be put in place, as well as awareness campaigns, infrastructure for sustainable recycling, collaborations, and acceptance of technical breakthroughs.

By doing this, we can reduce the negative effects of e-waste on the environment, safeguard public health, save precious resources, and pave the path for a more sustainable future.

In conclusion, a multifaceted strategy that incorporates legislative initiatives, public education campaigns, infrastructural improvement, and technical innovation is needed for successful management of e-waste. By using these tactics, we can lessen the negative consequences of e-waste, encourage responsible consumption and production, and aim to create a society that is more ecologically friendly and sustainable.

REFERENCES

Chinese Academy of Sciences. (2018). *The social and environmental impacts of e-waste recycling in China*. Author.

International Telecommunications Union. (2019). *The role of technology in e-waste management*. Author.

Kuehr, R., & Wang, F. (2015). *The social and economic impacts of e-waste management in developing countries*. United Nations University.

Kumar, S. K., Kumar, S., Kumar, P., & Kumar, A. (2021a). The application of artificial intelligence in e-waste management: A state-of-the-art review. *Sustainable Computing : Informatics and Systems*, *35*, 100680.

Kumar, S. K., Kumar, S., Kumar, P., & Kumar, A. (2021b). A deep learning approach for e-waste classification. *Materials Today: Proceedings*, *51*, 1246–1250.

Liu, Y., Zhao, G., Chen, Y., & Zhang, J. (2022). Image classification of e-waste using convolutional neural networks. *IEEE Access : Practical Innovations, Open Solutions*, *10*, 24433–24441. doi:10.1109/ACCESS.2022.3221820

Tiwari, A. K., Kumar, S., Kumar, P., & Kumar, A. (2020). Machine learning for the prediction of hazardous materials in e-waste. *Waste Management & Research*, *38*(10), 1088–1096.

Wang, Y., Li, H., Fu, S., & Wang, Y. (2021). An optimization model for the recycling of e-waste. *Journal of Cleaner Production*, *279*, 123872.

World Bank. (2020). *The economic and social impacts of technological advancements in e-waste management*. Author.

World Economic Forum. (2017). *The economic and social benefits of e-waste recycling*. Author.

Chapter 19
Technological Advancements in E-Waste Management Methods and Models

M. Rudra Kumar
G. Pullaiah College of Engineering and Technology, India

U. Sivaji
Institute of Aeronautical Engineering, Dundigal, India

N. Badrinath
https://orcid.org/0000-0003-4058-7136
Vellore Institute of Technology, India

B. Rupa Devi
https://orcid.org/0009-0005-1298-737X
Annamacharya Institute of Technology and Sciences, India

K. Dinesh Kumar
https://orcid.org/0000-0003-0843-1561
Amrita Vishwa Vidyapeetham University, India

V. Rajesh Kumar Korupalli
Woxsen University, India

ABSTRACT

In response to the rise of electronic trash on a worldwide scale, numerous new solutions have emerged. To recover valuable elements from electronic trash, many recycling processes have been developed, including mechanical, chemical, and biological methods. The use of artificial intelligence and machine learning algorithms in automated sorting systems has increased the operational efficacy of e-waste recycling dramatically. Urban mining has gained popularity as a potential method of obtaining precious metals from technological trash. Furthermore, the implementation of extended producer obligation (EPR) rules imposes obligation on producers for the correct management and disposal of their products. These technology advances collectively lead to more sustainable and effective electronic waste management, decreasing environmental impact and aiding the recovery of valuable materials.

DOI: 10.4018/979-8-3693-1018-2.ch019

INTRODUCTION

The term "e-waste" refers to a wide range of electronic and electrical equipment that has been discarded, including commonly used devices such as smartphones and laptops, as well as less noticeable household appliances like refrigerators and washing machines (Oswald & Reller, 2011). The 21st century has experienced an unprecedented proliferation of technology, resulting in the integration of electronic devices as an inseparable component of contemporary existence. Technology has significantly transformed various aspects of our lives, including communication, work, and daily living, through the utilization of microchips in smartphones and the operation of large-scale data centers that support the digital era. Nevertheless, the rapid and exponential proliferation of electronic devices in both production and consumption has concurrently engendered a significant and urgent global concern, commonly referred to as electronic waste or e-waste. These technological devices, previously lauded for their groundbreaking features and practicality, ultimately reach the conclusion of their operational cycle, necessitating their proper disposal. The increasing global demand for electronics has led to the emergence of a complex and multifaceted issue regarding the management of electronic waste(Tesfaye et al., 2017). This issue extends beyond environmental considerations and encompasses broader implications for health, economy, and society.

The exponential increase in electronic waste poses a significant and complex global issue, influenced by various factors that highlight the contradictory nature of technological advancement. Based on the findings of the Global E-waste Monitor 2020, it was determined that the global production of electronic waste (e-waste) amounted to approximately 53.6 million metric tons in the year 2019. It is projected that this volume will increase by 21% to reach 74.7 million metric tons by the year 2030 (Yu et al., 2010), provided that current trends continue. The current surge of electronic waste presents a multitude of significant challenges that require urgent attention and creative resolutions.

Figure 1. Global e-waste forecasting statistics
Source: https://ewastemonitor.info/gem-2020/

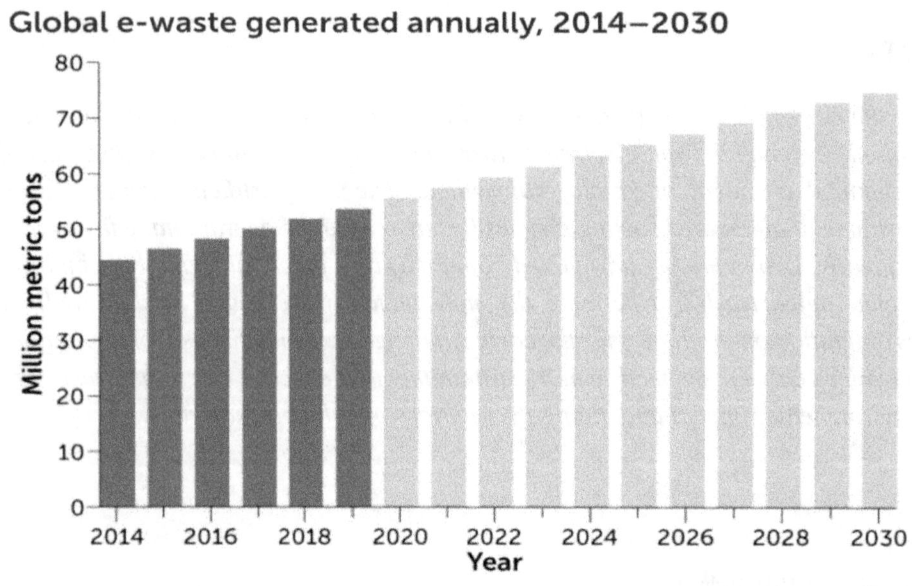

Primarily, the ecological ramifications of electronic waste are profoundly concerning. Electronic devices contain a combination of potentially harmful substances, such as lead, mercury, cadmium, and brominated flame retardants. Improper disposal of these toxic substances can result in their infiltration into the soil and water, thereby causing contamination of ecosystems and presenting a significant risk to both human and environmental well-being. The cumulative effect of electronic waste on the planet's intricate ecological equilibrium is a matter of significant apprehension.

In addition to its environmental consequences, the depletion of valuable resources represents another urgent concern. Electronic devices serve as valuable repositories of valuable resources, encompassing rare earth elements, gold, and silver. The conventional extraction methods employed for these materials result in the depletion of finite resources, contribute to environmental degradation, and frequently involve unethical mining practices. Electronic waste (e-waste) serves as an unexplored source of secondary raw materials (Tansel, 2017). Inadequate disposal of e-waste results in the squandering of valuable resources that have the potential to be recycled and reused. This, in turn, alleviates the strain on primary resource extraction. Moreover, the pervasive presence of informal recycling practices in numerous regions across the globe poses a significant risk to the health and well-being of workers and nearby communities involved in the dismantling and recycling of electronic waste. The act of inhaling toxic fumes, being exposed to hazardous chemicals, and handling electronic components without sufficient protective measures can result in various health complications, such as respiratory ailments, neurological disorders, and potentially cancerous conditions. The detrimental impact on human well-being caused by electronic waste serves as a poignant reminder of the pressing necessity for the implementation of responsible management strategies. In addition to the environmental and health aspects, data security emerges as a crucial concern. Electronic devices that have been discarded frequently contain data of a sensitive nature, including personal, corporate, and government information. Insufficient protocols for data erasure or disposal may lead to instances of data breaches, identity theft, and corporate espionage. The inadequate management of devices that store data can result in significant ramifications, jeopardizing privacy, and security on a broad scope. The absence of standardized regulations and enforcement mechanisms across various regions and countries presents a substantial obstacle to the efficient management of electronic waste. The lack of standardization in electronic waste (e-waste) policies and practices poses a significant obstacle to the effective coordination of global initiatives aimed at addressing this issue in a comprehensive and consistent manner (Bhat & Patil, 2014).

Given the complex nature of these challenges, finding effective solutions to the e-waste crisis necessitates the implementation of innovative approaches that go beyond traditional waste management methods. Technological advancements in the field of e-waste management assume a prominent role in addressing the challenges posed by electronic waste, presenting potential solutions to alleviate its impacts (Breivik et al., 2014). The present study focuses on the examination of technological advancements in the field of electronic waste (e-waste) management. Technological advancements have emerged as a promising solution in the field of electronic waste management, providing a comprehensive approach to tackle the environmental, economic, and social aspects associated with this issue. These innovations span across various disciplines, including robotics, artificial intelligence, blockchain technology, and circular economy models. As the examination of e-waste management unfolds, it becomes evident that technological advancements have emerged as significant catalysts for transformative shifts. An area of notable progress pertains to the automation of e-waste collection and sorting. The conventional approaches to the collection and sorting of electronic waste (e-waste) are frequently characterized by labor-intensive procedures, protracted time requirements, and susceptibility to errors. To address these constraints, sophisticated robotic systems and cameras powered by artificial intelligence have been implemented to effectively classify and organize electronic waste items according to their specific type and constituent elements. These technologies have the dual effect of expediting the collection process and improving precision, thereby safeguarding the retention of valuable materials during the sorting phase. In addition to this, the notion of electronic waste automated teller machines (E-Waste ATMs) has surfaced in certain areas. These novel machines enable individuals to engage in the act of depositing their obsolete electronic devices for the purpose of recycling, in return for receiving rewards or monetary incentives. This approach not only optimizes the procedure of electronic waste disposal but also promotes responsible recycling practices by offering tangible incentives to consumers (Van Yken et al., 2021).

Blockchain technology has become increasingly prominent in the domain of e-waste tracking and traceability. The utilization of blockchain technology, known for its attributes of transparency and immutability, is employed to establish a comprehensive and unchangeable log of electronic waste transactions. The blockchain ledger facilitates the monitoring and documentation of the entire lifecycle of electronic waste (e-waste) items, commencing from their initial collection and concluding with their recycling or disposal. This system guarantees a high level of transparency and accountability throughout each stage of the process. Furthermore, the adoption of QR codes and RFID tags has become increasingly prevalent, allowing for the assignment of distinctive identifiers to electronic devices. These identifiers greatly enhance the ability to monitor and trace devices throughout their entire lifespan (Forti et al., 2020). These technologies not only facilitate responsible disposal practices, but also offer consumers and regulators the ability to authenticate and validate the genuineness and legality of e-waste management procedures.

There has been notable advancement in the field of e-waste recycling, particularly in the realm of urban mining, which aims to extract value from discarded electronics. Conventional mining practices are characterized by their high resource consumption and adverse environmental impacts. In contrast, urban mining focuses on the retrieval of valuable metals and materials from electronic waste. Advancements in recycling methodologies have facilitated the secure and effective retrieval of precious metals, rare earth elements, and other valuable resources from electronic components. Moreover, sophisticated chemical methodologies are currently being utilized to extract and reclaim valuable substances from electronic waste. These processes not only mitigate the environmental consequences of resource extraction but also foster a more circular economy through the reuse of recovered materials. Circular economy models have emerged as a prominent approach for addressing the issue of electronic waste (e-waste) at its origin. The Product-as-a-Service (PaaS) (Forti et al., 2020; Sulaiman Zangina et al., 2023) model represents an innovative approach in which manufacturers investigate novel business models that redirect the focus from product ownership to product utilization. In the context of Platform as a Service (PaaS), consumers engage in subscription-based models to access and utilize products and services, as opposed to acquiring them through outright purchases. This transition provides manufacturers with an incentive to prioritize the design of products that possess qualities such as durability, repairability, and recyclability. Consequently, this approach effectively mitigates the production of electronic waste and fosters the notion of prolonging a product's lifespan. Similarly, there has been a global increase in the adoption of Extended Producer Responsibility (EPR) programs. The Extended Producer Responsibility (EPR) framework imposes responsibility on manufacturers for the complete life cycle of their products, encompassing activities such as product collection, recycling, and proper disposal. This approach incentivizes manufacturers to incorporate sustainability considerations into the design of their products, as they are accountable for the management of their products' end-of-life phase. The emergence of e-waste recycling robots represents a significant innovation within the recycling sector. Robotic systems are specifically designed for the purpose of disassembling and recycling electronic waste, thereby providing enhanced levels of efficiency, precision, and safety in comparison to manual procedures. The utilization of these robotic systems enables the efficient execution of complex operations, including the extraction of intricate components and the disposal of hazardous materials. Consequently, the recycling process is optimized, leading to a decrease in the potential risks associated with human exposure to harmful substances.

In the contemporary era characterized by the widespread use of mobile technology, the significance of e-waste tracking applications has witnessed a notable surge. These mobile applications enable consumers to conveniently find nearby e-waste drop-off locations, educate themselves about responsible disposal practices, and potentially receive incentives or rewards for recycling their outdated electronic devices. These applications serve to connect individuals with the e-waste recycling ecosystem, thereby enhancing the accessibility and convenience of responsible disposal practices. The significance of data wiping and secure disposal holds immense value within the realm of electronic waste management. To safeguard sensitive information stored on electronic devices, the field of data security has seen the development of sophisticated data wiping software and hardware solutions. These advancements aim to effectively erase data in a secure manner prior to recycling or disposal. These tools are crucial not only in ensuring the protection of data privacy but also in mitigating the risks of data breaches and identity theft. To simultaneously tackle the issue of waste and meet energy demands, certain pioneering technologies endeavor to transform electronic waste into energy utilizing methods such as incineration or pyrolysis. These methodologies not only mitigate the quantity of electronic waste but also produce sustainable energy, thereby contributing to a more environmentally conscious approach to waste management.

Community-based recycling models have recently emerged as viable grassroots solutions to address the issue of electronic waste. Efforts such as repair cafes and community e-waste collection centers serve to advance the principles of local, sustainable recycling and repair of electronic devices. These initiatives led by the community promote a collective sense of accountability and environmental guardianship, highlighting the significance of community-level practices such as minimizing, repurposing, and recycling electronic devices. The topic of global collaboration and standards is of significant importance in the academic discourse. It encompasses the exploration of how individuals and organizations from different parts of the world come together to work towards common goals and objectives.

The imperative of international collaboration and the establishment of shared standards is crucial in light of the worldwide scope of the electronic waste (e-waste) issue. The harmonization of e-waste management practices and regulations across borders necessitates collaborative endeavors involving governments, industries, and non-governmental organizations. The implementation of universal standards can effectively promote the adherence to responsible recycling practices on a global scale, thereby facilitating a comprehensive approach to addressing the e-waste problem. The administration of electronic waste finds itself at a pivotal moment in the course of human history. The rapid increase in electronic consumption and disposal necessitates the development of novel approaches that effectively tackle the complex issues associated with electronic waste, encompassing environmental and health risks, resource depletion, and data security vulnerabilities. The emergence of technological advancements in the field of e-waste management presents a promising prospect, as it provides a holistic strategy to address and minimize the environmental consequences of electronic waste. In the exploration of e-waste management, an examination of these advancements will elucidate the transformative impact they have on our strategies for handling electronic waste, thereby facilitating the progression towards a future characterized by enhanced sustainability and accountability (Ahirwar & Tripathi, 2021; Li & Achal, 2020).

LITERATURE REVIEW

Adverse Effects of E-Waste on Human Life and Environment

Electronic garbage (e-waste) contains heavy metals, persistent organic pollutants (POPs), and flame retardants, which pose long-term environmental risks. Pollution from formal and informal e-waste processing is high. Pollutants from these processes affect air, dust, soil, and water. These pollutants can enter the food chain and harm living things. Geochemical cycles and biological processes promote pollution buildup, movement, and transformation, creating long-term environmental hazards that can harm ecosystems and humans (Ahirwar & Tripathi, 2021; Bakhiyi et al., 2018). Health concerns from informal e-waste recycling are high for women and children. Food consumption, inhalation, and direct contact with polluted soil, dust, and skin can expose people to e-waste. This exposure can also cause respiratory and neurological troubles, tremors, pneumonia, and even death. Dioxins, furans, PCBs, and lead emitted during e-waste processing can cause cancer, nerve damage, and developmental difficulties in children. Dismantling and treatment workers risk occupational exposure, which can cause asthma, skin illnesses, neuritis, eye irritation, and gastrointestinal issues. E-waste recycling particles can cause inflammation, oxidative stress, and DNA damage, worsening health risks. E-waste releases toxic compounds, causing various environmental consequences. E-waste processing releases dioxins and other halogenated polymers that might cause cancer. E-waste exposure raises blood lead levels, which impair the neurological, circulatory, and renal systems and disrupt children's neurodevelopment. Microbial activities can convert inorganic mercury in e-waste into methylmercury, threatening the brain system. E-waste chromium compounds can damage DNA and cause asthma. Freon, used in air conditioners, can damage the ozone layer, worsening the greenhouse impact. Dioxins can be released through landfilling or illegally burning brominated flame retardants and chlorinated polymers. Due to its hazardous components and complex management, e-waste poses serious environmental and health risks. The threats to vulnerable populations, such as e-waste sector workers and neighboring communities, can affect future generations. Addressing the global environmental and health issues caused by e-waste requires managing and minimizing these risks (Frazzoli et al., 2010; Heacock et al., 2018).

Data Security and Cyber Crimes

The inadequate disposal of electronic equipment and their subsequent entry into the illegal e-waste market is a significant issue of concern, especially because it may lead to the exposure of sensitive data. Nevertheless, the existing body of studies pertaining to the magnitude of this matter is quite constrained. There are numerous examples that effectively illustrate the ease with which data can be extracted from secondary electronic and e-waste markets. A substantial amount of this data belongs to the categories of Personally Identifiable Information (PII) and Sensitive Personal Information (SPI), which are particularly coveted by individuals involved in cybercriminal activities. In a notable case study conducted by Simson Garfinkel at the Massachusetts Institute of Technology (MIT) in 2003 (Jan et al., 2022; Whitman & Mattord, 2021), an investigation was carried out to acquire a total of 158 hard drives through several channels. These sources included computer stores, small enterprises, thrift stores, and online auction platforms. The researchers made a significant finding that a considerable proportion of these discs still had retrievable data, even in instances when files had been erased, rewritten, or reformatted. The dataset included personal information, corporate data, credit card numbers, emails, and other types of sensitive data, highlighting the inherent risks associated with insufficient practices for disposing of data.

In 2009, the University of Vancouver conducted a significant case study that provided insights into the e-waste disposal sites in Ghana and revealed the widespread presence of hard drives being traded in local second-hand electronics markets. A significant proportion of these storage devices contained confidential financial information, including credit card numbers, bank account details, internet transaction logs, personal photographs, electronic mail correspondence, and even classified U.S. government agreements. These disclosures not only elicited apprehensions over personal privacy but also presented possible risks to the security of the nation. The prevalence of sensitive information being discovered in e-waste and second-hand markets is supported by a multitude of anecdotal reports. These examples involve the retrieval of military records, employment applications, medical history documents, tax information, and credit card details from previously owned computers and hard drives. Despite having been reported for over two decades, this issue continues to persist, emphasizing the urgent requirement for further extensive research, increased awareness, and strong data disposal processes to protect sensitive information from unauthorized access.

TRADITIONAL METHODS OF THE E-WASTE MANAGEMENT

Recycling

The section discusses the global management of electronic waste (e-waste) and its processing and recycling, focusing on metal recovery and non-metal component use. Manufacturers, importers, and distributors in Europe must manage electronic waste recycling and disposal costs under the WEEE and RoHS Directives (Daft, 2021; Feldman et al., 2016).

[1] Government-regulated recycling and processing systems have emerged, culminating in 500 EU e-waste treatment companies. Municipalities, recyclers, non-profits, and industry associations recycle e-waste in the US. Various governments have passed individual e-waste management laws to promote recycling and reduce landfill disposal.

[2] E-waste management in Japan has improved from collecting non-metal components to uniform treatment. This advancement has led to a solid industry chain for e-waste collection, transportation, and processing. Recycling and resource recovery have increased significantly.

[3] China's extended producer responsibility scheme and waste electrical appliance disposal fund encourage producers to pay for recycling and processing. This has increased the competitiveness of waste electrical appliance processing enterprises.

[4] Electronic waste disposal involves secure dismantling, component separation and enrichment, material regeneration, energy recovery, and pollutant treatment. These methods ensure the safe treatment and recycling of electronic trash, which may include harmful elements.

[5] E-waste recycling is being addressed in India through regulatory measures and infrastructural development. The nation's E-Waste (Management and Handling) Rules require producers to collect and recycle their electronic gadgets after use. India has established collection centers and authorized recyclers to manage e-waste. Additionally, e-waste campaigns and projects have been launched to educate individuals and organizations on correct disposal. The above initiatives show

India's commitment to sustainable electronic trash management and addressing its environmental and health risks.

[6] Electronic garbage recycling relies on metal recycling, which includes mechanical, physical, pyrometallurgy, hydrometallurgy, and bio-metallurgy methods. Each method has pros and cons, and their synergy is often used to maximize metal recovery.

[7] Electronic waste contains polymers, glass, and rubber, which can be recycled. Non-metallic materials are recycled using different methods depending on their properties and goals. Mechanical regeneration, modified regeneration, and energy recovery are examples.

Incineration

Incineration is a waste management technique characterized by the deliberate combustion of solid waste items at exceptionally elevated temperatures within specialized facilities referred to as incinerators. This approach is utilized to decrease the quantity of garbage, alleviate its ecological consequences, and harness energy from the process of combustion. Incinerators function within a temperature range of approximately 700 to 1,200 degrees Celsius, facilitating the efficient transformation of both organic and inorganic waste into ash, gases, and thermal energy. It is worth mentioning that incineration has the capability to produce electricity or heat, thereby making a valuable contribution to sustainable energy production via a process commonly referred to as waste-to-energy (WtE) (Chatterjee, 2012; Khaliq et al., 2014). Despite the potential advantages of volume reduction and energy recovery, incineration is also accompanied by environmental considerations pertaining to emissions, including dioxins and heavy metals. Consequently, the implementation of rigorous pollution control systems and compliance with regulatory standards becomes imperative. The meticulous sorting of garbage and the careful choice of appropriate waste streams are crucial factors in the effective and responsible execution of incineration as a strategy for waste management.

Landfills and Exports

Solid garbage is often dumped in landfills to control and contain it, reducing environmental contamination. Liners and monitoring systems are carefully designed into landfills to prevent chemical leakage into soil and groundwater. Garbage is compacted and covered with soil to reduce Oduors, pests, and land restoration. Landfilling is a practical waste management approach, although it emits methane and may harm ecosystems. Recycling, composting, and trash reduction are prioritized in modern waste management to reduce landfill use. Municipal solid waste is mostly produced by the US, China, and India (P. Kumar, 2012), with socioeconomic levels affecting garbage composition. Waste management methods include recycling, incineration, waste-to-energy conversion, composting, and landfill disposal. As ecological reactors, landfills alter waste material. Liner, soil cover, leachate collection, gas recovery, and flaring are essential to landfill sustainability. This study examines how landfill conditions affect gas and leachate production, focusing on bioreactor landfills as a future waste management option. This review covers waste volume reduction, resource recovery, waste valorization, environmental protection, and urban development site reclamation. This discussion covers landfill categorization, engineering, operations, mechanics, and landfill mining. Common waste management methods include landfilling to contain garbage and reduce environmental contamination. The US, China, and India generate the most garbage, with waste composition varying by socioeconomic status. Landfilling is one waste management option.

Complex processes occur in specific regions with this strategy. Sustainable landfilling requires liners, soil cover, leachate collection, and gas recovery. Bioreactor landfills can improve waste stabilization. Our study covers waste reduction, resource recovery, valorization, environmental protection, and site reclamation. This discussion covers landfill classifications, engineering principles, operational processes, mechanisms, and landfill mining.

Transporting abandoned electronics (e-waste) to developing nations has received interest. The scale of this challenge is hard to assess without a quantitative understanding of the amounts involved. This research aims to monitor the global transportation of electronic waste (e-waste) by consolidating and restricting current assessments of the amount of e-waste produced domestically in each country (MGEN), exported from OECD countries (MEXP), and imported from non-OECD countries. This study uses 2005 as the reference year and provides uncertainty ranges for all estimates. The estimation of MGEN, based on a global aggregate of 35,000 kt (with a range of 20,000–50,000 kt) according to a nation's GDP, matches independent assessments for specific countries. China, India, and five West African nations, the main beneficiaries of OECD e-waste exports, may have imported 5,000 kt (range: 3,600 to 7,300 kt) annually. This import volume accounts for 23% (with a range of 17% to 34%) of OECD domestic e-waste. Application of 23% to MGEN estimates MEXP for each OECD nation. MGEN (domestic e-waste production), MIMP (import), MEXP (export), and MNET (net e-waste flow) can be used to investigate and depict global e-waste generation and movement from OECD to non-OECD nations. Although there are still uncertainties, seven non-OECD nations' import predictions often match or exceed OECD export projections.

Manufacturer Take-Back Programs and Collection of Materials

India, like many other developed and developing nations, uses extended producer responsibility (EPR) for e-waste management under its E-waste (Management and Handling) Rules, 2011. Under these guidelines, producers must set up e-waste collecting centers and finance and organize an environmentally friendly management system. We analyze India's Rules' application in Ahmedabad, western India, in this post. Our case study uses interviews with regulated business entities, regulatory bodies enforcing the Rules, informal trash collectors and handlers, and publicly available implementation statistics. After the Rules were implemented, formal waste processing capacity increased, although only 5% to 15% of garbage generated is likely processed in formal facilities. The EPR requirement obliges businesses to comply with a few inexpensive rules, but the collection and recycling system has made it difficult for consumers to dump e-waste in formal centers. Our findings suggest that Indian EPR regulation should include deposit-refund systems and go beyond take-back obligations. In EPR laws, developing countries must carefully choose and create instruments. Collecting and categorizing abandoned electronic devices is a crucial step in managing electronic waste. The main goal is to recover valuable materials and components while minimizing environmental damage. This process entails setting up recycling centers, drop-off locations, and electronic waste disposal events(Brough & Jouhara, 2020; Khoshnevisan et al., 2021). E-waste collection aims to correctly process electrical gadgets to avoid landfill dumping. E-waste is categorized into metals, plastics, circuit boards, and batteries. Separating is essential for optimal recycling and recovery. Extraction of precious metals, copper, and polymers reduces the need for virgin resources and their environmental impact. Responsible e-waste collection also prevents inappropriate disposal, which can release harmful substances into the environment and harm humans and ecosystems. E-waste material and component collection helps conserve resources, reduce pollution, and handle electronic trash sustainably.

E-WASTE MINIMIZATION METHODS

The integration of advanced technology in the field of electronic waste (e-waste) management has brought forth a new era characterized by enhanced efficiency and sustainability. The proficient handling of electronic trash, which is an escalating global issue, relies on using the capabilities of artificial intelligence (AI), machine learning (ML), virtual reality (VR), blockchain, augmented reality (AR), the Internet of Things (IoT), and robotics. These revolutionary technologies operate in conjunction to fundamentally transform the processes involved in the management, recycling, and repurposing of electronic waste. Through the implementation of strategies aimed at increasing recycling rates, reducing environmental impacts, and promoting responsible disposal practices, this comprehensive approach lays the foundation for a future in e-waste management that is characterized by sustainability and effective utilization of resources. This investigation examines the various ways in which these advancements are transforming the domain of electronic waste management.

Artificial Intelligence (AI) for Sorting and Quality Control

The proper management of municipal solid waste relies heavily on the efficient classification of garbage, making it a crucial aspect. In this regard, the integration of robots presents a viable approach for enhancing the waste management process. Nevertheless, the implementation of robotic systems in such intricate settings requires sophisticated visual and operational functionalities due to the complex and unpredictable characteristics of industrial waste categorization. Current scholarly investigations have been focused on enhancing the precision and effectiveness of automated garbage sorting machines. This endeavor has propelled progress in sensor technology and waste recognition cameras, alongside the creation of intricate artificial intelligence (AI) algorithms. One potentially effective strategy entails the utilization of hyperspectral pictures for the purpose of identifying specific regions of significance within waste streams. Drawing on previous research, (Esmaeilian et al., 2018)have demonstrated the integration of simultaneous localization and mapping (SLAM) technology and instance segmentation methods into robots. This advancement enables these robots to effectively navigate complex field settings and independently gather construction and demolition waste. The utilization of deep learning methods, such as instance segmentation, enables robots to accurately perceive the boundaries of objects depicted in photos, hence enhancing their ability to recognize construction and demolition trash (Widmer et al., 2005). Within this context, numerous enterprises are currently leading the way in the adoption and application of artificial intelligence (AI)-based technologies for the sorting of electronic garbage (e-waste). Each of these organizations employs distinct methodologies and approaches that set them apart from one another. One example of a company utilizing advanced technology for efficient identification and segregation of various e-waste items is TOMRA, a prominent player in the worldwide sorting solutions industry. TOMRA utilizes state-of-the-art high-resolution cameras and machine learning algorithms in its operations. Zen Robotics use robotic devices that are equipped with artificial intelligence (AI) and sensors in order to identify and selectively retrieve recyclable materials from trash streams. The Bühler Group provides artificial intelligence (AI)-based sorting systems that aim to improve the retrieval of resources from electronic waste (e-waste), thereby making a substantial contribution to the concept of a circular economy. trash Robotics is a company that specializes in the development and implementation of robotic systems designed for trash sorting purposes, with a specific focus on electronic waste. These systems are equipped with artificial intelligence vision systems, which enable accurate identification and recognition of various materials (Fytili & Zabaniotou, 2008; Koul et al., 2022).

The use of robots specifically engineered for the purpose of sorting rubbish exhibits considerable promise in augmenting the efficiency of waste management processes, diminishing expenditures on labor, and raising the precision of waste classification. However, there are ongoing questions over the feasibility of waste-sorting robots, principally attributed to their elevated expenses in terms of installation and maintenance when compared to conventional waste-sorting techniques. Nevertheless, current research efforts are mostly directed towards investigating cost-efficient strategies for the development of such robots. This involves studying the utilization of less expensive materials and designing adaptive systems that may be employed in various operational environments. Furthermore, continuous endeavors are being made to improve the design of robot structures, enhance sensor technologies, optimize waste categorization algorithms, and advance the capabilities of robotic arms. These efforts aim to augment the overall efficacy and efficiency of robots in waste management tasks. Waste-sorting robots remain a topic of significant interest and have considerable potential for practical implementation in forthcoming real-world scenarios.

Machine Learning for Predictive Analysis

Machine learning is essential to transforming e-waste management since it solves the complex problems of the growing disposal of electronic devices. Electronic waste management can be improved by integrating data analysis, predictive modelling, and automation with cutting-edge technology. Collection, sorting, recycling, and sustainable strategy development are examples. This comprehensive study will examine how machine learning affects electronic waste management and provide insights into major organizations' strategies, focusing scientific and statistical features. Machine learning improves e-waste management through predictive analytics. This technology uses powerful algorithms to assess past data to predict electronic garbage development trends. Machine learning models can anticipate using consumer electronics sales, technology obsolescence, and product life cycles. These models help authorities and organizations plan electronic waste management. Data-driven methods optimize resource, employee, and recycling infrastructure distribution, reducing waste and improving collection and recycling. Classifying e-waste is another important part of its management. A major issue is the variety of electronic devices and components. Apple and Samsung are leading the way in using machine intelligence to solve this problem. These firms use machine learning methods to train models on enormous datasets of electronic components to automatically categorize and sort e-waste. Machine learning algorithms can distinguish cellphones, laptops, and circuit boards from other electronic waste using computer vision and sensor data. Recycling facilities are much more efficient with this capacity. This automation system improves sorting efficiency and material retrieval, contributing to sustainable e-waste treatment. Dell and HP rely on machine learning models to forecast demand for recovered electronic debris. To make intelligent manufacturing decisions, these models use market trends, economic factors, and customer behaviors. Machine learning algorithms reveal the quantities and types of recycled materials needed to meet market demand by analyzing prior market data (Al-Obaidi et al., 2022; Esmaeilian et al., 2018). Data-driven methods allow these firms to maximize recycled materials, reducing waste and complying with sustainability aims.

Machine learning helps identify and prevent e-waste fraud. Waste management companies like the one above use machine learning algorithms to detect and analyze electronic waste management process irregularities. This study analyses transportation routes, recycling facility methods, and waste disposal locations using algorithms. Machine learning-based fraud detection systems detect violations of compliance norms and laws to protect the e-waste recycling process. These technologies deter illegal e-waste dumping and misreporting. It is evident that many prominent organizations are adopting machine learning in e-waste management, however exact data is confidential. Machine learning algorithms have improved e-waste management efficiency, accuracy, and sustainability, according to these companies. Data-driven solutions can reduce environmental and health risks from improper electronic waste disposal, optimize resource use, and actively recycle and dispose of electronic equipment. Machine learning offers effective solutions for addressing the intricate issues associated with electronic waste management. The integration of data analysis, process automation, and predictive modelling enhances the efficiency of e-waste collection, sorting, and recycling. Although the exact deployment figures inside enterprises are not publicly released, several well-known companies have included machine learning in their projects for managing electronic trash. These organizations utilize data-driven approaches to reduce waste, promote sustainability, and advocate for appropriate disposal of electronic trash. The utilization of machine learning offers a viable and environmentally conscious approach to tackle the challenges associated with e-waste management, highlighting its increasing importance.

IoT's Smart Bins

IoT-connected smart bins address the growing problem of electronic waste (e-waste). IoT-powered e-waste bins are a pioneering use of technology. These clever bins include advanced sensors and networking capabilities. These characteristics streamline abandoned electronics collection and disposal. Electronics often contain hazardous and recyclable materials. Smart bins that use IoT require specialized sensors. The sensors are designed to measure electronic garbage factors such fill levels, temperature, humidity, and chemical composition. Fill-level sensors can properly measure electronic garbage in a container. Real-time fill level data is crucial for trash management and recycling operations. This data helps them optimize collection schedules, reducing unnecessary trips and carbon emissions. Data is sent instantly over Wi-Fi or cellular networks to a centralized management system or cloud computing platform. Connectivity enables continuous e-waste container monitoring and analysis. Processing this data often requires advanced data analytics and machine learning. The algorithms can predict when a smart bin would reach its maximum capacity based on past data, usage trends, and weather. This optimizes collection routes and resource allocation (Dhanwani et al., 2021; Guo et al., 2021). In addition to fill-level monitoring, IoT-enabled intelligent bins for electronic waste management often include picture recognition. The above technology may identify and classify electronic waste in designated containers. This technology helps classify and organize electronic trash and identify unapproved or hazardous substances, assuring safety and waste disposal compliance. These intelligent receptacles may also improve e-waste disposal traceability and accountability. E-waste may be tagged with RFID or QR codes during disposal. Tagging allows for complete tracking of each item from disposal to recycling or proper disposal. The e-waste management process becomes more transparent and accountable with this practice. GPS technology in IoT-enabled intelligent garbage containers is another benefit. GPS-equipped waste collection vehicles can obtain real-time e-waste bin locations and fill levels. This allows optimized collecting paths. Results include shorter collection times, lower fuel use, and improved e-waste management system efficiency.

IoT-enabled intelligent receptacles for electronic waste management provide a novel way to solve the problems of proper disposal. Modern sensors, wireless connectivity, data analytics, and image recognition enable real-time monitoring, efficient collection, and proper disposal of e-waste in these bins. Intelligent bins that provide accurate information about e-waste loading and composition help manage e-waste more sustainably and ecologically. These smart bins also increase electronic waste management safety and regulatory compliance.

RFID Technology Applications

RFID technology is a powerful tool for optimizing e-waste management by tracking, monitoring, and managing electronic trash throughout its lifecycle. RFID tags, readers, and a central database enable unique identification and effective data management of e-waste objects. RFID tags underpin this technology. At collection or disposal, these tags are attached to electronic equipment or components. They have a microchip and antenna, with passive, active, and semi-passive tags. Most e-waste management tags are passive RFID tags that use radio waves from RFID scanners to activate the microchip and communicate data. Active tags with power sources have a longer range than semi-passive tags. RFID tags are attached to e-waste objects to start RFID implementation. These tags provide item identification, manufacturer information, disposal date, and other important e-waste tracking data. The e-waste management system strategically places RFID readers at collection, recycling, and transfer stations. Radio waves from RFID readers trigger RFID tags on adjacent e-waste. After activation, RFID tags send their data to the reader, which sends it to a central database. The central database stores all RFID-generated data, enabling real-time visibility and full e-waste management insights. This database allows continuous tracking of tagged e-waste's location, processing, and disposal. RFID technology automates data acquisition, decreasing laborious record-keeping and inaccuracies (Condemi et al., 2019; Dhanwani et al., 2021; Esmaeilian et al., 2018; Guo et al., 2021).

RFID technology improves e-waste management in several ways. It greatly increases transparency and accountability throughout the disposal and recycling workflow, assuring environmental compliance and proper e-waste handling. RFID devices help identify and track dangerous items in e-waste, reducing health and environmental risks. RFID systems also optimize inventory management by giving real-time e-waste availability for recycling or refurbishing, reducing waste and optimizing resources. Ultimately, RFID technology improves e-waste management efficiency and accountability. E-waste is tracked from disposal to recycling or responsible disposal using RFID tags, readers, and a central database. RFID technology improves e-waste management transparency, safety, compliance, data-driven decision-making, and resource optimization.

Augmented Reality (AR) and Virtual Reality (VR)

VR and AR technologies are being used in e-waste management to streamline procedures, increase worker training, and boost recycling efficiency. These immersive technologies solve electronic waste issues using VR or AR. These are useful for e-waste management training and teaching. VR simulations allow staff to practice e-waste processing, sorting, and recycling in a safe, controlled environment. This method is useful for teaching personnel to recognize and handle dangerous electronic trash. However, AR can overlay digital data over workers' vision to provide real-time instruction and information. E-waste technicians can use AR-enabled equipment like smart glasses to obtain instructions, access reference materials, and validate safety protocols. Tracking and monitoring e-waste materials is another VR/AR use. RFID and GPS technologies combined with VR and AR systems offer real-time tracking and visualization of e-waste during disposal and recycling. Using AR devices, workers may scan RFID-tagged e-waste and instantaneously retrieve its origin, composition, and disposal requirements. This information ensures accurate sorting and disposal, reducing handling errors. VR and AR help design e-waste recycling facilities efficiently. Before construction, facility designers can generate VR prototypes and test layouts and workflows. This improves facility layout, equipment location, and workflow. AR can also guide workers through complex procedures in facilities, enhancing productivity and minimizing errors. In e-waste management, VR and AR are used for public awareness and education. These technologies can educate the public about appropriate e-waste disposal through immersive, interactive experiences. VR experiences can replicate the environmental impact of inappropriate disposal, while AR apps can provide real-time information about nearby e-waste collection and recycling centers to encourage recycling. E-waste data can also be visualized using VR and AR in data analytics platforms. Making data analysis more accessible and interactive helps decision-makers allocate resources, monitor compliance, and plan sustainability. Finally, VR and AR technologies provide intriguing prospects for improving numerous elements of e-waste management. They offer immersive training, real-time tracking, facility design optimization, public involvement, and data visualization. As these technologies evolve, they may make e-waste management more efficient, transparent, and sustainable (Das et al., 2021; Połomka et al., 2020; Yaqoob et al., 2023).

CURRENTLY AVAILABLE PRODUCTS

Tomra Autosort

Food sorting improves quality and safety by removing faults, unwanted objects, and undesired fractions from fruits, vegetables, nuts, and more. It efficiently sorts plastic types in recycling plants, improving recycling. Clear PET bottles can be sorted from colored ones and HDPE from LDPE, optimizing material recovery.

The AUTOSORT sorts ferrous and non-ferrous metals, which is important in recycling centers where aluminum cans must be separated from steel cans or copper wire from brass wire. Diamonds from gravel and gold from quartz are easily separated by the machine in mineral sorting, especially in mining. With throughput rates of 100 tons per hour and purity rates exceeding 99%, the TOMRA AUTOSORT is outstanding. It comes in many sizes and configurations to meet customer needs and operates continuously in harsh industrial conditions. Over 8,000 clients in 100 countries trust TOMRA AUTOSORT. It has recovered over 100 million tons of materials since its debut, helping reduce waste and recover resources. The TOMRA AUTOSORT improves product quality, throughput, operational costs, and sustainability. Its automation speeds up production and removes faults and foreign materials, improving product quality. Labor reduction reduces costs, and material recovery promotes sustainability, waste reduction, and the circular economy. This is a diverse and powerful sorting system that may improve product quality, operational efficiency, cost, and sustainability across industries (Bünger, 2021; Das et al., 2021; Patel et al., 2018; Połomka et al., 2020).

Predix Asset Insights

GE Digital's cloud-based asset management platform Predix Asset Insights analyses industrial sensor and device data using AI and ML. This system forecasts asset performance and maintenance needs to reduce operational interruptions, maximize asset utilization, and extend equipment lifespan. This technology could improve e-waste management. Predix Asset Insights predicts asset failures, a major benefit. This technology can foresee electrical device failures, allowing companies to schedule preventative maintenance. This method reduces unexpected inoperability, increasing the lifespan of electronic devices and reducing electronic waste. The program also optimizes maintenance schedules well. Predix Asset Insights helps enterprises optimize maintenance schedules by predicting asset repair scheduling and type. The optimization technique minimizes maintenance costs and enhances e-waste management efficiency by allocating resources more efficiently. In improving asset utilization, Predix Asset Insights excels. Finding underutilized equipment assets in an organization's fleet does this. This technology can detect malfunctioning machines or components in e-waste recycling. Organizations can reduce e-waste by eliminating or reassigning underused assets, reducing the need for new equipment. Practical examples demonstrate Predix Asset Insights' impact on e-waste management. Electronic manufacturers use software to predict component failures, enabling preventative maintenance and extending part lifespans. Waste management firms streamline e-waste processing equipment maintenance schedules to decrease costs and boost efficiency. The software's ability to identify underutilized assets in fleets helps e-waste recycling companies save unnecessary equipment purchases and e-waste generation (Patel et al., 2018; Sharan, 2002). Predix Asset Insights improves electronic waste management efficiency and effectiveness. AI and ML help companies predict asset performance and maintenance. They can reduce downtime, improve asset utilization, and extend equipment lifespan with these capabilities. Thus, these practices promote sustainable e-waste management.

Smartbins

Smart bins revolutionize e-waste management in an innovative and effective way. These containers can identify and sort e-waste like batteries, cell phones, computers, and TVs using sensors and AI. This innovative technology improves e-waste recycling accuracy and efficiency, reducing landfill trash. Smart bins can also collect data on e-waste types and volumes, helping to improve management policies and programs. A major benefit of smart bins in e-waste management is improved recycling. These bins make e-waste recycling easier, encouraging more people to recycle. Additionally, increased engagement can significantly reduce e-waste's environmental impact. E-waste contains lead, mercury, and cadmium, which can harm the environment if incorrectly disposed of operating efficiency is improved with smart bins. By automating sorting, they free up staff time for other important duties like e-waste collection and transportation. Smart bins also decrease manual labor in sorting and recycling, lowering e-waste management costs. Several global examples show how smart bins manage e-waste. The "Bin-e," developed by Clean Robotics in the US, can recognize and sort over 20 e-waste categories. These smart bins are in New York, San Francisco, and Los Angeles (Mart\'\inez et al., 2014; PR & Narula, 2018). Waste cycle's "e-waste bin," used in UK schools and businesses, can recognize and sort more than 10 e-waste categories. The "e-waste reverse vending machine" by E-Waste Exchange in India identifies and sorts over 15 e-waste types in public locations like airports and railway stations. Smart bins are a promising technological answer for trash problems. Smart bins make e-waste management more sustainable by automating sorting, encouraging recycling, and providing data for decision-making. They also raise awareness about e-waste and encourage recycling to create a greener electronics environment.

E-Cycle Tracker

EcycleTracker is a smartphone software that helps users track and manage their electronic waste (e-waste) and find nearby recycling facilities. The Electronics Takeback Coalition (ETC), a nonprofit dedicated to reducing electronic waste, produced the software to improve e-waste management. EcycleTracker's core feature is e-waste tracking. Individuals can track their electronic trash categories and volumes. This data can help design better electronic waste management programs and provide useful insights into usage habits. Additionally, the app makes finding nearby recycling sites easy. This technology simplifies finding nearby recycling facilities, making electronic garbage disposal easier for individuals. This trait makes recycling easier and more accessible. EcycleTracker also gives e-waste recycling advice. This guidance intends to make recycling electronic garbage environmentally sustainable and safe while minimizing hazardous material concerns. There are many benefits to using EcycleTracker. First, it promotes e-waste recycling to prevent the spread of harmful compounds that could poison the environment. It also streamlines recycling facility location and provides data on e-waste types and volumes, improving efficiency and efficacy. By eliminating labor-intensive operations like collecting and transporting electronic garbage, this technique may reduce e-waste management costs. EcycleTracker helps raise awareness of e-waste and encourages recycling. EcycleTracker is free to download from the App Store or Google Play(Aboelmaged, 2021; PR & Narula, 2018). After installing the app, users can create an account to track their electronic waste. The process of adding electronic garbage categories and quantities is simple. Additionally, the app lets users easily find nearby recycling facilities. By selecting the "Recycling Locations" tab in the app, users can view a list of nearby recycling sites. Specific addresses and operational schedules are included in this compilation. EcycleTracker promotes proper electronic waste management in a simple way. This effort helps people reduce their environmental impact and recycle technological waste, creating a more sustainable future.

VR E-Waste Education

VR e-waste education is a novel way to teach people about e-waste and sustainable management. This unique strategy strives to engage and teach people more interactively and captivatingly. Virtual e-waste education offers many engaging experiences. Virtual tours of e-waste recycling facilities can reveal the complex procedures involved in sorting and recycling the materials. VR programs can also show people how inappropriate e-waste disposal harms the environment. VR has various advantages over lectures and presentations for e-waste education. Immersive and captivating learning experiences boost engagement and motivation, improving learning outcomes. VR helps people to learn through personal experiences rather than passive viewing, improving comprehension and retention. VR e-waste education can also teach users empathy and compassion. VR can increase emotional response and motivation to address e-waste concerns by immersing people in real-world situations of its effects on communities and the environment. VR e-waste education is young but has great potential to change how people understand and manage e-waste. VR e-waste education programs may reach more people as VR technology becomes more affordable and accessible.

Several VR e-waste education programs exist:

[1] "Recycle Rush": UNEP created a VR game. Players learn about e-waste recycling and appropriate recycling in this game.
[2] WEF's "The Global e-Waste Challenge": This VR experience from the World Economic Forum shows how e-waste affects the environment and how individuals may help reduce it.
[3] European Commission's "The Circular Economy: A New Way to Make Products" VR experience: This VR experience helps users design and manufacture items sustainably to reduce e-waste.

These examples demonstrate the potential of VR e-waste teaching to raise awareness and promote ethical management. VR technology will certainly lead to more inventive and successful VR-based e-waste education programs.

RESEARCH AND DEVELOPMENT

Researchers need to focus on the following issues, while developing the methodologies on the e-waste minimization using sustainable solutions

[1] *Data collection:* The quality and quantity of data is essential for training AI and ML models. Data collection can be a challenge, especially for e-waste, as it is often difficult to track and collect data on the movement and disposal of electronic devices.
[2] *Model development:* There are many different algorithms and techniques that can be used to develop AI and ML models. The choice of algorithm and technique will depend on the specific problem that is being solved.
[3] *Model evaluation:* It is important to evaluate the performance of AI and ML models before they are deployed in the real world. This can be done by testing the models on a holdout dataset or by using metrics such as accuracy or precision.
[4] *Model deployment:* Once AI and ML models have been developed and evaluated, they need to be deployed in the real world. This may involve integrating them with existing systems or developing new applications.
[5] *Model monitoring:* Once AI and ML models are deployed, it is important to monitor their performance and make improvements as needed. This is important because the environment is constantly changing, and the models need to be able to adapt to these changes.
[6] *The security of data:* E-waste data often contains sensitive information, such as personal data or financial information. It is important to ensure that this data is secure and protected from unauthorized access.
[7] *The scalability of systems:* AI and ML systems can be complex and require a lot of data to train. It is important to ensure that the systems are scalable so that they can be used to process large amounts of data.
[8] *The sustainability of systems:* AI and ML systems can consume a lot of energy. It is important to design systems that are energy-efficient and sustainable.

DISCUSSIONS

Manufacturers can use AI and ML algorithms to create sustainable electronics. These technologies may help identify eco-friendly materials, energy-efficient components, and modular designs that simplify maintenance and recycling. This technique might significantly reduce electronic waste by promoting longer product lifecycles and reducing obsolescence. ioT-enabled intelligent devices can teach customers about e-waste management. Applications on smartphones and other devices can give recycling information, expedite trade-in programs, and incentivize recycling. These technical advances may encourage proper electronic device disposal by raising awareness and making recycling services more accessible. The reverse supply chain for electronic garbage can be optimized by machine learning algorithms for collecting and transporting abandoned gadgets. Predicting e-waste collection needs using historical data and real-time IoT sensor data can improve efficiency. Blockchain technology can establish and maintain transparency and accountability in this system, reducing the risks of illegal disposal.

AI and ML can help implement and improve circular economy techniques for e-waste management. These devices can track and analyze materials from manufacturing to recycling and reuse. Blockchain technology can immutably and securely document the whole lifecycle of materials, ensuring transparency and optimizing resource retrieval. AI-driven robotics and automation can identify and efficiently separate valuable components in electronic waste, improving recycling. Machine learning algorithms can learn and adapt to changes in electronic waste composition, improving sorting accuracy. Blockchain technology can validate proper recycling of diverse components and materials. Blockchain technology could create permanent eco-labels and certifications for electronic devices. This method ensures that consumers can easily confirm manufacturers' sustainability claims, encouraging the purchase of environmentally friendly products and reducing electronic waste over time.

International cooperation is needed to solve the global electronic waste problem. Blockchain technology could make global data exchange and collaboration safer and more efficient for governments, organizations, and recyclers. This technology could provide a global network for handling electronic waste, easing regional standardization and legislation. Researching and developing new technologies requires ongoing funding. Governments, companies, and universities can collaborate to find new ways to reduce electronic waste. This may involve developing better materials, recycling methods, or new artificial intelligence technology. Implementing proper approaches with AI, ML, IoT, blockchain, and other relevant technologies can significantly reduce electronic waste. These technological advances will encourage governments, industry, and individuals to work together to manage electronic trash more sustainably and ethically (Aboelmaged, 2021; Davis & Wolski, 2009; A. Kumar & Dixit, 2018; PR & Narula, 2018).

CONCLUSION

Technology has altered e-waste management and minimization. Blockchain, AI, ML, and IoT have transformed electronic waste sustainability and responsibility. New e-waste solutions benefit the environment, health, and economy. E-waste collection, sorting, and recycling benefit from AI and ML. These systems improve resource allocation and logistics by predicting e-waste creation patterns using large datasets and predictive analytics. Machine learning may reuse or recycle valuable electrical waste. E-waste disassembly is faster and more precise with AI-driven robotics and automation. Real-time e-waste tracking requires

IoT. Electronic bins and collection sites have IoT sensors that monitor e-waste container fill, temperature, and condition. Data-driven rubbish collection route optimization decreases operational expenses and environmental impact. IoT-enabled smart bins sort e-waste and alert authorities or recycling centers when full. Proactive e-waste management streamlines and prevents illicit dumping. E-waste lifecycle transparency and security are enabled by blockchain. Blockchain tracks and accounts for e-waste. This tool combats e-waste trafficking and enforces legislation. Recycling and electronic waste disposal can benefit from blockchain. It builds e-waste ecosystem trust and sustainability. Technology helps protect the environment. Untreated e-waste can pollute soil, water, and air with lead, mercury, and cadmium. AI and ML may categorize dangerous materials, minimizing pollution. Blockchain's transparency prohibits e-waste dumping and shipping to underdeveloped nations. These technologies improve health outside the environment. Workers can become very ill from unregulated e-waste recycling. AI-powered dismantling and recycling robots reduce health risks. E-waste storage and processing facilities can be remotely monitored by IoT devices for worker safety. E-waste recycling is rising due to technology. AI and ML sorting algorithms recycle rare metals, polymers, and minerals into industry. Job creation and value-added services in the circular economy minimize virgin material demand and boost economic growth. AI, ML, IoT, and blockchain have transformed e-waste minimization. These approaches assist us solve electronic waste's growing challenges. These solutions promote a sustainable and responsible future for electronic waste disposal by enhancing recycling efficiency, protecting the environment, improving human health, and boosting economic growth. These improvements help us manage electronic waste sustainably and reduce its harm.

REFERENCES

Aboelmaged, M. (2021). E-waste recycling behaviour: An integration of recycling habits into the theory of planned behaviour. *Journal of Cleaner Production, 278*, 124182. doi:10.1016/j.jclepro.2020.124182

Ahirwar, R., & Tripathi, A. K. (2021). E-waste management: A review of recycling process, environmental and occupational health hazards, and potential solutions. *Environmental Nanotechnology, Monitoring & Management, 15*, 100409.

Al-Obaidi, K. M., Hossain, M., Alduais, N. A. M., Al-Duais, H. S., Omrany, H., & Ghaffarianhoseini, A. (2022). A review of using IoT for energy efficient buildings and cities: A built environment perspective. *Energies, 15*(16), 5991. doi:10.3390/en15165991

Bakhiyi, B., Gravel, S., Ceballos, D., Flynn, M. A., & Zayed, J. (2018). Has the question of e-waste opened a Pandora's box? An overview of unpredictable issues and challenges. *Environment International, 110*, 173–192. doi:10.1016/j.envint.2017.10.021 PMID:29122313

Bhat, V., & Patil, Y. (2014). E-waste consciousness and disposal practices among residents of Pune city. *Procedia: Social and Behavioral Sciences, 133*, 491–498. doi:10.1016/j.sbspro.2014.04.216

Breivik, K., Armitage, J. M., Wania, F., & Jones, K. C. (2014). Tracking the global generation and exports of e-waste. Do existing estimates add up? *Environmental Science & Technology, 48*(15), 8735–8743.

Brough, D., & Jouhara, H. (2020). The aluminium industry: A review on state-of-the-art technologies, environmental impacts and possibilities for waste heat recovery. *International Journal of Thermofluids, 1*, 100007. doi:10.1016/j.ijft.2019.100007

Bünger, L. (2021). *Robotic waste sorting*. Worcester Polytechnic Institute.

Chatterjee, S. (2012). Sustainable electronic waste management and recycling process. *American Journal of Environmental Engineering, 2*(1), 23–33. doi:10.5923/j.ajee.20120201.05

Condemi, A., Cucchiella, F., & Schettini, D. (2019). Circular economy and E-waste: An opportunity from RFID TAGs. *Applied Sciences (Basel, Switzerland), 9*(16), 3422. doi:10.3390/app9163422

Daft, R. L. (2021). *Management*. Cengage Learning.

Das, P., Gabriel, J.-C. P., Tay, C. Y., & Lee, J.-M. (2021). Value-added products from thermochemical treatments of contaminated e-waste plastics. *Chemosphere, 269*, 129409. doi:10.1016/j.chemosphere.2020.129409 PMID:33388566

Davis, G., & Wolski, M. (2009). E-waste and the sustainable organisation: Griffith University's approach to e-waste. *International Journal of Sustainability in Higher Education, 10*(1), 21–32. doi:10.1108/14676370910925226

Dhanwani, R., Prajapati, A., Dimri, A., Varmora, A., & Shah, M. (2021). Smart Earth Technologies: A pressing need for abating pollution for a better tomorrow. *Environmental Science and Pollution Research International, 28*(27), 35406–35428. doi:10.100711356-021-14481-6 PMID:34018104

Esmaeilian, B., Wang, B., Lewis, K., Duarte, F., Ratti, C., & Behdad, S. (2018). The future of waste management in smart and sustainable cities: A review and concept paper. *Waste Management (New York, N.Y.), 81*, 177–195. doi:10.1016/j.wasman.2018.09.047 PMID:30527034

Feldman, J., Misenar, S., & Conrad, E. (2016). *Eleventh Hour CISSP®: Study Guide*. Syngress.

Forti, V., Baldé, C. P., Kuehr, R., & Bel, G. (2020). *The global e-waste monitor 2020. United Nations University (UNU), International Telecommunication Union (ITU) & International Solid Waste Association*. ISWA.

Frazzoli, C., Orisakwe, O. E., Dragone, R., & Mantovani, A. (2010). Diagnostic health risk assessment of electronic waste on the general population in developing countries' scenarios. *Environmental Impact Assessment Review, 30*(6), 388–399. doi:10.1016/j.eiar.2009.12.004

Fytili, D., & Zabaniotou, A. (2008). Utilization of sewage sludge in EU application of old and new methods—A review. *Renewable & Sustainable Energy Reviews, 12*(1), 116–140. doi:10.1016/j.rser.2006.05.014

Guo, H., Wu, S., Tian, Y., Zhang, J., & Liu, H. (2021). Application of machine learning methods for the prediction of organic solid waste treatment and recycling processes: A review. *Bioresource Technology, 319*, 124114. doi:10.1016/j.biortech.2020.124114 PMID:32942236

Heacock, M., Trottier, B., Adhikary, S., Asante, K. A., Basu, N., Brune, M.-N., Caravanos, J., Carpenter, D., Cazabon, D., Chakraborty, P., Chen, A., Barriga, F. D., Ericson, B., Fobil, J., Haryanto, B., Huo, X., Joshi, T. K., Landrigan, P., Lopez, A., ... Suk, W. (2018). Prevention-intervention strategies to reduce exposure to e-waste. *Reviews on Environmental Health*, *33*(2), 219–228. doi:10.1515/reveh-2018-0014 PMID:29750656

Jan, F., Min-Allah, N., Saeed, S., Iqbal, S. Z., & Ahmed, R. (2022). IoT-based solutions to monitor water level, leakage, and motor control for smart water tanks. *Water (Basel)*, *14*(3), 309. doi:10.3390/w14030309

Khaliq, A., Rhamdhani, M. A., Brooks, G., & Masood, S. (2014). Metal extraction processes for electronic waste and existing industrial routes: A review and Australian perspective. *Resources*, *3*(1), 152–179. doi:10.3390/resources3010152

Khoshnevisan, B., Duan, N., Tsapekos, P., Awasthi, M. K., Liu, Z., Mohammadi, A., Angelidaki, I., Tsang, D. C. W., Zhang, Z., Pan, J., Ma, L., Aghbashlo, M., Tabatabaei, M., & Liu, H. (2021). A critical review on livestock manure biorefinery technologies: Sustainability, challenges, and future perspectives. *Renewable & Sustainable Energy Reviews*, *135*, 110033. doi:10.1016/j.rser.2020.110033

Koul, B., Yakoob, M., & Shah, M. P. (2022). Agricultural waste management strategies for environmental sustainability. *Environmental Research*, *206*, 112285. doi:10.1016/j.envres.2021.112285 PMID:34710442

Kumar, A., & Dixit, G. (2018). An analysis of barriers affecting the implementation of e-waste management practices in India: A novel ISM-DEMATEL approach. *Sustainable Production and Consumption*, *14*, 36–52. doi:10.1016/j.spc.2018.01.002

Kumar, P. (2012). Managing E-Waste in India-A Review. *Academic Discourse*, *1*(1), 123–133.

Li, W., & Achal, V. (2020). Environmental and health impacts due to e-waste disposal in China—A review. *The Science of the Total Environment*, *737*, 139745. doi:10.1016/j.scitotenv.2020.139745 PMID:32516663

Martinez, M., Calvo-Parra, G., Gil, E., de la Rubia, O., Hillebrand, M., Rubio, F., Aipperspach, W., & Gombert, A. (2014). Environmental testing results over a tracker drive train. *AIP Conference Proceedings*, (1), 228–232.

Oswald, I., & Reller, A. (2011). E-waste: A story of trashing, trading, and valuable resources. *Gaia (Heidelberg)*, *20*(1), 41–47. doi:10.14512/gaia.20.1.9

Patel, P., Ali, M. I., & Sheth, A. (2018). From raw data to smart manufacturing: AI and semantic web of things for industry 4.0. *IEEE Intelligent Systems*, *33*(4), 79–86. doi:10.1109/MIS.2018.043741325

Połomka, J., Jędrczak, A., & Myszograj, S. (2020). Recovery of stabilizer glass in innovative MBT Installation—An analasys of new technological procedure. *Materials (Basel)*, *13*(6), 1356. doi:10.3390/ma13061356 PMID:32192113

PR, S. K., & Narula, D. S. (2018). Educational Strategies for all Eviating the Environmental Impact of E-Waste for Sustainable Development. *Cosmos An International Journal of Management*, *7*(2), 1–9.

Sharan, A. M. (2002). *Design of an Economical, and Mechanically Stable E-Cycle*. NECEC St. John's.

Sulaiman Zangina, A., Ahmed, A., Nuhu, Z., & Sani Alhassan, Z. (2023). e-Waste quantification and the associated valuable resources in Kano metropolis, Nigeria. *Journal of Material Cycles and Waste Management*, *25*(6), 1–10. doi:10.100710163-023-01793-7

Tansel, B. (2017). From electronic consumer products to e-wastes: Global outlook, waste quantities, recycling challenges. *Environment International*, *98*, 35–45. doi:10.1016/j.envint.2016.10.002 PMID:27726897

Tesfaye, F., Lindberg, D., Hamuyuni, J., Taskinen, P., & Hupa, L. (2017). Improving urban mining practices for optimal recovery of resources from e-waste. *Minerals Engineering*, *111*, 209–221. doi:10.1016/j.mineng.2017.06.018

Van Yken, J., Boxall, N. J., Cheng, K. Y., Nikoloski, A. N., Moheimani, N. R., & Kaksonen, A. H. (2021). E-waste recycling and resource recovery: A review on technologies, barriers and enablers with a focus on oceania. *Metals*, *11*(8), 1313. doi:10.3390/met11081313

Whitman, M. E., & Mattord, H. J. (2021). *Principles of information security*. Cengage Learning.

Widmer, R., Oswald-Krapf, H., Sinha-Khetriwal, D., Schnellmann, M., & Böni, H. (2005). Global perspectives on e-waste. *Environmental Impact Assessment Review*, *25*(5), 436–458. doi:10.1016/j.eiar.2005.04.001

Yaqoob, I., Salah, K., Jayaraman, R., & Omar, M. (2023). Metaverse applications in smart cities: Enabling technologies, opportunities, challenges, and future directions. *Internet of Things : Engineering Cyber Physical Human Systems*, *23*, 100884. doi:10.1016/j.iot.2023.100884

Yu, J., Williams, E., Ju, M., & Yang, Y. (2010). *Forecasting global generation of obsolete personal computers*. ACS Publications. doi:10.1021/es903350q

Chapter 20
Transitioning Toward a Circular Economy Through E-Waste Management

G. Raghunandan
Department of Commerce, Christ University (deemed), Bengaluru, India

Dhanabalan Thangam
https://orcid.org/0000-0003-1253-3587
Presidency Business School, Presidency College, Bengaluru, India

Shashikala Patil
International School of Management Excellence, Bengaluru, India

B. M. Rama Murthy
Acharya Institute of Management and Sciences, Bengaluru, India

Mohammad Juned
Presidency Business School, Presidency College, Bengaluru, India

Durairaj Duraisamy
https://orcid.org/0009-0000-6068-9928
School of Management, CMR University, Bangalore, India

Ramakrishna Narasimhaiah
https://orcid.org/0000-0002-4973-5775
Department of Economics, Jain University, Bengaluru, India

Nandeeswara Rao Ponnada
https://orcid.org/0000-0002-9292-9620
Department of Economics, Jain University, Bengaluru, India

Ravi V.
Department of Professional Studies, Christ University (deemed), Bengaluru, India

ABSTRACT

This chapter delves into the complexities and recommendations associated with the management of electronic waste, paving the way for a transition towards a circular economy. It underscores the significance of adopting sustainable practices, including eco-friendly product designs, extended producer responsibility, recycling, and remanufacturing. Several challenges are identified, such as the scarcity of data regarding material concentrations in specific electronic equipment and the absence of universally applicable standard methods for evaluating elemental concentrations in e-waste. The chapter underscores the value of integrating technical support and financial incentives to bolster e-waste manage-

DOI: 10.4018/979-8-3693-1018-2.ch020

ment efforts. Furthermore, it highlights the pivotal roles played by the private sector, non-governmental organizations, and government bodies in the effective implementation of these strategies. Collaboration between businesses, governments, and consumers is underscored as essential for establishing a robust circular economy.

INTRODUCTION

Electronic waste, often known as e-waste, comprises electronic devices powered by a battery or plug that have been discarded because they are no longer needed, functional, or have become obsolete. E-waste can be categorized into six main groups: lamps, small IT and telecom equipment, screens and monitors, temperature exchange equipment, large equipment, and small equipment. Among these categories, small equipment, which encompasses items like microwaves, vacuum cleaners, and kettles, represents the largest proportion of e-waste production by weight. In 2019, the global electronic waste production reached an astonishing 54 million metric tons. This surge in e-waste can be attributed to various factors, including rising disposable income and the widespread availability of electronic devices. Consequently, electronic waste has emerged as the fastest-growing waste category on a global scale. The "Global E-Waste Management Market Outlook, 2028" study report, released by Bonafide study, projects that the industry will grow from USD 60.27 billion in 2022 to USD 122.48 billion by 2028. Furthermore, from 2023 to 2028, the market is expected to expand at a 12.82% CAGR. A substantial amount of garbage has been released into the environment as a result of increased resource consumption brought on by population expansion, urbanisation, and rapid economic growth (Research and Markets, 2023). As of 2023 the total amount of global electronic waste is 61.3 million metric tons. This trajectory is expected to persist, with estimates suggesting that by 2030 it will be 74.7 million metric tons, and the annual worldwide generation of e-waste will expand by around 30 percent (statista.com, 2023). Figure 1 explains growth trend of Global E-waste till 2030.

Figure 1.

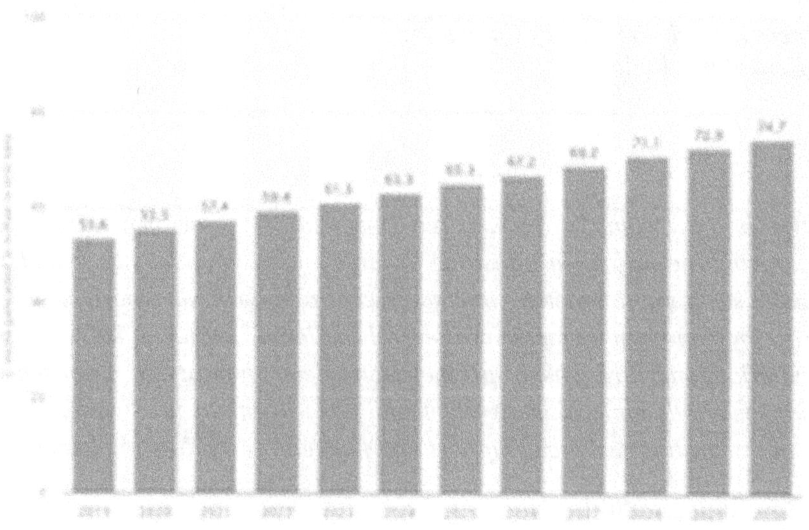

The digital revolution has transformed the way people live, communicate, and connect with others across the globe. We are currently living in a remarkable era of technological progress, characterized by a generation deeply enamored with technology. Our lives are filled with a multitude of devices, ranging from household appliances to computers, laptops, and ever-evolving smart phones. People frequently replace their gadgets, almost as often as they update their wardrobes. Each time a new and improved version becomes available, old devices are discarded in favor of new ones. Have you ever stopped to ponder the staggering amount of electronic waste generated by the global population? And, perhaps more importantly, where does all this electronic waste end up?

Electronic waste, often referred to as E-Waste, encompasses electronic devices and gadgets that have reached the end of their useful life. This category includes obsolete electronic devices, cellphones, notebooks, game consoles, and their various components. Here are some eye-opening and concerning facts about electronic waste: Globally, approximately 40 million tons of electronic waste is generated each year, equating to the disposal of 800 laptops every second. On average, cell phone users replace their devices roughly every 18 months, contributing to the growing e-waste problem. Within the total volume of E-Waste, a staggering 70% is considered toxic waste, yet only 12.5% of electronic waste is recycled. An alarming 85% of E-Waste ends up in landfills, where incineration often occurs, releasing harmful toxins into the air. These toxins can include lead, which poses risks to the central nervous system and kidneys, particularly affecting a child's mental development even at low-level exposure. Some of the most common hazardous electronic items include LCD desktop monitors, LCD televisions, plasma televisions, traditional Cathode Ray Tube (CRT) TVs, and computers with CRTs. Electronic waste contains a multitude of substances, many of which are toxic, such as mercury, lead, arsenic, cadmium, selenium, chromium, and flame retardants. Approximately 80% of E-Waste generated in the United States and many other countries is transported to Asia for processing and disposal. The global production of electronic devices is staggering, with 300 million computers and 1 billion cellphones manufactured annually. This production rate is expected to grow by 8% per year, further exacerbating the e-waste challenge (theworldcounts.com, 2023).

The proliferation of electronic waste, often referred to as Waste Electrical and Electronic Equipment (WEEE) or e-waste, is a rapidly escalating global issue. Predictions indicate that this waste stream will double by the year 2045 (Parajuly et al. 2019). This surge is primarily attributed to the ever-increasing demand for electronic products, placing substantial responsibilities on both advanced and developing economies worldwide. Electronic devices have become an integral part of our daily lives, experiencing a growth rate three times faster than that of other solid wastes. While these electronic products have evolved to play a complex and ubiquitous role in our daily routines, they contain valuable resources and hazardous materials that demand careful management. However, the systems for collecting and managing e-waste have not kept pace, resulting in inadequate handling and management. This situation poses various significant risks, including resource wastage and adverse effects on the environment and human health (Awasthi, Li, Koh, & Ogunseitan, 2019).

E-waste is a term subject to various definitions, occasionally causing controversy. According to a report by the United Nations Environment Programme (UNEP), e-waste encompasses a wide range of electrical and electronic equipment that has reached the end of its life cycle and holds no value for its owners. It is also known as Waste Electrical and Electronic Equipment (WEEE). Meanwhile, the European Directive 2002/96/EC defines e-waste as any waste electrical and electronic equipment, along with all its components, subassemblies, and consumables that are part of the product at the time of disposal (UNEP, 2023). The Organisation for Economic Co-operation and Development (OECD) describes e-waste

Table 1. Classification of e-waste

Sl.No	Category	Examples
1	Large household appliances, including cooling and freezing appliances	Televisions, Fans, Freezer, Refrigerator Water cooler, Ice maker, Washing machine Clothes dryer, Drying cabinet, Air-conditioners
2	Small household appliances	Juicers, electric mixers, meat grinders, coffee grinders, deep fryers, herb grinders, food processors, electric kettles, waffle irons, coffee makers, blenders, and dough blenders, as well as rice cookers, toasters, exhaust hoods, are yet another category of kitchen appliances.
3	IT equipment, including monitors	Computers, Telecommunications Equipment, Monitors, Keyboards, Printers, Servers, Drives, and Network HUBs
4	Consumer electronics	Mobile Devices, Smartphones, Wearables, Tablets, And Wireless Charging Platforms
5	Lamps and luminaires	Incandescent, Tungsten Halogen, Fluorescent Compact Fluorescent, Mercury Vapour Metal Halide, High-Pressure Sodium Vapour, Low-Pressure Sodium Vapour, LED
6	Toys	Alarm Clocks, Dance Mats, Digital Cameras, Electronic Pets, Handheld Games, Karaoke Machines, Plug & Play Video Games, Radios, MP3 & CD Players.
7	Tools	Multimeter, Soldering Wire, Soldering Iron, Solder Smoke Absorber, Alligator Clips, Hot Glue Gun, Heat Gun, and Heat Sink
8	Medical devices	Artificial cardiac pacemaker, Sphygmomanometer, Implantable cardioverter-defibrillator, Hearing aid, Defibrillation Glucose meter, Digital Stethoscope Thermometer, and Pulse oximetry
9	Monitoring and control instruments	indicator lights, dials, video displays, and audible alarms, buttons, keyboards, mice, trackballs, touch screens, and thumbwheels.
10	Automatic dispensers	Automatic Dispenser, Hand Sanitizer Dispenser, Automatic Soap Dispenser, Paper Towel Dispenser, Air Freshener Dispenser, and Automatic Tape Dispenser.

Source: (Wikipedia, 2023) .

as any household appliance that consumes electricity and has reached the end of its life cycle. E-waste, also known as electronic waste, is generated when electronic devices are disposed of at the end of their operational life. The proliferation of technology and our consumption-oriented culture has led to the significant accumulation of electronic waste. In the United States, the Environmental Protection Agency (EPA) categorizes e-waste into ten distinct classifications as mentioned in the Table 1(Wikipedia, 2023).

Waste Electrical and Electronic Equipment (WEEE) is a highly diverse and heterogeneous category, encompassing a wide range of constituents, some of which can be hazardous while others are not. Within these electronic devices lie valuable materials such as copper, aluminum, silver, and iron, which can be reclaimed through recycling processes? Therefore, recycling WEEE holds significant importance in waste management and the recovery of precious resources. To establish an efficient and environmentally friendly recycling method, it is essential to identify and quantify both valuable and hazardous components within this waste stream. This allows for a better understanding of its characteristics and facilitates the enhanced recovery of minerals, contributing to the conservation of natural resources and the development of sustainable e-waste management solutions (Ilyassova et al., 2021). Many companies manufacturing electrical and electronic devices frequently release product upgrades. Moreover, these products are often designed with a limited lifespan. The primary component in most electric and electronic equipment is the printed circuit board (PCB), which can contain approximately 20% copper and 250 g/t Au (gold), significantly higher than the concentrations found in copper or gold ore (Cocchiara et al, 2019). Various methods are available for recovering these valuable metals from electronic waste, including physical, pyrometallurgical, bio-metallurgical, and hydrometallurgical processes (Wang et al, 2021).

Despite having a reasonably established waste management infrastructure and concerted efforts, overall results in addressing the e-waste problem have been unsatisfactory. This could be attributed to the absence of design support evidence, well-developed policies, and research know-how concerning the recovery of e-waste products. To tackle these issues, the concept of a circular economy is regarded as a progressive approach. While an Extended Producer Responsibility (EPR)-based e-waste management system was initially considered a suitable plan, the implementation has not yielded the expected incentives. Individual operators and resource recovery have been limited to minimal collection and subsequent material recycling processes. Additionally, the potential for reusing discarded materials throughout the entire product lifecycle is almost nonexistent (Turaga, et al, 2019).

The circular economy represents a concept that aims to slow down the rate of electrical and electronic equipment (EEE) consumption by keeping waste materials within the system for as long as possible. This is achieved by minimizing and eliminating the need for their regeneration through smarter product design and business models (Ali & Shirazi, 2023). The approach involves the "7 R's" system, which includes reducing, reusing, refusing, recycling, recovering, rethinking, and redesigning. These aspects emphasize the social, economic, and environmental dimensions, aligning with India's E-waste management rule of 2016. However, various obstacles may impede the process, such as limited technology, inadequate collection systems, financial requirements, and a lack of training within the informal sector (Ekins et al, 2020).

The circular economy strives to promote cleaner and renewable technologies, innovative models, and developmental policies to eliminate waste by optimizing products and cycling materials. This ensures they remain in use for as long as possible, preserving their utility and value (Ellen MacArthur Foundation, 2019). This optimization can be observed through improved product design and forward-thinking waste management models, extending a product's lifespan and enabling resourceful material recovery. The general concept of a circular economy in e-waste management is illustrated in Figure 2.

Figure 2. The concept of 7Rs of the circular economy in e-waste management

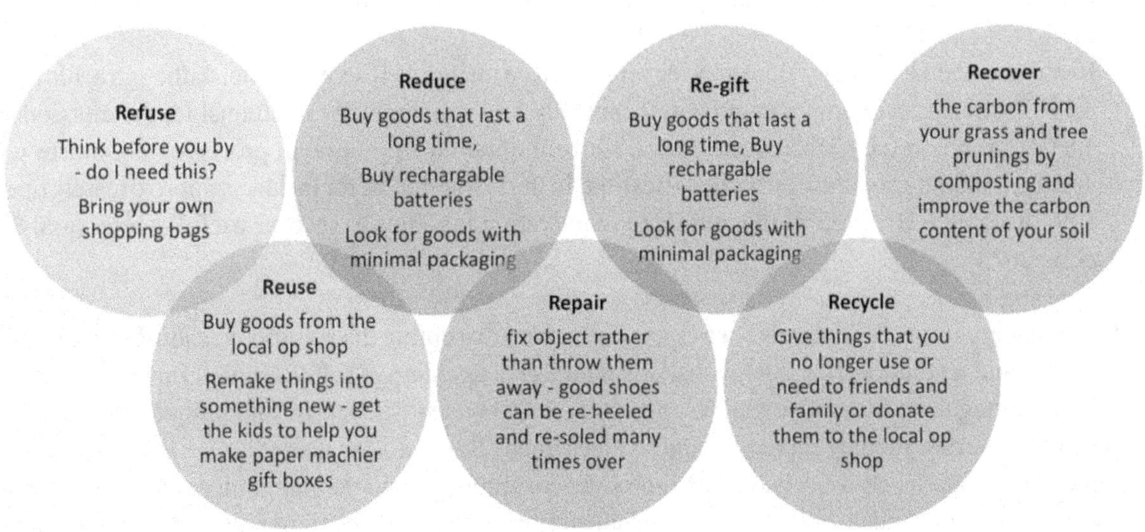

1. Reimagine solutions at various system levels by exploring alternatives and redefining challenges. As evident in the rapid rise of the sharing economy, a fundamental shift from ownership to sharing is already underway. Platforms like Airbnb facilitate home sharing, numerous websites promote car sharing, and Peerby encourages the sharing of various items. Sharing products beyond individual ownership significantly reduces the need for minerals and resources (goldenplains.vic.gov.au, 2023).
2. Minimize resource consumption by applying lean design principles and extending the lifespan of products. The integration of bionic algorithms into computer-aided design, coupled with 3D printing, can substantially reduce resource usage. For instance, Airbus APWorks recently unveiled the Light Rider, the world's first 3D-printed motorcycle with a frame weighing just six kilograms. This concept holds immense potential, especially in resource-intensive industries like construction (axil-is.com, 2023).
3. Repair components and parts to prolong product usage by the same owner. No one would discard a bicycle due to a flat tire. When the bicycle holds substantial value and tire repair is relatively simple, opting for a repair is the logical choice. However, for more affordable household appliances, people often dispose of items that could have been easily repaired (Palak Jain, 2020).
4. Reuse products by transferring them, either in their original form or with modifications, to another user. While second-hand marketplaces like eBay thrive on the reuse of consumer goods, industrial goods are gradually seeing an uptick in reuse. A railway depot in Groningen, for example, challenges contractors to repurpose items like sleepers, rails, clamps, railway switches, and overhead catenaries that are no longer needed elsewhere (Palak Jain, 2020).
5. Refurbish products by replacing malfunctioning components and parts with new ones. A noteworthy example is OUD NOW, which transforms old cabinets into modern designs. Designer Theo Herfkens replaces broken antique chassis with tailor-made, contemporary ones. This approach appeals to those who want to preserve their grandparents' heritage while enjoying a stylish interior (axil-is.com, 2023).
6. Recycle materials or resources by disassembling components and separating parts. Urban mining, where electronic waste is a valuable source of metals like gold, is gaining traction. Apple's recycling robot Liam disassembles old iPhones to recycle precious metals such as silver and platinum. Estimates suggest that urban mining can cover about one-third of the global demand for rare metals (axil-is.com, 2023).
7. Reclaim embedded energy from non-recyclable waste materials where feasible. Although achieving a 100% circular economy may remain technically and economically unattainable, the embedded energy from non-recyclable waste can be converted into energy through processes like waste-to-energy, including combustion and gasification. In the Netherlands, for instance, approximately one-fifth of waste is diverted from landfills to incineration facilities to recover useful energy (axil-is.com, 2023).

On the energy front, Tesla stands out as a game-changer, disrupting the automotive industry with electric vehicles and revolutionizing energy markets with its energy storage solutions. In terms of cascading materials, Freitag is an innovative player, upcycling well-traveled truck tarps into unique, high-quality bags and accessories, thereby increasing their value (Smart Product Manager, 2023). Exploring the social and behavioral aspects associated with e-product consumption and their alignment with the circular system is an ongoing endeavor. Beyond the technical intricacies and recycling rates within the e-waste circular economy, sustainable product manufacturing should also be prioritized. While the concept of a circular economy is considered crucial for the generation and management of e-products, it necessitates a collaborative effort across various sectors, including businesses, governments, and consumers (Henriques, Figueiredo, & Nunes, 2023). Both the technological and economic dimensions, as well as consumer behavior, play vital roles in defining the long-term sustainability of the circular economy. Therefore, there is a need to identify opportunities for targeted interventions aimed at enhancing e-waste management and achieving a robust circular economy.

E-WASTE SCENARIO AT THE GLOBAL LEVEL

Electric and electronic products are integral to global development, encompassing a wide range of items used in daily life, both in households and businesses worldwide. However, ownership per capita varies according to income levels. In 2019, the world generated approximately 53.6 million metric tons (Mt) of e-waste, excluding PV panels, equivalent to 7.3 kg per capita. Projections indicate that by 2030, the quantity of e-waste generated will surpass 74 Mt, signifying an alarming annual increase of nearly 2 Mt (Tub et al, 2020). In 2019, the documented formal collection and recycling of e-waste amounted to 9.3 Mt, representing only 17.4% of the total e-waste generated. Figure 3 has revealed that the Global E-waste had grown by 1.8 Mt since 2014, with an annual growth rate of around 0.4 Mt. However, during the same period, the overall e-waste generation increased by 9.2 Mt, at an annual growth rate of nearly 2 Mt. According to the E-waste Management - Global Strategic Business Report, The market for e-waste management is a rapidly expanding sector that is predicted to grow from a modest US$55.6 billion in 2022 to an anticipated US$158.9 billion by 2030. Making decisions in this quickly growing field requires a deep comprehension of market dynamics and trends. This market research study offers comprehensive information and important insights to help firms stay ahead of the curve in this exciting industry. This disparity underscores the inadequacy of recycling efforts in keeping pace with the global surge in e-waste (Forti et al, 2019).

Data on e-waste collection and recycling are derived from country-reported information. The most recent available data on formally collected and recycled e-waste worldwide generally pertain to the year

Figure 3. The e-waste scenario at the global level

2016. In 2019, the majority of e-waste generated (82.6%) likely evaded formal collection and environmentally sounds management practices. These unaccounted flows are typically not documented in a consistent or systematic manner. The absence of data on formally collected and recycled e-waste implies that a significant portion of the 2019 e-waste (44.3 Mt) was managed outside official collection systems and, in some instances, shipped to developing countries. In wealthier countries, small-sized electronic devices may end up in regular waste bins and be disposed of with municipal solid waste, bypassing proper recycling procedures and resulting in material losses. In EU countries alone, it is estimated that 0.6 Mt of e-waste finds its way into waste bins (Tub et al, 2020).

YEAR AND COMPONENT WISE E-WASTE GENERATED

According to the report "Global flows of e-waste 2019, by scenario" by statista, In 2019, 53.6 million metric tonnes of electronic garbage were produced globally. Merely 17. percent, or 9.3 million metric tonnes, of this total were recorded as being collected and recycled appropriately. However, almost 83% of e-waste was not recorded, which means it was probably recycled, burnt, exchanged, or disposed of in a way that did not protect the environment. In the nations of the European Union, an estimated 0.6 million metric tonnes of e-waste were disposed of in trash cans. The global e-waste volume in 2019 primarily consisted of Small equipment (17.4 Mt), Large equipment (13.1 Mt), and Temperature exchange equipment (10.8 Mt). Screens and monitors, Small IT and telecommunication equipment, and Lamps constituted smaller proportions of the e-waste generated in 2019, accounting for 6.7 Mt, 4.7 Mt, and 0.9 Mt, respectively. Since 2014, the e-waste categories experiencing the most significant growth (in terms of total weight of e-waste generated) are Temperature exchange equipment (with an annual average increase of 7%), Large equipment (+5%), and Lamps and Small equipment (+4%). This trend is primarily driven by the increased consumption of these products in lower-income countries, where these items contribute to improved living standards and the same explained in Figure 4. In contrast, Small IT and telecommunication equipment have been growing at a slower rate, while Screens and monitors have exhibited a slight decline (-1%). This decline can be attributed to the replacement of heavy CRT monitors and screens with lighter flat panel displays, resulting in a reduction in total weight despite an increase in the number of units in circulation (Forti et al, 2020).

REGION WISE E-WASTE GENERATED

In 2019, the majority of e-waste was generated in Asia, amounting to 24.9 Mt, while Europe had the highest per capita e-waste generation, averaging 16.2 kg per capita. Europe also boasted the highest documented formal e-waste collection and recycling rate, standing at 42.5%. Conversely, all other continents reported significantly lower quantities of e-waste that was formally collected and recycled compared to the estimated e-waste generation. Figure 5 explains the e-waste generated region wise.

According to e-waste monitor report, in 2019, Asia ranked second, accounting for 11.7% of global e-waste generation, followed by the Americas and Oceania at 9.4% and 8.8%, respectively. Africa had the lowest contribution, with just 0.9% of the total. It's worth noting that e-waste statistics can vary significantly among regions due to various factors such as income levels, existing policies, and the

Figure 4.

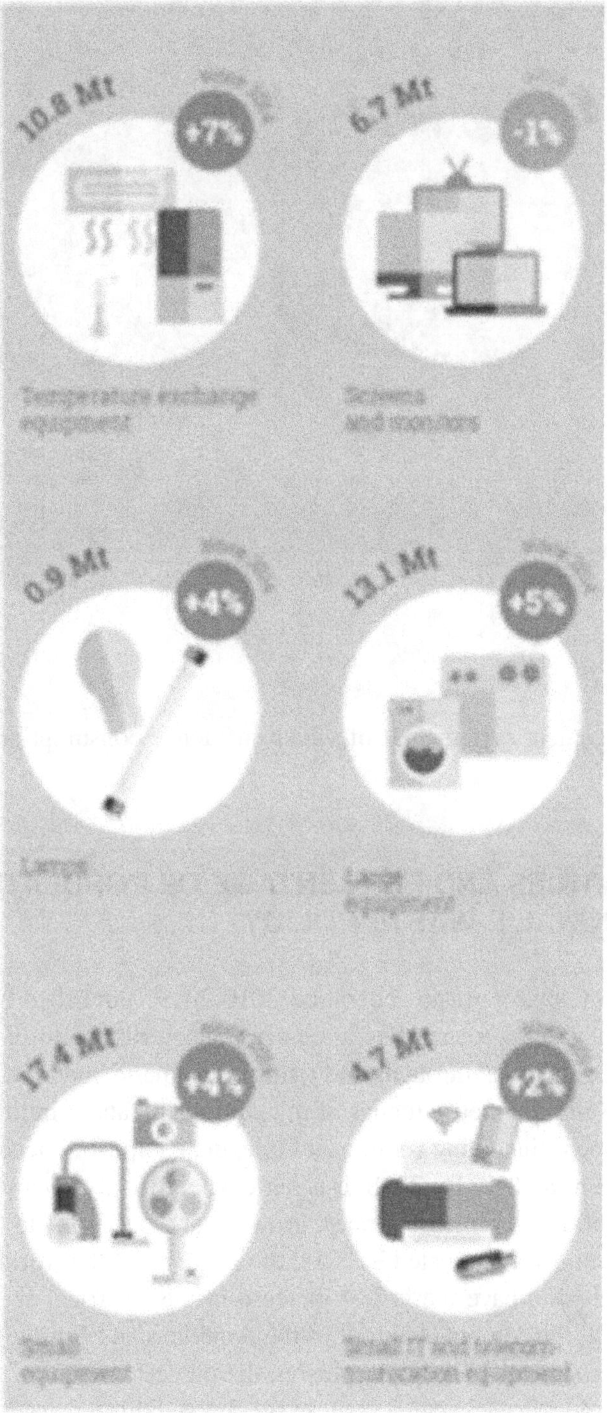
10.8 Mt
+7%
Temperature exchange equipment
6.7 Mt
-1%
Screens and monitors
0.9 Mt
+4%
Lamps
13.1 Mt
+5%
Large equipment
17.4 Mt
+4%
Small equipment
4.7 Mt
+2%

Figure 5.

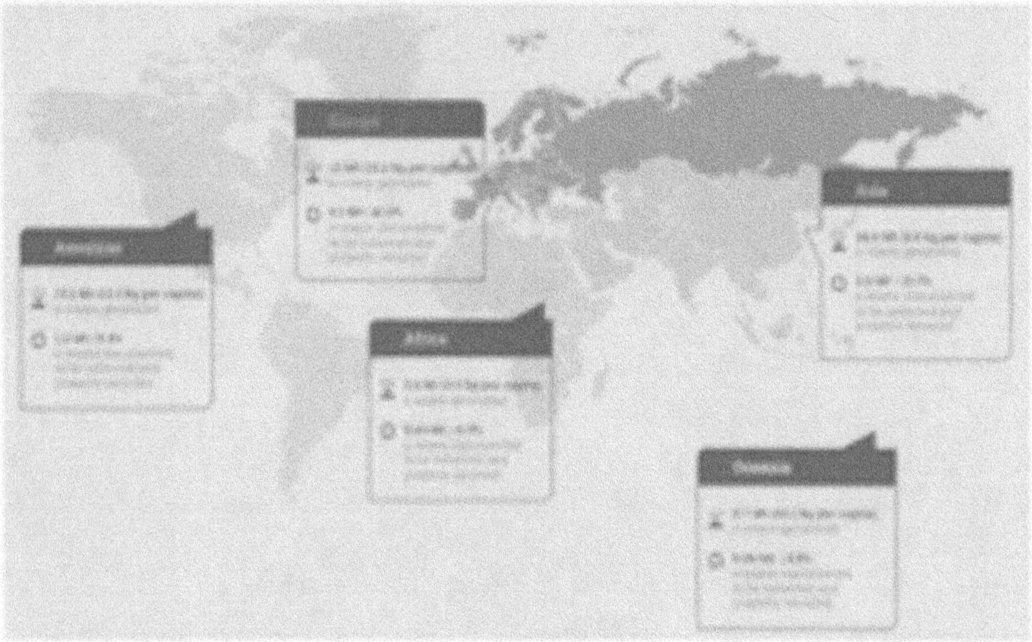

structure of waste management systems, all of which influence consumption and disposal behaviors (ewastemonitor.info, 2020).

NUMBER OF COUNTRIES AND PERCENTAGE OF POPULATION COVERED BY NATIONAL E-WASTE POLICY

According to the report “Global e-waste generation 2010-2019” published by Ian Tiseo (2023), The period in which electronics have emerged as a significant contributor to the waste stream has been characterised by technological breakthroughs and rising consumer demand. 12.8 million metric tonnes of small equipment, 11.8 million metric tonnes of big equipment, and 7 million metric tonnes of temperature exchange equipment (including cooling and freezing equipment) accounted for the majority of the world’s electronic trash in 2014. By 2018, it is anticipated that the overall volume of e-waste will have increased to about 50 million metric tonnes, growing at a pace of 4 to 5 percent annually. With 16 million metric tonnes of e-waste created in 2014, the majority of the world’s e-waste was produced in Asia, the nation expected to have the fastest increase in the electrical and electronics sector from 2014 to 2016. By contrast, the combined output of the Americas and Europe was around 11.6 million metric tonnes. That being said, at 15.6 kg per person, the biggest amount of e-waste per resident was in Europe. The Americas and Oceania came in close second and third, respectively, with 12.2 and 15.2 kilogrammes per person.

In October 2019, approximately 71% of the global population fell under the umbrella of a national e-waste policy, legislation, or regulation. This marks significant progress compared to 2014 when a mere 44% of the population enjoyed such coverage. Notably, the higher coverage rate is influenced by the

Figure 6.

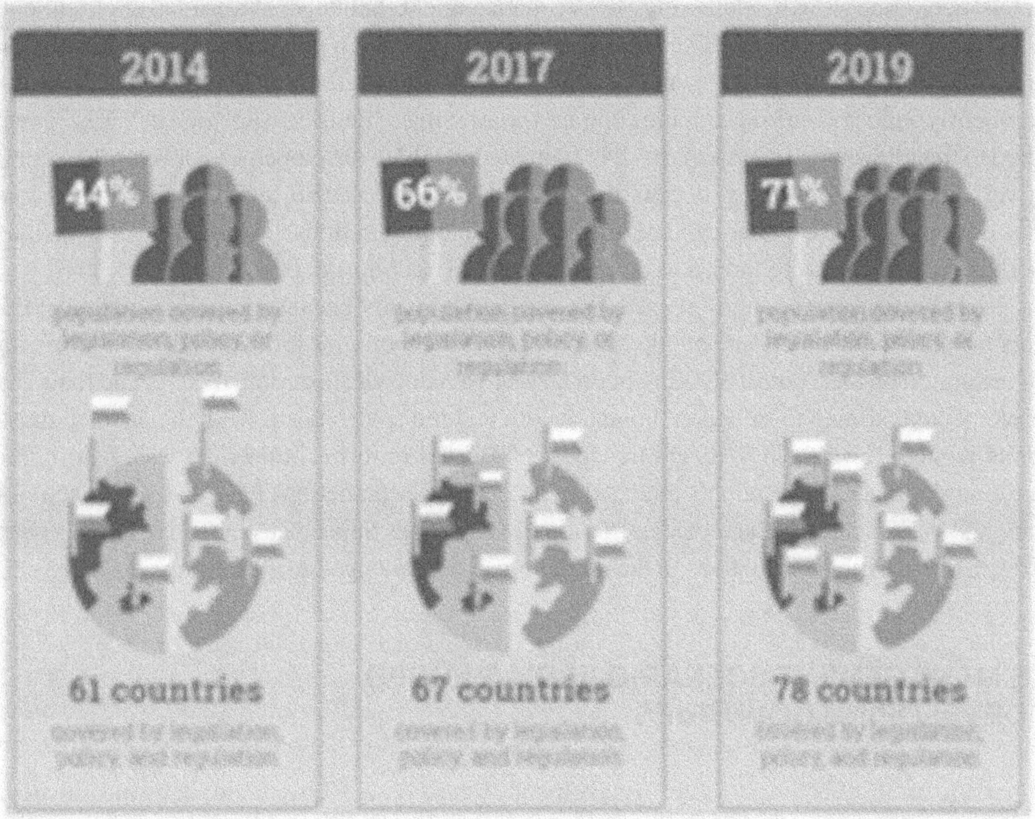

presence of national legal frameworks in the most populous countries, such as China and India. However, it's essential to recognize that this coverage extends to only 78 out of the 193 countries worldwide. Consequently, less than half of all nations currently benefit from the presence of e-waste policies, legislation, or regulations (Shahabuddin et al, 2023). Figure 6 explains the number of countries and percentage of population covered by national e-waste policy.

As far as India is concerned, only a meager 5% of e-waste is currently being recycled due to the absence of suitable infrastructure and the frailty of policies and institutional frameworks. This glaring shortfall in recycling results in a depletion of natural resources and heightened environmental degradation, which adversely affects those engaged in the recycling industry. Within India, for instance, 60% of the country's total e-waste emanates from 65 cities, with the highest contributors being Maharashtra, Tamil Nadu, and Andhra Pradesh (Manish & Chakraborty, 2019). Furthermore, e-waste poses significant health and environmental hazards due to the presence of toxic additives and hazardous substances such as mercury (which damages the human brain) and chlorofluorocarbons (CFCs). It is imperative to manage e-waste in an environmentally sound manner. Achieving effective waste management requires the design of an enhanced collection and recycling system aimed at accelerating waste recovery and treatment (Abalansa et al, 2021).

While several countries have collaborated to adopt national e-waste policies focusing on legislation and regulation of e-waste recycling, progress remains slow in numerous regions, marked by inadequate collection and management strategies. In developing nations, although improved living standards result from robust economic growth, addressing e-waste management challenges as a critical environmental and health issue has not yet been extensively considered, missing out on potential opportunities for product recovery. WEEE is predominantly generated by OECD (Organization for Economic Cooperation and Development) countries, with developing countries witnessing a significant upsurge in WEEE generation compared to their developed counterparts. It is estimated that both developed and developing nations will collectively dispose of an average of 550 million metric tons of WEEE by 2030. Developed countries typically deal with their e-waste by exporting it to developing nations, specifically China, India, and Africa, thereby posing severe threats to their ecosystems (weee-forum.org, 2023).

This cross-border movement of e-waste has escalated, mainly because importers and unauthorized recyclers face no levied duties and employ inappropriate technologies for disposal. To address this, the Basel Convention on the Control of Trans-boundary Movements of Hazardous Wastes and their Disposal was implemented, aiming to safeguard human health and the environment from the adverse impacts of hazardous wastes. Nevertheless, illegal transport of WEEE in forms other than waste has continued, undermining the Basel Convention's intended purpose. European nations have also taken action by introducing the WEEE directive and the Restriction of Hazardous Substances Directive (RoHS Directive, 2002/95/EC) as additional initiatives to tackle the e-waste challenge (eur-lex.europa.eu, 2023).

WEEE MANAGEMENT SCENARIO IN DEVELOPED AND DEVELOPING COUNTRIES

In the pursuit of an effective e-waste management system on a global scale, several critical factors come into play. These factors include waste collection, handling, recycling or recovery, and disposal, as outlined in the E-waste management rule of 2016. However, developing countries face significant challenges in managing the vast volumes of e-waste due to issues such as informal recycling practices and the absence of robust e-waste legislation. Additionally, low- and middle-income countries have a significant presence of informal sectors, which cannot be underestimated in this context (Mohammed, 2022). Figure 7 provides an overview of the current practices within the e-waste management system.

Despite slower technological advancement, developing countries have observed a faster consumption rate of Electrical and Electronic Equipment (EEE), resulting in a higher generation of e-waste, which is estimated to be double that of developed countries (Garlapati 2016). For example, by 2030, it is projected that developing countries will discard between 400-700 million dead computer components, compared to 200–300 million in developed countries. The technological lag and inadequate regulatory frameworks in developing nations amplify the challenges of e-waste management when compared to their developed counterparts. While e-waste management in developed nations receives more attention with respect to policy development and structural frameworks, developing countries face significant hurdles (Thakur & Kumar, 2022).

In comparing legislative policies between developed and developing countries, the former generally have stricter regulations and more effective enforcement mechanisms in place. Nonetheless, many developing countries are taking the initiative to formulate and implement specific e-waste management policies. Developed nations often employ national registry systems, encompassing proper collection and robust logistics systems (Sthiannopkao and Wong 2013). For instance, Germany has implemented programs like the Packaging Waste Program and Extended Producer Responsibility (EPR) to place

Figure 7.

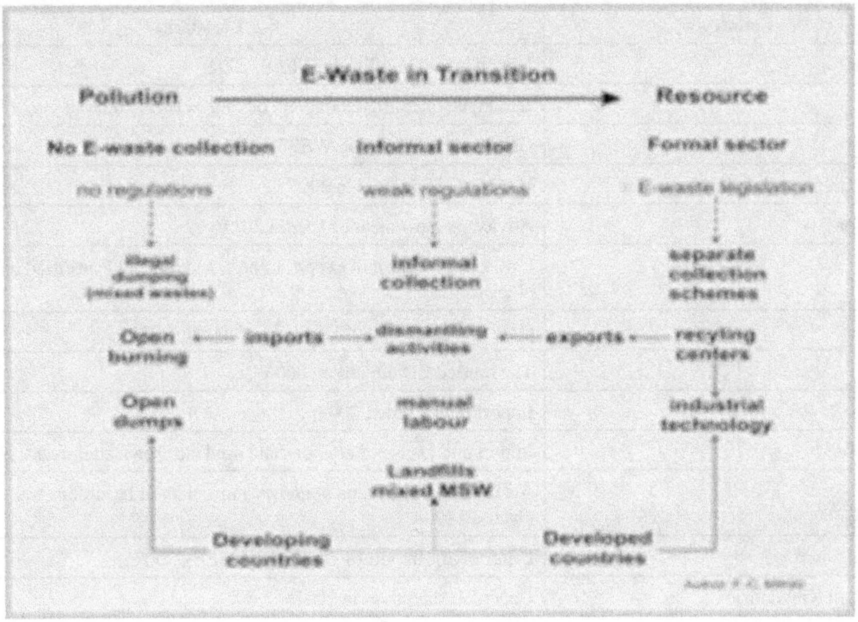

financial responsibilities on manufacturers for collecting and reducing packaging waste (rekart.co.in, 2023). The EPR concept has been adopted by countries such as Sweden, Switzerland, Norway, and Taiwan. Another significant EU legislation, the RoHS Directive, restricts the use of hazardous substances in e-wastes (Directive 2002/95/EC), endorsing their collection and recycling (Directive 2002/96/EC). Meanwhile, countries like China and India, despite the absence of a national registry, have implemented EPR legislation systems to monitor EEE production and establish manufacturer take-back systems.

One major contributor to the e-waste challenge in developing countries is the presence of thriving gray markets, where second-hand EEE is readily available, exacerbating the situation. Despite lower labor costs in countries like China, India, and Pakistan, the substantial quantities of e-waste (both imported and domestic) and the dominance of informal parties have been identified as major obstacles to effective e-waste management (Srivastava & Pathak, 2020).

Global e-waste recycling policies often rely on consumer behavior and the local market. However, research indicates that a significant portion of e-waste is generated by economically developed countries. Different regions have varying degrees of e-waste regulations and recycling infrastructure. Developed countries like Canada, Europe (with all 27 countries having their own regulations for WEEE recycling), and parts of the United States (where approximately half of the states have e-waste recycling laws) have well-established recycling and processing industries adhering to industry standards. In Latin America, several countries such as Colombia, Mexico, Costa Rica, Peru, and Brazil have e-waste policies in place, while Africa generally lacks comprehensive e-waste recycling regulations, leading to around 85% of surplus electronic imports being reclaimed rather than discarded (Shahabuddin et al, 2023). Egypt, a significant e-waste producer in Africa, lacks a robust e-waste management system. In Australia, the National Television and Computer Recycling Scheme utilize a combination of government directives and industry engagement for e-waste assembly and recycling. In Asia, South Korea, Japan, China, Taiwan, and India have e-waste policies, with Taiwan leading in e-waste recycling at approximately 82%,

Table 2. Legislative framework for WEEE management in different countries

Sl.No	Country	Legislation
1	Belgium	Directive 2002/96/EC on WEEE, 2002
2	China	Catalogue of restricted imports of solid wastes, 2008
3	Finland	Government Decree on WEEE, 2004
4	France, Germany and the Netherland	Under EU Directives in 2005
5	Hong Kong	Advice on movement of UEEE, 2011
6	Japan	Law for the control of export, import, and others of specified hazardous and other wastes
7	Nigeria	Guide for importers of UEEE into Nigeria, 2011
8	Norway	The revised EU directives, 2006
9	Pakistan	Import policy order, 2009
10	Singapore	Import and export of e-wastes and used electronic equipment, 2008
11	South Korea	Act on control of Trans-boundary movement of hazardous waste and their disposal, 1994
12	Thailand	Criterion for import of used EEE (UEEE), 2007
13	The United Kingdom	Under EU directives in 2007
14	The United States	HR 2284: Responsible electronics recycling act, 2011
15	Vietnam	Law on environmental protection, 2005

Source: (Barapatre & Rastogi, 2022)

followed by Japan and South Korea recycling about 75% of their e-waste. Although China has formal e-waste management regulations, it continues to work on expanding its recycling capacity and remains both the primary destination for e-waste dumping and the largest producer and consumer of electronics worldwide (Ahirwar & Tripathi, 2021). E-waste management protocols in many other countries worldwide are still in the developmental stages. Table 2 provides an overview of the legislative framework for WEEE management in different countries.

Forti, Baldé & Kuehr (2018) extensively details the operational practices employed by countries worldwide in the collection, preprocessing, and recycling of e-waste. Their study reveals that developed economies exhibit greater innovation in this realm compared to their developing counterparts. However, there remains substantial room for improvement in both categories. One noteworthy trend is that developed countries often export their e-waste to developing nations for disposal. This practice is driven by several factors, including the relatively high cost of labor and stringent environmental regulations governing hazardous waste disposal in developed nations (Abalansa et al, 2021). Despite the inadequacy of appropriate technologies, facilities, and resources for efficient e-waste recycling in developing countries, they continue to import e-waste from developed nations for short-term gains. These recyclers rely on rudimentary methods to extract precious materials from e-waste. However, a few countries, such as Tunisia, have begun efforts to formulate more effective approaches to e-waste management and recycling. A significant volume of e-waste is exported from European countries, with a substantial portion finding its way to Asia and Africa under the false pretense of being sent for reuse and refurbishment. This mislabeling often includes malfunctioning items being unlawfully classified as "used goods," resulting in significant harm to the environment and local populations (Sakhuja, 2022).

STRATEGIC APPROACHES FOR WEEE

Amidst growing concerns about recycling, the strategic approach of the circular economy (CE) is gaining recognition for its potential to alleviate environmental stress and promote economic balance (Akram et al. 2019). The waste management sector is more important than ever in this age of increased environmental awareness. Waste management's future presents a vibrant frontier full of opportunities for investment. With a current worth of an astounding US$987.1 billion, the business is poised to reach a far higher valuation of $1.4 trillion by 2030. Gaining a deep and comprehensive grasp of this market's moving parts is essential to capitalising on its growth (Research and Markets, 2023). The concept of CE has been carefully chosen in response to the relentless rise in electronic products and the recycling challenges posed by these end-of-life (EoL) products, as illustrated in Figure 8.

This strategic approach holds particular relevance for developing countries like India, which are emerging as potent economic systems. Implementing EoL product recycling practices can contribute significantly to sustainable development in such regions (Slaveykova et al. 2019) Sustainable development efforts in this context entail a focus on critical issues such as waste management, pollution reduction, e-waste management strategies, environmental sustainability, and protection. Among these, the "environmental management system" (EMS) stands out as a pivotal driving tool. EMS places emphasis on the creation of eco-friendly products, the establishment of stringent regulations, the cultivation of a green image, and support for producers in the effective implementation of CE principles. This approach empowers stakeholders and policymakers to reduce the environmental burden and build an efficient e-waste management system, with a specific focus on key enablers such as EMS and environmental partners (Krishnamurthy et al. 2019).

Figure 8.

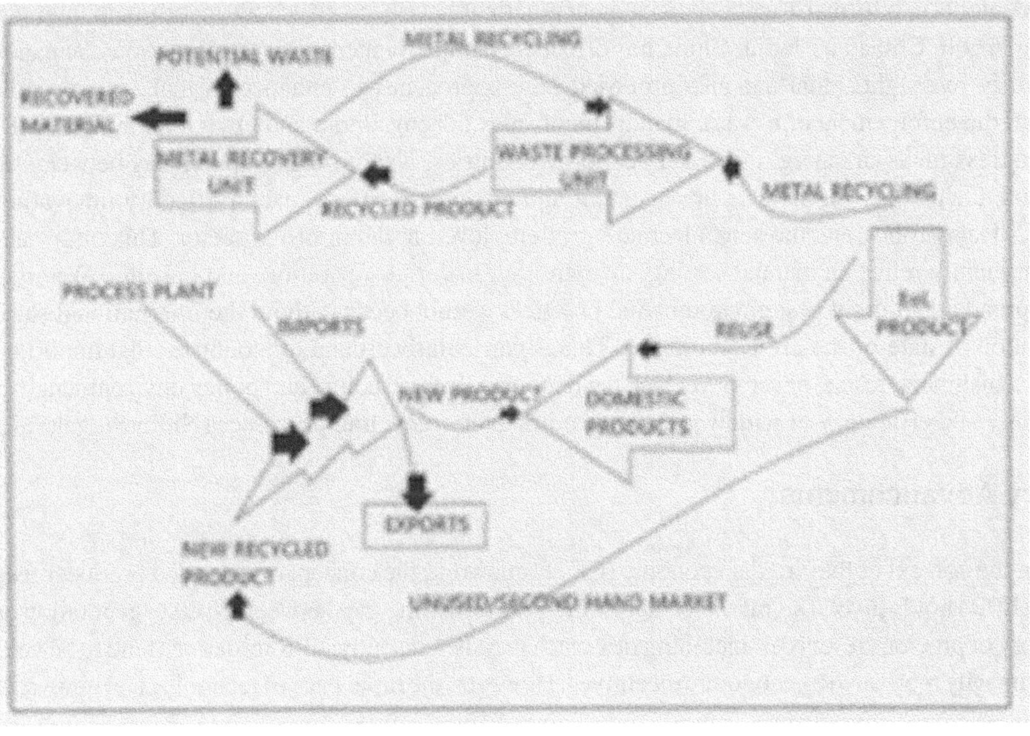

It is becoming increasingly evident that top-down regulatory approaches are insufficient for the international management of e-waste, given its high-value content that often undergoes inefficient and unsafe rudimentary recycling in emerging market economies. To address this challenge, it is essential to involve the private sector, including manufacturers, retailers, and labor investors, in the development of innovative solutions. These solutions should bring the extensive informal e-waste management sector under guidelines that safeguard environmental quality and human health while supporting workers whose livelihoods depend on discarded e-waste. In this context, we explore opportunities to create or adapt cost-effective innovations, such as Apple's Daisy robot recycler, for artisanal e-waste processing within the informal sectors (Blomsma and Brennan 2017). Thus there are strategies has been proposed with the aim to integrate e-waste management into a global circular economy (CE) agenda. This entails combining technical innovation for e-waste processing with financial incentives through collaborative efforts involving multiple agencies (Awasthi et al, 2019). The objective is to enhance rudimentary e-waste management practices in regions where waste accumulates, often at the expense of laborers' well-being and environmental quality due to the toxic components of electronic products.

Right-Scaling Technology

The expense associated with establishing and operating supply chains and equipment for effective e-waste management can be prohibitive due to stringent policies and regulations designed to protect the environment and human health. Consequently, rudimentary approaches to e-waste processing have proliferated, shifting the costs of human health impacts and environmental degradation to society. For instance, studies have shown alarming levels of lead in the blood of children residing near informal e-waste processing sites, such as Guiyu, China (Ogunseitan et al, 2009).

The informal rudimentary e-waste sector, comprised of small groups involved in collecting, transporting, and recycling, mainly emerged in Asia and Africa to meet unmet demands. However, recent advancements in informal e-waste management practices have been driven by government interventions, particularly in China and India. China has effectively implemented e-waste regulations, expanding its regulatory oversight. India has also proposed new approaches to enhance formal e-waste collection through the enforcement of e-waste management rules (Zheng Jinran and Qiu Quanlin, 2016). Despite the progress made in emerging market economy countries, a significant gap persists between the best available technology for e-waste management and resource recovery, exemplified by innovations like Apple's Daisy robot, and the actual technologies employed in the informal sector. This informal sector predominantly relies on manual sorting, dismantling, material separation, and burning. We argue that widespread adoption of best environmental practices cannot occur without the international sharing or adaptation of state-of-the-art technologies. This is particularly crucial for countries that import e-waste for refurbishment, resale, or scrap, yet lack the necessary infrastructure and policy environment to ensure the safety and efficiency of widely practiced rudimentary recycling processes (Nidhi Singal, 2023).

Policy Advancements

Within the context of the circular economy (CE) framework, the concept of extended producer responsibility (EPR) holds pivotal significance in bridging the widening gap between e-waste generation and the recovery of precious resources, including rare earth metals. Presently, EPR initiatives tend to be voluntary and primarily motivated by economic incentives. However, the rapid pace of technological innovation and the trend towards miniaturization in electronic products suggest that economic incentives alone may not suffice to justify EPR in the long term. To effectively manage various categories of e-waste, including those devoid of economically valuable materials but containing toxic components, mandatory EPR is essential. This approach aims not only to facilitate resource recovery but also to prevent environmental pollution and human health hazards resulting from toxic elements (Balde et al., 2015).

Implementing mandatory EPR will require policies that ensure equitable participation, particularly to prevent undue burdens on small manufacturers. To achieve this, a collective "superfund" mechanism can be established, whereby all electronic product manufacturers contribute financially. Manufacturers can also be incentivized by compensating them for adopting green chemistry or eco-design principles that eliminate the use of hazardous materials throughout their product life cycles. The EPR superfund should encompass the end-of-life recovery and material reuse aspects, with manufacturers contributing funds in proportion to their product sales. An independent agency would be responsible for designing and executing strategies for the collection, sorting, and recycling of obsolete electronic products on a global scale (D'Adamo et al, 2019).

This CE model is more attainable for larger electronic products that are easily traceable and for which widespread technical knowledge for repair or refurbishment is lacking in emerging market economies. For these product categories, preventing unsafe and unregulated rudimentary recycling is a priority for integration into the proposed EPR superfund and CE framework. However, for high-volume electronic products like mobile phones, laptops, and televisions, which dominate the e-waste stream, alternative e-waste management models are required. These models should involve local entrepreneurial stakeholders and employ minimally restrictive regulatory and policy instruments. The objective is to formalize rudimentary e-waste processing, integrating it into CE models with new technological tools and regulatory policies that ensure safe operations without compromising worker health or environmental integrity (Ibn-Mohammed et al, 2019). A noteworthy example is Guiyu, China, a formerly prominent e-waste management town. It transformed from housing around 5,000 informal sector operations to hosting 29 integrated formal companies in an industrial park. While this progress in environmental protection and occupational conditions is commendable, manual labor remains inefficient, and the lack of technical and financial infrastructure for EPR hinders the sustainable development of e-waste management CE in the region (Ogunseitan et al, 2009).

Economic Incentive Initiatives

Effective retrieval of old and obsolete electronic products from consumers poses a significant challenge in the realm of proper e-waste management worldwide. However, promising solutions are emerging, particularly in China, where web-based applications such as the Baidu and Recycling Brother (Huishouge) apps have gained popularity. These apps connect consumers seeking to dispose of e-waste with certified recycling and dismantling services. The Baidu app, in particular, was launched in collaboration with the United Nations Development Programme (UNDP) Asia–Pacific Innovation Fund (unep.org, 2017). There is growing interest in forming collaborative alliances involving various stakeholders, including electronics manufacturers and financial institutions, to establish comprehensive online e-waste management processes. This collaboration aims to integrate the experiences gained from informal e-waste recycling collectives in Chinese industrial parks with microfinance-driven economic incentives in India. This integration could potentially lead to the creation of self-help group (SHG) parks, operating under the oversight of environmental regulatory authorities (mca.gov.in, 2021).

In India, banks have taken the initiative to provide initial capital investments to promote sustainable rural development practices, particularly in rudimentary e-waste management. For instance, the National Bank for Agriculture and Rural Developments and the Small Industries Development Bank of India has collaborated to establish an SHG bank linkage scheme. This scheme offers short-term lending facilities to specific industrial sectors, especially micro, small, and medium enterprises (MSMEs). These micro-financial strategies are regulated and supervised by the Reserve Bank of India, and they also extend small-scale funding support to non-governmental organizations that promote SHG formation (Balde et al., 2015). To scale up this approach and support e-waste management at the national level in emerging economies worldwide, a similar scheme could be implemented. Another example is based on corporate social responsibility (CSR) funding for designated industrial sectors. Since 2014, the Indian Ministry of Corporate Affairs has mandated that industries allocate 2% of their average net profit from the previous three financial years to CSR activities. This framework could be adapted to transform e-waste management in transitioning economies (Turaga et al, 2019).

India has already established SHGs for various forms of community development, providing economic opportunities for low-income workers. An illustrative example is a solid waste collection and handling cooperative that transitioned from rudimentary procedures to a formal partnership with the municipality of Pune. This cooperative, comprising approximately 2,300 members, delivers formal collection and recycling services to about 400,000 households in Pune. Initially, local government subsidies covered administrative and equipment-related costs. More recently, the cooperative invested in resources such as a waste recycling storage site, protective equipment, and worker salaries (Awasthi, Li, Koh, & Ogunseitan, 2019). While these models offer promise, further research is necessary to assess their adaptability to processing hazardous solid waste like e-waste, which may contain toxic materials and potentially explosive components such as rechargeable batteries. Additionally, addressing concerns related to digital data and personal information security on discarded smartphones and computers is crucial in such contexts (WHO, 2021). Other examples of economic incentive initiatives in transitioning economies include the World Bank's pilot project on e-waste management in Egypt and Ethiopia. This project, conducted in collaboration with the Egyptian government, emphasizes economic incentives and employs a collection strategy involving non-governmental organizations and the private sector in e-waste management. Further research is required to compare these experiences and identify best practices that can be shared among countries grappling with mounting e-waste challenges, all while contributing to a sustainable circular economy on a global scale without compromising local employment opportunities (Balde et al., 2015).

RECOMMENDATIONS FOR ENHANCED INTEGRATION

In light of the rapid advancements in electronic product design and the widespread proliferation of mobile devices on a global scale, addressing the issue of e-waste, which represents the fastest-growing category of solid hazardous waste with international transport implications, demands innovative and cross-border management strategies. Particularly in countries undergoing economic transitions, various models for financial and technical support in e-waste management have emerged. Here, we put forth recommendations aimed at effectively integrating technical support and financial incentives to revolutionize rudimentary e-waste recycling and resource recovery across different scopes. To provide a comprehensive framework for such integration, we can draw upon the United Nations Sustainable Development Goals (SDGs), specifically Goals 8 and 12. These SDGs offer valuable guidance for exploring the multifaceted dimensions of collaboration required to transform informal e-waste management. Aligning with the principles of a Circular Economy (CE), these SDGs encourage policies that have the potential to generate employment through partnerships between public and private enterprises, diversify labor-intensive sectors with high added value, protect labor rights, and promote safe and secure working conditions for laborers while minimizing environmental and health impacts (unstats.un.org, 2023).

Recognizing the critical role of the information and communication technology sector, primarily driven by electronic products, in sustainable development, the United Nations Environment Programme (UNEP) highlights both the challenges posed by electronic waste and it's potential as a secondary resource material capable of meeting raw material demand for manufacturing new products, provided economic considerations justify it (Awasthi, Li, Koh, & Ogunseitan, 2019). In order to strengthen programmatic collaboration with national governments, United Nations agencies responsible for labor market integrity, economic development, environmental protection, and monitoring health risk factors can leverage their advanced knowledge and expertise in electronic product life-cycle stewardship. They can also facilitate the allocation of financial resources to address emerging e-waste challenges in many developing countries. Moreover, these agencies can support web-based platforms for knowledge-sharing related to e-waste (unstats.un.org, 2023).

Drawing from a past initiative by the UNDP Asia–Pacific Innovation Fund, which introduced an app-based approach that showed promise in China but faced scalability limitations, we propose adapting and tailoring similar initiatives for countries like Ghana and Nigeria. These initiatives should be locally customized, ensuring sustainability through engagement and investments by local or regional financial institutions, with support from local governments to provide security and infrastructure guarantees. Collaborative efforts with other UN agencies, including the World Health Organization, Environmental Programme, and International Labour Organization, will be essential to ensure the coherence of policies and actions across various sectors for effective global e-waste management (Awasthi, Li, Koh, & Ogunseitan, 2019). The multi-dimensional approach, guided by SDGs, CE principles, and UN agency collaboration, can pave the way for improved e-waste management and sustainable development, with a focus on emerging economies transitioning towards more effective practices.

E-WASTE AND THEIR IMPACT

According to the UNEP report on "Compendium of Technologies for the Recovery of Materials from WEEE / E-waste" One of the main effects of rising consumption is the rapid rise in garbage output. According to World Bank projections, 2.2 billion tonnes of municipal solid trash would be produced annually by 4.3 billion urban dwellers by 20251. The kind and makeup of trash have changed significantly throughout time as a result of to altered consumer behaviour and consumption. Consequently, many waste streams are developing, leading to issues with the sustainable management of local and national governments. Given that waste management has a significant impact on the environment, public health, and the negative effects of inappropriate garbage disposal management are well-documented and quite serious. In an era marked by rapid urbanization, electronic waste, or e-waste, has emerged as a paramount concern in the modernized world. This issue continually introduces toxic and hazardous elements, including lead, mercury, calcium, polybrominated biphenyls, and chromium, into our environment (Chen, et al., 2018). The primary apprehension surrounding e-waste is rooted in its composition of hazardous substances, posing threats to public health and environmental well-being. Improper handling of such waste leads to detrimental consequences, as perilous substances are released. E-waste comprises numerous toxic components that can inflict severe harm on the environment, human health, and animal well-being if not properly disposed of in an environmentally responsible manner (Yong, Lim & Ilankoon, 2019). Some of the adverse effects of the chemicals commonly found in e-waste on human health are outlined below:

Brominated Flame Retardants (BFRs): These chemicals have been routinely added to consumer products for decades to reduce fire-related injuries and property damage. Concern has arisen due to the occurrence of various classes of BFRs in the environment and humans. BFRs resist decomposition in the environment and can lead to impaired memory function, learning difficulties, and behavioral problems in babies born to exposed pregnant women. They interfere with estrogen and thyroid functioning (Kumar, 2021).

Lead: Present in most computer monitors and televisions, lead exposure, especially in children under six, can result in intellectual impairment, damage to the reproductive and nervous systems, and blood disorders. Lead poisoning can have multiple adverse health effects, particularly on the neurological development of children (Borthakur & Govind, 2017).

Cadmium: This element primarily targets the kidneys in both the general population and occupationally exposed individuals. Cadmium can accumulate in the human kidney for an extended period, causing kidney damage and affecting the respiratory system. Found in rechargeable batteries for laptops and other electronic devices, cadmium is highly toxic, especially to the kidneys and bones (Borthakur & Govind, 2017).

Mercury: Elemental and methyl mercury are toxic to the central and peripheral nervous systems. Inhalation of mercury vapor can harm the nervous, digestive, immune, respiratory, and renal systems and may even be fatal. Inorganic mercury salts are corrosive to the skin, eyes, and gastrointestinal tract, potentially inducing kidney toxicity. Mercury exposure can lead to neurological and behavioral disorders, including tremors, insomnia, memory loss, neuromuscular effects, headaches, and cognitive and motor dysfunction. It is used in lighting devices in flat-screen monitors and televisions and can affect breast milk (Kumar, 2021).

***Hexavalent Chromium Compounds*:** Hexavalent chromium, a known carcinogen, is employed in various industries, including electronics, for creating metal housings. Exposure to it can cause lung cancer and irritation or damage to the respiratory tract, eyes, throat, and skin (Kumar, 2021).

Plastic Compounds: Polyvinyl chloride (PVC) cabling is commonly used in printed circuit boards, connectors, plastic covers, and cables. When PVCs are incinerated or landfilled, they release dioxins that have harmful effects on human reproductive and immune systems (Kumar, 2021). E-waste contamination extends to air, soil, and water pollution, a stark reality observed during the recycling or landfilling of these hazardous and toxic materials. Hazardous organic compounds like PCBs (polychlorinated biphenyls), PBDEs (polybrominated diphenyl ethers), and PCDD/Fs (polychlorinated dibenzo-p-dioxins and dibenzofurans) contribute to air and water contamination, disrupting the food chain and adversely affecting human health (Borthakur and Singh, 2017).

Informal recycling exacerbates the e-waste crisis in developing countries such as India, Ghana, Nigeria, China, Thailand, Indonesia, Vietnam, Pakistan, and Bangladesh. Illicit and unregulated practices, including the unauthorized melting of circuit boards, open burning of wire piles, and disposal of metal-bearing acidic solutions, have become distressingly common (Borthakur and Singh, 2017). Dismantling, another inappropriate e-waste disposal method, often occurs under poor conditions, resulting in severe pollution and harm to public health. Generated waste is recklessly disposed of through open burning and dumping, representing the least desirable alternatives in countries lacking proper legislation and essential waste management infrastructure (Shittu, Williams & Shaw, 2021).

These practices typically involve the manual disassembly of e-waste using rudimentary tools such as hammers, chisels, and screwdrivers. Precious metals like gold are extracted by stripping them in open-pit acid baths, while copper recovery involves burning electric cables in open pits. Plastics are shredded and liquefied at high temperatures without adequate ventilation, releasing toxins like dioxins and furans. These rudimentary techniques pose threats to both environmental protection and worker health (Borthakur and Singh, 2017). The improper dumping of e-waste is also observed in developed countries, often attributed to inadequate implementation of environmental laws and WEEE management regulations by authorities.

All electronic devices contain printed circuit boards, which house toxic components like lead, brominates flame retardants, and antimony oxide. The heating of printed circuit boards (PCBs) for component removal is a common practice, releasing hazardous materials such as lead, mercury, and hexavalent chromium. These substances are found in cathode-ray tubes (CRTs), capacitors, mercury switches, batteries, LCDs, cartridges, and electrolytes. Cadmium and lead are prevalent in computer batteries and circuit boards, while cathode-ray tubes contain cadmium and lead oxide, and mercury is present in switches and flat-screen monitors (Borthakur and Singh, 2017). Polychlorinated biphenyls (PCBs) are identified in transformers and capacitors. These harmful substances contribute to various health disorders, including those related to growth, reproduction, thyroid function, lung health, and cellular abnormalities. Those

Table 3. Impacts associated with e-waste exposure

Pollutants or Toxins	Occurrences	Available Places
Heavy metals	Widespread	Found in many electronic components, including batteries, circuit boards, and cathode ray tubes
Polybrominated Diphenyl Ethers (PBDEs)	Common	Used as flame retardants in electronic devices, such as televisions, computers, and smart phones
Polychlorinated Biphenyls (PCBs)	Less common	Once used in a variety of electronic devices, but now banned in most countries
Polycyclic Aromatic Hydrocarbons (PAHs)	Common	Formed when organic matter is burned, such as in the open burning of e-waste
Hexavalent Chromium	Less common	Used in some electronic components, such as computer monitors and fluorescent light bulbs
Mercury	Widespread	Used in fluorescent light bulbs, thermostats, and some medical devices
Lead	Widespread	Used in batteries, solder, and some electronic components
Cadmium	Less common	Used in batteries, nickel-cadmium batteries, and some electronic components
Barium	Less common	Used in cathode ray tubes and some electronic components
Lithium	Common	Used in rechargeable batteries, such as those found in laptops and smart phones
Nickel	Common	Found in Blends, cells, switches, transistors, colorants
Selenium	Less common	Photovoltaic Cells, Colorants, Photocopiers, Fax Devices
Silver	Common	Condensers, switches, cells, resistors
Zinc	Widespread	Iron, copper-zinc alloy, mixed metals, single-use and rechargeable cells, luminescent materials

Source: (Kumar, 2021)

most affected are individuals directly exposed to these pollutants, including pregnant women, infants, and children. In light of these concerns, addressing the e-waste challenge is imperative for safeguarding both environmental sustainability and public health (Vanapalli et al, 2021).

One of the most concerning aspects of e-waste exposure is its impact on the development of the nervous system, particularly in children living in e-waste recycling areas. These children are exposed to extraordinarily high levels of hazardous substances over their lifetimes. Despite their smaller body weight compared to adults, the volume of toxicants they inhale far exceeds their body weight. These hazardous substances also have detrimental effects on various organ systems. The dangerous health impacts associated with e-waste exposure are summarized in Table 3.

EXPANDING PRODUCER RESPONSIBILITY (EPR) FOR ADVANCING A CIRCULAR ECONOMY

European nations have invested significant effort in reducing e-waste and its environmental impact. To achieve this goal, they have engaged manufacturers and expanded their responsibilities under an "extended producer responsibility (EPR) system." The Organization for Economic Co-operation and Development (OECD) defines EPR as an environmental approach where producers take on increased responsibility for their products at the post-consumer stage of their lifecycle (OECD 2001). EPR aims to incentivize manufacturers to consider environmental factors in their product design, shifting the responsibility for e-waste management from municipalities to producers. Manufacturers are tasked with handling recycling, reuse, and the final disposal of waste components, often by incorporating treatment and disposal costs into the product's selling price. These efforts also encourage producers to offer high-quality products in the global market.

The EPR system is a comprehensive policy framework that utilizes various tools such as advance recycling fees (ARF), waste collection charges, landfill bans, subsidies, and more. The central idea behind EPR for electronics is the acknowledgment that e-waste collection and management incur a net cost that must be addressed to prevent irresponsible and careless handling of these hazardous materials. EPR seeks to engage original equipment manufacturers (OEMs) by setting collection targets for e-waste accumulation and recovery, thereby preventing it from ending up in landfills. To meet these targets, manufacturers collaborate with compliance organizations in exchange for EPR fees (Shan, Xiao & Yang, 2021).However, India's diverse social structure, with a population ranging from lower to middle-income levels, has contributed to the growth of a gray market for used electrical and electronic equipment (UEEE), even after a product's end of life.

In Europe, the EPR concept has been applied to Waste Electrical and Electronic Equipment (WEEE) since 2003, following the first WEEE Directive (2002/96/EC) and its subsequent revision (the WEEE Recast Directive 2012/19/EU) in 2012. In December 2015, the European Commission launched the EU strategy for a circular economy, recognizing that EPR protocols play a vital role in efficient waste management systems and those incentives should be provided to producers for taking responsibility for recycling and reusing their products during the design phase. This circular economy concept aligns with the principles of eco-industrial development and aims to create a "win-win" scenario where economic growth and environmental protection can coexist. However, the transition from a linear economic model of "take, make, and dispose" to a circular one presents significant challenges (McDowall et al. 2017).

E-waste is considered a crucial resource within the circular economy framework. Circular economy seeks to reduce the consumption of new materials and minimize waste generation by establishing sustainable resource flow loops. It not only addresses resource scarcity and waste generation but also offers economic opportunities through recycling and recovering valuable materials. Circular economy encourages the reuse and recycling of e-waste, making it a promising sector for future development. To achieve this, environmental, health, economic, and technological considerations must be incorporated into the system (McDowall et al. 2017).

Implementing a single guiding principle may not be sufficient to achieve multiple environmental objectives, such as landfill diversion and reducing product toxicity. A Pigovian tax on market activities that pose threats to social order and well-being can effectively reduce landfills and promote the use of recycled materials in manufacturing processes, although illegal dumping remains a concern (Calcott and Walls, 2000). Alternatively, a deposit-refund system can address this issue, where a tax imposed on products is refunded when the product is returned for recycling. However, this system comes with implementation costs and complexities, leading to command-and-control strategies by governments. These strategies involve government-mandated reductions in pollution levels through the imposition of collection targets under EPR-based take-back regulations, with government oversight (Palmer and Walls 1999).

Recently, e-waste reverse logistics (RL) has gained momentum, involving the efficient and profitable management of raw materials, work-in-progress inventory, finished goods, and information in a reverse flow from consumption points back to the origin. The goal is to recover value and ensure appropriate disposal. Rapid technological advancements and market competition have shortened product lifecycles and accelerated product obsolescence, highlighting the need to incorporate RL into management practices (Calcott and Walls, 2000). The growth in e-waste, stricter environmental regulations, and increased consumer demand for corporate social responsibility have driven the adoption of reverse logistics. While various RL models have been developed in developed countries, developing nations are still in the early stages of implementation. In economies like India and China, challenges include inadequate legislation, limited economic incentives, low public awareness, the prevalence of unbranded products, and informal waste pickers who collect and treat e-waste without regard for health, safety, or environmental protection (Islam et al, 2020).

Several gaps hinder the effective implementation of reverse logistics. Producers often hesitate to invest in RL due to the associated infrastructure costs for collection, recycling, and reuse of e-waste. Difficulty in managing relationships with wholesalers, retailers, consumers, and recycling organizations also contributes to corporate reluctance toward RL (Kumar, 2019). Inadequate legislation can render RL processes ineffective. For example, in the European Directive, a shared responsibility model is used to define that producers are responsible for WEEE treatment, but the collection process is not clearly outlined. As a result, retailers may accept returned WEEE but refuse to cover transportation costs to producers or municipal collection sites. In some countries, the lack of economic incentives has been identified as a barrier to achieving set targets (Vidal, 2023). To overcome these challenges, both developed and developing countries are continually working to design more effective RL management models.

CHALLENGES IN THE MANAGEMENT OF ELECTRONIC WASTE RECYCLING

Countries worldwide are typically categorized as either developed or developing, and two distinct classification criteria are employed, resulting in different country groupings. The Basel Convention uses the Annex VII (developed countries) and non-Annex VII (developing) countries classification. This convention is a crucial international trade agreement designed to regulate the cross-border movement of hazardous waste (Yang, 2020). In contrast, the World Trade Organization (WTO) also classifies countries as developed or developing but does not provide a specific definition for this classification. While the Basel Convention designates the developed countries group to include the OECD, the European Union (EU), and Liechtenstein (Zhang, 2021), the WTO's classification includes the EU, North America (excluding Mexico), the European Free Trade Association (Iceland, Liechtenstein, Norway, and Switzerland), Japan, New Zealand, and Australia as developed countries. Developing countries, according to the WTO, encompass Africa, Mexico, South and Central America, the Caribbean, Europe (excluding the EU and EFTA), the Middle East, and Asia (excluding Australia, Japan, and New Zealand) (Shittu, Williams, & Shaw, 2021). These varying country classifications create significant challenges in the international trade of e-waste.

In practice, many developed economies act as e-waste exporters, while developing countries import these wastes. According to the Basel Convention Report, most hazardous e-waste is exchanged between developed countries, with minimal trade occurring with developing economies. However, the WTO claims that most developed countries send their e-waste to developing countries. India, China, and various African nations have become major destinations for the global dumping of WEEE. India grapples with this issue due to the growth of the UEEE market, the rise of the lower-middle class economy, the expansion of the upper-middle class economy through rapid industrialization, and the proliferation of the informal recycling sector (Cheshmeh et al, 2023). China, on the other hand, was one of the world's largest importers of e-waste but completely banned e-waste imports in 2000, with no reported imports of waste printed boards (WPBs) after that year, although certain countries reported exporting WPBs to China during the same period (Barapatre & Rastogi, 2022).

Due to technological, infrastructural, and financial constraints, developing economies continue to struggle with adopting recommended e-waste recycling and disposal techniques. A practical solution to this problem can be found in the "best-of-2-worlds" (Bo2W) theory (Barapatre & Rastogi, 2022). This theory suggests establishing collection and screening centers based on population size and adopting sustainable practices such as eco-friendly product designs, extended producer responsibility (EPR), product stewardship, recycling, and remanufacturing (Payne, 2022). The increasing global demand for WEEE production, coupled with its diverse composition, raises concerns about exposure to hazardous substances resulting from improper e-waste handling. This underscores the importance of adopting sustainable management practices, such as EPR, but there is still room for improvement in transforming these practices into a circular economy. To achieve this goal, it is crucial to identify and address the existing challenges.

DISCUSSIONS AND IMPLICATIONS

To develop effective management strategies for WEEE, a comprehensive understanding of its composition is essential, given the constant evolution of electronic devices driven by technological advancements.

Currently, there is a lack of information regarding material concentrations in specific equipment and limited data available on the components within these devices. Additionally, the absence of universally applicable standard methods for assessing elemental concentrations in WEEE complicates the analytical process. To address these challenges, the creation of an official database containing detailed information on WEEE components and standardized analytical protocols is necessary. Such a database would facilitate the characterization and comparison of results.

To establish a circular economy and close the loop, stringent policies and suitable infrastructure are required. These approaches should be tailored to the specific needs of different regions. While countries like the United States, Canada, India, and various European states have explicit legislation and adhere to EPR principles, there is room for improvement in their collection rates. One way to enhance these rates is by implementing competitive collection systems involving both public and private sectors (Compagnoni, 2022). Conversely, regions lacking specific regulations and handling capacities face even greater challenges due to the increasing production of WEEE. Addressing this issue necessitates the formulation of effective policies, taking into account region-specific conditions in each country.

The absence of adequate legislation often results in the emergence of informal WEEE management sectors that engage in illegal e-waste trading, involving both imported and domestic WEEE. This problem can be mitigated by banning WEEE imports and expanding formal collection rates. To achieve a circular economy, improvements in recycling methods are imperative, as they serve as the foundation of sustainable resource extraction from WEEE, thus alleviating environmental burdens. The recycling industry requires innovation and enhancement since current processes involve pretreatment, mechanical separation, and refining methods to obtain desired materials. Not all facilities possess the technical expertise required for efficient refining techniques, which hampers the recycling industry's growth. The intricate composition of WEEE renders conventional recycling methods inefficient, necessitating alternative approaches (Vaccari et al, 2019).

The circular economy concept emphasizes maximizing resource extraction from WEEE while minimizing environmental impacts. Although many hazardous substances have been phased out, heavy metals and organic pollutants persist. Restricting the use of hazardous materials in electronics can enhance recycling efficiency. Bio-based electronic components offer an alternative to eliminate the hazards associated with WEEE recycling (Cardamone, Ardolino & Arena, 2021). Furthermore, electronic product design should consider ease of disassembly during recycling and recovery processes, promoting reuse in line with circular economy principles. Environmental compatibility, profitability, and the efficient replacement of natural resources should guide recycling processes. The incorporation of bio-metallurgical techniques as alternatives to pyro-metallurgy and hydrometallurgy for refining is essential (Vaccari et al, 2019). Although biosorption and bioleaching are proven techniques, their limitations in terms of loading capacity and reaction time must be addressed.

Hazardous substances contained within WEEE gradually leach into the environment, causing contamination that adversely affects the health of humans and animals. This concern is particularly pronounced when linked to informal waste recycling processes. Workers engaged in such rudimentary and unsupervised practices face various occupational hazards. Additionally, populations residing in the vicinity are exposed to hazardous pollutant concentrations that contaminate the air, water, and land. Cesaro et al. (2018) proposed an approach to assess the potential relative impacts of different types of WEEE based on their metal content, chosen as target contaminants. This approach relies on prioritization criteria for WEEE management, with one of the main challenges being the computation of risks associated with potential contamination scenarios. To address this issue effectively, a comprehensive inventory of classified organic and inorganic pollutants resulting from the informal treatment of WEEE and the associated hazards can be compiled to evaluate risks to human health. Such an inventory would be valuable in formulating effective guidelines and recycling practices. Dealing with this issue is particularly challenging, especially when it comes to classifying WEEE based on material composition. The use of innovative materials in Electrical and Electronic Equipment (EEE) due to technological advancements makes it difficult to assess the potential long-term impacts of these diverse materials on health and the environment (Shaikh, Thomas, Zuhair, & Magalini, 2020).

Procedures followed by developed economies cannot be directly transplanted into underdeveloped or developing economies due to differences in socioeconomic conditions in various regions. An integrated approach that combines formal and informal systems supported by legal processes can offer a viable and constructive alternative. This can be achieved through a division of labor where the informal sector handles the collection and manual dismantling of valuable WEEE components, while the formal sector takes responsibility for metal refining and residue disposal. This approach would help reduce the on-site generation of WEEE and limit its entry into the informal sector. However, the feasibility of this approach hinges on the implementation of stringent policies regarding the illegal importation of WEEE, and developed countries play a crucial role in this regard (Vaccari et al, 2019).

The management of end-of-life products has become increasingly challenging for developing countries due to the surge in the quantity of e-waste being imported and the inadequacy of e-waste legislation. Furthermore, the general public often lacks awareness of the toxicity levels of hazardous e-waste and is compelled to work in unhealthy conditions during its recycling, as the government fails to provide safety measures in recycling facilities. This dire situation forces individuals to choose between poor working conditions and unemployment. Insufficient technical expertise, infrastructure, managed landfill sites, equipment, and recycling facilities compound the problem, exacerbated by the absence of well-regulated collection and preprocessing systems. Consequently, only a small fraction of the total volume of available e-waste is handled by collecting and disposal companies. Illicit dumping and incineration of this hazardous e-waste alongside other types of solid waste pose a severe threat to both human health and the environment.

CONCLUSION

The management of e-waste is a global concern and presents a significant challenge, particularly for developing countries that receive substantial amounts of this waste. To address this issue effectively, it is crucial to establish an appropriate legislative framework and policies tailored specifically for developed countries. Over the past few decades, numerous policies have been enacted, all of which have been rooted in the circular economy model. To enhance the recovery performance of e-waste, these legislations must be accompanied by investments in training campaigns and capacity building. Innovations in current processing techniques are needed to efficiently recover metals from the intricate e-waste stream. Additionally, there is a need for accurate databases and mapping of e-waste collection and treatment centers. To improve global e-waste management, developed countries should demonstrate a stronger commitment to technology development and the expansion of new recycling facilities. Simultaneously, developing countries should focus on implementing stringent legislation and enhancing the collection of WEEE (Waste Electrical and Electronic Equipment) to expand their recycling capacity. States with lower e-waste production can collaborate to establish shared treatment and recycling facilities.

E-waste recycling holds significant potential for urban mining, allowing the recovery of precious metals, which are found in much higher concentrations in e-waste compared to their natural ores. Circular economy principles and sustainability have already captured the attention of researchers, managers, and policymakers. Now, it's essential for the general public to recognize the moral obligation to transform the linear model into a circular one. An ideal version of a circular economy would be based on utilizing resources extracted from WEEEs to meet SDGs. A significant improvement in the e-waste management system, aligned with the principles of the circular economy, can be expected in the near future through the collaborative efforts of both formal and informal sectors in adopting integrated recycling practices.

REFERENCES

Abalansa, S., El Mahrad, B., Icely, J., & Newton, A. (2021). Electronic waste, an environmental problem exported to developing countries: The GOOD, the BAD and the UGLY. *Sustainability (Basel)*, *13*(9), 5302. doi:10.3390u13095302

Ahirwar, R., & Tripathi, A. K. (2021). E-waste management: A review of recycling process, environmental and occupational health hazards, and potential solutions. *Environmental Nanotechnology, Monitoring & Management*, *15*, 100409. doi:10.1016/j.enmm.2020.100409

Akram, R., Natasha, Fahad, S., Hashmi, M. Z., Wahid, A., Adnan, M., Mubeen, M., Khan, N., Rehmani, M. I. A., Awais, M., Abbas, M., Shahzad, K., Ahmad, S., Hammad, H. M., & Nasim, W. (2019). Trends of electronic waste pollution and its impact on the global environment and ecosystem. *Environmental Science and Pollution Research International*, *26*(17), 16923–16938. doi:10.100711356-019-04998-2 PMID:31025281

Ali, S., & Shirazi, F. (2023). The Paradigm of Circular Economy and an Effective Electronic Waste Management. *Sustainability (Basel)*, *15*(3), 1998. doi:10.3390u15031998

Alves, B. (2023, Apr 17). *Global flows of e-waste 2019, by scenario.* https://www.statista.com/statistics/1066948/share-of-electronic-waste-disposed-globally/

Awasthi, A. K., Li, J., Koh, L., & Ogunseitan, O. A. (2019). Circular economy and electronic waste. *Nature Electronics*, *2*(3), 86–89. doi:10.103841928-019-0225-2

Axil-is.com. (2023). *What is waste management hierarchy?* Available at: https://axil-is.com/blogs-articles/waste-management-hierarchy/

Barapatre, S., & Rastogi, M. (2022). e-Waste management: A transition towards a circular economy. In Handbook of Solid Waste Management: Sustainability through Circular Economy (pp. 1499-1521). Singapore: Springer Nature Singapore.

Blomsma, F., & Brennan, G. (2017). The emergence of circular economy: A new framing around prolonging resource productivity. *Journal of Industrial Ecology*, *21*(3), 603–614. doi:10.1111/jiec.12603

Cocchiara, C., Dorneanu, S. A., Inguanta, R., Sunseri, C., & Ilea, P. (2019). Dismantling and electrochemical copper recovery from Waste Printed Circuit Boards in H2SO4–CuSO4–NaCl solutions. *Journal of Cleaner Production*, *230*, 170–179. doi:10.1016/j.jclepro.2019.05.112

Compagnoni, M. (2022). Is Extended Producer Responsibility living up to expectations? A systematic literature review focusing on electronic waste. *Journal of Cleaner Production, 367*, 133101. doi:10.1016/j.jclepro.2022.133101

Ekins, P., Domenech Aparisi, T., Drummond, P., Bleischwitz, R., Hughes, N., & Lotti, L. (2020). *The circular economy: What, why, how and where*. Academic Press.

Ellen MacArthur Foundation. (2019). *What is a circular economy?* Available at: https://ellenmacarthurfoundation.org/topics/circular-economy-introduction/overview

ewastemonitor.info. (2020). *The Global E-waste Monitor 2020 – Quantities, flows, and the circular economy potential.* Available at: https://ewastemonitor.info/gem-2020/#:~:text=According%20to%20the%20report%2C%20Asia,Mt%20and%200.7%20Mt%20respectively

Florin-Constatin Mihai and Maria-Grazia Gnoni. (2016). *E-waste Management as a Global Challenge.* Available at: https://www.intechopen.com/chapters/51266

Forti, V., Baldé, C. P., Kuehr, R., & Bel, G. (2020). *The global e-waste monitor 2020. United Nations University (UNU), International Telecommunication Union (ITU) & International Solid Waste Association.* ISWA.

Forti, V., Baldé, K., & Kuehr, R. (2018). *E-waste statistics: Guidelines on classifications, reporting and indicators*. Academic Press.

Garlapati, V. K. (2016). E-waste in India and developed countries: Management, recycling, business and biotechnological initiatives. *Renewable & Sustainable Energy Reviews, 54*, 874–881. doi:10.1016/j.rser.2015.10.106

Goel, A. (2021). Application of circular economy in E-waste management–a review with an Indian perspective. *Strategies Toward Green Deal Implementation*, 70.

Henriques, R., Figueiredo, F., & Nunes, J. (2023). Product-Services for a Resource-Efficient and Circular Economy: An Updated Review. *Sustainability (Basel), 15*(15), 12077. doi:10.3390u151512077

Ilyassova, G., Nukusheva, A., Arenova, L., Karzhassova, G., & Akimzhanova, M. (2021). Prospects of legal regulation in the field of electronic waste management in the context of a circular economy. *International Environmental Agreement: Politics, Law and Economics, 21*(3), 367–388. doi:10.100710784-020-09514-3

Jain, P. (2020). *R's of Waste Management (Evolution).* Available at: https://indieseducation.com/rs-of-waste-management-evolution/

Krishnamurthy, S., Fudurich, G., & Rao, P. (2019). Circular economy for India: perspectives on stewardship principles, waste management, and energy generation. In *Modernization and accountability in the social economy sector* (pp. 272–298). IGI Global. doi:10.4018/978-1-5225-8482-7.ch015

Manish, A., & Chakraborty, P. (2019). E-waste management in India: Challenges and opportunities. *TerraGreen, 12*, 22–28.

Mohammed, S. I. (2022). E-Waste Management in Different Countries: Strategies, Impacts, and Determinants. In *Advances in Green Electronics Technologies*. IntechOpen.

Ngetich, M. J., Kiplagat, A. K., & Ng'etich, J. (2018). An Investigation into Domestic Water Consumption Patterns in Iten Town, Kenya. *Africa Environmental Review Journal, 3*(1), 200.

Ogunseitan, O. A., Schoenung, J. M., Saphores, J. D. M., & Shapiro, A. A. (2009). The electronics revolution: From e-wonderland to e-wasteland. *Science, 326*(5953), 670–671. doi:10.1126cience.1176929 PMID:19900918

Parajuly, K., Fitzpatrick, C., Muldoon, O., & Kuehr, R. (2020). Behavioral change for the circular economy: A review with focus on electronic waste management in the EU. *Resources, Conservation and Recycling. X, 6*, 100035. doi:10.1016/j.rcrx.2020.100035

Report, B. C. (2023, October 20). *E-waste collection: best practices and lessons learned in Indonesia, Trinidad and Tobago and in the European region, from PACE II pilot projects.* https://www.basel.int/Default.aspx?tabid=4210&meetId=4A404336-4967-EE11-90C9-005056A3140F&lang=en

Research and Markets. (2023a, December 20). *E-waste Management - Global Strategic Business Report.* https://www.researchandmarkets.com/report/e-waste#reld0-5774988

Research and Markets. (2023b, January 22). *Global E-Waste Management Market Outlook, 2028.* https://www.researchandmarkets.com/reports/5774988/global-e-waste-management-market-outlook

Sakhuja, D., Ghai, H., Bhatia, R. K., & Bhatt, A. K. (2022). Management of e-Waste: technological challenges and opportunities. Handbook of Solid Waste Management: Sustainability through Circular Economy, 1523-1557.

Shahabuddin, M., Uddin, M. N., Chowdhury, J. I., Ahmed, S. F., Uddin, M. N., Mofijur, M., & Uddin, M. A. (2023). A review of the recent development, challenges, and opportunities of electronic waste (e-waste). *International Journal of Environmental Science and Technology, 20*(4), 4513–4520. doi:10.100713762-022-04274-w

Shaikh, S., Thomas, K., Zuhair, S., & Magalini, F. (2020). A cost-benefit analysis of the downstream impacts of e-waste recycling in Pakistan. *Waste Management (New York, N.Y.), 118*, 302–312. doi:10.1016/j.wasman.2020.08.039 PMID:32919349

Slaveykova, V. I., Couture, P., Duquesne, S., D'Hugues, P., & Sánchez, W. (2019). Recycling, reuse, and circular economy: A challenge for ecotoxicological research. *Environmental Science and Pollution Research International, 26*(21), 1–4. doi:10.100711356-019-04626-z PMID:31104246

Smart Product Manager. (2023). *Tesla's Electric Revolution: Disrupting the Auto Industry with Innovation*. Available at: https://smartproductmanager.medium.com/teslas-electric-revolution-disrupting-the-auto-industry-with-innovation-8ed879f002eb

Srivastava, R. R., & Pathak, P. (2020). Policy issues for efficient management of E-waste in developing countries. In *Handbook of Electronic Waste Management* (pp. 81–99). Butterworth-Heinemann. doi:10.1016/B978-0-12-817030-4.00002-4

Statista.com. (2023). *Global E-Waste - Statistics & Facts*. Available at: https://www.statista.com/topics/3409/electronic-waste-worldwide/#topicOverview

Sthiannopkao, S., & Wong, M. H. (2013). Handling e-waste in developed and developing countries: Initiatives, practices, and consequences. *The Science of the Total Environment*, *463*, 1147–1153. doi:10.1016/j.scitotenv.2012.06.088 PMID:22858354

Thakur, P., & Kumar, S. (2022). Evaluation of e-waste status, management strategies, and legislations. *International Journal of Environmental Science and Technology*, *19*(7), 6957–6966. doi:10.100713762-021-03383-2

Theworldcounts.com. (2023). *Electronic Waste Facts*. Available at: https://www.theworldcounts.com/stories/electronic-waste-facts

Tiseo, I. (2023, July 18). *Global e-waste generation 2010-2019*. https://www.statista.com/statistics/499891/projection-ewaste-generation-worldwide/

Turaga, R. M. R., Bhaskar, K., Sinha, S., Hinchliffe, D., Hemkhaus, M., Arora, R., & Sharma, H. (2019). E-waste management in India: Issues and strategies. *Vikalpa*, *44*(3), 127–162. doi:10.1177/0256090919880655

UNE. (2023). *Sustainable Future of E-waste*. https://www.unep.org/ietc/news/story/sustainable-future-e-waste#:~:text=E%2Dwaste%20is%20becoming%20one,the%20amount%20produced%20in%202014

Vaccari, M., Vinti, G., Cesaro, A., Belgiorno, V., Salhofer, S., Dias, M. I., & Jandric, A. (2019). WEEE treatment in developing countries: Environmental pollution and health consequences - An overview. *International Journal of Environmental Research and Public Health*, *16*(9), 1595. doi:10.3390/ijerph16091595 PMID:31067685

Wang, J., Chen, S., Zeng, X., Huang, J., Liang, Q., Shu, J., & Sun, Z. (2021). Recovery of high purity copper from waste printed circuit boards of mobile phones by slurry electrolysis with ammonia-ammonium system. *Separation and Purification Technology*, *275*, 119180. doi:10.1016/j.seppur.2021.119180

Weee-forum.org. (2023). *International E-Waste Day: 57.4M Tonnes Expected in 2021*. Available at: https://weee-forum.org/ws_news/international-e-waste-day-2021/

Wikipedia. (2023). *United States Environmental Protection Agency*. Available at: https://en.wikipedia.org/wiki/United_States_Environmental_Protection_Agency

Compilation of References

Abalansa, S., El Mahrad, B., Icely, J., & Newton, A. (2021). Electronic waste, an environmental problem exported to developing countries: The GOOD, the BAD and the UGLY. *Sustainability (Basel), 13*(9), 5302. doi:10.3390u13095302

Abanades, S., Abbaspour, H., Ahmadi, A., Das, B., Ehyaei, M. A., Esmaeilion, F., El Haj Assad, M., Hajilounezhad, T., Jamali, D. H., Hmida, A., & Ozgoli, H. A. (2021). A critical review of biogas production and usage with legislations framework across the globe. *International Journal of Environmental Science and Technology*, 1–24. PMID:34025745

Abdelbasir, S. M., Hassan, S. S., Kamel, A. H., & El-Nasr, R. S. (2018). Status of electronic waste recycling techniques: A review. *Environmental Science and Pollution Research International, 25*(17), 16533–16547. doi:10.100711356-018-2136-6 PMID:29737485

Abdussalam-Mohammed, W. (2019). Review of therapeutic applications of nanotechnology in medicine field and its side effects. *Journal of Chemical Reviews, 1*(3), 243–251. doi:10.33945/SAMI/JCR.2019.3.5

Aboelmaged, M. (2021). E-waste recycling behavior: An integration of recycling habits into the theory of planned behavior. *Journal of Cleaner Production, 278*, 124182. doi:10.1016/j.jclepro.2020.124182

Adanu, S., Francis, G., & Attah, M. (2020). Challenges of adopting sustainable technologies in e-waste management at Agbogbloshie, Ghana. *Heliyon, 6*(8), e04548. doi:10.1016/j.heliyon.2020.e04548 PMID:32802976

Agbeshie, A. A., Adjei, R., Anokye, J., & Banunle, A. (2020). Municipal waste dumpsite: Impact on soil properties and heavy metal concentrations, Sunyani, Ghana. *Scientific African, 8*, e00390. doi:10.1016/j.sciaf.2020.e00390

Ahirwar, R. (n.d.). *E-waste management: A review of recycling process, environmental and occupational health hazards, and potential solutions.* Elsevier. Retrieved November 20, 2023, from https://www.sciencedirect.com/science/article/pii/S2215153220303925

Ahirwar, R., & Tripathi, A. K. (2021). E-waste management: A review of recycling process, environmental and occupational health hazards, and potential solutions. *Environmental Nanotechnology, Monitoring & Management, 15*, 100409.

Ahirwar, R., & Tripathi, A. K. (2021). E-waste management: A review of recycling process, environmental and occupational health hazards, and potential solutions. *Environmental Nanotechnology, Monitoring & Management, 15*, 100409. doi:10.1016/j.enmm.2020.100409

Ahlers, J., Hemkhaus, M., Hibler, S., & Hannak, J. (2021). *Analysis of Extended Producer Responsibility Schemes.* Academic Press.

Ahluwalia, P. K., & Nema, A. K. (2007). A life cycle based multi-objective optimization model for the management of computer waste. *Resources, Conservation and Recycling, 51*(4), 792–826. doi:10.1016/j.resconrec.2007.01.001

Ahmad, S., Akram, M., Husain, D., Ahmad, A., Sharma, M., Prakash, R., & Ahmed, M. (2023). Ecological Footprint Assessment of e-Waste Recycling. In *Environmental Assessment of Recycled Waste* (pp. 67–83). Springer Nature Singapore. doi:10.1007/978-981-19-8323-8_5

Ahmad, S., Jha, S., Eljialy, A. E. M., & Khan, S. (2021). A Systematic Review on e-Wastage Frameworks. *International Journal of Advanced Computer Science and Applications*, *12*(12), 701–709. doi:10.14569/IJACSA.2021.0121287

Airehrour, D., Cherrington, M., Madanian, S., & Singh, J. (2019). Reducing ICT carbon footprints through adoption of green computing. Academy of Economic Studies in Bucharest. Department of Economic Informatics. doi:10.12948/ie2019.04.17

Akram, R. (2019). *Trends of electronic waste pollution and its impact on the global environment and ecosystem.* Springer. doi:10.1007/s11356-019-04998-2

Alabi, A. O., & Bakare, A. A. (2017). Genetic damage induced by electronic waste leachates and contaminated underground water in two prokaryotic systems. *Toxicology Mechanisms and Methods*, *27*(9), 657–665. doi:10.1080/15376516.2017.1349228 PMID:28670935

Alam, M. M. (2021). Voice-based information access services for low-literate users in rural Bangladesh. *Information Technologies and International Development*, *17*, 1–21.

Alam, S., Bhardwaj, L. K., Mallick, R., & Rai, S. (2023). Estimation of Heavy Metals and Fluoride Ion in Vegetables Grown Nearby the Stretch of River Yamuna, Delhi (NCR), India. *Indian Journal of Environmental Protection*, *43*(1), 64–73.

Al-Fuqaha, A., Guizani, M., Mohammadi, M., Aledhari, M., & Ayyash, M. (2015). Internet of Things: A Survey on Enabling Technologies, Protocols, and Applications. *IEEE Communications Surveys and Tutorials*, *17*(4), 2347–2376. doi:10.1109/COMST.2015.2444095

Alghazo, J., & Ouda, O. K. (2016, October). Electronic waste management and security in GCC countries: A growing challenge. In *Proceedings of the ICIEM International Conference, Sousse, Tunisia* (pp. 27-30). Academic Press.

Ali, A., Malik, S. A., Shafiullah, M., Malik, M. Z., & Zahir, M. H. (2023). Policies and regulations for solar photovoltaic end-of-life waste management: Insights from China and the USA. *Chemosphere*, *340*, 139840. doi:10.1016/j.chemosphere.2023.139840 PMID:37597624

Ali, S., & Shirazi, F. (2023). The Paradigm of Circular Economy and an Effective Electronic Waste Management. *Sustainability (Basel)*, *15*(3), 1998. doi:10.3390u15031998

Almulhim, A. I. (2022). Household's awareness and participation in sustainable electronic waste management practices in Saudi Arabia. *Ain Shams Engineering Journal*, *13*(4), 101729. doi:10.1016/j.asej.2022.101729

Al-Obaidi, K. M., Hossain, M., Alduais, N. A. M., Al-Duais, H. S., Omrany, H., & Ghaffarianhoseini, A. (2022). A review of using IoT for energy efficient buildings and cities: A built environment perspective. *Energies*, *15*(16), 5991. doi:10.3390/en15165991

AlQattan, N., Acheampong, M., Jaward, F. M., Ertem, F. C., Vijayakumar, N., & Bello, T. (2018). Reviewing the potential of Waste-to-Energy (WTE) technologies for Sustainable Development Goal (SDG) numbers seven and eleven. *Renewable Energy Focus*, *27*, 97–110. doi:10.1016/j.ref.2018.09.005

Alsheyab, M. A. (2015). Potential recovery of precious metals from waste laptops in Jordan. *Rare Metals*, *34*(7), 517–521. doi:10.100712598-014-0337-9

Alves, B. (2023, Apr 17). *Global flows of e-waste 2019, by scenario.* https://www.statista.com/statistics/1066948/share-of-electronic-waste-disposed-globally/

Amankwah-Amoah, J. (2016). Global business and emerging economies: Towards a new perspective on the effects of e-waste. *Technological Forecasting and Social Change, 105*, 20–26. doi:10.1016/j.techfore.2016.01.026

Ambitionbox.com. (2022). *462 Chemicals Companies in India | AmbitionBox*. AmbitionBox. Available at: https://www.ambitionbox.com/chemicals-companies-in-india

Amri, F., Ben Zaied, Y., & Ben Lahouel, B. (2019). ICT, total factor productivity, and carbon dioxide emissions in Tunisia. *Technological Forecasting and Social Change, 146*, 212–217. doi:10.1016/j.techfore.2019.05.028

Anand, A. (2022). *Tackling e-waste is the need of the hour and here is how you can chip in.* CNBCTV18.

Ananno, A. A., Masud, M. H., Dabnichki, P., Mahjabeen, M., & Chowdhury, S. A. (2021). Survey and analysis of consumers' behaviour for electronic waste management in Bangladesh. *Journal of Environmental Management, 282*, 111943. doi:10.1016/j.jenvman.2021.111943 PMID:33465720

Andeobu, L., Wibowo, S., & Grandhi, S. (2021). An assessment of e-waste generation and environmental management of selected countries in Africa, Europe and North America: A systematic review. *The Science of the Total Environment, 792*, 148078. doi:10.1016/j.scitotenv.2021.148078 PMID:34147806

Andeobu, L., Wibowo, S., & Grandhi, S. (2021, August 27). A Systematic Review of E-Waste Generation and Environmental Management of Asia Pacific Countries. *International Journal of Environmental Research and Public Health, 18*(17), 9051. doi:10.3390/ijerph18179051 PMID:34501640

Anh Khoa, T., Phuc, C. H., Lam, P. D., Nhu, L. M. B., Trong, N. M., Phuong, N. T. H., Van Dung, N., Tan-Y, N., Nguyen, H. N., & Duc, D. N. M. (2020). Waste management system using IoT-based machine learning in university. *Wireless Communications and Mobile Computing, 2020*, 1–13. doi:10.1155/2020/6138637

Annamalai, J. (2015). Occupational health hazards related to informal recycling of E-waste in India: An overview. *Indian Journal of Occupational and Environmental Medicine, 19*(1), 61. doi:10.4103/0019-5278.157013 PMID:26023273

Anwar, S., Ghaffar, M., Razzaq, F., & Bibi, B. (2018). E-waste reduction via virtualization in green computing. *American Scientific Research Journal for Engineering, Technology, and Sciences, 41*(1), 1–11.

Appolloni, A., D'Adamo, I., Gastaldi, M., Santibanez-Gonzalez, E. D., & Settembre-Blundo, D. (2021). Growing e-waste management risk awareness points towards new recycling scenarios: The view of the Big Four's youngest consultants. *Environmental Technology & Innovation, 23*, 101716. doi:10.1016/j.eti.2021.101716

Ardente, F., Latunussa, C. E., & Blengini, G. A. (2019). Resource efficient recovery of critical and precious metals from waste silicon PV panel recycling. *Waste Management (New York, N.Y.), 91*, 156–167. doi:10.1016/j.wasman.2019.04.059 PMID:31203937

Arya, S., & Kumar, S. (2020). E-waste in India at a glance: Current trends, regulations, challenges and management strategies. *Journal of Cleaner Production, 271*, 122707. doi:10.1016/j.jclepro.2020.122707

Asante, K. A., Adu-Kumi, S., Nakahiro, K., Takahashi, S., Isobe, T., Sudaryanto, A., Devanathan, G., Clarke, E., Ansa-Asare, O. D., Dapaah-Siakwan, S., & Tanabe, S. (2011). Human exposure to PCBs, PBDEs and HBCDs in Ghana: Temporal variation, sources of exposure and estimation of daily intakes by infants. *Environment International, 37*(5), 921–928. doi:10.1016/j.envint.2011.03.011 PMID:21470682

Asante, K. A., Agusa, T., Biney, C. A., Agyekum, W. A., Bello, M., Otsuka, M., Itai, T., Takahashi, S., & Tanabe, S. (2012). Multi-trace element levels and arsenic speciation in urine of e-waste recycling workers from Agbogbloshie, Accra in Ghana. *The Science of the Total Environment, 424*, 63–73. doi:10.1016/j.scitotenv.2012.02.072 PMID:22446112

Asibey, M. O., Lykke, A. M., & King, R. S. (2022). Understanding the factors for increased informal electronic waste recycling in Kumasi, Ghana. *International Journal of Environmental Health Research, 32*(2), 305–320. doi:10.1080/09603123.2020.1755016 PMID:32393050

Asokan, V. A., Teah, H. Y., Kawazu, E., & Hotta, Y. (2023). Ambitious EV policy expedites the e-waste and socio-environmental impacts in India. *Resources, Conservation and Recycling, 190*, 106829. doi:10.1016/j.resconrec.2022.106829

Atia, T. A., Elia, G., Hahn, R., Altimari, P., & Pagnanelli, F. (2019). Closed-loop hydrometallurgical treatment of end-of-life lithium ion batteries: Towards zero-waste process and metal recycling in advanced batteries. *Journal of Energy Chemistry, 35*, 220–227. doi:10.1016/j.jechem.2019.03.022

Atzori, L., Iera, A., & Morabito, G. (2010). The Internet of Things: A survey. *Computer Networks, 54*(15), 2787–2805. doi:10.1016/j.comnet.2010.05.010

Avis, W. (2021). *Drivers, barriers, and opportunities of e-waste management in Africa*. Academic Press.

Awasthi, A. K., Awasthi, M. K., Mishra, S., Sarsaiya, S., & Pandey, A. K. (2022). Evaluation of E-waste materials linked potential consequences to environment in India. *Environmental Technology & Innovation, 28*, 102477. doi:10.1016/j.eti.2022.102477

Awasthi, A. K., Iacovidou, E., Awasthi, M. K., Johnson, M., Parajuly, K., Zhao, M., Mishra, S., & Pandey, A. K. (2023). Assessing Strategic Management of E-Waste in Developing Countries. *Sustainability (Basel), 15*(9), 7263. doi:10.3390u15097263

Awasthi, A. K., Li, J., Koh, L., & Ogunseitan, O. A. (2019). Circular economy and electronic waste. *Nature Electronics, 2*(3), 86–89. doi:10.103841928-019-0225-2

Awasthi, A. K., Wang, M., Awasthi, M. K., Wang, Z., & Li, J. (2018). Environmental pollution and human body burden from improper recycling of e-waste in China: A short-review. *Environmental Pollution, 243*, 1310–1316. doi:10.1016/j.envpol.2018.08.037 PMID:30268981

Awasthi, A. K., Zeng, X., & Li, J. (2016). Comparative examining and analysis of e-waste recycling in typical developing and developed countries. *Procedia Environmental Sciences, 35*, 676–680. doi:10.1016/j.proenv.2016.07.065

Axil-is.com. (2023). *What is waste management hierarchy?* Available at: https://axil-is.com/blogs-articles/waste-management-hierarchy/

Azevedo, L. P., Araújo, F. G. D. S., Lagarinhos, C. A. F., Tenório, J. A. S., & Espinosa, D. C. (2019). Resource recovery from e-waste for environmental sustainability: a case study in Brazil. In *Electronic waste management and treatment technology* (pp. 175–200). Butterworth-Heinemann. doi:10.1016/B978-0-12-816190-6.00008-X

Babatunde, D. E., Denwigwe, I. H., Babatunde, O. M., Gbadamosi, S. L., Babalola, I. P., & Agboola, O. (2019). Environmental and societal impact of nanotechnology. *IEEE Access : Practical Innovations, Open Solutions, 8*, 4640–4667. doi:10.1109/ACCESS.2019.2961513

Babu, S. C.V. (2023). Introduction to IoT and its Applications. Anniyappa Publications.

Bagwan, W. A. (2024). Electronic waste (E-waste) generation and management scenario of India, and ARIMA forecasting of E-waste processing capacity of Maharashtra state till 2030. *Waste Management Bulletin, 1*(4), 41–51. doi:10.1016/j.wmb.2023.08.002

Bakhiyi, B., Gravel, S., Ceballos, D., Flynn, M. A., & Zayed, J. (2018). Has the question of e-waste opened a Pandora's box? An overview of unpredictable issues and challenges. *Environment International, 110*, 173–192. doi:10.1016/j.envint.2017.10.021 PMID:29122313

Balde, C. P. (2015). *The global e-waste monitor – 2014.* United Nations University, IAS – SCYCLE.

Balde, C. P., & Kuehr, R. (2015). E-waste statistics-Guidelines on classification, reporting and indicators. United Nations University.

Baldé, C. P., Forti, V., Gray, V., Kuehr, R., & Stegmann, P. (2017). The global e-waste monitor – 2017. United Nations University (UNU), International Telecommunication Union (ITU) and International Solid Waste Association, Bonn/Geneva/Vienna.

Balde, C. P., Wang, F., Kuehr, R., & Huisman, J. (2015). The global e-waste monitor – 2014. United Nations University, IAS – SCYCLE. Bonn, Germany.

Baldé, K., D'Angelo, E., Luda, V., Deubzer, O., & Kühr, R. (2022). *Global transboundary e-waste flows monitor 2022.* United Nations Institute for Training and Research (UNITAR).

Baldé, C. P., Forti, V., & Kuehr, R. (2020). *The Global E-waste Monitor 2020: Quantities, flows, and the circular economy potential. United Nations University (UNU)/International Telecommunication Union (ITU)/International Solid Waste Association.* ISWA.

Barapatre, S., & Rastogi, M. (2021). e-Waste Management: A Transition Towards a Circular Economy. In S. Barapatre & M. Rastogi (Eds.), Handbook of Solid Waste Management. https://doi.org/ doi:10.1007/978-981-15-7525-9_68-

Barapatre, S., & Rastogi, M. (2022). e-Waste management: A transition towards a circular economy. In Handbook of Solid Waste Management: Sustainability through Circular Economy (pp. 1499-1521). Singapore: Springer Nature Singapore.

Barba-Gutiérrez, Y., Adenso-Díaz, B., & Hopp, M. (2008). An analysis of some environmental consequences of European electrical and electronic waste regulation. *Resources, Conservation and Recycling*, *52*(3), 481–495. doi:10.1016/j.resconrec.2007.06.002

Basel Convention – Controlling transboundary movements of hazardous wastes and their disposal. (2017). https://www.basel.int/TheConvention/Overview/tabid/1271/Default.aspx

Baumann, M., Moody, T. S., Smyth, M., & Wharry, S. (2020). A perspective on continuous flow chemistry in the pharmaceutical industry. *Organic Process Research & Development*, *24*(10), 1802–1813. doi:10.1021/acs.oprd.9b00524

Bazargan, A., Lam, K. F., & McKay, G. (2012). Challenges and opportunities of e-waste management. *Nova Scientia.*

Bekabil, U. T. (2020). Industrialization and environmental pollution in Africa: An empirical review. *J. Resour. Dev. Manag*, *69*, 18–21.

Belboom, S., Renzoni, R., Deleu, X., Digneffe, J.-M., & Leonard, A. (2011). Electrical waste management effects on environment using life cycle assessment methodology: the fridge case study. *SETAC EUROPE 17th LCA Case Study Symposium Sustainable Lifestyles*, 2.

Berger, W., Simon, F. G., Weimann, K., & Alsema, E. A. (2010). A novel approach for the recycling of thin film photovoltaic modules. *Resources, Conservation and Recycling*, *54*(10), 711–718. doi:10.1016/j.resconrec.2009.12.001

Bhardwaj, L. K., Sharma, S., & Jindal, T. (2023b). Estimation of Physico-Chemical and Heavy Metals in the Lakes of Grovnes & Broknes Peninsula, Larsemann Hill, East Antarctica. *Chemistry Africa*, 1-18.

Bhardwaj, L. K. (2022). Evaluation of bis (2-ethylhexyl) phthalate (DEHP) in the PET bottled mineral water of different brands and impact of heat by GC–MS/MS. *Chemistry Africa*, *5*(4), 929–942. doi:10.100742250-022-00385-6

Bhardwaj, L. K., & Jindal, T. (2020). Persistent organic pollutants in lakes of Grovnes Peninsula at Larsemann Hill area, East Antarctica. *Earth Systems and Environment*, *4*(2), 349–358. doi:10.100741748-020-00154-w

BhardwajL. K.KumarD.KumarA. (2023c). Phytoremediation Potential of Ocimum Sanctum: A Sustainable Approach for Remediation of Heavy Metals. doi:10.20944/preprints202308.0593.v1

BhardwajL. K.RathP.ChoudhuryM. (2023a). Assessment of E-Waste Pollution Level in Soil and Water and Its Impact on Human Health: A Review. doi:10.20944/preprints202311.1714.v1

Bhardwaj, L. K., & Sharma, A. (2021). Estimation of physico-chemical, trace metals, microbiological and phthalate in PET bottled water. *Chemistry Africa*, *4*(4), 981–991. doi:10.100742250-021-00267-3

Bhardwaj, L. K., Sharma, S., & Jindal, T. (2021). Occurrence of polycyclic aromatic hydrocarbons (PAHs) in the Lake water at Grovnes Peninsula Over East Antarctica. *Chemistry Africa*, *4*(4), 965–980. doi:10.100742250-021-00278-0

BhardwajL. K.SinghV. V.DwivediK.RaiS. A. (2023d). Comprehensive Review on Sources, Types, Impact and Challenges of Air Pollution. https://doi.org/ doi:10.20944/preprints202311.1245.v1

BhardwajL. K.VikramV. (2023). Air Pollution and Its Effect on Human Health. https://doi.org/ doi:10.20944/preprints202307.1691.v1

Bhaskar, K., & Kumar, B. (2019). Electronic waste management and sustainable development goals: Is there a business case for linking the two? *Journal of Indian Business Research*, *11*(2), 120–137. doi:10.1108/JIBR-01-2018-0051

Bhaskar, K., & Turaga, R. M. R. (2018). India's E-Waste Rules and Their Impact on E-Waste Management Practices: A Case Study. *Journal of Industrial Ecology*, *22*(4), 930–942. doi:10.1111/jiec.12619

Bhat, V. (2023, April). E-waste management and Achieving SDG-Challenges in Indian Context. *IOP Conference Series. Earth and Environmental Science*, *1161*(1), 012008. doi:10.1088/1755-1315/1161/1/012008

Bhat, V., & Patil, Y. (2014). E-waste consciousness and disposal practices among residents of Pune city. *Procedia: Social and Behavioral Sciences*, *133*, 491–498. doi:10.1016/j.sbspro.2014.04.216

Binelli, A., Sarkar, S. K., Chatterjee, M., Riva, C., Parolini, M., Bhattacharya, B., Bhattacharya, A. K., & Satpathy, K. K. (2007). Concentration of polybrominated diphenyl ethers (PBDEs) in sediment cores of Sundarban mangrove wetland, northeastern part of Bay of Bengal (India). *Marine Pollution Bulletin*, *54*(8), 1220–1229. doi:10.1016/j.marpolbul.2007.03.021 PMID:17512955

Bisschop, L. (2012). Is it all going to waste? Illegal transport of e-waste in a European trade hub. *Crime, Law, and Social Change*, *58*(3), 221–249. doi:10.100710611-012-9383-0

Blomsma, F., & Brennan, G. (2017). The emergence of circular economy: A new framing around prolonging resource productivity. *Journal of Industrial Ecology*, *21*(3), 603–614. doi:10.1111/jiec.12603

Borthakur, A., & Singh, P. (2016). Researches on informal E-waste recycling sector: It's time for a 'Lab to Land' approach. *Journal of Hazardous Materials, 323*(Pt B), 730-732.

Borthakur, A. (2014). Generation and management of electronic waste in the city of Pune, India. *Bulletin of Science, Technology & Society*, *34*(1-2), 43–52. doi:10.1177/0270467614541242

Borthakur, A. (2023). Design, adoption and implementation of electronic waste policies in India. *Environmental Science and Pollution Research International*, *30*(4), 8672–8681. doi:10.100711356-022-18836-5 PMID:35141823

Borthakur, A., & Govind, M. (2017). Emerging trends in consumers' E-waste disposal behaviour and awareness: A worldwide overview with special focus on India. *Resources, Conservation and Recycling*, *117*, 102–113. doi:10.1016/j.resconrec.2016.11.011

Borthakur, A., & Govind, M. (2017). How well are we managing E-waste in India: Evidences from the city of Bangalore. *Energy, Ecology & Environment, 2*(4), 225–235. doi:10.100740974-017-0060-0

Borthakur, A., & Govind, M. (2018, February). Management of the challenges of electronic waste in India: An analysis. *Proceedings of the Institution of Civil Engineers-Waste and Resource Management, 171*(1), 14–20. doi:10.1680/jwarm.17.00035

Borthakur, A., & Singh, P. (2012). Electronic waste in India: Problems and policies. *International Journal of Environmental Sciences, 3*(1), 353–362.

Borthakur, A., & Singh, P. (2016). E-waste in the era of a digitally empowered India: A real challenge (A glance at the world). *Waste Management (New York, N.Y.), 58*, II.

Braverman, H. (2020). Scientific management. In *Industrial Work and Life* (pp. 111–127). Routledge. doi:10.4324/9781003085706-9

Breivik, K., Armitage, J. M., Wania, F., & Jones, K. C. (2014). Tracking the global generation and exports of e-waste. Do existing estimates add up? *Environmental Science & Technology, 48*(15), 8735–8743.

Brewer, A., Dror, I., & Berkowitz, B. (2022). Electronic waste as a source of rare earth element pollution: Leaching, transport in porous media, and the effects of nanoparticles. *Chemosphere, 287*, 132217. doi:10.1016/j.chemosphere.2021.132217 PMID:34826916

Brigden, K. (2005). *Recycling of electronic wastes in China and India: workplace and environmental contamination.* http://www. greenpeace. org/raw/content/china/en/press/reports/recycling-of-electronic-wastes. pdf

Brigden, K., Labunska, I., Santillo, D., & Johnston, P. (2008). Chemical contamination at e-waste recycling and disposal sites in Accra and Korforidua. Academic Press.

Brough, D., & Jouhara, H. (2020). The aluminium industry: A review on state-of-the-art technologies, environmental impacts and possibilities for waste heat recovery. *International Journal of Thermofluids, 1*, 100007. doi:10.1016/j.ijft.2019.100007

Buet, C. (2018). *Assessment of Generation of E-Waste, its Impacts on Environment and Resource Recovery Potential in Bangladesh.* Academic Press.

Bungadaeng, S., Prueksasit, T., & Siriwong, W. (2019). Inhalation exposure to respirable particulate matter among workers in relation to their e-waste open burning activities in Buriram Province, Thailand. *Sustainable Environment Research, 29*(1), 1–12. doi:10.118642834-019-0030-7

Bünger, L. (2021). *Robotic waste sorting*. Worcester Polytechnic Institute.

Bushehri, F. I. (2010). *UNEP's role in promoting environmentally sound management of e-waste.* In 5th ITU Symposium on ''ICTs, the Environment and Climate Change, Cairo, Egypt.

Camilleri, M. A. (2022). Strategic attributions of corporate social responsibility and environmental management: The business case for doing well by doing good! *Sustainable Development (Bradford), 30*(3), 409–422. doi:10.1002d.2256

Canning, L. (2006). Rethinking market connections: Mobile phone recovery, reuse and recycling in the UK. *Journal of Business and Industrial Marketing, 21*(5), 320–329. doi:10.1108/08858620610681623

Cayumil, R., Khanna, R., Rajarao, R., Mukherjee, P. S., & Sahajwalla, V. (2016). Concentration of precious metals during their recovery from electronic waste. *Waste Management (New York, N.Y.), 57*, 121–130. doi:10.1016/j.wasman.2015.12.004 PMID:26712661

Central Pollution Control Board (CPCB). (2018). *E-waste*. Available at: https://cpcb.nic.in/e-waste/

Cerchier, P., Dabalà, M., Pezzato, L., Tammaro, M., Zucaro, A., Fiorentino, G., ... Brunelli, K. (2022). Silicon-PV panels recycling: Technologies and perspectives. *La Metallurgia Italiana*, 16.

Cesaro, A., Marra, A., Kuchta, K., Belgıorno, V., & Van Hullebusch, E. D. (2018). WEEE management in a circular economy perspective: An overview. *Global NEST Journal*, *20*(4), 743–750. doi:10.30955/gnj.002623

Chakraborty, M., Kettle, J., & Dahiya, R. (2022). Electronic Waste Reduction Through Devices and Printed Circuit Boards Designed for Circularity. *IEEE Journal on Flexible Electronics*, *1*(1), 4–23. doi:10.1109/JFLEX.2022.3159258

Chakraborty, S. C., Qamruzzaman, M., Zaman, M. W. U., Alam, M. M., Hossain, M. D., Pramanik, B. K., Nguyen, L. N., Nghiem, L. D., Ahmed, M. F., Zhou, J. L., Mondal, M. I. H., Hossain, M. A., Johir, M. A. H., Ahmed, M. B., Sithi, J. A., Zargar, M., & Moni, M. A. (2022). Metals in e-waste: Occurrence, fate, impacts and remediation technologies. *Process Safety and Environmental Protection*, *162*, 230–252. doi:10.1016/j.psep.2022.04.011

Chancerel, P. (2009). Export of e-waste. Review of the international legal framework. *Environmental Science and Pollution Research International*, *16*(2), 131–141.

Chandra, M. (2015). Environmental concerns in India: Problems and solutions. *J. Int'l Bus. & L.*, *15*, 1.

Chandrasekaran, S. R., Avasarala, S., Murali, D., Rajagopalan, N., & Sharma, B. K. (2018). Materials and energy recovery from e-waste plastics. *ACS Sustainable Chemistry & Engineering*, *6*(4), 4594–4602. doi:10.1021/acssuschemeng.7b03282

Chan, J. K. Y., Xing, G. H., Xu, Y., Liang, Y., Chen, L. X., Wu, S. C., Wong, C. K. C., Leung, C. K. M., & Wong, M. H. (2007). Body loadings and health risk assessment of polychlorinated dibenzo-p-dioxins and dibenzofurans at an intensive electronic waste recycling site in China. *Environmental Science & Technology*, *41*(22), 7668–7674. doi:10.1021/es071492j PMID:18075072

Chan, J. K., & Wong, M. H. (2013). A review of environmental fate, body burdens, and human health risk assessment of PCDD/Fs at two typical electronic waste recycling sites in China. *The Science of the Total Environment*, *463*, 1112–1122. doi:10.1016/j.scitotenv.2012.07.098 PMID:22925483

Charnes, A., Cooper, W. W., & Rhodes, E. (1978). Measuring the of decision making units. *European Journal of Operational Research*, *2*(6), 429–444. doi:10.1016/0377-2217(78)90138-8

Chatterjee, S. (2012). Sustainable electronic waste management and recycling process. *American Journal of Environmental Engineering*, *2*(1), 23–33. doi:10.5923/j.ajee.20120201.05

Chauhan, G., Madou, M. J., Kalra, S., Chopra, V., Ghosh, D., & Martinez-Chapa, S. O. (2020). Nanotechnology for COVID-19: Therapeutics and vaccine research. *ACS Nano*, *14*(7), 7760–7782. doi:10.1021/acsnano.0c04006 PMID:32571007

Cheema, S. M., Hannan, A., & Pires, I. M. (2022). Smart Waste Management and Classification Systems Using Cutting Edge Approach. *Sustainability (Basel)*, *14*(16), 10226. doi:10.3390u141610226

Chen, M., Zhang, F.-S., & Zhu, J. (2011). Leaching toxicity of Pb and Ba containing in cathode ray tube glasses by SEP-TCLP. In *Recycling of Electronic Waste II, Proceedings of the Second Symposium, TMS.* The Minerals, Metals & Materials Society. 10.1002/9781118086391.ch15

Chen, A. (2019). Environmental impact of electronic waste recycling in China: A case study of Guiyu. *The Science of the Total Environment*, *654*, 585–590.

Chen, A., Dietrich, K. N., Huo, X., & Ho, S. M. (2011). Developmental neurotoxicants in e-waste: An emerging health concern. *Environmental Health Perspectives*, *119*(4), 431–438. doi:10.1289/ehp.1002452 PMID:21081302

Chen, J., Huang, S., BalaMurugan, S., & Tamizharasi, G. S. (2021). Artificial intelligence based e-waste management for environmental planning. *Environmental Impact Assessment Review*, *87*, 106498. doi:10.1016/j.eiar.2020.106498

Chen, L., Msigwa, G., Yang, M., Osman, A. I., Fawzy, S., Rooney, D. W., & Yap, P. S. (2022). Strategies to achieve a carbon neutral society: A review. *Environmental Chemistry Letters*, *20*(4), 2277–2310. doi:10.100710311-022-01435-8 PMID:35431715

Chen, L., Yu, C., Shen, C., Zhang, C., Liu, L., Shen, K., Tang, X., & Chen, Y. (2010). Study on adverse impact of e-waste disassembly on surface sediment in East China by chemical analysis and bioassays. *Journal of Soils and Sediments*, *10*(3), 359–367. doi:10.100711368-009-0176-8

Chen, S. J., Tian, M., Wang, J., Shi, T., Luo, Y., Luo, X. J., & Mai, B. X. (2011). Dechlorane Plus (DP) in air and plants at an electronic waste (e-waste) site in South China. *Environmental Pollution*, *159*(5), 1290–1296. doi:10.1016/j.envpol.2011.01.026 PMID:21324409

Chinese Academy of Sciences. (2018). *The social and environmental impacts of e-waste recycling in China*. Author.

Chi, X., Streicher-Porte, M., Wang, M. Y., & Reuter, M. A. (2011). Informal electronic waste recycling: A sector review with special focus on China. *Waste Management (New York, N.Y.)*, *31*(4), 731–742. doi:10.1016/j.wasman.2010.11.006 PMID:21147524

Choi, B.-C., Shin, H.-S., Lee, S.-Y., & Hur, T. (2006). Life cycle assessment of a personal computer and its effective recycling rate. *The International Journal of Life Cycle Assessment*, *11*(2), 122–128. doi:10.1065/lca2004.12.196

Chowdhury, M. S., Rahman, K. S., Chowdhury, T., Nuthammachot, N., Techato, K., Akhtaruzzaman, M., Tiong, S. K., Sopian, K., & Amin, N. (2020). An overview of solar photovoltaic panels' end-of-life material recycling. *Energy Strategy Reviews*, *27*, 100431. doi:10.1016/j.esr.2019.100431

Claude, L. (2017). Tracking e-waste: The development of an electronic waste tracking system. *Resources, Conservation and Recycling*, *120*, 222–229.

Cocchiara, C., Dorneanu, S. A., Inguanta, R., Sunseri, C., & Ilea, P. (2019). Dismantling and electrochemical copper recovery from Waste Printed Circuit Boards in H2SO4–CuSO4–NaCl solutions. *Journal of Cleaner Production*, *230*, 170–179. doi:10.1016/j.jclepro.2019.05.112

Compagnoni, M. (2022). Is Extended Producer Responsibility living up to expectations? A systematic literature review focusing on electronic waste. *Journal of Cleaner Production*, *367*, 133101. doi:10.1016/j.jclepro.2022.133101

Condemi, A., Cucchiella, F., & Schettini, D. (2019). Circular economy and E-waste: An opportunity from RFID TAGs. *Applied Sciences (Basel, Switzerland)*, *9*(16), 3422. doi:10.3390/app9163422

CPCB Estimates E-Waste Generation in India and Ministry Introduces New Rules for Better Management. (n.d.). https://legalitysimplified.com/2023/07/24/cpcb-estimates-e-waste-generation-in-india-and-ministry-introduces-new-rules-for-better-management/#:~:text=According%20to%20the%20data%20collected,and%202021%2D22%2C%20respectively

CPCB. (2021). *CPCB on behalf of Member Secretary. Annual Report 2020-21*. CPCB.

Cucchiella, F., D'Adamo, I., Koh, S. C. L., & Rosa, P. (2015). Recycling of WEEEs: An economic assessment of present and future e-waste streams. *Renewable & Sustainable Energy Reviews*, *51*, 263–272. doi:10.1016/j.rser.2015.06.010

Cui, J., & Røven, H. J. (2011). Electronic Waste. In Waste: A Handbook for Management (pp. 281-296). Elsevier. doi:10.1016/B978-0-12-381475-3.10020-8

Daft, R. L. (2021). *Management*. Cengage Learning.

Darby, L., & Obara, L. (2005). Household recycling behaviour and attitudes towards the disposal of small electrical and electronic equipment. *Resources, Conservation and Recycling*, *44*(1), 17–35. doi:10.1016/j.resconrec.2004.09.002

Dasgupta, D., Debsarkar, A., Hazra, T., Bala, B. K., Gangopadhyay, A., & Chatterjee, D. (2017). Scenario of future e-waste generation and recycle-reuse-landfill-based disposal pattern in India: A system dynamics approach. *Environment, Development and Sustainability*, *19*(4), 1473–1487. doi:10.100710668-016-9815-6

Das, P., Gabriel, J.-C. P., Tay, C. Y., & Lee, J.-M. (2021). Value-added products from thermochemical treatments of contaminated e-waste plastics. *Chemosphere*, *269*, 129409. doi:10.1016/j.chemosphere.2020.129409 PMID:33388566

Davis, G., & Wolski, M. (2009). E-waste and the sustainable organisation: Griffith University's approach to e-waste. *International Journal of Sustainability in Higher Education*, *10*(1), 21–32. doi:10.1108/14676370910925226

Davis, J. M., & Garb, Y. (2015). A model for partnering with the informal e-waste industry: Rationale, principles and a case study. *Resources, Conservation and Recycling*, *105*, 73–83. doi:10.1016/j.resconrec.2015.08.001

Dawar, I., Gupta, S. D., Singh, R., Kothari, Y., & Layek, S. (2023). Forest Fire Detection using Deep Learning Techniques. *2023 2nd International Conference on Vision Towards Emerging Trends in Communication and Networking Technologies (ViTECoN)*, 1–6.

Dawson, L. (2019). 'Our Waste, our Resources; A Strategy for England'–Switching to a circular economy through the use of extended producer responsibility. *Environmental Law Review*, *21*(3), 210–218. doi:10.1177/1461452919851943

Deathe, A. L., MacDonald, E., & Amos, W. (2008). E-waste Management Programmes and the Promotion of Design for the Environment: Assessing Canada's Contributions. *Review of European Community & International Environmental Law*, *17*(3), 319–334. doi:10.1111/j.1467-9388.2008.00610.x

Debnath, B., & Baidya, R. (2015). E-waste recycling as criteria for green computing approach: analysis by QFD tool. *Computational Advancement in Communication Circuits and Systems: Proceedings of ICCACCS 2014*, 139–144.

Debnath, B., Roychoudhuri, R., & Ghosh, S. K. (2016). E-waste management–a potential route to green computing. *Procedia Environmental Sciences*, *35*, 669–675. doi:10.1016/j.proenv.2016.07.063

Deng, R., Chang, N. L., Ouyang, Z., & Chong, C. M. (2019). A techno-economic review of silicon photovoltaic module recycling. *Renewable & Sustainable Energy Reviews*, *109*, 532–550. doi:10.1016/j.rser.2019.04.020

Devika, S. (2010, November). Environmental impact of improper disposal of electronic waste. In *Recent Advances in Space Technology Services and Climate Change 2010 (RSTS & CC-2010)* (pp. 29–31). IEEE. doi:10.1109/RSTSCC.2010.5712793

Dhanwani, R., Prajapati, A., Dimri, A., Varmora, A., & Shah, M. (2021). Smart Earth Technologies: A pressing need for abating pollution for a better tomorrow. *Environmental Science and Pollution Research International*, *28*(27), 35406–35428. doi:10.100711356-021-14481-6 PMID:34018104

Dharini, K., Cynthia, J. B., Kamalambikai, B., Arul, J. P., Celestina, S., & Muthu, D. (2017). Hazardous E-waste and its impact on soil structure. *Iopscience*. 10.1088/1755-1315/80/1/012057

Dhir, A., Koshta, N., Goyal, R. K., Sakashita, M., & Almotairi, M. (2021). Behavioral reasoning theory (BRT) perspectives on E-waste recycling and management. *Journal of Cleaner Production*, *280*, 124269. doi:10.1016/j.jclepro.2020.124269

Dinesh Kumar, K., & Umamaheswari, E. (2023). An efficient proactive VM consolidation technique with improved LSTM network in a cloud environment. *Computing*, 1–28. doi:10.100700607-023-01214-5

Dissanayake, D. G. K., & Weerasinghe, D. (2021). Towards circular economy in fashion: Review of strategies, barriers and enablers. *Circular Economy and Sustainability*, 1–21.

Dixit, S., Arora, R., Kumar, K., Bansal, S., Vatin, N., Araszkiewicz, K., & Epifantsev, K. (2022). Replacing E-waste with coarse aggregate in architectural engineering and construction industry. *Materials Today: Proceedings*, *56*, 2353–2358. doi:10.1016/j.matpr.2021.12.154

Doan, L. T. T., Amer, Y., Lee, S. H., & Phuc, P. N. K. (2019). Strategies for E-waste management: A literature review. *International Journal of Energy and Environmental Engineering*, *13*(3), 157–162.

Dong, L. (2009). *Research on waste crystalline silicon solar panels resource recovery*. Southwest Jiaotong University.

Duan, H., Eugster, M., Hischier, R., Streicher-Porte, M., & Li, J. (2009). Life cycle assessment study of a Environment Canada, 2000. *Understanding the Environmental Aspects of Electronic Products: A Life Cycle assessment Case Study of a Business Telephone*.

Dutta, D., & Goel, S. (2021). Understanding the gap between formal and informal e-waste recycling facilities in India. *Waste Management (New York, N.Y.)*, *125*, 163–171. doi:10.1016/j.wasman.2021.02.045 PMID:33706255

Dutta, D., Rautela, R., Gujjala, L. K. S., Kundu, D., Sharma, P., Tembhare, M., & Kumar, S. (2023). A review on recovery processes of metals from E-waste: A green perspective. *The Science of the Total Environment*, *859*, 160391. doi:10.1016/j.scitotenv.2022.160391 PMID:36423849

Eguchi, A., Nomiyama, K., Devanathan, G., Subramanian, A., Bulbule, K. A., Parthasarathy, P., Takahashi, S., & Tanabe, S. (2012). Different profiles of anthropogenic and naturally produced organohalogen compounds in serum from residents living near a coastal area and e-waste recycling workers in India. *Environment International*, *47*, 8–16. doi:10.1016/j.envint.2012.05.003 PMID:22717641

Ekins, P., Domenech Aparisi, T., Drummond, P., Bleischwitz, R., Hughes, N., & Lotti, L. (2020). *The circular economy: What, why, how and where*. Academic Press.

Ellen MacArthur Foundation. (2019). *What is a circular economy?* Available at: https://ellenmacarthurfoundation.org/topics/circular-economy-introduction/overview

Environment Canada. (n.d.). Chinese desktop personal computer. *The Science of the Total Environment*, *407*, 1755–1764.

ESDO. (2010). *Study on E-waste: Bangladesh Situation*. Environment and Social DevelopmentOrganization.

Esmaeilian, B., Wang, B., Lewis, K., Duarte, F., Ratti, C., & Behdad, S. (2018). The future of waste management in smart and sustainable cities: A review and concept paper. *Waste Management (New York, N.Y.)*, *81*, 177–195. doi:10.1016/j.wasman.2018.09.047 PMID:30527034

European Union (EU)-India Technical Cooperation Project: Energy. (2021). *PV waste management in India: Comparative analysis of state of play and recommendations*. Available at https://www.cecp-eu.in/uploads/documents/events/pv-waste-management-report-25-01-2021.pdf

E-waste roadmap 2023 for India: Daily Current Affairs. (n.d.). https://www.dhyeyaias.com/current-affairs/daily-current-affairs/e-waste-roadmap-2023-for-india

ewastemonitor.info. (2020). *The Global E-waste Monitor 2020 – Quantities, flows, and the circular economy potential*. Available at: https://ewastemonitor.info/gem-2020/#:~:text=According%20to%20the%20report%2C%20Asia,Mt%20and%200.7%20Mt%20respectively

Ezeah, C. (2015). Environmental justice: The key to sustainable waste management in developing countries. *International Journal of Environmental Research and Public Health*, *12*(6), 6635–6657.

Fadaei, A. (2022). E-waste management status worldwide: Major challenges and solutions. *Proceedings of the International Academy of Ecology and Environmental Sciences, 12*(4), 281.

Faist Emmenegger, M., Frischknecht, R., Stutz, M., Guggisberg, M., Witschi, R., & Otto, T. (2006). Life cycle assessment of the mobile communication system UMTS: Towards eco-efficient systems. *The International Journal of Life Cycle Assessment, 11*(4), 265–276. doi:10.1065/lca2004.12.193

Fang, B., Yu, J., Chen, Z., Osman, A. I., Farghali, M., Ihara, I., ... Yap, P. S. (2023). Artificial intelligence for waste management in smart cities: A review. *Environmental Chemistry*. PMID:37362015

Fang, J., Zhang, L., Rao, S., Zhang, M., Zhao, K., & Fu, W. (2022). Spatial variation of heavy metals and their ecological risk and health risks to local residents in a typical e-waste dismantling area of southeastern China. *Environmental Monitoring and Assessment, 194*(9), 604. doi:10.100710661-022-10296-1 PMID:35867165

Farrell, C. C., Osman, A. I., Doherty, R., Saad, M., Zhang, X., Murphy, A., Harrison, J., Vennard, A. S. M., Kumaravel, V., Al-Muhtaseb, A. H., & Rooney, D. W. (2020). Technical challenges and opportunities in realising a circular economy for waste photovoltaic modules. *Renewable & Sustainable Energy Reviews, 128*, 109911. doi:10.1016/j.rser.2020.109911

Farrell, M. J. (1957). The measurement of productive efficiency. *Journal of the Royal Statistical Society. Series A (General), 120*(3), 253–281. doi:10.2307/2343100

Feldman, J., Misenar, S., & Conrad, E. (2016). *Eleventh Hour CISSP®: Study Guide*. Syngress.

Fernández, L. J., Ferrer, R., Aponte, D. F., & Fernandez, P. (2011). Recycling silicon solar cell waste in cement-based systems. *Solar Energy Materials and Solar Cells, 95*(7), 1701–1706. doi:10.1016/j.solmat.2011.01.033

Fiksel, J., Sanjay, P., & Raman, K. (2021). Steps toward a resilient circular economy in India. *Clean Technologies and Environmental Policy, 23*(1), 203–218. doi:10.100710098-020-01982-0 PMID:33199980

Florek, M. (2017). Quantitative assessment of illegal e-waste imports in the port of Gdansk, Poland. *Waste Management (New York, N.Y.), 70*, 19–28.

Florin-Constatin Mihai and Maria-Grazia Gnoni. (2016). *E-waste Management as a Global Challenge*. Available at: https://www.intechopen.com/chapters/51266

Forbes. (2023). *What is Organizational Behavior?* https://www.forbes.com/advisor/business/what-is-organizational-behavior

Forti, V., Balde, C. P., Kuehr, R., & Bel, G. (2020). *The Global E-waste Monitor 2020: Quantities, flows and the circular economy potential*. Academic Press.

Forti, V., Baldé, K., & Kuehr, R. (2018). *E-waste statistics: Guidelines on classifications, reporting and indicators*. Academic Press.

Forti, V., Baldé, C. P., Kuehr, R., & Bel, G. (2020). *The global e-waste monitor 2020. United Nations University (UNU), International Telecommunication Union (ITU) & International Solid Waste Association*. ISWA.

Forti, V., Balde, C. P., Kuehr, R., & Bel, G. (2020). *The Global E-waste Monitor 2020: Quantities, flows and the circular economy potential. United Nations University/United Nations Institute for Training and Research: Bonn, Germany; International Telecommunication Union: Geneva, Switzerland*. International Solid Waste Association.

Frazzoli, C. (2016). Integrated health and environmental monitoring for e-waste workers in Agbogbloshie, Ghana, and findings from early investigations. *Reviews on Environmental Health, 31*(1-2), 141–146.

Frazzoli, C., Orisakwe, O. E., Dragone, R., & Mantovani, A. (2010). Diagnostic health risk assessment of electronic waste on the general population in developing countries' scenarios. *Environmental Impact Assessment Review*, *30*(6), 388–399. doi:10.1016/j.eiar.2009.12.004

Fu, J., Zhang, H., Zhang, A., & Jiang, G. (2018). *E-waste recycling in China: a challenging field.* Academic Press.

Fytili, D., & Zabaniotou, A. (2008). Utilization of sewage sludge in EU application of old and new methods—A review. *Renewable & Sustainable Energy Reviews*, *12*(1), 116–140. doi:10.1016/j.rser.2006.05.014

Gaidajis, G., Angelakoglou, K., & Aktsoglou, D. (2010). E-waste: Environmental problems and current management. *Journal of Engineering Science and Technology Review*, *3*(1), 193–199. doi:10.25103/jestr.031.32

Ganguly, S. (2014). E-Waste, generation, disposal and its associated environmental impact: A Review. *Indian Journal of Scientific Research and Technology*, *2*(5), 5–6.

Gangwar, C., Choudhari, R., Chauhan, A., Kumar, A., Singh, A., & Tripathi, A. (2019). Assessment of air pollution caused by illegal e-waste burning to evaluate the human health risk. *Environment International*, *125*, 191–199. doi:10.1016/j.envint.2018.11.051 PMID:30721825

Garcia-Segura, S., Qu, X., Alvarez, P. J., Chaplin, B. P., Chen, W., Crittenden, J. C., Feng, Y., Gao, G., He, Z., Hou, C. H., Hu, X., Jiang, G., Kim, J.-H., Li, J., Li, Q., Ma, J., Ma, J., Nienhauser, A. B., Niu, J., ... Westerhoff, P. (2020). Opportunities for nanotechnology to enhance electrochemical treatment of pollutants in potable water and industrial wastewater–a perspective. *Environmental Science. Nano*, *7*(8), 2178–2194. doi:10.1039/D0EN00194E

Garlapati, V. K. (2016). E-waste in India and developed countries: Management, recycling, business and biotechnological initiatives. *Renewable & Sustainable Energy Reviews*, *54*, 874–881. doi:10.1016/j.rser.2015.10.106

Garvin Heath. (2019). *PV Modules End of Life Management Setting the Stage. NREL.* National Renewable Energy Laboratory.

Gautam, A., Shankar, R., & Vrat, P. (2022). Managing end-of-life solar photovoltaic e-waste in India: A circular economy approach. *Journal of Business Research*, *142*, 287–300. doi:10.1016/j.jbusres.2021.12.034

Gautam, P., Behera, C. K., Sinha, I., Gicheva, G., & Singh, K. K. (2022). High added-value materials recovery using electronic scrap-transforming waste to valuable products. *Journal of Cleaner Production*, *330*, 129836. doi:10.1016/j.jclepro.2021.129836

Ghaleb Al-Aqqad, A. O., & Menyhart, J. (2023). Solar panels problem after end-of-life and waste management (SWOT analysis for the global directives of PV's waste management). *International Review of Applied Sciences and Engineering*, *14*(2), 220–229. doi:10.1556/1848.2022.00511

Gharaei, A., Karimi, M., & Shekarabi, S. A. H. (2019). An integrated multi-product, multi-buyer supply chain under penalty, green, and quality control polices and a vendor managed inventory with consignment stock agreement: The outer approximation with equality relaxation and augmented penalty algorithm. *Applied Mathematical Modelling*, *69*, 223–254. doi:10.1016/j.apm.2018.11.035

Ghiasi, T. S., Kaverzin, A. A., Dismukes, A. H., de Wal, D. K., Roy, X., & van Wees, B. J. (2021). Electrical and thermal generation of spin currents by magnetic bilayer graphene. *Nature Nanotechnology*, *16*(7), 788–794. doi:10.103841565-021-00887-3 PMID:33958763

Ghisellini, P., Cialani, C., & Ulgiati, S. (2016). A review on circular economy: The expected transition to a balanced interplay of environmental and economic systems. *Journal of Cleaner Production*, *114*, 11–32. doi:10.1016/j.jclepro.2015.09.007

Ghosh, B. K., Mekhilef, S., Ahmad, S., & Ghosh, S. K. (2022). A review on global emissions by E-products based waste: Technical management for reduced effects and achieving sustainable development goals. *Sustainability (Basel)*, *14*(7), 4036. doi:10.3390u14074036

Ghosh, S. K., Debnath, B., Baidya, R., De, D., Li, J., Ghosh, S. K., Zheng, L., Awasthi, A. K., Liubarskaia, M. A., Ogola, J. S., & Tavares, A. N. (2016). Waste electrical and electronic equipment management and Basel Convention compliance in Brazil, Russia, India, China and South Africa (BRICS) nations. *Waste Management & Research*, *34*(8), 693–707. doi:10.1177/0734242X16652956 PMID:27422616

Ghulam, S. T., & Abushammala, H. (2023). Challenges and Opportunities in the Management of Electronic Waste and Its Impact on Human Health and Environment. *Sustainability (Basel)*, *15*(3), 1837. doi:10.3390u15031837

Giovanardi, M., Konstantinou, T., Pollo, R., & Klein, T. (2023). Internet of Things for building façade traceability: A theoretical framework to enable circular economy through life-cycle information flows. *Journal of Cleaner Production*, *382*, 135261. doi:10.1016/j.jclepro.2022.135261

Global e-waste generation outlook 2030. (n.d.). *Statista*. Retrieved December 9, 2023, from https://www.statista.com/statistics/1067081/generation-electronic-waste-globally-forecast/

Goel, A. (2021). Application of circular economy in E-waste management–a review with an Indian perspective. *Strategies Toward Green Deal Implementation*, 70.

Gomez, J. (2018). Smart City Architecture Based on IoT: Case of Barcelona. *IEEE Access : Practical Innovations, Open Solutions*, *6*, 7606–7616. doi:10.1109/ACCESS.2017.2781224

Gottardo, S., Mech, A., Drbohlavová, J., Małyska, A., Bøwadt, S., Sintes, J. R., & Rauscher, H. (2021). Towards safe and sustainable innovation in nanotechnology: State-of-play for smart nanomaterials. *NanoImpact*, *21*, 100297. doi:10.1016/j.impact.2021.100297 PMID:33738354

Grant, K., & Goldizen, F. C. (2019). Human health and ecological assessment associated with electronic waste recycling: An overview. *Reviews on Environmental Health*, *34*(3), 255–264.

Grant, K., Goldizen, F. C., Sly, P. D., Brune, M. N., Neira, M., van den Berg, M., & Norman, R. E. (2013). Health consequences of exposure to e-waste: A systematic review. *The Lancet. Global Health*, *1*(6), e350–e361. doi:10.1016/S2214-109X(13)70101-3 PMID:25104600

Greenpeace - Where Does E-Waste End Up? - English Version. (n.d.). Retrieved November 21, 2023, from https://media.greenpeace.org/archive/Where-Does-E-Waste-End-Up----English-Version-27MZIF2D3F8X.html

Gui, L., Atasu, A., Ergun, Ö., & Toktay, L. B. (2013). Implementing extended producer responsibility legislation: A multi-stakeholder case analysis. *Journal of Industrial Ecology*, *17*(2), 262–276. doi:10.1111/j.1530-9290.2012.00574.x

Guo, H., Wu, S., Tian, Y., Zhang, J., & Liu, H. (2021). Application of machine learning methods for the prediction of organic solid waste treatment and recycling processes: A review. *Bioresource Technology*, *319*, 124114. doi:10.1016/j.biortech.2020.124114 PMID:32942236

Guo, Y., Huo, X., Li, Y., Wu, K., Liu, J., Huang, J., ... Xu, X. (2010). Monitoring of lead, cadmium, chromium and nickel in placenta from an e-waste recycling town in China. *The Science of the Total Environment*, *408*(16), 3113–3117. doi:10.1016/j.scitotenv.2010.04.018 PMID:20451954

Gupta, D., & Dash, S. (2023). Challenges of implementing extended producer responsibility for plastic-waste management: Lessons from India. *Social Responsibility Journal*, *19*(9), 1595–1612. doi:10.1108/SRJ-08-2022-0326

GuptaP.SharmaA.BhardwajL. K. (2023). Solid Waste Management (SWM) and Its Effect on Environment & Human Health. doi:10.20944/preprints202309.0384.v1

Gupta, S. (2011). E-waste management: Teaching how to reduce, reuse and recycle for sustainable development-need of some educational strategies. *Journal of Education and Practice*, *2*(3), 1735–2222.

Gupta, S. (2019). E-waste management and health outcomes: A review. *Environmental Research*, *179*, 108–120.

Gupta, S. (2021). A review on potential of IoT in solid waste management: Opportunities, challenges, and future directions. *Environmental Science and Pollution Research International*, *28*(22), 28146–28162. PMID:33527248

Gupta, S. K. (2020). Study of Nanotechnology and Its Application. *J Phys Opt Sci*, *2*(1), 1–7.

Guterres, A. (2020). Sustainable Development Goals Report 2020—Goal 12—Electronic Waste. UNStats.

Ha, N. N., Agusa, T., Ramu, K., Tu, N. P. C., Murata, S., Bulbule, K. A., Parthasaraty, P., Takahashi, S., Subramanian, A., & Tanabe, S. (2009). Contamination by trace elements at e-waste recycling sites in Bangalore, India. *Chemosphere*, *76*(1), 9–15. doi:10.1016/j.chemosphere.2009.02.056 PMID:19345395

Heacock, M., Kelly, C. B., Asante, K. A., Birnbaum, L. S., Bergman, Å. L., Bruné, M. N., Buka, I., Carpenter, D. O., Chen, A., Huo, X., Kamel, M., Landrigan, P. J., Magalini, F., Diaz-Barriga, F., Neira, M., Omar, M., Pascale, A., Ruchirawat, M., Sly, L., ... Suk, W. A. (2016). E-waste and harm to vulnerable populations: A growing global problem. *Environmental Health Perspectives*, *124*(5), 550–555. doi:10.1289/ehp.1509699 PMID:26418733

Heacock, M., Trottier, B., Adhikary, S., Asante, K. A., Basu, N., Brune, M.-N., Caravanos, J., Carpenter, D., Cazabon, D., Chakraborty, P., Chen, A., Barriga, F. D., Ericson, B., Fobil, J., Haryanto, B., Huo, X., Joshi, T. K., Landrigan, P., Lopez, A., ... Suk, W. (2018). Prevention-intervention strategies to reduce exposure to e-waste. *Reviews on Environmental Health*, *33*(2), 219–228. doi:10.1515/reveh-2018-0014 PMID:29750656

Heath, G. A., Silverman, T. J., Kempe, M., Deceglie, M., Ravikumar, D., Remo, T., Cui, H., Sinha, P., Libby, C., Shaw, S., Komoto, K., Wambach, K., Butler, E., Barnes, T., & Wade, A. (2020). Research and development priorities for silicon photovoltaic module recycling to support a circular economy. *Nature Energy*, *5*(7), 502–510. doi:10.103841560-020-0645-2

Heinrich, A. J., Oliver, W. D., Vandersypen, L. M., Ardavan, A., Sessoli, R., Loss, D., Jayich, A. B., Fernandez-Rossier, J., Laucht, A., & Morello, A. (2021). Quantum-coherent nanoscience. *Nature Nanotechnology*, *16*(12), 1318–1329. doi:10.103841565-021-00994-1 PMID:34845333

Henriques, R., Figueiredo, F., & Nunes, J. (2023). Product-Services for a Resource-Efficient and Circular Economy: An Updated Review. *Sustainability (Basel)*, *15*(15), 12077. doi:10.3390u151512077

He, P., Wang, C., & Zuo, L. (2018). The present and future availability of high-tech minerals in waste mobile phones: Evidence from China. *Journal of Cleaner Production*, *192*, 940–949. doi:10.1016/j.jclepro.2018.04.222

Herat, S. (2008). Contamination of solid waste from toxic materials in electronic waste (E-waste). *Journal of Solid Waste Technology Management*, *34*(4), 1–18.

Herat, S., & Agamuthu, P. (2012). E-waste: A problem or an opportunity? Review of issues, challenges, and solutions in Asian countries. *Waste Management & Research*, *30*(11), 1113–1129. doi:10.1177/0734242X12453378 PMID:22851536

Hicks, C., Dietmar, R., & Eugster, M. (2005). The recycling and disposal of electrical and electronic waste in China-legislative and market responses. *Environmental Impact Assessment Review*, *25*(5), 459–471. doi:10.1016/j.eiar.2005.04.007

Hieronymi, K., Kahhat, R., & Williams, E. (Eds.). (2012). *E-waste management: From waste to resource*. Routledge. doi:10.4324/9780203116456

Hischier, R., Wäger, P., & Gauglhofer, J. (2005). Does WEEE recycling make sense from an environmental perspective? The environmental impacts of the Swiss takeback and recycling systems for waste electrical and electronic equipment (WEEE). *Environmental Impact Assessment Review, 25*(5), 525–539. doi:10.1016/j.eiar.2005.04.003

Hossain, S., Sulatan, S., Shahnaz, F., Akram, A. B., Nesa, M., & Happell, J. (2010). E-waste: Bangladesh situation. Study Report Environment and Social Development Organization-Environmental and Social Development Organisation Dhaka. Bangladesh.

Hossain, M. S., Al-Hamadani, S. M., & Rahman, M. T. (2015). E-waste: A challenge for sustainable development. *Journal of Health & Pollution, 5*(9), 3–11. doi:10.5696/2156-9614-5-9.3 PMID:30524771

Hossain, S., Sulatan, S., Shahnaz, F., & Hossain, M. L. (2011). *Illegal import and trade off of E-waste in Bangladesh. Environment and Social Development Organization*. ESDO.

HR, G., & Aithal, P.S. (2022). Why is it Called Doctor of Philosophy and Why Choosing Appropriate Research Philosophical Paradigm is Indispensable During Ph. D. Program in India? *International Journal of Philosophy and Languages, 1*(1), 42–58.

Hsu, E., Barmak, K., West, A. C., & Park, A. H. A. (2019). Advancements in the treatment and processing of electronic waste with sustainability: A review of metal extraction and recovery technologies. *Green Chemistry, 21*(5), 919–936. doi:10.1039/C8GC03688H

Huang, J., Nkrumah, P. N., Anim, D. O., & Mensah, E. (2014). E-waste disposal effects on the aquatic environment: Accra, Ghana. *Reviews of Environmental Contamination and Toxicology*, 19–34. PMID:24515808

Huo, X., Peng, L., Xu, X., Zheng, L., Qiu, B., Qi, Z., Zhang, B., Han, D., & Piao, Z. (2007). Elevated blood lead levels of children in Guiyu, an electronic waste recycling town in China. *Environmental Health Perspectives, 115*(7), 1113–1117. doi:10.1289/ehp.9697 PMID:17637931

Ignatuschtschenko, E. (2018). Electronic waste in China, Japan, and vietnam: A comparative analysis of waste management strategies. *Vienna Journal of East Asian Studies, 9*(1), 29–58. doi:10.2478/vjeas-2017-0002

Ikhlayel, M. (2018). An integrated approach to establish e-waste management systems for developing countries. *Journal of Cleaner Production, 170*, 119–130. doi:10.1016/j.jclepro.2017.09.137

Ilankoon, I. M. S. K., Ghorbani, Y., Chong, M. N., Herath, G., Moyo, T., & Petersen, J. (2018). E-waste in the international context–A review of trade flows, regulations, hazards, waste management strategies and technologies for value recovery. *Waste Management, 82*, 258-275.

Ilankoon, I. M. S. K., Ghorbani, Y., Chong, M. N., Herath, G., Moyo, T., & Petersen, J. (2018). E-waste in the international context–A review of trade flows, regulations, hazards, waste management strategies and technologies for value recovery. *Waste Management (New York, N.Y.), 82*, 258–275. doi:10.1016/j.wasman.2018.10.018 PMID:30509588

Ilyassova, G., Nukusheva, A., Arenova, L., Karzhassova, G., & Akimzhanova, M. (2021). Prospects of legal regulation in the field of electronic waste management in the context of a circular economy. *International Environmental Agreement: Politics, Law and Economics, 21*(3), 367–388. doi:10.100710784-020-09514-3

Impacts of E-Waste on Health—Blog. (n.d.). Retrieved December 9, 2023, from https://www.greenit.co.in/impacts-of-e-waste-on-health.html

India e-Waste Management Market Report 2023: India has Become the Largest Producer of e-Waste after China and the United States - Analysis & Forecasts 2018-2022 & 2023-2027. (n.d.). https://www.globenewswire.com/en/news-release/2023/03/01/2618469/28124/en/India-e-Waste-Management-Market-Report-2023-India-has-Become-the-Largest-Producer-of-e-Waste-after-China-and-United-States-Analysis-Forecasts-2018-2022-2023-2027.html

International Energy Agency Photovoltaic Power Systems Programme (IEA PVPS). (2016). *Task12 and the International Renewable Energy Agency (IRENA), End of Life Management of Photovoltaic Panels Trends in PV Module Recycling Technologies*. Available at: https://iea-pvps.org/key-topics/end-of-life-management-of-photovoltaic-panels-trends-in-pv-module-recycling-technologies-by-task-12/

International Telecommunications Union. (2019). *The role of technology in e-waste management*. Author.

Islam, A., Ahmed, T., Awual, M. R., Rahman, A., Sultana, M., Abd Aziz, A., ... Hasan, M. (2020). Advances in sustainable approaches to recover metals from e-waste-A review. *Journal of Cleaner Production, 244*, 118815. doi:10.1016/j.jclepro.2019.118815

Islam, M. N. (2016). E-waste management of Bangladesh. *International Journal of Innovative Human Ecology and Nature Studies, 4*(2), 1–12.

Islam, M. T., Huda, N., Baumber, A., Shumon, R., Zaman, A., Ali, F., Hossain, R., & Sahajwalla, V. (2021). A global review of consumer behavior towards e-waste and implications for the circular economy. *Journal of Cleaner Production, 316*, 128297. doi:10.1016/j.jclepro.2021.128297

Ismail, H., & Hanafiah, M. M. (2020). A review of sustainable e-waste generation and management: Present and future perspectives. *Journal of Environmental Management, 264*, 110495. doi:10.1016/j.jenvman.2020.110495 PMID:32250915

Ismail, H., & Hanafiah, M. M. (2021). Evaluation of e-waste management systems in Malaysia using life cycle assessment and material flow analysis. *Journal of Cleaner Production, 308*, 127358. doi:10.1016/j.jclepro.2021.127358

Jain, A. (2008). Global e-waste growth. *E-waste: Implication, Regulations, and Management in India and Current Global Best Practices*, 3-22.

Jain, P. (2020). *R's of Waste Management (Evolution)*. Available at: https://indieseducation.com/rs-of-waste-management-evolution/

Jain, M., Kumar, D., Chaudhary, J., Kumar, S., Sharma, S., & Verma, A. S. (2023). *Review on E-waste management and its impact on the environment and society*. Waste Management Bulletin. doi:10.1016/j.wmb.2023.06.004

Janani, S., Preethi, K., Gowtham, S., & Kulanthaivel, P. (2023). A sustainable reuse of e waste as a partial replacement material for aggregate. *Materials Today: Proceedings*. Advance online publication. doi:10.1016/j.matpr.2023.05.740

Jan, F., Min-Allah, N., Saeed, S., Iqbal, S. Z., & Ahmed, R. (2022). IoT-based solutions to monitor water level, leakage, and motor control for smart water tanks. *Water (Basel), 14*(3), 309. doi:10.3390/w14030309

Jangre, J., Prasad, K., & Patel, D. (2022). Analysis of barriers in e-waste management in developing economy: An integrated multiple-criteria decision-making approach. *Environmental Science and Pollution Research International, 29*(48), 72294–72308. doi:10.100711356-022-21363-y PMID:35696062

Javed, R., Zia, M., Naz, S., Aisida, S. O., Ain, N. U., & Ao, Q. (2020). Role of capping agents in the application of nanoparticles in biomedicine and environmental remediation: Recent trends and future prospects. *Journal of Nanobiotechnology, 18*(1), 1–15. doi:10.118612951-020-00704-4 PMID:33225973

Jayarajan, P., Thenmozhi, S., Maheswar, R., Malathy, S., & Udaiyakumar, R. (2018). Smart cloud enabled E-waste management system. *2018 International Conference on Computer Communication and Informatics (ICCCI)*, 1–5.

Jayaraman, K., Vejayon, S., Raman, S., & Mostafiz, I. (2019). The proposed e-waste management model from the conviction of individual laptop disposal practices-An empirical study in Malaysia. *Journal of Cleaner Production*, *208*, 688–696. doi:10.1016/j.jclepro.2018.10.125

Jian, L., & Zhang, S. (2010). Study e-waste management based on EPR system. *2010 International Conference on E-Business and E-Government*, 820–823. 10.1109/ICEE.2010.213

Jin, X. (2018). A review on data-driven approaches for the prognosis of critical equipment in large-scale mechanical systems. *Mechanical Systems and Signal Processing*, *100*, 165–193.

Jones, R., & Brown, A. (2020). Electronic waste: Impacts on the environment and human health. *Environmental Science & Technology*, *54*(7), 3981–3992.

Jonkers, N., Krop, H., van Ewijk, H., & Leonards, P. E. (2016). Life cycle assessment of flame retardants in an electronics application. *The International Journal of Life Cycle Assessment*, *21*(2), 146–161. doi:10.100711367-015-0999-z

Jordan, D. C., & Kurtz, S. R. (2013). Photovoltaic degradation rates—An analytical review. *Progress in Photovoltaics: Research and Applications*, *21*(1), 12–29. doi:10.1002/pip.1182

Joshi, S., Sharma, M., Ekren, B. Y., Kazancoglu, Y., Luthra, S., & Prasad, M. (2023). Assessing Supply Chain Innovations for Building Resilient Food Supply Chains: An Emerging Economy Perspective. *Sustainability (Basel)*, *15*(6), 4924. doi:10.3390u15064924

Jugal, K. (2010). E-waste management: As a challenge to public health in India. *Indian Journal of Community Medicine*, *35*(3), 382–385. doi:10.4103/0970-0218.69251 PMID:21031101

Kamal, S. (2021). Internet of Things (IoT)-based healthcare monitoring: A review of applications, challenges, and future perspectives. *Journal of King Saud University. Computer and Information Sciences*. Advance online publication. doi:10.1016/j.jksuci.2021.02.004

Kandel, A. (2022). *A REPORT On E-waste Management E-waste Management*. Sharda University.

Kang, H., & Schoenung, J. M. (2005). Electronic waste recycling: A review of U.S. infrastructure and technology options. *Resources, Conservation and Recycling*, *45*(4), 368–400. doi:10.1016/j.resconrec.2005.06.001

Kannan, D., Govindan, K., & Shankar, M. (2016). Formalize recycling of electronic waste. *Nature*, *530*(7590), 281–281. doi:10.1038/530281b PMID:26887484

Karim, R., Bari, N., & Amin, M. (2014). E-waste management in Bangladesh. *2nd International Conference on Green Energy and Technology*.

Kaur, B., Kaur, K., & Kaur, H. (2023). E-waste Management: A Transition Towards a Circular Economy. *The Impact of Climate Change and Sustainability Standards on the Insurance Market*, 403-416.

Kaya, M. (2016). Recovery of metals and nonmetals from electronic waste by physical and chemical recycling processes. *Waste Management (New York, N.Y.)*, *57*, 64–90. doi:10.1016/j.wasman.2016.08.004 PMID:27543174

Kazancoglu, Y., Ozbiltekin-Pala, M., & Ozkan-Ozen, Y. D. (2021). Prediction and evaluation of greenhouse gas emissions for sustainable road transport within Europe. *Sustainable Cities and Society*, *70*, 102924. doi:10.1016/j.scs.2021.102924

Kellner, R., & Goe, M. (2016). Technological trends in e-waste management. *Waste Management & Research*, *34*(7), 570–577.

Khaliq, A., Rhamdhani, M., Brooks, G., & Masood, S. (2014). *Metal Extraction Processes for Electronic Waste and Existing Industrial Routes: A Review and Australian Perspective*. Academic Press.

Khaliq, A., Rhamdhani, M. A., Brooks, G., & Masood, S. (2014). Metal extraction processes for electronic waste and existing industrial routes: A review and Australian perspective. *Resources*, *3*(1), 152–179. doi:10.3390/resources3010152

Khan, A., Hussain, S., & Sampene, A. K. (2023). Investing in green intellectual capital to enhance green corporate image under the Influence of green innovation climate: A Case of Chinese Entrepreneurial SMEs. *Journal of Cleaner Production*, *418*, 138177. doi:10.1016/j.jclepro.2023.138177

Khan, K., Su, C.-W., Umar, M., & Yue, X.-G. (2021). Do crude oil price bubbles occur? *Resources Policy*, *71*, 101936. doi:10.1016/j.resourpol.2020.101936

Khatiwada, B., Jariyaboon, R., & Techato, K. (2023). E-waste management in Nepal: A case study overcoming challenges and opportunities. *e-Prime-Advances in Electrical Engineering, Electronics and Energy*, *4*, 100155.

Khetriwal, D. S., Kraeuchi, P., & Widmer, R. (2009). Producer responsibility for e-waste management: Key issues for consideration–learning from the Swiss experience. *Journal of Environmental Management*, *90*(1), 153–165. doi:10.1016/j.jenvman.2007.08.019 PMID:18162284

Khoshnevisan, B., Duan, N., Tsapekos, P., Awasthi, M. K., Liu, Z., Mohammadi, A., Angelidaki, I., Tsang, D. C. W., Zhang, Z., Pan, J., Ma, L., Aghbashlo, M., Tabatabaei, M., & Liu, H. (2021). A critical review on livestock manure biorefinery technologies: Sustainability, challenges, and future perspectives. *Renewable & Sustainable Energy Reviews*, *135*, 110033. doi:10.1016/j.rser.2020.110033

Kibert, C. J. (2004). Green buildings: An overview of progress. *Journal of Land Use & Environmental Law*, *19*(2), 491–502.

Kiddee, P., Naidu, R., & Wong, M. H. (2013). Electronic waste management approaches: An overview. *Waste Management (New York, N.Y.)*, *33*(5), 1237–1250. doi:10.1016/j.wasman.2013.01.006 PMID:23402807

Kiddee, P., Naidu, R., & Wong, M. H. (2013). E-waste management in India: A strategy for the attainment of SDGs 2030. *Materials Today: Proceedings*, *60*, 811–814.

Kiehn, K., & Weller Vojkovic, A. (2018). *Millennials motivations for shopping second-hand clothing as part of a sustainable consumption practice*. Academic Press.

Kirtane, A. R., Verma, M., Karandikar, P., Furin, J., Langer, R., & Traverso, G. (2021). Nanotechnology approaches for global infectious diseases. *Nature Nanotechnology*, *16*(4), 369–384. doi:10.103841565-021-00866-8 PMID:33753915

Kitila, A. W., & Woldemikael, S. M. (2021). Electronic waste management in Addis Ababa: The case of Bole and Nefas Silk Lafto sub-cities. *African Journal of Science, Technology, Innovation and Development*, *13*(2), 235–246. doi:10.1080/20421338.2020.1712014

Komoto, K., Lee, J. S., Zhang, J., Ravikumar, D., Sinha, P., Wade, A., & Heath, G. A. (2018). *End-of-life management of photovoltaic panels: Trends in PV module recycling technologies (No. NREL/TP-6A20-73847). National Renewable Energy Lab.* NREL. doi:10.2172/1561523

Koul, B., Yakoob, M., & Shah, M. P. (2022). Agricultural waste management strategies for environmental sustainability. *Environmental Research*, *206*, 112285. doi:10.1016/j.envres.2021.112285 PMID:34710442

Krishnamurthy, S., Fudurich, G., & Rao, P. (2019). Circular economy for India: perspectives on stewardship principles, waste management, and energy generation. In *Modernization and accountability in the social economy sector* (pp. 272–298). IGI Global. doi:10.4018/978-1-5225-8482-7.ch015

Kuehr, R. (2019). E-waste seen from a global perspective. In Waste Electrical and Electronic Equipment (WEEE) Handbook (pp. 1-16). Woodhead Publishing. doi:10.1016/B978-0-08-102158-3.00001-X

Kuehr, R., & Wang, F. (2015). *The social and economic impacts of e-waste management in developing countries*. United Nations University.

Kumar, K. D., & Umamaheswari, E. (2018). Efficient Cloud Resource Scaling Based on Prediction Approaches. *International Journal of Engineering & Technology, 7*(4.10).

Kumar, A., & Dixit, G. (2018). An analysis of barriers affecting the implementation of e-waste management practices in India: A novel ISM-DEMATEL approach. *Sustainable Production and Consumption, 14*, 36–52. doi:10.1016/j.spc.2018.01.002

Kumar, A., & Holuszko, M. (2016). Electronic waste and existing processing routes: A Canadian perspective. *Resources, 5*(4), 35. doi:10.3390/resources5040035

Kumar, A., Holuszko, M., & Denise, C. R. E. (2017). E-waste: An overview on generation, collection, legislation and recycling practices. *Resources, Conservation and Recycling, 122*, 32–42. doi:10.1016/j.resconrec.2017.01.018

Kumar, H., Singh, M. K., Gupta, M. P., & Madaan, J. (2020). Moving towards smart cities: Solutions that lead to the Smart City Transformation Framework. *Technological Forecasting and Social Change, 153*, 119281. doi:10.1016/j.techfore.2018.04.024

Kumari, N., & Lenka, U. (2023). Scientometric review of disability inclusion in the workplace.

Kumari, N., & Lenka, U. (2023). Employment and Retention of Differently-abled People in the Workplace Through Assistive Technologies. *International Journal of Digital Technologies, 2*(1).

Kumar, K. D., & Umamaheswari, E. (2018). Prediction methods for effective resource provisioning in cloud computing: A survey. *Multiagent and Grid Systems, 14*(3), 283–305. doi:10.3233/MGS-180292

Kumar, K. D., & Umamaheswari, E. (2019). Ewptnn: An efficient workload prediction model in cloud computing using two-stage neural networks. *Procedia Computer Science, 165*, 151–157. doi:10.1016/j.procs.2020.01.097

Kumar, K. D., & Umamaheswari, E. (2020). Hpcwmf: A hybrid predictive cloud workload management framework using improved LSTM neural network. *Cybernetics and Information Technologies, 20*(4), 55–73. doi:10.2478/cait-2020-0047

Kumar, K. V. R., Kumar, K. D., Poluru, R. K., Basha, S. M., & Reddy, M. P. K. (2020). Internet of things and fog computing applications in intelligent transportation systems. In *Architecture and security issues in fog computing applications* (pp. 131–150). IGI Global. doi:10.4018/978-1-7998-0194-8.ch008

Kumar, N., & Jha, A. (2019). Application of principles of supply chain management to the pharmaceutical good transportation practices. *International Journal of Pharmaceutical and Healthcare Marketing, 13*(3), 306–330. doi:10.1108/IJPHM-09-2017-0048

Kumar, P. (2012). Managing E-Waste in India-A Review. *Academic Discourse, 1*(1), 123–133.

Kumar, P., & Fulekar, M. H. (2019). Multivariate and statistical approaches for the evaluation of heavy metals pollution at e-waste dumping sites. *SN Applied Sciences, 1*(11), 1–13. doi:10.100742452-019-1559-0

Kumar, S. K., Kumar, S., Kumar, P., & Kumar, A. (2021a). The application of artificial intelligence in e-waste management: A state-of-the-art review. *Sustainable Computing : Informatics and Systems, 35*, 100680.

Kumar, S. K., Kumar, S., Kumar, P., & Kumar, A. (2021b). A deep learning approach for e-waste classification. *Materials Today: Proceedings, 51*, 1246–1250.

Kumar, S., & Rawat, S. (2018). Future e-Waste: Standardisation for reliable assessment. *Government Information Quarterly, 35*(4), S33–S42. doi:10.1016/j.giq.2015.11.006

Kumar, V., Garg, R., Rahman, Z., & Kazmi, A. A. (2011). Sustainability and E-waste management scenario in India. *The First International Conference on Interdisciplinary Research and Development, 31*, 31.

Kumhar, B. L. (2020). Concept and Approaches of Nano Technology. *Agri Mirror: Future India, 1*(2), 19–23.

Kurniawan, K., Soefihara, M. D. A., Nababan, D. C., & Kim, S. (2022). Current status of the recycling of e-waste in Indonesia. *Geosystem Engineering, 25*(3-4), 83–94. doi:10.1080/12269328.2022.2142856

Kurniawan, T. A., Othman, M. H. D., Hwang, G. H., & Gikas, P. (2022). Unlocking digital technologies for waste recycling in Industry 4.0 era: A transformation towards a digitalization-based circular economy in Indonesia. *Journal of Cleaner Production, 357*, 131911. doi:10.1016/j.jclepro.2022.131911

Lambert, F., Gaydardzhiev, S., Leonard, G., Lewis, G., Bareel, P.-F., & Bastin, D. (2015). Copper leaching from waste electric cables by biohydrometallurgy. *Minerals Engineering, 76*, 38–46. doi:10.1016/j.mineng.2014.12.029

Larsson, S., Jansson, M., & Boholm, Å. (2019). Expert stakeholders' perception of nanotechnology: Risk, benefit, knowledge, and regulation. *Journal of Nanoparticle Research, 21*(3), 1–17. doi:10.100711051-019-4498-1

Lebbie, T. S., Moyebi, O. D., Asante, K. A., Fobil, J., Brune-Drisse, M. N., Suk, W. A., Sly, P. D., Gorman, J., & Carpenter, D. O. (2021). E-waste in Africa: A serious threat to the health of children. *International Journal of Environmental Research and Public Health, 18*(16), 8488. doi:10.3390/ijerph18168488 PMID:34444234

Leccisi, E., Raugei, M., & Fthenakis, V. (2016). The energy and environmental performance of ground-mounted photovoltaic systems—A timely update. *Energies, 9*(8), 622. doi:10.3390/en9080622

Lee, D., Offenhuber, D., Duarte, F., Biderman, A., & Ratti, C. (2018). Monitour: Tracking global routes of electronic waste. *Waste Management, 72*, 362-370.

Lee, C., & Williams, E. (2021). The global trade in electronic waste: A research review. *Environmental Science & Technology, 55*(1), 417–429.

Lee, J. (2020). IoT-based environmental monitoring for e-waste exposure in urban areas: A case study in Seoul, Korea. *Environmental Research, 183*, 109205.

Lee, S., & Na, S. (2010). E-waste recycling systems and sound circulative economies in East Asia: A comparative analysis of systems in Japan, South Korea, China and Taiwan. *Sustainability (Basel), 2*(6), 1632–1644. doi:10.3390u2061632

Leng, J., Ruan, G., Jiang, P., Xu, K., Liu, Q., Zhou, X., & Liu, C. (2020). Blockchain-empowered sustainable manufacturing and product lifecycle management in industry 4.0: A survey. *Renewable & Sustainable Energy Reviews, 132*, 110112. doi:10.1016/j.rser.2020.110112

Lepawsky, J. (2015). The changing geography of global trade in electronic discards: Time to rethink the e-waste problem. *The Geographical Journal, 181*(2), 147–159. doi:10.1111/geoj.12077

Leung, A. O. W. (2019). Environmental contamination and health effects due to E-waste recycling. *Electronic waste management and treatment technology*, 335-362.

Leung, A. O., Zheng, J., Yu, C. K., Liu, W. K., Wong, C. K., Cai, Z., & Wong, M. H. (2011). Polybrominated diphenyl ethers and polychlorinated dibenzo-p-dioxins and dibenzofurans in surface dust at an E-waste processing site in Southeast China. *Environmental Science & Technology, 45*(13), 5775–5782. doi:10.1021/es103915w PMID:21639085

Leung, A., Cai, Z. W., & Wong, M. H. (2006). Environmental contamination from electronic waste recycling at Guiyu, southeast China. *Journal of Material Cycles and Waste Management, 8*(1), 21–33. doi:10.100710163-005-0141-6

Li, J. (2020). E-waste recycling induced polybrominated diphenyl ethers (PBDEs) exposure in human tissues in Ghana, West Africa. *Environmental Science and Pollution Research International*, *27*(26), 32888–32899. PMID:32524403

Li, J., Duan, H., & Shi, P. (2011). Heavy metal contamination of surface soil in electronic waste dismantling area: Site investigation and source-apportionment analysis. *Waste Management & Research*, *29*(7), 727–738. doi:10.1177/0734242X10397580 PMID:21406506

Li, J., Zeng, X., Chen, M., Ogunseitan, O. A., & Stevels, A. (2015). "Control-Alt-Delete": Rebooting solutions for the e-waste problem. *Environmental Science & Technology*, *49*(12), 7095–7108. doi:10.1021/acs.est.5b00449 PMID:26007633

Lindeman, M. (2018). A survey of data mining and knowledge discovery process models and methodologies. *Knowledge and Information Systems*, *54*(2), 227–307.

Lin, S., Ali, M. U., Zheng, C., Cai, Z., & Wong, M. H. (2022). Toxic chemicals from uncontrolled e-waste recycling: Exposure, body burden, health impact. *Journal of Hazardous Materials*, *426*, 127792. doi:10.1016/j.jhazmat.2021.127792 PMID:34802823

Liu, K., Tan, Q., Yu, J., & Wang, M. (2023). A global perspective on e-waste recycling. *Circular Economy*, 100028.

Liu, Q., Li, K. Q., Zhao, H., Li, G., & Fan, F. Y. (2009). The global challenge of electronic waste management. *Environmental Science and Pollution Research International*, *16*(3), 248–249. doi:10.100711356-009-0131-7 PMID:19308473

Liu, Q., Tang, Y., & Chen, S. (2013). E-waste recycling induced polybrominated diphenyl ethers, polychlorinated biphenyls, and polycyclic aromatic hydrocarbons contamination in the ambient air. *Journal of Environmental Sciences (China)*, *25*(1), 83–92.

Liu, Y., Zhao, G., Chen, Y., & Zhang, J. (2022). Image classification of e-waste using convolutional neural networks. *IEEE Access : Practical Innovations, Open Solutions*, *10*, 24433–24441. doi:10.1109/ACCESS.2022.3221820

Livemint. (2023). *Indian chemical industry: A growing force and its prospects mint.* Available at: https://www.livemint.com/market/stock-market-news/indian-chemical-industry-a-growing-force-and-its-prospects-11695551490833.html#:~:text=Chemical%20industry%20revenue%20has%20been,in%20the%20last%205%20years.&text=The%20Indian%20chemicals%20sector%20stands,of%20%24178%20billion%20in%202021

Li, W., & Achal, V. (2020). Environmental and health impacts due to e-waste disposal in China–A review. *The Science of the Total Environment*, *737*, 139745. doi:10.1016/j.scitotenv.2020.139745 PMID:32516663

Li, Z., Qiu, F., Tian, Q., Yue, X., & Zhang, T. (2022). Production and recovery of tellurium from metallurgical intermediates and electronic waste-A comprehensive review. *Journal of Cleaner Production*, *366*, 132796. doi:10.1016/j.jclepro.2022.132796

Lu, C., Lin, Z., Zhong, Y., Ren, W., Tobias, M., Mu, Z., Ma, Z., Geng, Y., & Xue, B. (2015). An overview of e-waste management in China. *Journal of Material Cycles and Waste Management*, *17*(1), 1–12. doi:10.100710163-014-0256-8

Lunardi, M. M., Alvarez-Gaitan, J. P., Bilbao, J. I., & Corkish, R. (2018). A review of recycling processes for photovoltaic modules. In B. Zaidi (Ed.), *Solar panels and photovoltaic materials* (pp. 9–27). Intechopen. doi:10.5772/intechopen.74390

Luo, C., Liu, C., Wang, Y., Liu, X., Li, F., Zhang, G., & Li, X. (2011). Heavy metal contamination in soils and vegetables near an e-waste processing site, south China. *Journal of Hazardous Materials*, *186*(1), 481–490. doi:10.1016/j.jhazmat.2010.11.024 PMID:21144651

Luo, Q., Cai, Z. W., & Wong, M. H. (2007). Polybrominated diphenyl ethers in fish and sediment from river polluted by electronic waste. *The Science of the Total Environment*, *383*(1-3), 115–127. doi:10.1016/j.scitotenv.2007.05.009 PMID:17560632

Luu, D. H. (2021). The impact of capital structure on firm value: A case study in Vietnam. *The Journal of Asian Finance. Economics and Business*, *8*(5), 287–292.

Lu, Y., & Xu, Z. (2016). Precious metals recovery from waste printed circuit boards: A review for current status and perspective. *Resources, Conservation and Recycling*, *113*, 28–39. doi:10.1016/j.resconrec.2016.05.007

Magalini, F., Thiébaud, E., & Kaddouh, S. (2019). Quantifying WEEE in Romania 2019 vs 2015. Sofies UK.

Mairaj, S., Nagar, R. D., Bhardwaj, L. K., Rehman, F., Punnakal, A., Chauhan, A., & Jindal, T. (2022). Antioxidant responses of arsenite-induced oxidative stress in rice (Oryza sativa L.) and its modulation by eugenol (extracted from Ocimum sanctum). *Indian Journal of Agricultural Research*, *56*(4), 396–400.

MAIT. (2016). *Salient Features of the E-Waste (Management) Rules, 2016 and its likely implication.* http://www.mait.com/assets/india-weee-rev—salient-features.pdf

Maitre-Ekern, E. (2021). Re-thinking producer responsibility for a sustainable circular economy from extended producer responsibility to pre-market producer responsibility. *Journal of Cleaner Production*, *286*, 125454. doi:10.1016/j.jclepro.2020.125454

Ma, J., Cheng, J., Wang, W., Kunisue, T., Wu, M., & Kannan, K. (2011). Elevated concentrations of polychlorinated dibenzo-p-dioxins and polychlorinated dibenzofurans and polybrominated diphenyl ethers in hair from workers at an electronic waste recycling facility in Eastern China. *Journal of Hazardous Materials*, *186*(2-3), 1966–1971. doi:10.1016/j.jhazmat.2010.12.091 PMID:21255916

Majewski, P., Al-shammari, W., Dudley, M., Jit, J., Lee, S. H., Myoung-Kug, K., & Sung-Jim, K. (2021). Recycling of solar PV panels-product stewardship and regulatory approaches. *Energy Policy*, *149*, 112062. doi:10.1016/j.enpol.2020.112062

Mallapragada, D. S., Dvorkin, Y., Modestino, M. A., Esposito, D. V., Smith, W. A., Hodge, B. M., Harold, M. P., Donnelly, V. M., Nuz, A., Bloomquist, C., Baker, K., Grabow, L. C., Yan, Y., Rajput, N. N., Hartman, R. L., Biddinger, E. J., Aydil, E. S., & Taylor, A. D. (2023). Decarbonization of the chemical industry through electrification: Barriers and opportunities. *Joule*, *7*(1), 23–41. doi:10.1016/j.joule.2022.12.008

Mandeep, & Shukla, P. (2020). Microbial nanotechnology for bioremediation of industrial wastewater. *Frontiers in Microbiology*, *11*, 590631. doi:10.3389/fmicb.2020.590631 PMID:33224126

Mangaonkar, N., Sabhani, P., & Sirsat, S. (2014). Green Technology for avoiding E-waste with the help of GIS. *2014 International Conference on Advances in Communication and Computing Technologies (ICACACT 2014)*, 1–6. 10.1109/EIC.2015.7230750

Manikandan, S., Inbakandan, D., & Nachiyar, C. V. (2023). Towards sustainable metal recovery from e-waste: A Mini Review. *Sustainable Chemistry for the Environment*, 100001.

Manish, A., & Chakraborty, P. (2019). E-waste management in India: Challenges and opportunities. *TerraGreen*, *12*, 22–28.

Manomaivibool, P. (2009). Extended producer responsibility in a non-OECD context: The management of waste electrical and electronic equipment in India. *Resources, Conservation and Recycling*, *53*(3), 136–144. doi:10.1016/j.resconrec.2008.10.003

María Fernández-González, J., Díaz-López, C., Martín-Pascual, J., & Zamorano, M. (2020). *Recycling organic fraction of municipal solid waste: Systematic literature review and bibliometric analysis of research trends.* doi:10.3390/su12114798

Marimuthu, V., & Hazardous, A. R.-J. (2022). Investigation of the mechanical properties of M40-grade concrete with PCB fiber from recycled electronic waste. *Journal of Hazardous, Toxic and Radioactive Waste*, *27*(1), 04022034. Advance online publication. doi:10.1061/(ASCE)HZ.2153-5515.0000725

Marks, D., Miller, M. A., & Vassanadumrongdee, S. (2023). Closing the loop or widening the gap? The unequal politics of Thailand's circular economy in addressing marine plastic pollution. *Journal of Cleaner Production, 391*, 136218. doi:10.1016/j.jclepro.2023.136218

Martinelli, C., Pucci, C., Battaglini, M., Marino, A., & Ciofani, G. (2020). Antioxidants and nanotechnology: Promises and limits of potentially disruptive approaches in the treatment of central nervous system diseases. *Advanced Healthcare Materials, 9*(3), 1901589. doi:10.1002/adhm.201901589 PMID:31854132

Martinez, M., Calvo-Parra, G., Gil, E., de la Rubia, O., Hillebrand, M., Rubio, F., Aipperspach, W., & Gombert, A. (2014). Environmental testing results over a tracker drive train. *AIP Conference Proceedings*, (1), 228–232.

Masuduzzaman, M., Shishir, K. S. A., & Alauddin, M. (2018). Utilization of E-waste in Concrete and its Environmental Impact-A Review. In *2018 international conference on smart city and emerging technology (ICSCET)* (pp. 1–4). IEEE. doi:10.1109/ICSCET.2018.8537301

Mayanti, B., & Helo, P. (2023). Circular economy through waste reverse logistics under extended producer responsibility in Finland. *Waste Management & Research.*

Minashkina, D., & Happonen, A. (2022). Analysis of the Past Seven Years of Waste-Related Doctoral Dissertations: A Digitalization and Consumer e-Waste Studies Mystery. *Energies, 15*(18), 6526. doi:10.3390/en15186526

Ministry of Environment, Forest and Climate Change. (2022). Annual Report of 2021-22. Government of India.

Mishra, D. N., & Panda, R. K. (2021). How delightful is Indian wellness tourism? A netnographic study. [AHTR]. *Advances in Hospitality and Tourism Research, 9*(1), 132–156. doi:10.30519/ahtr.784232

Mishra, D. N., & Panda, R. K. (2023). Evaluating visitor–therapist relationship in Indian spa and wellness resorts. *Journal of Hospitality and Tourism Insights, 6*(5), 2433–2461. doi:10.1108/JHTI-06-2022-0246

Misra, N. R., Kumar, S., & Jain, A. (2021). A review on E-waste: Fostering the need for green electronics. *2021 International Conference on Computing, Communication, and Intelligent Systems (ICCCIS)*, 1032–1036. 10.1109/ICCCIS51004.2021.9397191

MoEF. (2011). *Implementation of e-waste rules in India*. CPCB.

MoEF. (2016). *E-Waste (Management) Rules, 2016.* http://www.moef.gov.in/sites/default/files/EWM%20Rules%202016%20english%2023.03.2016.pdf

Mohammed, S. I. (2022). E-Waste Management in Different Countries: Strategies, Impacts, and Determinants. In *Advances in Green Electronics Technologies.* IntechOpen.

Moheb-Alizadeh, H., Sadeghi, A. H., Sahebi Fakhrabad, C. A., Jaunich, M. K., Kemahlioglu-Ziya, E., & Handfield, R. B. (2023). *Reverse Logistics Network Design to Estimate the Economic and Environmental Impacts of Take-back Legislation: A Case Study for E-waste Management System in Washington State.* https://arxiv.org/abs/2301.09792v1

Moriguchi, Y. (2006). *Establishing a sound material cycle society in Asia.* In The Asia 3R Conference, Tokyo, Japan.

Morlet, A. (2018). *The Circular Economy Opportunity for Urban and Industrial Innovation in China.* Ellen MacArthur Foundation.

Mor, R. S., Sangwan, K. S., Singh, S., Singh, A., & Kharub, M. (2021). E-waste management for environmental sustainability: An exploratory study. *Procedia CIRP, 98*, 193–198. doi:10.1016/j.procir.2021.01.029

Morris, A., & Metternicht, G. (2016). Assessing effectiveness of WEEE management policy in Australia. *Journal of Environmental Management, 181*, 218–230. doi:10.1016/j.jenvman.2016.06.013 PMID:27353372

Mourshed, M., Masud, M. H., Rashid, F., & Joardder, M. U. H. (2017). Towards the effective plastic waste management in Bangladesh: A review. *Environmental Science and Pollution Research International, 24*(35), 27021–27046. doi:10.100711356-017-0429-9 PMID:29079979

Mousavi, S., Hosseinzadeh, A., & Golzary, A. (2023). Challenges, recent development, and opportunities of smart waste collection: A review. *The Science of the Total Environment, 886*, 163925. Advance online publication. doi:10.1016/j.scitotenv.2023.163925 PMID:37164079

Moyen Massa, G., & Archodoulaki, V.-M. (2023). Electrical and Electronic Waste Management Problems in Africa: Deficits and Solution Approach. *Environments (Basel, Switzerland), 10*(3), 44. doi:10.3390/environments10030044

Mudali, U. K., Patil, M., Saravanabhavan, R., & Saraswat, V. K. (2021). Review on e-waste recycling: Part I—a prospective urban mining opportunity and challenges. *Transactions of the Indian National Academy of Engineering : an International Journal of Engineering and Technology, 6*(3), 547–568. doi:10.100741403-021-00216-z

Mukherjee, A. (2021). Harnessing digital technologies for environmental sustainability: A systematic review. *Journal of Cleaner Production, 280*, 124388.

Mukta, T. A., & Ahmed, I. (2020). Review on E-waste management strategies for implementing green computing. *Int. J. Comput. Appl, 177*, 45–52.

Munir, S. (2020). Towards smart sustainable cities: A review of the role digital innovation can play in humanizing cities. *Sustainable Cities and Society, 61*, 102365. doi:10.1016/j.scs.2020.102365

Muñoz, I., Gazulla, C., Bala, A., Puig, R., & Fullana, P. (2009). LCA and ecodesign in the toy industry: Case study of a teddy bear incorporating electric and electronic components. *The International Journal of Life Cycle Assessment, 14*(1), 64–72. doi:10.100711367-008-0044-6

Murad, W., & Siwar, C. (2007). Waste management and recycling practices of the urban poor: A case study in Kuala Lumpur city, Malaysia. *Waste Management & Research, 25*(1), 3–13. doi:10.1177/0734242X07070766 PMID:17346002

Murphy, D., Meyers, C., Oelson, S., McKean, K., & Custer, S. (1998). *Exceptions*. Sopris West CO.

Murthy, V., & Ramakrishna, S. (2022). A review on global E-waste management: Urban mining towards a sustainable future and circular economy. *Sustainability (Basel), 14*(2), 647. doi:10.3390u14020647

Muszyńska, B., Rojowski, J., Dobosz, K., & Opoka, W. (2015). Biological and physico-chemical properties of thallium. *Medicina Internacia Revuo*, (105), 180–185.

Naik, S., & Eswari, J. S. (2022). Electrical waste management: Recent advances challenges and future outlook. *Total Environment Research Themes*, 100002.

Naik, S., & Eswari, J. S. (2022). Electrical waste management: Recent advances challenges and outlook. *Total Environment Research Themes*, 100002.

Nakamura, S., & Kondo, Y. (2006). A waste input-output life-cycle cost analysis of the recycling of end-of-life electrical home appliances. *Ecological Economics, 57*(3), 494–506. doi:10.1016/j.ecolecon.2005.05.002

Namo eWaste. (2022). *Namo eWaste—E-Waste Recycling & Management Solution in India*. Namo eWaste.

Nancy, A. A., Ravindran, D., Raj Vincent, P. M. D., Srinivasan, K., & Gutierrez Reina, D. (2022). IoT-Cloud-Based Smart Healthcare Monitoring System for Heart Disease Prediction via Deep Learning. *Electronics (Basel), 11*(15), 2292. doi:10.3390/electronics11152292

Nande, A., Raut, S., Swart, H. C., & Dhoble, S. J. (2023). Heavy metal waste management—Side products of industries and electronic waste. *360-Degree Waste Management, Volume 2. Biomedical, Pharmaceutical, Industrial Waste, and Remediation, 2*, 203–219. doi:10.1016/B978-0-323-90909-9.00012-5

Naphade, M. (2013). Smart cities and Internet of Things. *IEEE Transactions on Automation Science and Engineering, 10*(3), 1443–1454.

Narayanan, A. (2019). Harnessing IoT and mobile technologies for women's safety. *Sustainable Cities and Society, 46*, 101410.

Nazir, Majeed, & Birjais. (2018). Contribution of green computing towards it for providing sustainable environment. *International Journal of Advanced Research in Computer Science, 9*(2).

Needhidasan, S., Samuel, M., & Chidambaram, R. (2014). Electronic waste–an emerging threat to the environment of urban India. *Journal of Environmental Health Science & Engineering, 12*(1), 36. doi:10.1186/2052-336X-12-36 PMID:24444377

Ngetich, M. J., Kiplagat, A. K., & Ng'etich, J. (2018). An Investigation into Domestic Water Consumption Patterns in Iten Town, Kenya. *Africa Environmental Review Journal, 3*(1), 200.

Ni, H. G., Zeng, H., Tao, S., & Zeng, E. Y. (2010). Environmental and human exposure to persistent halogenated compounds derived from e-waste in China. *Environmental Toxicology and Chemistry, 29*(6), 1237–1247. doi:10.1002/etc.160 PMID:20821565

Nnorom, I. C., & Osibanjo, O. (2008). Overview of electronic waste (e-waste) management practices and legislations, and their poor applications in the developing countries. *Resources, Conservation and Recycling, 52*(6), 843–858. doi:10.1016/j.resconrec.2008.01.004

Nnorom, I. C., & Osibanjo, O. (2008). Sound management of brominated flame retarded (BFR) plastics from electronic wastes: State of the art and options in Nigeria. *Resources, Conservation and Recycling, 52*(12), 1362–1372. doi:10.1016/j.resconrec.2008.08.001

Noel-Brune, M., Goldizen, F. C., Neira, M., van den Berg, M., Lewis, N., King, M., Suk, W. A., Carpenter, D. O., Arnold, R. G., & Sly, P. D. (2013). Health effects of exposure to e-waste. *The Lancet. Global Health, 1*(2), e70. doi:10.1016/S2214-109X(13)70020-2 PMID:25104155

Norouzi, N., & Fani, M. (2020). Black gold falls, black plague arise-An Opec crude oil price forecast using a gray prediction model. *Upstream Oil and Gas Technology, 5*, 100015. doi:10.1016/j.upstre.2020.100015

Nowakowski, P., & Pamuła, T. (2020). Application of deep learning object classifier to improve e-waste collection planning. *Waste Management (New York, N.Y.), 109*, 1–9. doi:10.1016/j.wasman.2020.04.041 PMID:32361385

Nuwematsiko, R., Oporia, F., Nabirye, J., Halage, A. A., Musoke, D., & Buregyeya, E. (2021). Knowledge, perceptions, and practices of electronic waste management among consumers in Kampala, Uganda. *Journal of Environmental and Public Health, 2021*. doi:10.1155/2021/3846428 PMID:33628280

Oglu Karimli, V. M., & Oglu Baghirov, H. S. (2020). *Administrative management theory at high educational institutions and its following stages*. Administrative Management Theory at High Educational Institutions and Its Following Stages.

Ogunseitan, O. A., Schoenung, J. M., Saphores, J. D., & Shapiro, A. A. (2009). The electronics revolution: From e-wonderland to e-wasteland. *Science, 326*(5953), 670–671. doi:10.1126cience.1176929 PMID:19900918

Olla, P., & Toth, J. (2010). E-waste education strategies: Teaching how to reduce, reuse and recycle for sustainable development. *International Journal of Environment and Sustainable Development*, *9*(1–3), 294–309. doi:10.1504/IJESD.2010.029977

Ongondo, F. O., Williams, I. D., & Cherrett, T. J. (2011). How are WEEE doing? A global review of the management of electrical and electronic wastes. *Waste Management, 31*(4), 714-730.

Ongondo, F. O., Williams, I. D., & Cherrett, T. J. (2011). How are WEEE doing? A global review of the management of electrical and electronic wastes. *Waste Management (New York, N.Y.)*, *31*(4), 714–730. doi:10.1016/j.wasman.2010.10.023 PMID:21146974

Orisakwe, O. E., Frazzoli, C., Ilo, C. E., & Oritsemuelebi, B. (2019). Public health burden of e-waste in Africa. *Journal of Health & Pollution*, *9*(22), 190610. doi:10.5696/2156-9614-9.22.190610 PMID:31259086

Orlins, S., & Guan, D. (2016). China's toxic informal e-waste recycling: Local approaches to a global environmental problem. *Journal of Cleaner Production*, *114*, 71–80. doi:10.1016/j.jclepro.2015.05.090

Osibanjo, O., & Nnorom, I. C. (2007). The challenge of electronic waste (e-waste) management in developing countries. *Waste Management & Research*, *25*(6), 489–501. doi:10.1177/0734242X07082028 PMID:18229743

Oswald, I., & Reller, A. (2010). Electronic waste–a challenge for sustainable development. *Environmental Sciences Europe*, *22*(1), 1–9.

Oswald, I., & Reller, A. (2011). E-waste: A story of trashing, trading, and valuable resources. *Gaia (Heidelberg)*, *20*(1), 41–47. doi:10.14512/gaia.20.1.9

Oteng-Ababio, M. (2012). Electronic waste management in Ghana-issues and practices. *Sustainable Development*, *20*(4), 293–307.

Oteng, D., Zuo, J., & Sharifi, E. (2022). An expert-based evaluation on end-of-life solar photovoltaic management: An application of Fuzzy Delphi Technique. *Sustainable Horizons*, *4*, 100036. doi:10.1016/j.horiz.2022.100036

Otto, S., Kibbe, A., Henn, L., Hentschke, L., & Kaiser, F. G. (2018). The economy of E-waste collection at the individual level: A practice oriented approach of categorizing determinants of E-waste collection into behavioral costs and motivation. *Journal of Cleaner Production*, *204*, 33–40. doi:10.1016/j.jclepro.2018.08.293

Panchal, R., Singh, A., & Diwan, H. (2021). Economic potential of recycling e-waste in India and its impact on import of materials. *Resources Policy*, *74*, 102264. doi:10.1016/j.resourpol.2021.102264

Panda, R. (2013). E-waste management: A step towards green computing. *International Journal of Environmental Engineering and Management*, *4*(5), 2231–1319.

Pandey, P., & Govind, M. (2014). Social repercussions of e-waste management in India: A study of three informal recycling sites in Delhi. *The International Journal of Environmental Studies*, *71*(3), 241–260.

Pani, S. K., & Pathak, A. A. (2021). Managing plastic packaging waste in emerging economies: The case of EPR in India. *Journal of Environmental Management*, *288*, 112405. doi:10.1016/j.jenvman.2021.112405 PMID:33780821

Panjaitan, T. W., Dargusch, P., Wadley, D., & Aziz, A. A. (2023). A study of management decisions to adopt emission reduction measures in heavy industry in an emerging economy. *Scientific Reports*, *13*(1), 1–15. doi:10.103841598-023-28417-2 PMID:36697460

Pan, X., Wong, C. W., & Li, C. (2022). Circular economy practices in the waste electrical and electronic equipment (WEEE) industry: A systematic review and future research agendas. *Journal of Cleaner Production*, *365*, 132671. doi:10.1016/j.jclepro.2022.132671

Parajuly, K., Kuehr, R., Awasthi, A. K., Fitzpatrick, C., Lepawsky, J., Smith, E., ... & Zeng, X. (2019). *Future e-waste scenarios*. Academic Press.

Parajuly, K., Kuehr, R., Awasthi, A. K., Fitzpatrick, C., Lepawsky, J., Smith, E., Widmer R., & Zeng, X. (2019). *Future e-waste scenarios*. Academic Press.

Parajuly, K., Fitzpatrick, C., Muldoon, O., & Kuehr, R. (2020). Behavioral change for the circular economy: A review with focus on electronic waste management in the EU. *Resources, Conservation and Recycling. X*, *6*, 100035. doi:10.1016/j.rcrx.2020.100035

Parajuly, K., Kuehr, R., Awasthi, A. K., Fitzpatrick, C., Lepawsky, J., Smith, E., & Zeng, X. (2019). *Future e-waste scenarios. StEP Initiative*. UNU ViE-SCYCLE, and UNEP IETC.

Pardini, K., Rodrigues, J. J. P. C., Kozlov, S. A., Kumar, N., & Furtado, V. (2019). IoT-based solid waste management solutions: A survey. *Journal of Sensor and Actuator Networks*, *8*(1), 5. doi:10.3390/jsan8010005

Pariatamby, A., & Victor, D. (2013). Policy trends of e-waste management in Asia. *Journal of Material Cycles and Waste Management*, *15*(4), 411–419. doi:10.100710163-013-0136-7

Park, P.-J., Tahara, K., Jeong, I.-T., & Lee, K.-M. (2006). Comparison of four methods for integrating environmental and economic aspects in the end-of-life stage of a washing machine. *Resources, Conservation and Recycling*, *48*(1), 71–85. doi:10.1016/j.resconrec.2006.01.001

Parmar, Pandya, & Kumbharana. (2011). Optimization of Energy Usage for Computer Systems by Effective Implementation of Green Computing. *International Journal of Advanced Networking Applications*, 87–90.

Parvez, S. M., Hasan, S. S., Knibbs, L. D., Jahan, F., Rahman, M., Raqib, R., Islam, N., Aich, N., Moniruzzaman, M., Islam, Z., Fujimura, M., & Sly, P. D. (2022). Ecological Burden of e-Waste in Bangladesh—an Assessment to Measure the Exposure to e-Waste and Associated Health Outcomes: Protocol for a Cross-sectional Study. *JMIR Research Protocols*, *11*(8), e38201. doi:10.2196/38201 PMID:35972788

Patel, P., Ali, M. I., & Sheth, A. (2018). From raw data to smart manufacturing: AI and semantic web of things for industry 4.0. *IEEE Intelligent Systems*, *33*(4), 79–86. doi:10.1109/MIS.2018.043741325

Patra, B. C., & Lenka, U. (2021). An AHP analysis of scientometrically derived factors of entrepreneurial intentions of women and constructing a conceptual research framework. *Journal of Research in Marketing and Entrepreneurship*, *23*(1), 139–158. doi:10.1108/JRME-02-2020-0017

Patra, B. C., & Lenka, U. (2022). Barriers to entrepreneurial intentions of women: Nominal group technique, analytic hierarchy process, and scientometric approach instigating the necessity of policy intervention. *Journal of Public Affairs*, *22*(2), e2429.

Patra, B. C., & Lenka, U. (2023). Scientometric, fuzzy NGT and DEMATEL analysis for determining sustainable business practices for entrepreneurial firms. *Benchmarking*.

Patra, B. C., & Mishra, D. N. (2023). Boosting entrepreneurial intentions: A novel EEG-guided protocol. *MethodsX*, *11*, 102358. doi:10.1016/j.mex.2023.102358 PMID:37701733

Pendharkar, V. (2018). *Indian cities stare at a mountain of e-waste, with little idea of how to manage it*. Citizen Matters.

Perkins, D. N., Kjellén, M., & Nehler, T. (2017). The digital divide: Information technology, computerization, obsolescence, and electronic waste. *Cognitive Technologies*, 173-210.

Pickren, G. (2014). Political ecologies of electronic waste: Uncertainty and legitimacy in the governance of e-waste geographies. *Environment & Planning A*, *46*(1), 26–45. doi:10.1068/a45728

Pinto, V. N. (2008). E-waste hazard: The impending challenge. *Indian Journal of Occupational and Environmental Medicine*, *12*(2), 65. doi:10.4103/0019-5278.43263 PMID:20040981

Pollock, D., & Coulon, R. (1996). Life cycle assessment of an inkjet print cartridge. *Proceedings of the 1996 IEEE International Symposium on Electronics and the Environment, 1996, ISEE-1996*, 154–160. 10.1109/ISEE.1996.501870

Połomka, J., Jędrczak, A., & Myszograj, S. (2020). Recovery of stabilizer glass in innovative MBT Installation—An analasys of new technological procedure. *Materials (Basel)*, *13*(6), 1356. doi:10.3390/ma13061356 PMID:32192113

Porcari, A., Borsella, E., Benighaus, C., Grieger, K., Isigonis, P., Chakravarty, S., Kines, P., & Jensen, K. A. (2019). From risk perception to risk governance in nanotechnology: A multi-stakeholder study. *Journal of Nanoparticle Research*, *21*(11), 1–19. doi:10.100711051-019-4689-9

Pouikli, K. (2020). Concretising the role of extended producer responsibility in European Union waste law and policy through the lens of the circular economy. Springer Berlin Heidelberg.

PR, S. K., & Narula, D. S. (2018). Educational Strategies for all Eviating the Environmental Impact of E-Waste for Sustainable Development. *Cosmos An International Journal of Management*, *7*(2), 1–9.

Pradhan, J. (2014). *Informal e-waste recycling: environmental risk assessment of heavy metal contamination in Mandoli industrial area, Delhi, India.* doi:10.1007/s11356-014-2713-2

Pramila, S., Fulekar, M., & Bhawana, P. (2012). E-waste- A challenge for tomorrow. *Research Journal of Recent Sciences ISSN*, *2277*, 2502.

Prasad, K. D., Krishna Murthy, P., Hima Gireesh, Ch., Prasad, M., & Sravani, K. (2021). Prioritization of e-waste management strategies towards green computing using AHP-QFD approach. *Proc. Eng*, *3*, 33–40.

Preeti, M., & Sayali, A. (2021). Scientometric analysis of research on end-of-life electronic waste and electric vehicle battery waste. *J. Scientometr. Res*, *10*(1), 37–46. doi:10.5530/jscires.10.1.5

Prek, M. (2004). Environmental impact and life cycle assessment of heating and air conditioning systems, a simplified case study. *Energy and Building*, *36*(10), 1021–1027. doi:10.1016/j.enbuild.2004.06.008

Priya, A., & Hait, S. (2018). Comprehensive characterization of printed circuit boards of various end-of-life electrical and electronic equipment for beneficiation investigation. *Waste Management (New York, N.Y.)*, *75*, 103–123. doi:10.1016/j.wasman.2018.02.014 PMID:29454818

Prusty, B. (2020). ICT enabled low-cost rural women entrepreneurship development: A framework for sustainable rural development. *Computers, Materials & Continua*, *63*(1), 381–399.

Puckett, J., Westervelt, S., Gutierrez, R., & Davis, S. (2002). *Exporting harm: The high-tech trashing of Asia.* The Basel Action Network.

Purchase, C. K., Al Zulayq, D. M., O'Brien, B. T., Kowalewski, M. J., Berenjian, A., Tarighaleslami, A. H., & Seifan, M. (2021). Circular economy of construction and demolition waste: A literature review on lessons, challenges, and benefits. *Materials (Basel)*, *15*(1), 76. doi:10.3390/ma15010076 PMID:35009222

Purchase, D., Abbasi, G., Bisschop, L., Chatterjee, D., Ekberg, C., Ermolin, M., Fedotov, P., Garelick, H., Isimekhai, K., Kandile, N. G., Lundström, M., Matharu, A., Miller, B. W., Pineda, A., Popoola, O. E., Retegan, T., Ruedel, H., Serpe, A., Sheva, Y., ... Wong, M. H. (2020). Global occurrence, chemical properties, and ecological impacts of e-wastes (IUPAC Technical Report). *Pure and Applied Chemistry*, *92*(11), 1733–1767. doi:10.1515/pac-2019-0502

Qin, X., Qin, Z., Li, Y., Zhao, Y., Xia, X., Yan, S., Tian, M., Zhao, X., Xu, X., & Yang, Y. (2011). Polybrominated diphenyl ethers in chicken tissues and eggs from an electronic waste recycling area in southeast China. *Journal of Environmental Sciences (China)*, *23*(1), 133–138. doi:10.1016/S1001-0742(10)60384-1 PMID:21476352

Quinn, M., Jordan, H., & Lacy-Nichols, J. (2021). Upstream and downstream explanations of the harms of ultra-processed foods in national dietary guidelines. *Public Health Nutrition*, *24*(16), 5426–5435. doi:10.1017/S1368980021003505 PMID:34392856

Qu, W., Bi, X., Sheng, G., Lu, S., Fu, J., Yuan, J., & Li, L. (2007). Exposure to polybrominated diphenyl ethers among workers at an electronic waste dismantling region in Guangdong, China. *Environment International*, *33*(8), 1029–1034. doi:10.1016/j.envint.2007.05.009 PMID:17618686

Raharjo, S., Wulandari, S., & Fitriani, S. (2021, June). Waste bank system improvement for electronic waste recycling in Indonesia: A case study of Padang City. *IOP Conference Series. Earth and Environmental Science*, *802*(1), 012053. doi:10.1088/1755-1315/802/1/012053

Rajesh, R., Kanakadhurga, D., & Prabaharan, N. (2022). Electronic waste: A critical assessment on the unimaginable growing pollutant, legislations and environmental impacts. *Environmental Challenges*, *7*, 100507. doi:10.1016/j.envc.2022.100507

Raj, P. (2020). IoT-Driven Healthcare Resource Allocation in Rural India. *Journal of Medical Systems*, *44*, 100. doi:10.100710916-020-01566-7

Rajput, R., & Nigam, N. A. (2021). An overview of E-waste, its management practices and legislations in present Indian context. *Journal of Applied and Natural Science*, *13*(1), 34–41. doi:10.31018/jans.v13i1.2440

Ramachandra, T. V., & Saira-Varghese, K. (2004). Environmentally sound options for e-wastes management. *Envies Journal of Human Settlements*.

Ramasubramanian, B., Tan, J., Chellappan, V., & Ramakrishna, S. (2023). Recent Advances in Extended Producer Responsibility Initiatives for Plastic Waste Management in Germany and UK. *Materials Circular Economy*, *5*(1), 6. doi:10.100742824-023-00076-8

Ramesh, M. V., Paramasivan, M., Akshay, P., & Jarin, T. (2023). A review on electric and electronic waste material management in 21st century. *Materials Today: Proceedings*. Advance online publication. doi:10.1016/j.matpr.2023.01.057

Ramos-Ruiz, A., Wilkening, J. V., Field, J. A., & Sierra-Alvarez, R. (2017). Leaching of cadmium and tellurium from cadmium telluride (CdTe) thin-film solar panels under simulated landfill conditions. *Journal of Hazardous Materials*, *336*, 57–64. doi:10.1016/j.jhazmat.2017.04.052 PMID:28472709

Ranbhise, P. (2014). Green computing a way towards environmentally sustainable future. *2014 International Conference on Contemporary Computing and Informatics (IC3I)*, 1094–1100. 10.1109/IC3I.2014.7019596

Rathore, N., & Panwar, N. L. (2022). Strategic overview of management of future solar photovoltaic panel waste generation in the Indian context. *Waste Management & Research*, *40*(5), 504–518. doi:10.1177/0734242X211003977 PMID:33885343

Ratti, R. (2020). Industrial applications of green chemistry: Status, Challenges and Prospects. *SN Applied Sciences*, *2*(2), 263. doi:10.100742452-020-2019-6

Rautela, R., Arya, S., Vishwakarma, S., Lee, J., Kim, K.-H., & Kumar, S. (2021). E-waste management and its effects on the environment and human health. *The Science of the Total Environment*, *773*, 145623. doi:10.1016/j.scitotenv.2021.145623 PMID:33592459

Ren, M., Zhang, S., Liu, X., Feng, G., Xu, Z., & Fang, J. (2010). Concentrations and profiles of polychlorinated dioxins and furans in a discarded electronic waste open burning site. In *2010 4th International Conference on Bioinformatics and Biomedical Engineering* (pp. 1-4). IEEE.

Report, B. C. (2023, October 20). *E-waste collection: best practices and lessons learned in Indonesia, Trinidad and Tobago and in the European region, from PACE II pilot projects.* https://www.basel.int/Default.aspx?tabid=4210&meetId=4A404336-4967-EE11-90C9-005056A3140F&lang=en

Research and Markets. (2023a, December 20). *E-waste Management - Global Strategic Business Report.* https://www.researchandmarkets.com/report/e-waste#reld0-5774988

Research and Markets. (2023b, January 22). *Global E-Waste Management Market Outlook, 2028.* https://www.researchandmarkets.com/reports/5774988/global-e-waste-management-market-outlook

Reynolds, O. (2021). The World's Fastest Growing Economies. *Focus Economics, 16.*

Rim, K. T. (2020). In silico prediction of toxicity and its applications for chemicals at work. *Toxicology and Environmental Health Sciences*, *12*(3), 191–202. doi:10.100713530-020-00056-4 PMID:32421081

Rizzi, F., Bartolozzi, I., Borghini, A., & Frey, M. (2013). Environmental management of end-of-life products: Nine factors of sustainability in collaborative networks. *Business Strategy and the Environment*, *22*(8), 561–572. doi:10.1002/bse.1766

Robinson, B. H. (2009). E-waste: An assessment of global production and environmental impacts. *The Science of the Total Environment*, *408*(2), 183–191. doi:10.1016/j.scitotenv.2009.09.044 PMID:19846207

Roy, H., Rahman, T. U., Suhan, M. B. K., Al-Mamun, M. R., Haque, S., & Islam, M. S. (2022). A comprehensive review on hazardous aspects and management strategies of electronic waste: Bangladesh perspectives. *Heliyon*, *8*(7), e09802. doi:10.1016/j.heliyon.2022.e09802 PMID:35815143

Roy, J. J., Rarotra, S., Krikstolaityte, V., Zhuoran, K. W., Cindy, Y. D.-I., Tan, X. Y., Carboni, M., Meyer, D., Yan, Q., & Srinivasan, M. (2022). Green recycling methods to treat lithium-ion batteries E-waste: A circular approach to sustainability. *Advanced Materials*, *34*(25), 2103346. doi:10.1002/adma.202103346 PMID:34632652

Rucevska, I., Nellemann, C., Isarin, N., Yang, W., Liu, N., Yu, K., Sandnæs, S., Olley, K., McCann, H., Devia, L., Bisschop, L., Soesilo, D., Schoolmeester, T., Henriksen, R., & Nilsen, R. (2015). *Waste crime – waste risks: Gaps in meeting the global waste challenge. A UNEP Rapid Response Assessment.* United Nations Environment Programme and GRID-Arendal, Nairobi and Arendal. http://www.grida.no

Ruggiano, N., & Perry, T. E. (2019). Conducting secondary analysis of qualitative data: Should we, can we, and how? *Qualitative Social Work: Research and Practice*, *18*(1), 81–97. doi:10.1177/1473325017700701 PMID:30906228

Ruiz, A. (2022). *Latest global E-waste statistics and what they tell us.* TheRoundup.

Sadiku, Ampah, & Musa. (2018). Green computing: a primer. *Journal of Scientific and Engineering Research, 5*(4), 247–251.

Saeed, B. B., Afsar, B., Hafeez, S., Khan, I., Tahir, M., & Afridi, M. A. (2019). Promoting employee's proenvironmental behavior through green human resource management practices. *Corporate Social Responsibility and Environmental Management*, *26*(2), 424–438. doi:10.1002/csr.1694

Saha, L., Kumar, V., Tiwari, J., Rawat, S., Singh, J., & Bauddh, K. (2021). Electronic waste and their leachates impact on human health and environment: Global ecological threat and management. *Environmental Technology & Innovation, 24*, 102049. doi:10.1016/j.eti.2021.102049

Sahoo, M., Vishwakarma, S., Panigrahi, C., & Kumar, J. (2021). Nanotechnology: Current applications and future scope in food. *Food Frontiers, 2*(1), 3–22. doi:10.1002/fft2.58

Saini, S., Chauhan, A., Thakur, G., & Sapra, L. (2023). Challenges and Opportunities in Secure Smart Cities for Enhancing the Security and Privacy. *Enabling Technologies for Effective Planning and Management in Sustainable Smart Cities*, 1–27.

Sakhuja, D., Ghai, H., Bhatia, R. K., & Bhatt, A. K. (2022). Management of e-Waste: technological challenges and opportunities. Handbook of Solid Waste Management: Sustainability through Circular Economy, 1523-1557.

Samsami, S., Mohamadizaniani, M., Sarrafzadeh, M. H., Rene, E. R., & Firoozbahr, M. (2020). Recent advances in the treatment of dye-containing wastewater from textile industries: Overview and perspectives. *Process Safety and Environmental Protection, 143*, 138–163. doi:10.1016/j.psep.2020.05.034

Saphores, J. D., Nixon, H., Ogunseitan, O. A., & Shapiro, A. A. (2006). Household willingness to recycle electronic waste: An application to California. *Environment and Behavior, 38*(2), 183–208. doi:10.1177/0013916505279045

Sasidharan, B. (2019). Internet of Things (IoT) for Smart Solid Waste Management: A review. *Waste Management (New York, N.Y.), 94*, 267–282.

Savvilotidou, V., Hahladakis, J. N., & Gidarakos, E. (2015). Leaching capacity of metals–metalloids and recovery of valuable materials from waste LCDs. *Waste Management (New York, N.Y.), 45*, 314–324. doi:10.1016/j.wasman.2015.05.025 PMID:26087646

Schischke, K., & Spielmann, M. (2001). Environmental assessment in production of electronic components – possibilities and obstacles of LCA methodology. *13th Discussion Forum on Life Cycle Assessment*, 9.

Schluep, M., Hagelüken, C., Kuehr, R., Magalini, F., Maurer, C., Meskers, C., & Wang, F. (2013). *Recycling–from e-waste to resources*. United Nations Environment Programme & United Nations University.

Schwarzer, S., De Bono, A., Giuliani, G., Kluser, S., & Peduzzi, P. (2005). E-waste, the hidden side of IT equipment's manufacturing and use. *Environment Alert Bulletin, 5*.

Schwarzer, R., Rochau, U., Saverno, K., Jahn, B., Bornschein, B., Muehlberger, N., Flatscher-Thoeni, M., Schnell-Inderst, P., Sroczynski, G., Lackner, M., Schall, I., Hebborn, A., Pugner, K., Fehervary, A., Brixner, D., & Siebert, U. (2015). Systematic overview of cost–effectiveness thresholds in ten countries across four continents. *Journal of Comparative Effectiveness Research, 4*(5), 485–504. doi:10.2217/cer.15.38 PMID:26490020

Secretariat, R. S. (2011). *E-waste in India. India Research Unit (Larrdis)*. Rajya Sabha Secretariat.

Seif, R., Salem, F. Z., & Allam, N. K. (2023). E-waste recycled materials as efficient catalysts for renewable energy technologies and better environmental sustainability. *Environment, Development and Sustainability*. Advance online publication. doi:10.100710668-023-02925-7 PMID:36691418

Seow, Y. (2020). *Singapore's smart container initiative enhances cross-border e-waste transport*. Waste Management World. Retrieved from https://www.mdpi.com/2071-1050/14/2/647

Sepúlveda, A., Schluep, M., Renaud, F. G., Streicher, M., Kuehr, R., Hagelüken, C., & Gerecke, A. C. (2010). A review of the environmental fate and effects of hazardous substances released from electrical and electronic equipments during recycling: Examples from China and India. *Environmental Impact Assessment Review*, *30*(1), 28–41. doi:10.1016/j.eiar.2009.04.001

Sethurajan, M., van Hullebusch, E. D., Fontana, D., Akcil, A., Deveci, H., Batinic, B., Leal, J. P., Gasche, T. A., Ali Kucuker, M., Kuchta, K., Neto, I. F. F., Soares, H. M. V. M., & Chmielarz, A. (2019). Recent advances on hydrometallurgical recovery of critical and precious elements from end of life electronic wastes-a review. *Critical Reviews in Environmental Science and Technology*, *49*(3), 212–275. doi:10.1080/10643389.2018.1540760

Shahabuddin, M., Uddin, M. N., Chowdhury, J. I., Ahmed, S. F., Uddin, M. N., Mofijur, M., & Uddin, M. A. (2023). A review of the recent developments, challenges, and opportunities of electronic waste (e-waste). *International Journal of Environmental Science and Technology*, *20*(4), 4513–4520. doi:10.100713762-022-04274-w

Shaikh, S., Thomas, K., Zuhair, S., & Magalini, F. (2020). A cost-benefit analysis of the downstream impacts of e-waste recycling in Pakistan. *Waste Management (New York, N.Y.)*, *118*, 302–312. doi:10.1016/j.wasman.2020.08.039 PMID:32919349

Shang, Y., Hasan, M. K., Ahammed, G. J., Li, M., Yin, H., & Zhou, J. (2019). Applications of nanotechnology in plant growth and crop protection: A review. *Molecules (Basel, Switzerland)*, *24*(14), 2558. doi:10.3390/molecules24142558 PMID:31337070

Sharan, A. M. (2002). *Design of an Economical, and Mechanically Stable E-Cycle*. NECEC St. John's.

Sharma, H., Rawal, N., & Mathew, B. B. (2015). The characteristics, toxicity and effects of cadmium. *International Journal of Nanotechnology and Nanoscience*, *3*(10).

Sharma, A., Mahajan, P., & Garg, R. (2023). End-of-life solar photovoltaic panel waste management in India: Forecasting and environmental impact assessment. *International Journal of Environmental Science and Technology*, 1–20. doi:10.100713762-023-04953-2 PMID:37360560

Sharma, H. B., Vanapalli, K. R., Samal, B., Cheela, V. R. S., Dubey, B. K., & Bhattacharya, J. (2021). Circular economy approach in solid waste management system to achieve UN-SDGs: Solutions for post-COVID recovery. *The Science of the Total Environment*, *800*, 149605. doi:10.1016/j.scitotenv.2021.149605 PMID:34426367

Sheng, W., & Yan, G., JingGuang, L., TingMing, S., YunFeng, Z., & YongNing, W. (2011). Particle-bound PCDD/Fs in the atmosphere of an electronic waste dismantling area in China. *Biomedical and Environmental Sciences*, *24*(2), 102–111. PMID:21565680

Shen, H., Ding, G., Han, G., Wang, X., Xu, X., Han, J., Lou, X., Xu, C., Cai, D., Song, Y., & Lu, W. (2010). Distribution of PCDD/Fs, PCBs, PBDEs and organochlorine residues in children's blood from Zhejiang, China. *Chemosphere*, *80*(2), 170–175. doi:10.1016/j.chemosphere.2010.03.012 PMID:20378148

Shen, Y., & Dierking, I. (2019). Perspectives in liquid-crystal-aided nanotechnology and nanoscience. *Applied Sciences (Basel, Switzerland)*, *9*(12), 2512. doi:10.3390/app9122512

Sheoran, M., & Das Gupta, D. (2023). *International best practices for e-waste take back and policy interventions for India*. Facilities. doi:10.1108/F-03-2023-0027

Sheoran, M., Kumar, P., Sharma, S., & Bukya, M. (2022). Current situation analysis of solar PV waste management in India. *Materials Today: Proceedings*, *58*, 773–782. doi:10.1016/j.matpr.2022.03.118

Sheoran, M., Sharma, S., & Kumar, P. (2020). Compatible standard policy measure to tackle solar photovoltaic waste in Indian scenario. *Journal of Physics: Conference Series*, *1504*(1), 012012. doi:10.1088/1742-6596/1504/1/012012

Shittu, O. S., Williams, I. D., & Shaw, P. J. (2021). Global E-waste management: Can WEEE make a difference? A review of e-waste trends, legislation, contemporary issues and future challenges. *Waste Management (New York, N.Y.)*, *120*, 549–563. doi:10.1016/j.wasman.2020.10.016 PMID:33308953

Shreyas Madhav, A. V., Rajaraman, R., Harini, S., & Kiliroor, C. C. (2022). Application of artificial intelligence to enhance collection of E-waste: A potential solution for household WEEE collection and segregation in India. *Waste Management & Research*, *40*(7), 1047–1053. doi:10.1177/0734242X211052846 PMID:34726090

Shubbak, M. H. (2019). Advances in solar photovoltaics: Technology review and patent trends. *Renewable & Sustainable Energy Reviews*, *115*, 109383. doi:10.1016/j.rser.2019.109383

Sileyew, K. J. (2019). *Research design and methodology*. Cyberspace.

Silva, R. J., Bidoglio, G., Robouch, P. B., Puigdomenech, I., Wanner, H., & Rand, M. H. (2012). *Chemical thermodynamics of americium*. Newnes.

Sim, S., & Wong, N. K. (2021). Nanotechnology and its use in imaging and drug delivery. *Biomedical Reports*, *14*(5), 1–9. doi:10.3892/br.2021.1418 PMID:33728048

Singh, A., & Sharma, R. (n.d.). *Health risk assessment of heavy metals via dietary intake of foodstuffs from the wastewater irrigated site of a dry tropical area of India*. Elsevier. Retrieved November 19, 2023, from https://www.sciencedirect.com/science/article/pii/S0278691509005572

Singh, A., Dubey, S., & Dubey, H. K. (2019). Nanotechnology: The future engineering. *Nanotechnology*, *6*(2), 230–233.

Singh, A., Sharma, R. K., Agrawal, M., & Marshall, F. M. (2010). Health risk assessment of heavy metals via dietary intake of foodstuffs from the wastewater irrigated site of a dry tropical area of India. *Food and Chemical Toxicology*, *48*(2), 611–619. doi:10.1016/j.fct.2009.11.041 PMID:19941927

Singhal, D., Tripathy, S., & Jena, S. K. (2019). Sustainability through remanufacturing of e-waste: Examination of critical factors in the Indian context. *Sustainable Production and Consumption*, *20*, 128–139. doi:10.1016/j.spc.2019.06.001

Singh, T., Sharma, S., & Nagesh, S. (2017). Socio-economic status scales updated for 2017. *International Journal of Research in Medical Sciences*, *5*(7), 3264–3267. doi:10.18203/2320-6012.ijrms20173029

Sinha, P. (2019). Assessment of environmental impacts and human health risks associated with e-waste handling processes in India: A review. *Environment International*, *132*, 105079. doi:10.1016/j.envint.2019.105079

Sinha, S. (2011). *Waste Electrical and Electronic Equipment: The EU and India, Sharing Best Practices*. Toxics Link.

Sivaramanan, S. (2013). E-Waste Management, Disposal and Its Impacts on the Environment. *Universal Journal of Environmental Research and Technology*, *3*(5).

Skinner, M. K., Manikkam, M., & Guerrero-Bosagna, C. (2010). Epigenetic transgenerational actions of environmental factors in disease etiology. *Trends in Endocrinology and Metabolism*, *21*(4), 214–222. doi:10.1016/j.tem.2009.12.007 PMID:20074974

Slaveykova, V. I., Couture, P., Duquesne, S., D'Hugues, P., & Sánchez, W. (2019). Recycling, reuse, and circular economy: A challenge for ecotoxicological research. *Environmental Science and Pollution Research International*, *26*(21), 1–4. doi:10.100711356-019-04626-z PMID:31104246

Smart Product Manager. (2023). *Tesla's Electric Revolution: Disrupting the Auto Industry with Innovation.* Available at: https://smartproductmanager.medium.com/teslas-electric-revolution-disrupting-the-auto-industry-with-innovation-8ed879f002eb

Smith, J. (2018). E-waste: Environmental problems and current management. *Journal of Environmental Management, 195*, 73–82.

Socolof, M. L., Overly, J. G., & Geibig, J. R. (2005). Environmental life-cycle impacts of CRT and LCD desktop computer displays. *Journal of Cleaner Production, 13*(13-14), 1281–1294. doi:10.1016/j.jclepro.2005.05.014

Sodhi, M. S., & Reimer, B. (2001). Models for recycling electronics end-of-life products: Modelle zum Recycling von Altprodukten aus dem Elektronikbereich. *OR-Spektrum, 23*(1), 97–115. doi:10.1007/PL00013347

Song, X., Yang, J., & Liu, G. (2015). Stakeholder coordination in municipal solid waste management: A critical review. *Waste Management (New York, N.Y.), 38*, 157–168.

Song, X., Zhang, C., Yuan, W., & Yang, D. (2018). Life-cycle energy use and GHG emissions of waste television treatment system in China. *Resources, Conservation and Recycling, 128*, 470–478. doi:10.1016/j.resconrec.2016.09.004

SOTI. (2022). *Reduce, reuse, rethink: From discard mentality to tech sustainability.* Available at https://soti.net/media/6647/industry-report-reduce-reuse-rethink-from-discard-mentality-to-tech-sustainability-letter.pdf

So, W. W. M., Lee, J. C. K., & Chow, C. F. (2019). *Environmental sustainability and education for waste management.* Springer Singapore. doi:10.1007/978-981-13-9173-6

Sportelli, M. C., Izzi, M., Kukushkina, E. A., Hossain, S. I., Picca, R. A., Ditaranto, N., & Cioffi, N. (2020). Can nanotechnology and materials science help the fight against SARS-CoV-2? *Nanomaterials (Basel, Switzerland), 10*(4), 802. doi:10.3390/nano10040802 PMID:32326343

Sridhar, A., Ponnuchamy, M., Kumar, P. S., & Kapoor, A. (2021). Food preservation techniques and nanotechnology for increased shelf life of fruits, vegetables, beverages and spices: A review. *Environmental Chemistry Letters, 19*(2), 1715–1735. doi:10.100710311-020-01126-2 PMID:33192209

Srivastava, R., & Sharma, D. (2015). Factors affecting e-waste management: an interpretive structural modeling approach. *2015 Fifth international conference on communication systems and network technologies*, 1307–1312. 10.1109/CSNT.2015.158

Srivastav, A. L., Markandeya, Patel, N., Pandey, M., Pandey, A. K., Dubey, A. K., Kumar, A., Bhardwaj, A. K., & Chaudhary, V. K. (2023). Concepts of circular economy for sustainable management of electronic wastes: Challenges and management options. *Environmental Science and Pollution Research International, 30*(17), 48654–48675. doi:10.100711356-023-26052-y PMID:36849690

Srivastava, R. R., & Pathak, P. (2020). Policy issues for efficient management of E-waste in developing countries. In *Handbook of Electronic Waste Management* (pp. 81–99). Butterworth-Heinemann. doi:10.1016/B978-0-12-817030-4.00002-4

Statista.com. (2023). *Global E-Waste - Statistics & Facts*. Available at: https://www.statista.com/topics/3409/electronic-waste-worldwide/#topicOverview

Stefansdotter, A., Knusden, J. S., Flack, M., & Hansen, P. G. (2016). *Nudging for sustainable consumption of electronics. Summary Report.* Denmark: Nordic Council of Ministers. Available online at: https://norden.diva-portal.org/smash/get/diva

StEP (Solving the e-Waste Problem) Annual Report, 2015/2016.

StEP (Solving the e-Waste Problem). (2018). http://www.step-initiative.org

StEP Green Paper (Solving the e-Waste Problem). (2015). *E-waste prevention, take-back system design and policy approaches*. Academic Press.

Sthiannopkao, S., & Wong, M. H. (2013). Handling e-waste in developed and developing countries: Initiatives, practices, and consequences. *The Science of the Total Environment*, *463*, 1147–1153. doi:10.1016/j.scitotenv.2012.06.088 PMID:22858354

Stone, S. (2020). *The 4Rs: Reduce, Reuse, Recycle and Recover with the Gloucestershire Energy from Waste Facility*. Gloucester Energy from Waste Facility.

Strijker, D., Bosworth, G., & Bouter, G. (2020). Research methods in rural studies: Qualitative, quantitative and mixed methods. *Journal of Rural Studies*, *78*, 262–270. doi:10.1016/j.jrurstud.2020.06.007

Sudas, P., Saini, A., & Taneja, S. (2012). Implementation of cloud computing in SMEs in India. *Recent Trends in Computer Science & Information Technology, 100.*

Sulaiman Zangina, A., Ahmed, A., Nuhu, Z., & Sani Alhassan, Z. (2023). e-Waste quantification and the associated valuable resources in Kano metropolis, Nigeria. *Journal of Material Cycles and Waste Management*, *25*(6), 1–10. doi:10.100710163-023-01793-7

Sum, E. Y. (1991). The recovery of metals from electronic scrap. *Journal of the Minerals Metals & Materials Society*, *43*(4), 53–61. doi:10.1007/BF03220549

Sun, X., Atiyeh, H. K., Huhnke, R. L., & Tanner, R. S. (2019). Syngas fermentation process development for production of biofuels and chemicals: A review. *Bioresource Technology Reports*, *7*, 100279. doi:10.1016/j.biteb.2019.100279

Sun, Y. (2017). The applications of IoT in healthcare: A systematic review. *Journal of King Saud University. Computer and Information Sciences*. Advance online publication. doi:10.1016/j.jksuci.2017.09.001

Suresh Babu, C. V., & Sivaneshwaran, J. (2023). Controlling Computer Features Through Hand Gesture. In V. Thayananthan (Ed.), AI-Based Digital Health Communication for Securing Assistive Systems (pp. 85-113). IGI Global. doi:10.4018/978-1-6684-8938-3.ch005

Suresh Babu, C. V., Akshayah, N. S., Maclin Vinola, P., & Janapriyan, R. (2023). IoT-Based Smart Accident Detection and Alert System. In P. Swarnalatha & S. Prabu (Eds.), *Handbook of Research on Deep Learning Techniques for Cloud-Based Industrial IoT* (pp. 322–337). IGI Global. doi:10.4018/978-1-6684-8098-4.ch019

Suresh, S., Singhvi, S., & Rustagi, V. (2019). *Managing India's PV module waste*. Available at: https://bridgetoindia.com/backend/wp-content/uploads/2019/04/BRIDGE-TO-INDIA-Managing-Indias-Solar-PV-Waste-1.pdf

Suri, H. (2020). Ethical considerations of conducting systematic reviews in educational research. Systematic reviews in educational research: Methodology, perspectives and application, 41-54. doi:10.1007/978-3-658-27602-7_3

Syafa Bakri, S. N., Surif, S., & Ramasamy, R. K. 2008. A case study of life cycle assessment (LCA) on ballast for fluorescent lamp in Malaysia. *IEEE International Symposium on Electronics and the Environment*, 1–4.

Syaifullah, D. H., Tjahjono, B., McIlhatton, D., & Zagloel, T. Y. M. (2022). The impacts of safety on sustainable production performance in the chemical industry: A systematic review of literature and conceptual framework. *Journal of Cleaner Production, 366*, 132876. doi:10.1016/j.jclepro.2022.132876

Taherdoost, H. (2022). What are different research approaches? Comprehensive review of qualitative, quantitative, and mixed method research, their applications, types, and limitations. *Journal of Management Science & Engineering Research*, *5*(1), 53–63. doi:10.30564/jmser.v5i1.4538

Tang, X., Shen, C., Chen, L., Xiao, X., Wu, J., Khan, M. I., Dou, C., & Chen, Y. (2010). Inorganic and organic pollution in agricultural soil from an emerging e-waste recycling town in Taizhou area, China. *Journal of Soils and Sediments, 10*(5), 895–906. doi:10.100711368-010-0252-0

Tang, X., Shen, C., Shi, D., Cheema, S. A., Khan, M. I., Zhang, C., & Chen, Y. (2010). Heavy metal and persistent organic compound contamination in soil from Wenling: An emerging e-waste recycling city in Taizhou area, China. *Journal of Hazardous Materials, 173*(1-3), 653–660. doi:10.1016/j.jhazmat.2009.08.134 PMID:19783093

Tan, L. (2020). Smart Waste Management: A Case Study in Singapore. *Journal of Sustainable Cities and Society, 58*, 102137. doi:10.1016/j.scs.2020.102137

Tansel, B. (2017). From electronic consumer products to e-wastes: Global outlook, waste quantities, recycling challenges. *Environment International, 98*, 35–45. doi:10.1016/j.envint.2016.10.002 PMID:27726897

Tasnim, S. S., Rahman, M. M., Hasan, M. M., Shammi, M., & Tareq, S. M. (2022). Current challenges and future perspectives of solar-PV cell waste in Bangladesh. *Heliyon, 8*(2), e08970. doi:10.1016/j.heliyon.2022.e08970 PMID:35243090

Tesfaye, F., Lindberg, D., Hamuyuni, J., Taskinen, P., & Hupa, L. (2017). Improving urban mining practices for optimal recovery of resources from e-waste. *Minerals Engineering, 111*, 209–221. doi:10.1016/j.mineng.2017.06.018

Thakur, P., & Kumar, S. (2022). Evaluation of e-waste status, management strategies, and legislations. *International Journal of Environmental Science and Technology, 19*(7), 6957–6966. doi:10.100713762-021-03383-2

Themelis, N. J. (2019). *Recovery of materials and energy from urban wastes*. Springer Berlin Heidelberg. doi:10.1007/978-1-4939-7850-2

Theworldcounts.com. (2023). *Electronic Waste Facts*. Available at: https://www.theworldcounts.com/stories/electronic-waste-facts

Thiounn, T., & Smith, R. C. (2020). Advances and approaches for chemical recycling of plastic waste. *Journal of Polymer Science, 58*(10), 1347–1364. doi:10.1002/pol.20190261

Tiseo, I. (2023, July 18). *Global e-waste generation 2010-2019*. https://www.statista.com/statistics/499891/projection-ewaste-generation-worldwide/

Tiwari, A. K., Kumar, S., Kumar, P., & Kumar, A. (2020). Machine learning for the prediction of hazardous materials in e-waste. *Waste Management & Research, 38*(10), 1088–1096.

Tiwari, D., & Dhawan, N. G. (2014). E-waste management: An emerging challenge to manage and recover valuable resources. *International Journal of Environmental Research and Development, 4*(3), 253–260.

Tran, H. P., Wang, F., Dewulf, J., Huynh, T. H., & Schaubroeck, T. (2016). Estimation of the unregistered inflow of electrical and electronic equipment to a domestic market: A case study on televisions in Vietnam. *Environmental Science & Technology, 50*(5), 2424–2433. doi:10.1021/acs.est.5b01388 PMID:26820611

Tsitomeneas, Kokkosis, & Charitopoulos. (2014). *Legislation, design and management of the electrical and electronic waste (e-waste) procedures.* Academic Press.

Tsydenova, O., & Bengtsson, M. (2011). Chemical hazards associated with the treatment of waste electrical and electronic equipment. *Waste Management (New York, N.Y.), 31*(1), 45–58. doi:10.1016/j.wasman.2010.08.014 PMID:20869229

Tue, N. M., Sudaryanto, A., Minh, T. B., Isobe, T., Takahashi, S., Viet, P. H., & Tanabe, S. (2010). Accumulation of polychlorinated biphenyls and brominated flame retardants in breast milk from women living in Vietnamese e-waste recycling sites. *The Science of the Total Environment, 408*(9), 2155–2162. doi:10.1016/j.scitotenv.2010.01.012 PMID:20116831

Tunali, M., Tunali, M. M., & Yenigun, O. (2021). Characterization of different types of electronic waste: Heavy metal, precious metal and rare earth element content by comparing different digestıon methods. *Journal of Material Cycles and Waste Management, 23*(1), 149–157. doi:10.100710163-020-01108-0

Turaga, R. M. R., Bhaskar, K., Sinha, S., Hinchliffe, D., Hemkhaus, M., Arora, R., Chatterjee, S., Khetriwal, D. S., Radulovic, V., Singhal, P., & Sharma, H. (2019). E-waste management in India: Issues and strategies. *Vikalpa, 44*(3), 127–162. doi:10.1177/0256090919880655

UNE. (2023). *Sustainable Future of E-waste*. https://www.unep.org/ietc/news/story/sustainable-future-e-waste#:~:text=E%2Dwaste%20is%20becoming%20one,the%20amount%20produced%20in%202014

UNEP. (2006). *Basel conference addresses electronic wastes challenge*. United Nations Environment Programme.

UNEP. (2007). *E-waste-Volume I: Inventory assessment manual*. United Nations Environmental Programme.

UNEP. (2022). *Sustainable Consumption and Production Policies*. United Nations Environment Programme.

UNITAR. (2020). *Global E-waste Statistics (2020)—Sustainable Cycles (SCYCLE) Programme Map*. The United Nations Institute for Training and Research. Available online: https://globalewaste.org/map/

US EPA. (2014). *Regulatory and Guidance Information by Topic: Waste – Guide for Industrial Waste Management*. Retrieved from https://www.epa.gov/regulatory-information-topic/regulatory-and-guidance-information-topic-waste#solid

US Government Accountability Office. (2008). *Electronic Waste EPA Needs to Better Control Harmful U.S. Exports through Stronger Enforcement and More Comprehensive Regulation*. United States Government Accountability Office.

Uwizeyimana, T. (2019). Enhancing E-Waste Management through IoT: A Case Study in Rwanda. *Waste Management (New York, N.Y.), 87*, 526–534. doi:10.1016/j.wasman.2019.03.040

Vaccari, M., Vinti, G., Cesaro, A., Belgiorno, V., Salhofer, S., Dias, M. I., & Jandric, A. (2019). WEEE treatment in developing countries: Environmental pollution and health consequences - An overview. *International Journal of Environmental Research and Public Health, 16*(9), 1595. doi:10.3390/ijerph16091595 PMID:31067685

Van Eygen, E., De Meester, S., Tran, H. P., & Dewulf, J. (2016). Resource savings by urban mining: The case of desktop and laptop computers in Belgium. *Resources, Conservation and Recycling, 107*, 53–64. doi:10.1016/j.resconrec.2015.10.032

Van Yken, J., Boxall, N. J., Cheng, K. Y., Nikoloski, A. N., Moheimani, N. R., & Kaksonen, A. H. (2021). E-waste recycling and resource recovery: A review on technologies, barriers and enablers with a focus on oceania. *Metals, 11*(8), 1313. doi:10.3390/met11081313

Vasanthi, P., Kaliappan, S., & Srinivasaraghavan, R. (2008). Impact of poor solid waste management on ground water. *Environmental Monitoring and Assessment, 143*(1), 227–238. doi:10.100710661-007-9971-0 PMID:17999155

Vats, M. C., & Singh, S. K. (2014). E-Waste characteristic and its disposal. *International Journal of Ecological Science and Environmental Engineering, 1*(2), 49-61.

Vats, M. C., & Singh, S. K. (2014). Status of e-waste in India-A review. *transportation, 3*(10), 16917-16931.

Veenstra, A., Wang, C., Fan, W., & Ru, Y. (2010). An analysis of E-waste flows in China. *International Journal of Advanced Manufacturing Technology, 47*(5-8), 449–459. doi:10.100700170-009-2356-5

Vega, F., Baena-Moreno, F. M., Fernández, L. M. G., Portillo, E., Navarrete, B., & Zhang, Z. (2020). Current status of CO2 chemical absorption research applied to CCS: Towards full deployment at industrial scale. *Applied Energy, 260*, 114313. doi:10.1016/j.apenergy.2019.114313

Vijayan, R. V., Krishnan, M. M., Parayitam, S., Duraisami, S. P. A., & Saravanaselvan, N. R. (2023). Exploring e-waste recycling behaviour intention among the households: Evidence from India. *Cleaner Materials*, *7*, 100174. doi:10.1016/j.clema.2023.100174

Vishwakarma, A., Kanaujia, K., & Hait, S. (2023). *Global E-Waste Management Strategies and Future Implications.* Elsevier., doi:10.1016/B978-0-323-99919-9.00013-1

Vishwakarma, S., Kumar, V., Arya, S., Tembhare, M., Dutta, D., & Kumar, S. (2022). E-waste in Information and Communication Technology Sector: Existing scenario, management schemes and initiatives. *Environmental Technology & Innovation*, *27*, 102797. doi:10.1016/j.eti.2022.102797

Voskergian, D., & Ishaq, I. (2023). Smart e-waste management system utilizing Internet of Things and Deep Learning approaches. *Journal of Smart Cities and Society*, *2*(2), 77–98. doi:10.3233/SCS-230007

Wäger, P. A., Hischier, R., & Eugster, M. (2011). Environmental impacts of the Swiss collection and recovery systems for waste electrical and electronic equipment (WEEE): A follow-up. *The Science of the Total Environment*, *409*(10), 1746–1756. doi:10.1016/j.scitotenv.2011.01.050 PMID:21342702

Wang, D., Cai, Z., Jiang, G., Leung, A., Wong, M. H., & Wong, W. K. (2005). Determination of polybrominated diphenyl ethers in soil and sediment from an electronic waste recycling facility. *Chemosphere*, *60*(6), 810–816. doi:10.1016/j.chemosphere.2005.04.025 PMID:15946727

Wang, F., Huisman, J., Baldé, K., & Stevels, A. (2012, September). *A systematic and compatible classification of WEEE. In 2012 Electronics Goes Green 2012*. IEEE.

Wang, F., Leung, A. O. W., Wu, S. C., Yang, M. S., & Wong, M. H. (2009a). Chemical and ecotoxicological analyses of sediments and elutriates of contaminated rivers due to e-waste recycling activities using a diverse battery of bioassays. *Environmental Pollution*, *157*(7), 2082–2090. doi:10.1016/j.envpol.2009.02.015 PMID:19299057

Wang, H., Han, M., Yang, S., Chen, Y., Liu, Q., & Ke, S. (2011b). Urinary heavy metal levels and relevant factors among people exposed to e-waste dismantling. *Environment International*, *37*(1), 80–85. doi:10.1016/j.envint.2010.07.005 PMID:20678797

Wang, H., Lv, S., Li, F., Liu, Q., & Ke, S. (2010). Study on the changes of urinary 8-hydroxydeoxyguanosine levels and burden of heavy metal around e-waste dismantling site. *The Science of the Total Environment*, *408*(24), 6092–6099. doi:10.1016/j.scitotenv.2010.09.019 PMID:20889187

Wang, J., Chen, S., Zeng, X., Huang, J., Liang, Q., Shu, J., & Sun, Z. (2021). Recovery of high purity copper from waste printed circuit boards of mobile phones by slurry electrolysis with ammonia-ammonium system. *Separation and Purification Technology*, *275*, 119180. doi:10.1016/j.seppur.2021.119180

Wang, J.-P., & Guo, X.-K. (2006). Impact of electronic wastes recycling on environmental quality. *Biomedical and Environmental Sciences*, *19*, 137–142. PMID:16827186

Wang, J., & Xu, Z. (2015). Disposing and recycling waste printed circuit boards: Disconnecting, resource recovery, and pollution control. *Environmental Science & Technology*, *49*(2), 721–733. doi:10.1021/es504833y PMID:25525865

Wang, T., Fu, J., Wang, Y., Liao, C., Tao, Y., & Jiang, G. (2009b). Use of scalp hair as indicator of human exposure to heavy metals in an electronic waste recycling area. *Environmental Pollution*, *157*(8-9), 2445–2451. doi:10.1016/j.envpol.2009.03.010 PMID:19346038

Wang, Y., Li, H., Fu, S., & Wang, Y. (2021). An optimization model for the recycling of e-waste. *Journal of Cleaner Production*, *279*, 123872.

Wang, Y., Luo, C., Li, J., Yin, H., Li, X., & Zhang, G. (2011a). Characterization of PBDEs in soils and vegetations near an e-waste recycling site in South China. *Environmental Pollution, 159*(10), 2443–2448. doi:10.1016/j.envpol.2011.06.030 PMID:21763040

Wang, Y., Peris, A., Rifat, M. R., Ahmed, S. I., Aich, N., Nguyen, L. V., Urík, J., Eljarrat, E., Vrana, B., Jantunen, L. M., & Diamond, M. L. (2020). Measuring exposure of e-waste dismantlers in Dhaka Bangladesh to organophosphate esters and halogenated flame retardants using silicone wristbands and T-shirts. *The Science of the Total Environment, 720*, 137480. doi:10.1016/j.scitotenv.2020.137480 PMID:32146393

Wang, Z., Guo, D., & Wang, X. (2016). Determinants of residents' e-waste recycling behavior intentions: Evidence from China. *Journal of Cleaner Production, 137*, 850–860. doi:10.1016/j.jclepro.2016.07.155

Wath, S. B., Dutt, P. S., Chakrabarti, T., Wath, S. B., & Dutt, P. S. (2011). *E-waste scenario in India, its management and implications*. doi:10.1007/s10661-010-1331-9

Wath, S. B., Vaidya, A. N., Dutt, P. S., & Chakrabarti, T. (2010). A roadmap for development of sustainable E-waste management system in India. *The Science of the Total Environment, 409*(1), 19–32. doi:10.1016/j.scitotenv.2010.09.030 PMID:20951410

Weckend, S., Wade, A., & Heath, G. A. (2016). *End-of-life management: solar photovoltaic panels (No. NREL/TP-6A20-73852). National Renewable Energy Lab.* NREL.

Weee-forum.org. (2023). *International E-Waste Day: 57.4M Tonnes Expected in 2021.* Available at: https://weee-forum.org/ws_news/international-e-waste-day-2021/

Weerasundara, L., & Mahatantila, K. (2020). *E-waste as a challenge for public and ecosystem health.* Elsevier. doi:10.1016/B978-0-12-817030-4.00003-6

Weiss, C., Carriere, M., Fusco, L., Capua, I., Regla-Nava, J. A., Pasquali, M., Scott, J. A., Vitale, F., Unal, M. A., Mattevi, C., Bedognetti, D., Merkoçi, A., Tasciotti, E., Yilmazer, A., Gogotsi, Y., Stellacci, F., & Delogu, L. G. (2020). Toward nanotechnology-enabled approaches against the COVID-19 pandemic. *ACS Nano, 14*(6), 6383–6406. doi:10.1021/acsnano.0c03697 PMID:32519842

Whitman, M. E., & Mattord, H. J. (2021). *Principles of information security.* Cengage Learning.

Widmer, R., & Lombard, R. (2005). *E-waste Assessment in South Africa a Case Study of the Gauteng Province.* EMPA.

Widmer, R., Oswald-Krapf, H., Sinha-Khetriwal, D., Schnellmann, M., & Böni, H. (2005). Global perspectives on e-waste. *Environmental Impact Assessment Review, 25*(5), 436–458. doi:10.1016/j.eiar.2005.04.001

Wikipedia. (2023). *United States Environmental Protection Agency.* Available at: https://en.wikipedia.org/wiki/United_States_Environmental_Protection_Agency

Williams, E., Kahhat, R., Allenby, B., Kavazanjian, E., Kim, J., & Xu, M. (2008). Environmental, social, and economic implications of global reuse and recycling of personal computers. *Environmental Science & Technology, 42*(17), 6446–6454. doi:10.1021/es702255z PMID:18800513

Wilson, D. C., Velis, C., & Cheeseman, C. (2006). Role of informal sector recycling in waste management in developing countries. *Habitat International, 30*(4), 797–808. doi:10.1016/j.habitatint.2005.09.005

Wilson, J. (2009). *Electronics: A history of everyday things.* Cambridge University Press.

Wingfield, S., & Lim, M. (2022). *The United Nations Basel convention's global plastic waste partnership: history, evolution and progress. In Microplastic Environment Pattern Process.* Springer.

Wong, M. H., Wu, S. C., Deng, W. J., Yu, X. Z., Luo, Q., Leung, A. O. W., Wong, C. S. C., Luksemburg, W. J., & Wong, A. S. (2007). Export of toxic chemicals–a review of the case of uncontrolled electronic-waste recycling. *Environmental Pollution, 149*(2), 131–140. doi:10.1016/j.envpol.2007.01.044 PMID:17412468

World Bank. (2020). *The economic and social impacts of technological advancements in e-waste management.* Author.

World Economic Forum. (2017). *The economic and social benefits of e-waste recycling*. Author.

World Health Organization (WHO). (2020). *Electronic waste and health.* Retrieved from https://www.who.int/news-room/fact-sheets/detail/electronic-waste-(e-waste)

Wu, D., Zhou, J., Creyer, M. N., Yim, W., Chen, Z., Messersmith, P. B., & Jokerst, J. V. (2021). Phenolic-enabled nanotechnology: Versatile particle engineering for biomedicine. *Chemical Society Reviews, 50*(7), 4432–4483. doi:10.1039/D0CS00908C PMID:33595004

Wu, K., Xu, X., Liu, J., Guo, Y., Li, Y., & Huo, X. (2010). Polybrominated diphenyl ethers in umbilical cord blood and relevant factors in neonates from Guiyu, China. *Environmental Science & Technology, 44*(2), 813–819. doi:10.1021/es9024518 PMID:20000818

Wu, Q., Du, Y., Huang, Z., Gu, J., Leung, J. Y., Mai, B., Xiao, T., Liu, W., & Fu, J. (2019). Vertical profile of soil/sediment pollution and microbial community change by e-waste recycling operation. *The Science of the Total Environment, 669*, 1001–1010. doi:10.1016/j.scitotenv.2019.03.178 PMID:30970449

Xavier, L. H., Giese, E. C., Ribeiro-Duthie, A. C., & Lins, F. A. F. (2021). Sustainability and the circular economy: A theoretical approach focused on e-waste urban mining. *Resources Policy, 74*, 101467. doi:10.1016/j.resourpol.2019.101467

Xavier, L. H., Ottoni, M., & Lepawsky, J. (2021). Circular economy and e-waste management in the Americas: Brazilian and Canadian frameworks. *Journal of Cleaner Production, 297*, 126570. doi:10.1016/j.jclepro.2021.126570

Xing, G. H., Chan, J. K. Y., Leung, A. O. W., Wu, S. C., & Wong, M. H. (2009). Environmental impact and human exposure to PCBs in Guiyu, an electronic waste recycling site in China. *Environment International, 35*(1), 76–82. doi:10.1016/j.envint.2008.07.025 PMID:18804865

Xing, G. H., Liang, Y., Chen, L. X., Wu, S. C., & Wong, M. H. (2011). Exposure to PCBs, through inhalation, dermal contact and dust ingestion at Taizhou, China–A major site for recycling transformers. *Chemosphere, 83*(4), 605–611. doi:10.1016/j.chemosphere.2010.12.018 PMID:21295325

Xu, Y., Li, J., Tan, Q., Peters, A. L., & Yang, C. (2018). Global status of recycling waste solar panels: A review. *Waste Management, 75*, 450-458.

Xu, Y., Yeh, C. H., Liu, C. G., Ramzan, S., & Zhang, L. (2021a). Evaluating and managing interactive barriers for sustainable e-waste management in China. *The Journal of the Operational Research Society, 72*(9), 2018–2031. doi:10.1080/01605682.2020.1759381

Yabe, J., Ishizuka, M., & Umemura, T. (2010). Current levels of heavy metal pollution in Africa. *The Journal of Veterinary Medical Science, 72*(10), 1257–1263. doi:10.1292/jvms.10-0058 PMID:20519853

Yadav, R., & Pathak, G. S. (2016). Young consumers' intention towards buying green products in a developing nation: Extending the theory of planned behavior. *Journal of Cleaner Production, 135*, 732–739. doi:10.1016/j.jclepro.2016.06.120

Yadav, V., Karmakar, S., Dikshit, A. K., & Vanjari, S. (2016). A feasibility study for the locations of waste transfer stations in urban centers: A case study on the city of Nashik, India. *Journal of Cleaner Production, 126*, 191–205. doi:10.1016/j.jclepro.2016.03.017

Yanagitani, K., & Kawahara, K. (2000). LCA study of air conditioners with an alternative refrigerant. *The International Journal of Life Cycle Assessment*, *5*(5), 287–290. doi:10.1007/BF02977581

Yan, C. H., Xu, J., & Shen, X. M. (2013). Childhood lead poisoning in China: Challenges and opportunities. *Environmental Health Perspectives*, *121*(10). Advance online publication. doi:10.1289/ehp.1307558 PMID:24218672

Yaqoob, I., Salah, K., Jayaraman, R., & Omar, M. (2023). Metaverse applications in smart cities: Enabling technologies, opportunities, challenges, and future directions. *Internet of Things : Engineering Cyber Physical Human Systems*, *23*, 100884. doi:10.1016/j.iot.2023.100884

Ylä-Mella, J., Keiski, R. L., & Pongrácz, E. (2015). Electronic waste recovery in Finland: Consumers' perceptions towards recycling and re-use of mobile phones. *Waste Management (New York, N.Y.)*, *45*, 374–384. doi:10.1016/j.wasman.2015.02.031 PMID:25797074

Yoon, H., & Jang, Y. C. (2006, May). The practice and challenges of electronic waste recycling in Korea with emphasis on extended producer responsibility (EPR). In *Proceedings of the 2006 IEEE International Symposium on Electronics and the Environment* (pp. 326-330). IEEE. 10.1109/ISEE.2006.1650085

Yu, J., Williams, E., Ju, M., & Yang, Y. (2010). *Forecasting global generation of obsolete personal computers*. ACS Publications. doi:10.1021/es903350q

Yu, Z., Loisel, J., Brosseau, D. P., Beilman, D. W., & Hunt, S. J. (2010). Global peatland dynamics since the Last Glacial Maximum. *Geophysical Research Letters*, *37*(13), 2010GL043584. doi:10.1029/2010GL043584

Yu, Z., Lu, S., Gao, S., Wang, J., Li, H., Zeng, X., Sheng, G., & Fu, J. (2010). Levels and isomer profiles of Dechlorane Plus in the surface soils from e-waste recycling areas and industrial areas in South China. *Environmental Pollution*, *158*(9), 2920–2925. doi:10.1016/j.envpol.2010.06.003 PMID:20630636

Zarei, M., Taghipour, H., & Hassanzadeh, Y. (2018). Survey of quantity and management condition of end-of-life tires in Iran: A case study in Tabriz. *Journal of Material Cycles and Waste Management*, *20*(2), 1099–1105. doi:10.100710163-017-0674-5

Zeng, X., Duan, H., Wang, F., & Li, J. (2017). Examining environmental management of e-waste: China's experience and lessons. *Renewable & Sustainable Energy Reviews*, *72*, 1076–1082. doi:10.1016/j.rser.2016.10.015

Zeng, X., Gong, R., Chen, W. Q., & Li, J. (2016). Uncovering the recycling potential of "new" WEEE in China. *Environmental Science & Technology*, *50*(3), 1347–1358. doi:10.1021/acs.est.5b05446 PMID:26709550

Zeng, X., & Li, J. (2016). Measuring the recyclability of e-waste: An innovative method and its implications. *Journal of Cleaner Production*, *131*, 156–162. doi:10.1016/j.jclepro.2016.05.055

Zeng, X., Mathews, J. A., & Li, J. (2018). Urban mining of e-waste is becoming more cost-effective than virgin mining. *Environmental Science & Technology*, *52*(8), 4835–4841. doi:10.1021/acs.est.7b04909 PMID:29616548

Zeng, X., Yang, C., Chiang, J. F., & Li, J. (2017). Innovating e-waste management: From macroscopic to microscopic scales. *The Science of the Total Environment*, *575*, 1–5. doi:10.1016/j.scitotenv.2016.09.078 PMID:27723459

Zhang, J., Jiang, Y., Zhou, J., Wu, B., Liang, Y., Peng, Z., Fang, D., Liu, B., Huang, H., He, C., Wang, C., & Lu, F. (2010). Elevated body burdens of PBDEs, dioxins, and PCBs on thyroid hormone homeostasis at an electronic waste recycling site in China. *Environmental Science & Technology*, *44*(10), 3956–3962. doi:10.1021/es902883a PMID:20408536

Zhang, L., Qu, J., Sheng, H., Yang, J., Wu, H., & Yuan, Z. (2019). Urban mining potentials of university: In-use and hibernating stocks of personal electronics and students' disposal behaviors. *Resources, Conservation and Recycling*, *143*, 210–217. doi:10.1016/j.resconrec.2019.01.007

Zhang, P., Guo, Z., Ullah, S., Melagraki, G., Afantitis, A., & Lynch, I. (2021). Nanotechnology and artificial intelligence to enable sustainable and precision agriculture. *Nature Plants*, *7*(7), 864–876. doi:10.103841477-021-00946-6 PMID:34168318

Zhang, Q., Zhou, T., Xu, X., Guo, Y., Zhao, Z., Zhu, M., Li, W., Yi, D., & Huo, X. (2011). Downregulation of placental S100P is associated with cadmium exposure in Guiyu, an e-waste recycling town in China. *The Science of the Total Environment*, *410*, 53–58. doi:10.1016/j.scitotenv.2011.09.032 PMID:21992841

Zhang, T., Lowry, G. V., Capiro, N. L., Chen, J., Chen, W., Chen, Y., Dionysiou, D. D., Elliott, D. W., Ghoshal, S., Hofmann, T., Hsu-Kim, H., Hughes, J., Jiang, C., Jiang, G., Jing, C., Kavanaugh, M., Li, Q., Liu, S., Ma, J., ... Alvarez, P. J. J. (2019). In situ remediation of subsurface contamination: Opportunities and challenges for nanotechnology and advanced materials. *Environmental Science. Nano*, *6*(5), 1283–1302. doi:10.1039/C9EN00143C

Zhang, X., Zhong, T., Liu, L., & Ouyang, X. (2015). Impact of soil heavy metal pollution on food safety in China. *PLoS One*, *10*(8), e0135182. Advance online publication. doi:10.1371/journal.pone.0135182 PMID:26252956

Zhan, L., Xia, F., Ye, Q., Xiang, X., & Xie, B. (2015). Novel recycle technology for recovering rare metals (Ga, In) from waste light-emitting diodes. *Journal of Hazardous Materials*, *299*, 388–394. doi:10.1016/j.jhazmat.2015.06.029 PMID:26150281

Zhao, G., Wang, Z., Dong, M. H., Rao, K., Luo, J., Wang, D., Zha, J., Huang, S., Xu, Y., & Ma, M. (2008). PBBs, PBDEs, and PCBs levels in hair of residents around e-waste disassembly sites in Zhejiang Province, China, and their potential sources. *The Science of the Total Environment*, *397*(1-3), 46–57. doi:10.1016/j.scitotenv.2008.03.010 PMID:18439655

Zhao, Y., Peng, B., Elahi, E., & Wan, A. (2021). Does the extended producer responsibility system promote the green technological innovation of enterprises? An empirical study based on the difference-in-differences model. *Journal of Cleaner Production*, *319*, 128631. doi:10.1016/j.jclepro.2021.128631

Zheng, G., Xu, X., Li, B., Wu, K., Yekeen, T. A., & Huo, X. (2013). Association between lung function in school children and exposure to three transition metals from an e-waste recycling area. *Journal of Exposure Science & Environmental Epidemiology*, *23*(1), 67–72. doi:10.1038/jes.2012.84 PMID:22854517

Zheng, J., Chen, K., Yan, X., Chen, S.-J., Hu, G.-C., Peng, X.-W., Yuan, J., Mai, B.-X., & Yang, Z.-Y. (2013). Heavy metals in food, house dust, and water from an e-waste recycling area in South China and the potential risk to human health. *Ecotoxicology and Environmental Safety*, *96*, 205–212. doi:10.1016/j.ecoenv.2013.06.017 PMID:23849468

Zheng, J., Luo, X. J., Yuan, J. G., He, L. Y., Zhou, Y. H., Luo, Y., Chen, S.-J., Mai, B.-X., & Yang, Z. Y. (2011). Heavy metals in hair of residents in an e-waste recycling area, South China: Contents and assessment of bodily state. *Archives of Environmental Contamination and Toxicology*, *61*(4), 696–703. doi:10.100700244-011-9650-6 PMID:21360078

Zheng, L., Wu, K., Li, Y., Qi, Z., Han, D., Zhang, B., Gu, C., Chen, G., Liu, J., Chen, S., Xu, X., & Huo, X. (2008). Blood lead and cadmium levels and relevant factors among children from an e-waste recycling town in China. *Environmental Research*, *108*(1), 15–20. doi:10.1016/j.envres.2008.04.002 PMID:18514186

Zhou, J., Kroll, A. V., Holay, M., Fang, R. H., & Zhang, L. (2020). Biomimetic nanotechnology toward personalized vaccines. *Advanced Materials*, *32*(13), 1901255. doi:10.1002/adma.201901255 PMID:31206841

About the Contributors

Rajesh Kumar K. V. is currently working as Assistant Professor in School of Business, Woxsen University, Hyderabad, Telangana. He earned his M. Tech in Embedded Systems from JNTU, Kakinada, and AMIE in ECE from the Institution of Engineers, Kolkata. He is pursuing his Ph.D. from VIT University, Chennai. He was earlier associated with VIT as Junior Research Fellow and upGrad Educational Private Limited where he was guiding students in selecting research problems and in writing a thesis, to deal with AI/ML models, for those who are pursuing M.Sc. in Data Science / AI ML courses from LIVERPOOL JOHN MOORES University through upGrad. His research specialization is signals and sensor-based Data Analytics for Bio-Medical Applications.

Hemachandran Kannan holds the Area Chair of the Analytics Department at Woxsen University. Dr. Hemachandran is an Advisory board member at APT AI Labs, LLC. He has been a passionate teacher with 15 years of teaching experience and 5 years of research experience. He served as an effective resource person at various national and international scientific conferences and also gave lectures on topics related to Artificial Intelligence. He was bestowed as Best faculty at Woxsen University in 2021-2022 and also in Ashoka Institute of Engineering & Technology in 2019 – 2020. He is having rich working experience in Natural Language Processing, Computer Vision, Building Video recommendation systems, Building Chatbots for HR policies and Education Sector, Automatic Interview processes, and Autonomous Robots. He is working on various real-time use cases and projects in collaboration with Industries such as Advertflair, LLC, Course5i and Apstek Corp. He has organized many International Conferences, Hackathons and Ideathon. An open-ended positive person who has a stupendous peer-reviewed publication record with more than 35 journals and international conference publications.

* * *

Afaq Ahmad is an accomplished communication researcher and educator. He currently holds a faculty position at the School of Liberal Arts and Humanities at Woxsen University in Hyderabad, India. Dr. Ahmad specializes in the field of journalism and media studies and has an impressive track record of guiding numerous PhD scholars in various academic institutions. His research expertise spans a broad spectrum of areas, including convergence media, digital media, the role of media in promoting human rights, film communication, and broadcast media. Dr. Ahmad brings a diverse skill set to the table, which includes proficiency in research design, lecturing, curriculum assessment, and the editing and proofreading of a wide range of materials, including manuscripts, research proposals, and doctoral theses.

Ramamurthy B. M. is an MBA, M.Phil, Ph.D in Finance, passionate about exploring new pedagogical tools for teaching Accounting and Finance courses with two and half decades of experience in teaching, research and administration. He is currently serving as an Associate Professor and Principal – UG Programs at AIMS B-School, Peenya, Bengaluru. He has completed Simulation Certificate in Entrepreneurship Strategy from Wharton Interactive, Wharton University of Pennsylvania. He has been instrumental in introducing "Simulations" as a new pedagogical tool to enhance experiential teaching and learning processes at AIMS B-School. He has organized International Conference on "Transformative Business Innovation" in collaboration with University of Cyberjaya at Malaysia in the year 2022. He has received "Dr. Radhakrishna Shikshana Ratna National Award" for the distinguished contributions to the development of the Nation and achieving outstanding excellence in the field of Teaching/Training/Research and Administration presented by International Institute for Social and Economic Reforms(R), Bengaluru on 13 th January 2014. He has published patent on "Monetary Incentives and Employee Performance" in 2023.

C. V. Suresh Babu is a pioneer in content development. A true entrepreneur, he founded Anniyappa Publications, a company that is highly active in publishing books related to Computer Science and Management. Dr. C.V. Suresh Babu has also ventured into SB Institute, a center for knowledge transfer. He holds a Ph.D. in Engineering Education from the National Institute of Technical Teachers Training & Research in Chennai, along with seven master's degrees in various disciplines such as Engineering, Computer Applications, Management, Commerce, Economics, Psychology, Law, and Education. Additionally, he has UGC-NET/SET qualifications in the fields of Computer Science, Management, Commerce, and Education. Currently, Dr. C.V. Suresh Babu is a Professor in the Department of Information Technology at the School of Computing Science, Hindustan Institute of Technology and Science (Hindustan University) in Padur, Chennai, Tamil Nadu, India. For more information, you can visit his personal blog at .

Jyoti Chawla is currently working as Professor in the Department of Applied Sciences (Chemistry), School of Engineering and Technology, Manav Rachna International Institute of Research and Studies, Faridabad. She received her Ph.D. in Chemical Sciences from Guru Nanak Dev University (GNDU) where she conducted research as junior research fellow under collaborative IUC-DAE project of GNDU, Amritsar and Bhabha Atomic Research Centre (BARC) Mumbai. She is NET qualified. Her research interest includes Water Treatment, Surface Chemistry, Risk assessment of Nanomaterials. She has contributed more than 50 papers in journals and conferences proceedings. She has also contributed 5 chapters in Books. She had guided two Ph.D. scholars. Recently she is guiding two research scholars.

Bhagya D. is working as Assistant Professor at the Department of Information Science and Engineering, Sapthagiri College of Engineering, Bengaluru. She has obtained PhD degree in Electronics and Communication Engineering from Vellore Institute of Technology (VIT) University, Chennai during March 2022. She completed BTech in Electronics and Communication Engineering and M.Tech. in Communication Engineering from Mahatma Gandhi University, Kottayam, Kerala. She has a teaching experience of 3 years. Her areas of interest include analog and digital electronics, signal processing and Information theory. Her research is mainly focussed on image processing, biomedical signal processing, and machine learning techniques for medical applications. Her research works till date are published in reputed SCI/Scopus-indexed IEEE journals.

Jagadeesan D. obtained his Bachelor's degree in Computer Science and Engineering from Anna University in 2005 at Chennai, Tamil Nadu, India. He obtained his Master's degree in Computer Science and Engineering from Dr. M.G.R University in 2007 at Chennai, Tamil Nadu, India and his Ph. D in Computer Science and Engineering at Sri Chandrasekharendra Saraswathi Viswa Mahavidyalaya (SCSVMV University), Kanchipuram, Tamil Nadu, India. He is Life Member of the ISTE and a Member of the IEEE. He is Microsoft Certified Power BI Data Analyst Associate. Currently, he working as Associate Professor in Department of Computer Science and Engineering, School of Technology at The Apollo University, Murukambattu, Chittoor. He is specialized in Networking, Mobile Ad-hoc Network, Compiler Design and Computation Theory. His current research interests are Mobile Routing Protocols, Heterogeneous Network, Network Security, e-Agriculture, Healthcare Data Analytics.

Raghunandan G. is working as an Assistant Professor in the School of Commerce, Finance and Accountancy, CHRIST (Deemed to be University). He has to his credit a Bachelor of Engineering in Chemical Engineering, MBA in Finance, PhD in the area of Management Information Systems and over 15 years of Industry and Teaching experience. His areas of Interest are International Finance, Strategic Finance, Quantitative Techniques, Financial Risk Management and Wealth Management. Currently, he has been working on two externally funded projects one by ICSSR, New Delhi on the topic "Entrepreneurship Ecosystem in North East Region of India" and the other on "Tech4all for Special Students" by Cheshire Disability Trust.

Sam Goundar is an Editor-in-Chief of the International Journal of Blockchains and Cryptocurrencies (IJFC) – Inderscience Publishers, Editor-in-Chief of the International Journal of Fog Computing (IJFC) – IGI Publishers, Section Editor of the Journal of Education and Information Technologies (EAIT) – Springer and Editor-in-Chief (Emeritus) of the International Journal of Cloud Applications and Computing (IJCAC) – IGI Publishers. He is also with more than 20 high impact factor journals. As a researcher, apart from Blockchains, Cryptocurrencies, Fog Computing, Mobile Cloud Computing and Cloud Computing, Dr. Sam Goundar also researches in Educational Technology, Artificial Intelligence, ICT in Climate Change, ICT Devices in the Classroom, Using Mobile Devices in Education, and e-Government. He has published on all these topics. He was a Research Fellow with the United Nations University. He is a Senior Lecturer in CS at British University Vietnam, Adjunct Senior Lecturer at The University of the South Pacific, Affiliate Professor of Information Technology at Pontificia Universidad Catolica Del Peru, and Adjunct Professor at The University of Fiji.

Adip Krishna Guduru is a student at Woxsen University currently pursuing his Bachelor's of Technology in Computer Science Engineering, with a specialisation in Artificial Intelligence and Machine Learning. He is a Tech-Enthusiast who enjoys exploring all parts of the field ranging from networking and web development, to most recently, machine learning, computer vision, and natural language processing. His latest projects include a web-based chatting interface named JollyChat, a computer vision based pothole detector, and an NLP based legal case prioritisation system called VerdictAI. He is passionate about advancing computer science and looks forward to working on innovative technologies.

A. K. M. Jawadul Haque is working in the private Bank in Bangladesh. He completed MS in Environmental Science in 2021. His research interest in sustainability issues especially on e-waste management to have a look a better world in future. Despite his MBA and banking profession for last twenty two years Mr. Haque is enthusiast about environment. He envisioned sustainability practices to save energy, water and reduce waste. He is also concerned about climate action by engaging himself at the Rotary Club of Shyamoli in Bangladesh. At present, he is pursuing professional Masters in Peace Conflict and Human Rights at the University of Dhaka.

Md. Shafiqul Islam specializes in sustainable development issues including sustainable waste management, e-waste management, waste management principles and value creation, environmental hazard & disaster management, organic farming with particular expertise on tea garden, NGOs and civil society. Other interests include rural development, livelihood, food security, humanitarian assistance, maternal & child health and nutrition, organizational issues in development agencies, and bottom up approaches of development. An Environmentalist by education and doing, he has conducted extensive field research in cross range of field including climate change adaptation, disaster risk reduction, ecological resilience, and has advised and consulted for a wide range of national international development agencies, universities NGOs and private sectors. Shafiqul Islam is an Associate Professor of Environmental Science at the International University of Business Agriculture and Technology. He is a specialist on Disaster Management, with a regional focus on Drought in Barind Tract in Bangladesh. Dr. Islam has published widely on the Climate Change Adaptation. His recent work includes Cultivation and Drought Management in Agriculture (Springer Nature, 2022). His two chapters published in two renowned books from IGI Global. After studying in Chittagong University, Shafiqul Islam was awarded a doctorate in Sustainable Approaches to Drought Management: Exploratory Study on selected regions of Northwest in Bangladesh' from the Institute of Disaster Management and Vulnerability Studies, University of Dhaka, Bangladesh. He worked as a Post-Doctoral Research Fellow at the London School of Economics and Political Science (LSE), UK and conducted a research on "Exploring Social Ecological Resilience of Fisher Folks to Coastal Disasters". He worked for 13 years with WB DFID, USAID, EU, GWAB, NETZ, DanChurchAid and ADD funded projects for research, collaboration, partnership, implementation, network linkages and dissemination. He returned to academia as an Assistant Professor at the Center for Sustainable Development in the University of Liberal Arts Bangladesh in 2010. Dr Islam is a Forester with broad interests in the ecology, silviculture, and conservation of biodiversity. He was a semester Faculty at the United International University in 2019. He was the Associate Professor and Chairman at the Department of Environmental Science (Graduate Program) in State University of Bangladesh from 2019-2022. He supervised more than 80 Master's student theses as the major Professor from 2019 to 2022. After receiving his doctoral degree in Disaster Management from the University of Dhaka, Bangladesh. His expertise includes community-based disaster risk management, climate change adaptation, and disaster and environmental education. Prof. Islam has published more than 50 academic papers and book chapters. Dr. Islam has been involved in research in the areas of gender, disasters, environment, climate change, education, social inclusion, indigenous community and other areas of social development since the late nineties.

Avula Jhansy is a dedicated student currently pursuing a B-Tech degree in Data Science and Artificial Intelligence at WOXSEN UNIVERSITY in Hyderabad. Her primary interests are data analysis and Machine learning. She is familiar with programming languages like Python, Java (object-oriented programming), SQL, HTML, and JavaScript. As a beginner in the field, she is committed to independent work and exploration of various topics.

Mohammad Juned is currently working as an Assistant Professor at Presidency College, Bangalore. He is a graduate in Information Technology with an MBA and a doctorate in Supply Chain Management. His teaching and research interests include supply chain management, operations management, quantitative methods and statistics, business analytics, and total quality management. He presented and published papers in related subjects.

Prakhyath Kandula is working as a Research Assistant at Woxsen University in the Department of Operations.

K. Dinesh Kumar working as an Assistant Professor in the School of Computing, Amrita Vishwa Vidyapeetham University. Ph.D. degree received from the School of Computer Science and Engineering, VIT University. Published international publications and participated in various international and national conferences. Active review member for various reputed leading research journals like IEEE transactions, Springer, Wiley, IGI Global, Taylor & Francis Online, Hindawi etc. and also acted as review member for international conferences like MBDAS-2021, Shenzhen, China and ICMEMSCE-2018, Xiamen, China. Co-author of three textbooks. Certified AWS Solution Architect (Associate). Research area is Cloud computing and Machine learning also have interests in Fog computing and Computer networks.

Rajeev Kumar is currently working as Associate Professor in the Department of Applied Sciences (Chemistry), School of Engineering and Technology, Manav Rachna International Institute of Research and Studies, Faridabad. He obtained his Ph.D. degree from Indian Institute of Technology, Roorkee (IIT Roorkee). He had done M.Sc. from CCS University, Meerut. He had qualified NET-JRF from CSIR-UGC. He has more than 14 years of teaching and research experience. He has published 32 research papers in national and international reputed journals. He has also contributed 5 chapters in Books. He had co-guided two Ph.D. scholars. Recently he is guiding two research scholars. His current research interests include Environmental Sciences, Orgeno-metallic synthesis, Analytical Chemistry, Water Treatment, and Optimization techniques.

Vijay Kumar received his Ph.D. in Statistics from Maharishi Dayanand University, Rohtak. Presently, he is working as a Professor& Head of Department, Applied Sciences, Manav Rachna International Institute of Research & Studies, Faridabad. His research interest includes Decision Making and Machine Learning. He has authored more than 65 research publications including research papers, conference papers, and chapters in Books. He has authored 4 books, out of which 3 are research books. He gave several invited talks in various seminars and conferences. Prof. Kumar is a member of editorial boards of several International/National journals and member of various academic advisory committees.

A. Muralidhar obtained his Bachelor's degree from Sri Krishna Devaraya University and Master's degree from Jawaharlal Nehru Technological University Hyderabad. He has awarded Doctor of Philosophy in Computer science from Vellore Institute of Technology – Chennai. He has a total Professional experience of more than 18 years working in various prestigious institutions. He has published 20 papers in various National and International peer reviewed journals and conferences. He is currently working as Associate Professor at Vellore Institute of Technology - Chennai Campus, India. His teaching and research expertise covers a wide range of subject area including Machine Learning, Knowledge Discovery and Data mining, Database Technologies, Big Data Analytics, etc.

Badrinath N. is working as Associate Professor in Department of Computer Science and Engineering, Vellore Institute of Technology, Vellore. He has 18 years of opulent experience in Teaching, research and Industrial Exposure with Research activities. He has published several papers in international & national journals and conferences and delivered guest lectures in many national & international seminars. He has completed BE (CSE) from Bharathidasan University, M. Tech (Advanced Computing) from SASTRA university and PhD in soft computing from Bharathidasan University.

P. V. Satya Prasad is an erudite academician with over 15 years of teaching and research experience in the domain of Journalism and Media Studies having taught at the under-graduate as well as Graduate level in higher education institutions in India as well as in the Gulf region. Dr Satya Prasad worked as a senior journalist in leading print media publications like The Hindu, in India and Oman Daily Observer, Muscat, Oman and Gulf Daily News, Bahrain, in a career spanning over two decades. Dr Satya Prasad is currently working as Dean and Associate Professor, School of Liberal Arts and Humanities, Woxsen University, Hyderabad, India.

Dipalee D. Rane received her ME from University of Pune and Bachelor's degree from J.T. Mahajan Faizpur, Maharashtra, India. Currently she is pursuing PhD from Vellore Institute of Technology (VIT) Chennai Campus. Since last 16 years she is working with D. Y. Patil College of Engineering, Akurdi, Pune- 44, Maharashtra, India.

Jayaditya Reddy is an undergraduate student pursuing a B.Tech in Computer Science with a specialization in Artificial Intelligence and Machine Learning at Woxsen University. He has prior experience working with computer vision and is very interested in the world of machine learning and data analytics. A passionate, curious and enthusiastic individual, Jayaditya leaves no chance to learn. Jayaditya has worked on several projects in the field of machine learning, including developing a computer vision model to detect and classify objects in images and a natural language processing model to generate text. He is also interested in the applications of machine learning to real-world problems. His latest projects include a computer vision based pothole detector, and an NLP based legal case prioritisation system called VerdictAI.

Oladri Renuka is an emerging talent in the field of Data Science and Artificial Intelligence. Currently in the third year of her B.Tech program. With a strong foundation in Machine Learning, Python programming, and Data Analysis, HTML, CSS, data visualization, she is well-equipped to delve deep into the fascinating world of Data Science. Her academic journey has honed her skills and deepened her understanding of these critical domains. Renuka's primary area of interest lies in the fusion of Data Science and Machine Learning. She aspires to leverage her knowledge to contribute to the ongoing advancements in this exciting field. Her dedication to research and her enthusiasm for tackling complex data-driven challenges make her a promising prospect in the world of academia and industry. As she continues her educational journey, Renuka eagerly anticipates exploring new frontiers in Data Science and Machine Learning. Her commitment to pushing the boundaries of knowledge in these fields underscores her determination to make a meaningful impact in the world of data-driven technologies.

B. Rupa Devi, Head and Associate Professor, Department of MCA in AITS Tirupati. She is pursuing her PhD in Department of CSE, SVUCE, SVU, Tirupati. Her areas on interest are Data Mining, Machine

Learning and Data Analytics. She has published 15 International journals. She is ratified as Associate Professor by JNTUA, Anantapuramu in 2010. She has 20 years of teaching experience.

Bhanu Prakash Saripalli serves as an Assistant Professor at the School of Technology, Woxsen University, bringing more than 13 years of valuable experience in teaching and research. Throughout his academic journey, he has made significant contributions to the field, publishing papers in esteemed national and international conferences and peer-reviewed journals. In 2022, he earned his Ph.D. from DIT University, India, focusing on supercapacitor-based smoothing of Solar Photovoltaic power generation fluctuations. His research interests center around Mitigating fluctuations in solar photovoltaic power generation, Energy Management, Power electronics for renewable energy, and Supercapacitor Energy storage systems. With four peer-reviewed publications in reputed journals, two conference papers, and two book chapters on Smart and Intelligent Systems in Proceedings of SIS 2021, his expertise is well-established.

Arti Saxena is currently working as Associate Professor in the Department of Applied Sciences (Maths), School of Engineering and Technology, Manav Rachna International Institute of Research and Studies, Faridabad. She received her Ph.D. from Dayalbagh Educational Institute, Deemed to be University, Agra. She has contributed more than 30 papers in journals and conferences proceedings. She has also contributed 5 chapters in Books and 4 Research Books. She had guided two Ph.D. scholars. Recently she is guiding two research scholars. She got DIRECTOR'S MEDAL for securing highest marks in M. Sc. Mathematics specialization in COMPUTER APPLICATIONS. She got many appreciations during my research and afterwards. She got a position in top 32% in NIIT National IT Aptitude test (NITAT) 2004. Successfully organised a National Conference on "Recent Trends in Mathematical Modeling and Soft Computing Techniques" Department of Mathematics, on 29th March 2014, FET, MRIU. Many more educational workshops and seminars organised in department.

Bodapatla Sindhu Priya is an undergraduate student currently pursuing B. Tech 3rd year in Data Science and Artificial Intelligence at Woxsen University. She possesses a diverse set of skills, including proficiency in HTML, CSS, JavaScript, Python, and Data Analysis. Her academic journey reflects a dedication to contemporary knowledge, training, and practical applications in the field of Data Science. Data Science equips learners with the skills to deploy and model data-based solutions, making it a fitting choice for someone.

Sivaji U. received the B.Tech. and M.Tech. degrees in computer science and engineering from JNTU Hyderabad, in 1999 and 2003, respectively, and the Ph.D. degree from the Computer Science engineering, Andhra University, Andhra Pradesh, India, in 2022. He is currently an Associate Professor with the Department of Information Technology, Institute of Aeronautical Engineering, Dundigal, Hyderabad, Telangana. He has teaching experience of more than 20 years in the areas of computer science and information technology. He has published more than 30 number of research papers/articles. He has also written a number of books in the area of computer science engineering. His current research interests include software Engineering, Machine Learning, networks, cyber security, software Automation, and cloud computing.

Ravi V. is working as Assistant Professor in CHRIST (Deemed to be University) having teaching experience of 15 years. The specialised areas are Accounts, Finance and general management. Published various research articles in National and international journals with high index. Attended National and International conference at various countries to present papers.

Index

G

H

I

J

M

T

U

V

W

www.ingramcontent.com/pod-product-compliance
Lightning Source LLC
LaVergne TN
LVHW062115300926
849161LV00010B/2467